Contemporary Clinical Neuroscience

Series editor

Mario Manto
Division of Neurosciences, Department of Neurology
CHU-Charleroi, Charleroi, Belgium
University of Mons, Mons, Belgium

Contemporary Clinical Neurosciences bridges the gap between bench research in the neurosciences and clinical neurology work by offering translational research on all aspects of the human brain and behavior with a special emphasis on the understanding, treatment, and eradication of diseases of the human nervous system. These novel, state-of-the-art research volumes present a wide array of preclinical and clinical research programs to a wide spectrum of readers representing the diversity of neuroscience as a discipline. Volumes in the series have focused on Attention Deficit Hyperactivity Disorder, Neurodegenerative diseases, G Protein Receptors, Sleep disorders, and Addiction issues.

More information about this series at http://www.springer.com/series/7678

Aasef Shaikh • Fatema Ghasia
Editors

Advances in Translational Neuroscience of Eye Movement Disorders

 Springer

Editors
Aasef Shaikh
Neurological Institute, University Hospitals
Neurology Service, Louis Stokes Cleveland
VA Medical Center
Department of Neurology, Case Western
Reserve University
Cleveland, OH, USA

Fatema Ghasia
Pediatric Ophthalmology
Cole Eye Institute, Cleveland Clinic
Cleveland, OH, USA

Contemporary Clinical Neuroscience
ISBN 978-3-030-31409-5   ISBN 978-3-030-31407-1   (eBook)
https://doi.org/10.1007/978-3-030-31407-1

This Springer imprint is published by the registered company Springer Nature Switzerland AG
The registered company address is: Gewerbestrasse 11, 6330 Cham, Switzerland

Preface

Eye movements are the window to the brain and mind. They serve as an important diagnostic marker, surrogate to modern imaging technology. The study of ocular motor behavior can examine the physiology of the human brain. This book, *Advances in Neuroscience of Eye Movements*, is about basic, translational, and clinical science of eye movements and is dedicated to David A. Robinson – a pioneer in the field – who invented one of the reliable method for measuring eye movements and systematically studied and quantified the electrophysiological properties of eye movement-related neurons from the frontal eye fields to the superior colliculus to ocular motoneurons. Dr. Robinson was one of the first to interpret behavioral and electrophysiological properties of the ocular motor system using mathematical models. He was also a translational scientist before that term was coined, successfully bringing basic science and computational techniques to account for clinical disorders.

Eye movements have been studied for at least two centuries, before the time of David Robinson. However, the primary focus of eye movement literature has changed over time. Early literature emphasized disease phenomenology and subjective descriptions of various ocular motor abnormalities. Several decades later, scientists and engineers, such as David Robinson, were enticed by the simplicity and elegance of ocular motor physiology. Computational rigor, physiological elegance, and clinical utility of the eye movement literature attracted neurologists, otolaryngologists, and ophthalmologists. Application of ocular motor physiology in the clinical arena grew exponentially in the last two decades of the twentieth century. Many diseases of the human brain that affect eye movement were readily explained by computational and physiological concepts of ocular motor control models. The neuroscience of eye movement, depth of physiological understanding, and computational explanation for almost every aspect of these movements had a natural next step to extend these concepts and study neural control of other motor behavior and dysfunction. Simultaneously, there was an unprecedented growth of portable and user-friendly technology to precisely and cost-effectively measure the eye movements. With the development of data acquisition systems, along with the growing interest of non-motor scientists, the study of eye movements made yet another leap into human cognition and behavior applications. The interaction and common interest of engineers, physicians, biologists, and psychologists were instrumental in the multidisciplinary growth of ocular motor research. This volume, *Advances in Neuroscience of Eye Movements*, reflects the truly multidisciplinary nature of the ocular motor field.

Cleveland, OH, USA   Aasef Shaikh
Cleveland, OH, USA   Fatema Ghasia

Contents

Contributions of David A. Robinson (1924–2016) to Understanding Eye Movements: An Appreciation

David S. Zee, Mark J. Shelhamer, R. John Leigh, and Lance M. Optican

Abstract In this appreciation of the scientific contributions of David A. Robinson by his former students, examples are provided of his seminal work in the fields of ocular motor and vestibular research. Drawing on a mathematical and engineering background, Robinson first invented a reliable method for measuring eye movements and then employed a bottom-up approach, defining the mechanical forces acting on the eyeballs, quantifying the neural code by which ocular motoneurons program eye movements. He then moved on to elucidate the premotor neural processing required to generate eye movements that respond to visual and vestibular needs. Each of his many contributions has had a sustained impact within and beyond his field – so that the neural control of eye movements has become a microcosm for motor control in general. Examples include understanding the mechanical properties of the orbital tissues, which has influenced treatment of strabismus; identification of mathematical integration of sensory signals by networks of neurons to hold gaze steady; central interaction of vestibular and visual signals to generate appropriate eye movements during self-rotation; adaptive properties of eye movements to optimize their contributions to clear vision; and internal interactions of motor and feedback signals to guide rapid eye movements (saccades). Robinson promoted translational research by proposing models based on normal biology that made specific, testable predictions about how abnormal eye movements, such as nystagmus, could arise. His pioneering contributions continue to have relevance in neurophysiology, computation neuroscience, neurology, ophthalmology, and otolaryngology.

D. S. Zee
Department of Neurology, Johns Hopkins University, Baltimore, MD, USA

M. J. Shelhamer
Department of Otolaryngology – Head & Neck Surgery, Johns Hopkins University, Baltimore, MD, USA

R. J. Leigh
Department of Neurology, Case Western Reserve University, Cleveland, OH, USA

L. M. Optican (✉)
Laboratory of Sensorimotor Research, National Eye Institute, NIH, Bethesda, MD, USA
e-mail: lmo@lsr.nei.nih.gov

© Springer Nature Switzerland AG 2019
A. Shaikh, F. Ghasia (eds.), *Advances in Translational Neuroscience of Eye Movement Disorders*, Contemporary Clinical Neuroscience,
https://doi.org/10.1007/978-3-030-31407-1_1

Keywords Eye movements · Vestibuloocular reflex · Visual · Orbital · Strabismus · Nystagmus · Adaptation · Saccades · Magnetic search coil

1 Introduction

David A. Robinson was a pioneer in the application of engineering analysis to bio-medical systems. His primary focus was on motor systems. He asked how the brain generated the innervation patterns needed to cause muscles to contract appropri-ately for a movement. He studied vestibular eye movements (due to head motion), optokinetic movements (response to large moving fields), smooth pursuit move-ments (tracking small moving targets), vergence movements (changes in the align-ment of the eyes between viewing distant and close targets), and saccadic movements (voluntary fast, brief eye movements used to change version – the direction of view of both eyes together). Robinson made significant contributions to understanding all of these movements because he started with a bottom-up approach: first understand the mechanical properties of the orbital tissues (globe, extraocular muscles, liga-ments, etc.) and then determine the innervation needed by this *oculomotor plant* to generate each type of eye movement.

Robinson was a Renaissance man and grasped the potential significance of his work to neuroscience and medicine in general; thus, he was a pioneer in bench-to-bedside, or translational, research. He cleverly applied quantitative hypotheses (models) of normal behavior, based on engineering principles, to interpret abnormal behavior in patients. These simple, accessible models provided clinician-scientists with new insights into the mechanisms of a variety of clinical phenomena, ranging from bedside signs (such as Alexander's law and the Bielschowsky head-tilt test) to effects of cerebellar lesions, and several forms of nystagmus, and strabismus.

Here, we – his students and colleagues, who called him "Dave," – will each pres-ent a few, representative examples of Robinson's major contributions to understand-ing eye movements in health and disease. We hope that these will also exemplify how he changed both clinical and basic research for the better.

2 Lance M. Optican: Robinson's Contribution to Eye Movements

David A. Robinson's contributions to our understanding of the mechanics, neuro-physiology, and behavior of eye movements are vast. Here, I will focus on his life-long effort to understand the mechanics of the oculomotor plant, which formed the basis for many of his other contributions.

Dave brought his expertise and experience as an engineer to biomedical research into eye movements in the early 1960s. At that time, it was thought that the location of a target to be brought to the fovea of the retina (where it could be seen best) was directly translated into an innervational command encoding the desired eye orientation (e.g., a step change for saccades). Eye movements were then the result of passing this innervation change to a slightly underdamped second-order system (representing the orbital tissues, globe, and muscles, called the *oculomotor plant*) (Westheimer 1954). In his seminal studies, Robinson measured the forces on the eyeball during saccades and pursuit (Robinson 1964, 1965). He found that the plant was, in fact, heavily overdamped, and the briskness of eye movements could only be achieved by crafting innervational signals that would compensate for the rather slow dynamics of the plant. For example, a saccade required a high-frequency burst of innervation (*pulse*) to overcome the viscous drag of the eye muscles, a tonic increase in firing (*step*) to hold the eye in its final orientation against the elastic restoring forces of the orbit, and a gradual transition between the pulse and step (*slide*) to compensate for long time-constant changes in muscle force. At the time, very little was known about how neurons in the oculomotor system behaved, but later, neurophysiological experiments found that innervation during saccades followed a pulse-slide-step pattern (Fuchs and Luschei 1970; Robinson 1970; Goldstein 1983).

This discovery embodies one of the central contributions that Dave made to eye movement research, which we might call the *outside→in* or *downstream→upstream* approach. Robinson emphasized that before you could understand what innervation the brain needed to generate to make the eyes move, you had to understand how the plant would react to those commands. Just as this approach applied to the motor neurons innervating the eye muscles, it also applied to the previous stage, the premotor neurons, and so on, deeper and deeper into the brain. Thus, Dave's approach taught us that to understand the brain we had to walk inward from the periphery.

The original plant model was a lumped, linear model, representing all the details of the orbit in terms of basic viscoelastic elements. Robinson realized that to make a more detailed model of the plant, it was important to measure the properties of the orbital tissues, particularly the extraocular muscles (EOMs). This led to a landmark study of orbital properties in human patients during strabismus surgery (Robinson et al. 1969). The lateral and medial recti were detached from the globe so that the length-tension relationship of lateral recti muscles under various levels of innervation, the passive length-tension relationship, and the passive restoring force on the globe could be measured (Fig. 1). This study improved the models of the oculomotor plant by allowing different properties to be ascribed to active and passive tissues in the orbit.

The next step in understanding eye movement mechanics was to consider the geometry of the orbital tissues. Robinson realized that the orientation of the eye was determined by the balance of passive and active forces acting on the globe. However, it was not sufficient to know the force exerted by the muscle. Because the eye rotates in three dimensions, it is the torque around an axis, and not the force alone, that is relevant. Dave determined how the geometry of the globe led to the conversion of forces to torques in a seminal paper on strabismus (or squint, i.e., the misalignment of the two eyes, Robinson 1975). The eye's orientation could be determined from

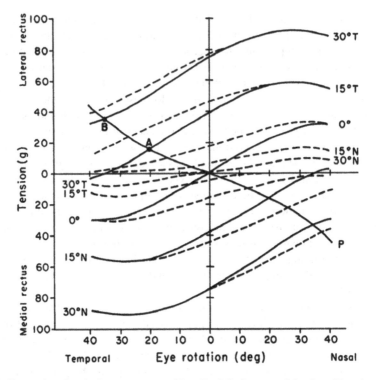

Fig. 1 Composite of static forces on eye position. Partially innervated developed length-tension curves for lateral and medial recti are shown above and below (respectively) the dashed lines. Their sums for gaze efforts of 0, 15, and 30° nasally (N) and temporally (T) are in solid lines. Curve P is the combined force of the passive muscle components and the globe suspensory tissues. (Reproduced, with permission, from Robinson et al. 1969)

the mechanical properties of the EOMs and the passive orbital tissues, the muscle's force, length and innervation, and the torque vector. This work created the field of computational analysis of strabismus and its surgical correction. Interestingly, Dave found that the path of the EOM from the origin to its insertion on the globe was not constrained by geometry. For example, if the lateral rectus muscle simply followed the shortest path (a great circle) from the insertion to the origin, when the eye turned, the muscle would slip over the globe, crossing the cornea. Obviously, something must prevent the muscle from taking such absurd positions. This insight from modeling led eventually to the finding that muscles passed through pulleys before inserting on the globe (Demer et al. 1995) (Fig. 2).

Even in his early work, Westheimer (1954) realized that the eye plant was nonlinear and that his lumped, linear second-order model was only an approximation. Indeed, others had shown that both length-tension and force-velocity relationships were nonlinear. Given Robinson's model of the mechanics of the orbit, it should have been possible to build a more accurate model of the plant. However, even by

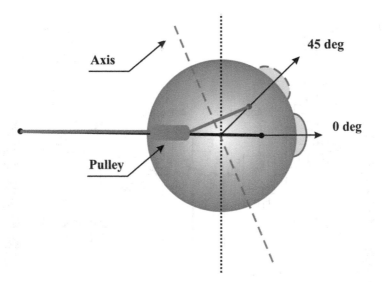

Fig. 2 Effect of pulleys on the axis of rotation of the lateral rectus muscle. As the eye elevates from 0° to 45°, the axis tips from vertical (black dotted line) to about 22.5° back (green dashed line). This allows innervation to the muscle to have the same effect, regardless of the elevation of the eye

today, this has not been done. Dave recognized that making a distributed, nonlinear model of the plant was prevented by our lack of understanding of the dynamic behavior of orbital elements. In a landmark paper, Robinson laid out the implications of nonlinear length-tension and force-velocity curves (which were also dependent upon innervation and active state tension) (Fig. 3). Given these dependencies, Dave concluded that more facts, rather than more modeling, were needed (Robinson 1981).

That study highlights what made Dave such a superb and influential modeler: his recognition of the primacy of the data and his unwillingness to be led astray by theoretical concepts or principles that, while elegant, might have no relation to nature's solutions. Thanks to Robinson's work, and his outside→in approach, we now have a remarkably detailed understanding of how the brain processes visual information into the innervations needed to make an eye movement. Dave always emphasized that the reason so much could be understood about eye movements was because of the simplicity of the system: it had a clear goal (align the fovea with the image of the target), it used only cranial nerves to innervate the EOMs (i.e., no spinal cord processing was needed), and the globe essentially rotated as a sphere around a fixed point in the orbit (i.e., only three degrees of freedom). This knowledge of one system can now be extended to clinical disorders (see below) and may be used to guide our understanding of other motor control systems.

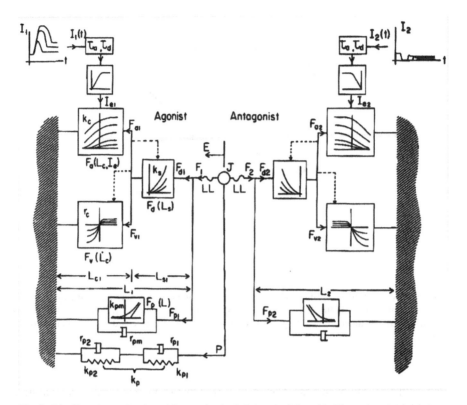

Fig. 3 Distributed organization of the mechanical elements of the orbit. The series-elastic element ($F_d(L_s)$) and the force-velocity relationship ($F_v\left(\dot{L}_c\right)$) are parametrically modulated by active state tension (F_a). The length-tension-innervation curves are denoted by $F_a(L_c, I_*)$. Waveforms for the innervations I_1 and I_2 are shown for saccades of different sizes. Time constants τ_a and τ_d represent activation and deactivation time constants. The stiffnesses k and viscosities r with various subscripts indicate mean or local slopes of functions often used in linear approximations. The curves in each box represent the nonlinearity appropriate for the state of innervation and active state tension. (Reproduced, with permission, from Robinson 1981)

3 Mark Shelhamer: Robinson's Contributions to the Vestibular System

3.1 Engineering Outlook

David Robinson made significant contributions to the understanding of vestibular and vestibuloocular functions. His work in this area was inspired in part by his observation, upon first seeing a diagram of the vestibular labyrinth, that it looked like a system that was designed by an engineer. In other words, that was a system whose structure directly related to its function and therefore one to which he could apply his engineering insights. In fact, one cannot fully grasp the basis of his contributions to neurophysiology – including the vestibular system – without understanding that he came from a solid background in electrical engineering and was a practicing engineer in industry before becoming an academic neuroscientist. The magnetic search coil system was a direct result of this experience, enabling precision measurement of eye movements with excellent temporal resolution (Robinson 1963). Likewise, his development of a method to measure single-unit neural responses in awake, behaving monkeys was also a key breakthrough (Robinson 1970), which meshed well with his contention that an appreciation of the purpose of a motor system is essential if one wants to understand its function. In that sense, the function of eye movements in animals that are not awake and free to move their heads is a moot point and perhaps not worthy of serious investigation. The third direct consequence of his engineering viewpoint is that it provided key insights stemming from a systems approach: an appreciation of how different physiological systems work together to accomplish functional goals, driven by the biological needs of the organism. Finally, the use of mathematical models was clearly influenced by this engineering background, especially as these models dealt directly with signals (position, velocity, etc.) familiar to any engineer of that era.

3.2   Identification of the Neural Integrator

In all his research, he brought an engineering sensibility that helped to organize data and guide future work. This is seen perhaps most clearly in the research that led to the insight that a central neural integrator had to exist – neural machinery that converts eye-velocity signals to the eye-position signals that are needed to maintain eccentric eye positions. This integrator was postulated very clearly in work that characterized the responses of abducens motoneurons during several types of eye movements (Skavenski and Robinson 1973). (This integrator is different from the velocity storage integrator described below. However, it was also noted in this same publication that relative phase considerations pointed to the likelihood of this other integrator to extend the low-frequency response of the vestibuloocular reflex (VOR).)

The reasoning as described in this seminal 1973 publication is very clear and in retrospect holds up well as a demonstration of the value of the systems approach. The semicircular canals (SCC) of the vestibular system are angular-acceleration sensors, yet there is an eye-position signal on the abducens motoneurons. Therefore, two mathematical integrations (in the sense of calculus) must take place, to convert acceleration to velocity and velocity to position. The first of these integrations is provided by the canals themselves, which act as angular-velocity sensors over the physiological range of head movements due to their highly damped dynamics. Experimental and theoretical considerations showed that the mechanics of the ocular orbit itself could not provide the second integration, at least over the frequency range of interest. Therefore, a distinct central velocity-to-position integrator had to exist, which was later found and studied extensively (Fig. 4).

Also noted in this work was the distributed nature of some of the signals on the extraocular muscles: different cells had slightly different combinations of velocity

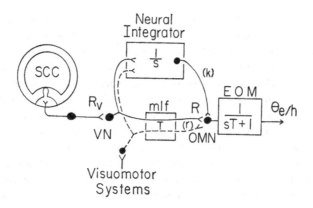

Fig. 4 Early "wiring diagram" indicating the necessity for a central neural integrator to provide the second integration that finally converts head angular acceleration sensed by the vestibular semicircular canals (SCC), via the vestibular nuclei (VN), to the eye-position signal found on abducens ocular motoneurons (OMN; discharge rate, R). A parallel path, here through the medial longitudinal fasciculus (mlf), with gain T, provides the accompanying velocity component that is needed to drive eye velocity and position correctly. θ_e/h: eye position with respect to the head. (Reproduced, with permission, Skavenski and Robinson 1973)

and position parameters, suggesting a mixing in this final common path of preliminary signals that were subject to slightly different processing via internal circuitry. This presaged later work in neural networks (see below).

3.3 Symbiosis in Visual-Vestibular Interaction

A perfect example of the modeling aspect of Robinson's work is a classic paper that not only made a significant contribution to the understanding of visual-vestibular interaction but also elucidated the philosophy behind the use of models in physiology (Robinson 1977b). The model in this case (Fig. 5) represents a beautiful accumulation of the known facts at the time regarding the main properties of optokinetic nystagmus (OKN), which is generated by the movement of a visual field, the vestibuloocular reflex (VOR), which is stimulated by head movement and their combination. The model begins with the observation that the semicircular canals of the vestibular system are well-suited to the transduction of transient head movements, while the visual system is better suited to transduction of sustained head motion that results in homogeneous motion of the visual field. Information on head motion from these two systems with complementary dynamics is combined in a model that leads to a parsimonious explanation of how the systems work together to maintain compensatory eye velocity (the VOR) during prolonged head movements. Several other observations are explained by this model; these emergent properties support the validity of the model and suggest further experiments. One of these is the prolonged

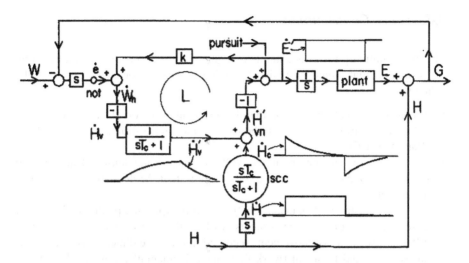

Fig. 5 Diagram representing vestibular-optokinetic symbiosis: the combining of vestibular and visual information about head movement to produce correct compensatory eye movements. The loop L instantiates, through positive feedback of the eye-velocity command Ė, a central leaky integrator or low-pass filter. This is one of the earliest presentations of what came to be known as velocity storage: the prolongation of the semicircular canal time constant to improve the low-frequency response to sustained head movements. (Reproduced, with permission, from Robinson 1977a)

time constant of vestibular nystagmus compared to that of the cupula and first-order vestibular afferents (velocity storage; Raphan et al. 1979). The model also confirms an earlier prediction of the presence of an efference copy signal of the oculomotor velocity command.

This work not only beautifully laid out the mathematics and the modeling of velocity storage but also made the case perfectly clear for why modeling is so important. Modeling is a hypothesis made concrete: "To ask if modelling is useful in oculomotor physiology is to ask if hypotheses are useful in science. It is an absurd question." This came at a time when many studies in neurophysiology were merely cataloging varieties of neural responses with no attempt to explain them with an overarching concept or model. This kind of interpretation was perceived by some as outside the realm of scientific investigation: teleology was not allowed. "Unfortunately, some of the entrenched establishment have even come to believe that the whole purpose of science is to collect data and actually oppose all attempts to explain them by models." Again, Robinson's view was always guided by the functional biological needs of the organism. "The purpose of the system being modelled must be clear, it must fit in with the animal's natural behavior and perform a useful function. It must be compatible with other functional subsystems and with phylogeny."

3.4 Other Contributions

This brief summary neglects a number of other contributions to the understanding of vestibuloocular function, with an emphasis on function. These include the following:

- The very practical observation that the instantaneous gain of the VOR is under voluntary control dependent on the mental state and instructions to the subject (Barr et al. 1976).
- Early work on adaptation of the VOR to magnifying lenses (Gauthier and Robinson 1975), including the observation that VOR gain can be state dependent.
- The use of matrices (Robinson 1982) to help interpret the plethora of data that were being generated by experiments in which the subjects were moved around different axes and eye movements recorded in the three dimensions of horizontal, vertical, and torsional (work that, not coincidentally, was enabled by the search coil system for measuring eye movements).
- The finding that the VOR can be turned off at the neural and functional levels during the generation of saccadic eye movements (Laurutis and Robinson 1986), which resolved an ongoing controversy and again demonstrated the tight integration of multiple systems to accomplish functional goals.

Another notable example of this type of functional systems thinking is in a model of the generation of VOR fast phases (Chun and Robinson 1978). It was found that the timing and eye positions of the beginning and end of nystagmus fast phases can be adequately described by random processes with appropriate probability distributions. However, the eye position at the end of each fast phase had a specific constraint, which explains the overall pattern of this nystagmus: the eye movements are not symmetric about the midline but instead are biased in the direction of the fast phases. This is also the direction in which the head is moving during the head rotation that generates vestibular nystagmus, and thus the interpretation of this constraint is that it represents an area of space in which something important (or harmful) might appear. In other words, it is not the case that the head movement drives the eyes as far as they can go in one direction, with this smooth tracking movement interrupted by a fast phase only because the eyes have reached the limit of their excursion in the head. Rather, the fast phase purposefully moves the eyes so that they can look at a location that has some possible relevance, after which the slow tracking movement maintains gaze on that location. Again, it is the function of the entire system that must be understood, and the mathematical model in this case meshes perfectly with that understanding.

3.5 Neural Networks and Farewell to Black Boxes

As noted above, ideas about distributed processing of oculomotor signals were present already in work as early as that with Skavenski in 1973. This was made explicit in later work on neural networks (Cannon et al. 1983; Arnold and Robinson 1991), where Robinson came to think of his earlier modeling as of lesser significance, because the newer network approaches showed that there are no black boxes (as in the early "black-box" models) but rather that signals are represented in a distributed manner in the brain. There is no single eye-velocity signal that is converted to a single eye-position signal, as might be performed by a discrete well-defined integrator (Anastasio and Robinson 1989); the presence of various amounts of position and velocity dependence in different brainstem neurons made this clear.

The contributions of this work in neural networks came not only from the neurophysiology but also the bridging of fields by which he could again explain the need for the math to physiologists and the nature of the physiology (including teleology) to engineers: "It is argued that trying to explain how any real neural network works on a cell-by-cell, reductionist basis is futile and we may have to be content with trying to understand the brain at higher levels of organization" (Robinson 1992). This line nicely summarizes Robinson's views on neuroscience. Part of his genius was in asking the right question. One aspect of this is limiting the scope of the problem under study, so that it could be understood with straightforward input-output approaches, investigated with the available technology (some of which he invented), and analyzed with the systems theory with which he was so intimately familiar. With the move into neural networks and more complex functional questions (selection of which visual target to track, cognitive influences on oculomotor behavior, etc.), he was led to admit that there are limits to the reductionist approach, such that understanding "higher levels of organization" will be necessary for further progress. Work on these "higher levels," however, must still be based on the foundations provided by the reductionist approach that he championed.

3.6 Translational Influences

It would be hard to identify a specific contribution in the vestibular area that has had a distinct translational impact on clinical care. Nevertheless, the work in this area has helped provide the foundation for some of the current clinical approaches and helped solidify their scientific and neurophysiological foundations. This work has influenced how the clinical-research community thinks about the VOR and vestibular contributions to eye movements. The flexibility of the VOR, modifiable as needed to serve the functional needs of the organism as part of a larger multisensory system, is a prevalent theme. This is reflected in current approaches to vestibular rehabilitation that work to modify not only VOR gain, per se, but also the combined gaze-control system including preprogrammed saccades. The systems view and the needs of the organism are again paramount.

4   R. John Leigh: Robinson's Contribution to Clinical Medicine

David A. Robinson was a pioneer in translational research before that term came into vogue. He was unusual in being a basic scientist who did not hesitate to apply his knowledge of computational neurobiology to the investigation and interpretation of clinical disorders of eye movements. He was unafraid of "getting his hands dirty" with clinical problems, being curious and confident that a mathematical approach could provide insights, if applied intelligently. When he first entered biomedicine, he realized that he needed more than his electrical engineering background could provide and took medical school classes in physiology and anatomy, including human dissection. He relished solving clinical mysteries, a good example being periodic alternating nystagmus (PAN) (Leigh et al. 1981; Garbutt et al. 2004). Individuals affected by this rare clinical disorder show continuous, spontaneous nystagmus that reverses direction approximately every 2 minutes (the period of oscillations tends to be consistent for each individual but varies between individuals, typically being 3–4 minutes).

At the time that Robinson and his clinical colleagues encountered an individual with PAN, his interests lay in developing models to account for normal interactions between vestibular and optokinetic systems. As mentioned in the prior section, Raphan et al. (1979) had shown that, when monkeys were rotated in darkness, vestibular nucleus neurons encoded a temporally enhanced (partially integrated) signal compared with primary afferents from the semicircular canals of the vestibular labyrinth. They coined the term "velocity storage" to describe this central perseveration of the raw vestibular signal. Henn, Young, Waespe, and their colleagues in Zurich demonstrated that, when monkeys were rotated in an illuminated environment, neurons in the vestibular nuclei (VN) encoded both inputs from the semicircular canals of the vestibular labyrinth and visual (optokinetic) inputs (Henn et al. 1974; Waespe and Henn 1977). Independently, Young and Oman (1969) and Malcolm and Melvill Jones (1970) had characterized an adaptive mechanism that nulled and even reversed the direction of nystagmus induced by unnaturally sustained rotations.

Prompted by Kornhuber's observation that PAN had similarities to the reversal phase of vestibular nystagmus that occurs during sustained rotations (Kornhuber 1959), Robinson developed a model for the vestibuloocular reflex (Fig. 6) that incorporated both velocity storage and an adaptation mechanism (Leigh et al. 1981). He reasoned that, if velocity storage became unstable (causing unnaturally prolonged vestibular responses), then the adaptation mechanism would act to null and then reverse the direction of the nystagmus. Assigning parameter values to the

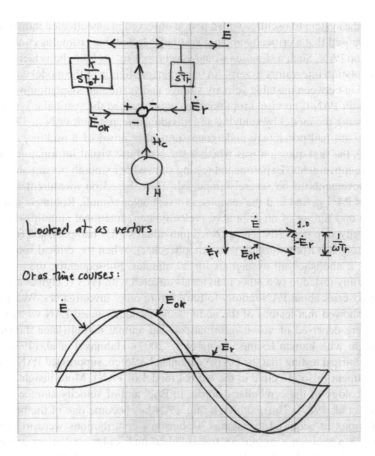

Fig. 6 Copy of the first page of Robinson's notes in which he applied a small-signal linear approximation to model periodic alternating nystagmus. In the top diagram, head velocity ($\dot{H}$) is transduced by the vestibular semicircular canals into a signal $\dot{H}_c$, which is passed on, after a sign change in the vestibular nucleus (summing junction), to become an eye-velocity command ($\dot{E}$); this is a simplified version of the vestibuloocular pathway. The signal $\dot{E}$ is also passed to the velocity storage integrator, which perseverates both the vestibular and optokinetic inputs (not shown) with a time constant T_o and gain k to generate a signal $\dot{E}_{ok}$. A central adaptation, or repair, operator (with time constant T_r) monitors the slow-phase eye-velocity command, $\dot{E}$, and acts to eliminate persistent nystagmus by sending a signal back ($\dot{E}_r$) to the vestibular nucleus, to cancel the source of vestibular imbalance. Robinson postulated that if the value of the gain k of the velocity storage mechanism, which was normally <1.0, increased, the system would become unstable. Then the adaptation operator would act to cancel the increasing signal, causing it to run away in the opposite direction – causing oscillations. At the bottom, Robinson schematizes these oscillations either as the vector sum of outputs from the velocity storage ($\dot{E}_{ok}$) and adaptation ($\dot{E}_r$) integrators or as the timed sum of two corresponding sine waves

velocity storage and adaptation integrators based on experimental work, he found that the model could generate PAN with a period like that observed in affected individuals. Applying a small-signal, linear approximation approach, Robinson went on to derive equations specifying parameter values for the system to become unstable and for the system to oscillate with a period observed in any affected individual. He also proposed that a critically applied vestibular (rotational) stimulus could temporarily stop PAN. Such a stimulus would synchronously set both the velocity storage and adaptation integrators to zero. An important prediction of the model was that the interaction between unstable velocity storage and vestibular adaptation would only account for PAN if no visual (optokinetic) input reached the vestibular system.

In testing the model by studying an individual affected with PAN (a 39-year-old woman who had previously undergone surgical removal of a midline, cerebellar abscess), the first question was whether she could use visual information to influence eye movements. Experimental testing showed her smooth pursuit and optokinetic movements to be severely impaired. Then, based on measurements of the patient's PAN period and the responses to vestibular stimuli, Robinson calculated the size and timing of an impulsive (velocity-step) rotational stimulus that could temporarily stop her PAN. The affected individual's PAN had persisted for 5 years (preventing her from working as a schoolteacher). When the critical rotation was applied, it abolished all nystagmus for 20 minutes. Subsequently, the model was successfully tested in two other individuals affected with PAN (Leigh et al. 1981).

Discoveries about PAN rapidly followed from other investigators. Waespe et al. (1985) showed that lesions of the nodulus in macaque caused PAN when the animals were deprived of visual information – an animal model (later shown to be consistent with human lesions; Jeong et al. 2007). Halmagyi et al. (1980) found from empirical testing that the $GABA_B$ agonist, baclofen, suppressed PAN – a clinical treatment (also effective in the animal model of PAN). More insights into the pharmacology of the cerebellar disorder in PAN and of velocity storage followed (Cohen et al. 1987). Thus, although rare, PAN has become one of the best understood forms of acquired nystagmus. Robinson's contributions were to grasp the relationship between a mysterious form of abnormal ocular oscillations and current concepts of vestibular physiology and then to develop a model that could be tested quantitatively in a specific way. Could more modern approaches to modeling PAN, such as using a neural network or a Bayesian approach, be developed? This seems probable, but any such model would have to live up to Robinson's intuitively appealing model that makes precise predictions.

On a more general level, David Robinson taught clinicians that they must formulate their ideas into clear, preferably quantitative, hypotheses that can be tested experimentally. And he taught basic scientists not to be afraid of applying current concepts of basic science to understanding clinical disorders. In promoting translational research, he was aware of Francis Bacon's observation that corroborative evidence, emanating from different approaches, often converges on the truth. Such research often requires the development of collaborations, trust, and friendships between investigators from diverse backgrounds – one of the most important attributes of the research community.

5 David S. Zee: Concluding Remarks

The breadth of Dave Robinson's contributions to clinical medicine and basic neuroscience is nicely exemplified in the discussions above. Importantly, one can see that his impact in so many fields, neurology, ophthalmology, otolaryngology, neuropsychology, biomedical engineering, and neurophysiology, is still obvious and continuing to influence new areas of scientific and clinical inquiry (Shadmehr 2017; Shaikh et al. 2016). Dave believed fervently in the scientific method and quantitative hypothesis testing. He abhorred "hand waving" to explain behavior. He believed that science was for the good of humanity and that scientific advances should be openly shared, without attention to patents or personal profit. Dave believed deeply in evolution and Mother Nature's "design." He emphasized that to understand eye movements, one must begin with the phylogenetically oldest forms – the slow and quick phases of vestibular responses – because their anatomical substrate was the scaffolding upon which all subsequent subtypes of eye movements developed. Indeed, so much of our current understanding of how fast and slow eye movements are generated grew out of Dave's studies of the vestibuloocular reflex.

As noted above, Dave attacked problems in a striking "bottom-up" approach. For example, to understand eye movements, you need to measure and quantify them. So he developed the magnetic field search coil technique, an "obvious" choice, as he put it, based on his experience in industry with motors and magnetic fields. This, of course, is a classic example of cross-fertilization between different fields. Dave also knew that to understand what type of problem the brain was trying to solve when generating eye movements, one must know the mechanical properties of the object (globe) to be moved. So he began measuring the properties of the muscles and orbital tissues. Next, he knew that to confirm what he had inferred from his studies of ocular muscles about the control signals to generate eye movements, he had to measure neural activity during eye movements. So he first developed and studied the properties of a microelectrode with which he could listen to the brain and relate the number of spikes to ongoing behavior. He and Albert Fuchs then developed ways of measuring eye movements in monkeys and immobilizing their heads (using the human orthopedic "crown" technique). One of the most exciting public presentations, which was so stimulating to many neuroscientists especially early in their careers, was the movie Dave made in which he showed a two-dimensional recording of monkey's eye movements on an oscilloscope and at the same time one could hear, from the implanted microelectrode, how the neural discharge of a single oculomotor neuron would change as the animal moved its eyes around the orbit using different types of eye movements (see Leigh and Zee 2015, for a video clip). The famous "pulse-step" change of innervation emerged from these studies of the mechanical properties of eye muscles and the corresponding neural activity in ocular motoneurons.

Dave next moved up the neural tree to the areas in the brainstem where the pulse-step might be generated. This is where I came into the picture with the study of a patient whose saccades were slow. Because we showed that her saccades could be modified in flight, their ballistic nature, which then was the current understanding, came into question, and the internal feedback, "bang-bang," control model for generation of saccades emerged. This model survives, with only minor modifications, to this day. But, as Dave said, anybody can make a model, but making a physiological and anatomically realistic model was another issue. So the next step was to find neurons that behaved as if they were in a network in which internal feedback was being used to guide a saccade in flight, and he and his colleagues did just that.

Dave often asked how does the brain manage to keep such neural networks working properly in the long-term as we develop or age or are faced with the accidents of nature that cause disease and trauma. Hence, Dave's interest in neural plasticity and the "repair shop" occupied so much of his scientific career. This issue, of course, directly relates to problems that neurologists face when trying to diagnose patients with a more chronic lesion that has been modified by the brain's attempt to repair it.

My desire to work in Dave's laboratory came when I was a neurology resident and had an epiphany while listening to Dave's inspiring lecture on how models and mathematics could explain a "simple" neuro-ophthalmological syndrome called internuclear ophthalmoplegia in which the fibers carrying signals from the abducens nucleus to the oculomotor nuclei are interrupted. I approached Dave after his talk and said I wanted to work with him. He said fine and "I have been waiting for a neurologist to come and work with me for many years." As John Leigh has emphasized here, Dave realized early on that much could be learned about how the normal brain functions by studying patients who had been afflicted due to unfortunate accidents of nature, from disease and trauma. I joined Dave's lab as a third-year neurology resident, and he sat down with me almost daily, one on one, trying to teach me some control systems. Out of these sessions grew his "control systems for neurologists" course through which so many fellows in his lab and our clinical fellows learned so much.

Dave was rigorous, uncompromisingly analytical, and could make you feel worthless at times with his famous red pen which scorched so many of our early (and late) drafts of papers and grants. But he gave of his time generously, helping so many of us with our careers and never asking to be a coauthor or coinvestigator. Dave truly was an intellectual giant in neuroscience, and his discoveries are a driving force for so many clinical and basic neuroscientists. One might even say that Dave also provided an intellectual "pulse-step," pushing us forward and not allowing us to fall back, so that we all might succeed in our careers. We are deeply thankful to him in so many ways.

Acknowledgments We thank Dr. Christian Quaia for comments on the manuscript.

References

Anastasio, T. J., & Robinson, D. A. (1989). The distributed representation of vestibulo-oculomotor signals by brain-stem neurons. *Biological Cybernetics, 61*(2), 79–88.

Arnold, D. B., & Robinson, D. A. (1991). A learning network model of the neural integrator of the oculomotor system. *Biological Cybernetics, 64*(6), 447–454.

Barr, C. C., Schultheis, L. W., & Robinson, D. A. (1976). Voluntary, non-visual control of the human vestibulo-ocular reflex. *Acta Oto-Laryngologica, 81*(5–6), 365–375.

Cannon, S. C., Robinson, D. A., & Shamma, S. (1983). A proposed neural network for the integrator of the oculomotor system. *Biological Cybernetics, 49*(2), 127–136.

Chun, K. S., & Robinson, D. A. (1978). A model of quick phase generation in the vestibuloocular reflex. *Biological Cybernetics, 28*(4), 209–221.

Cohen, B., Helwig, D., & Raphan, T. (1987). Baclofen and velocity storage – A model of the effects of the drug on the vestibuloocular reflex in the Rhesus-Monkey. *Journal of Physiology (London), 393*, 703–725.

Demer, J. L., Miller, J. M., Poukens, V., Vinters, H. V., & Glasgow, B. J. (1995). Evidence for fibromuscular pulleys of the recti extraocular muscles. *Investigative Ophthalmology & Visual Science, 36*(6), 1125–1136.

Fuchs, A. F., & Luschei, E. S. (1970). Firing patterns of abducens neurons of alert monkeys in relationship to horizontal eye movement. *Journal of Neurophysiology, 33*(3), 382–392.

Garbutt, S., Thakore, N., Rucker, J., Han, Y. N., Kumar, A. N., & Leigh, R. J. (2004). Effects of visual fixation and convergence on periodic alternating nystagmus due to MS. *Neuro-Ophthalmology, 28*(5–6), 221–229. https://doi.org/10.1080/01658100490889650.

Gauthier, G. M., & Robinson, D. A. (1975). Adaptation of the human vestibuloocular reflex to magnifying lenses. *Brain Research, 92*(2), 331–335.

Goldstein, H. P. (1983). *The neural encoding of saccades in the rhesus monkey*. Baltimore: Johns Hopkins University.

Halmagyi, G. M., Rudge, P., Gresty, M. A., Leigh, R. J., & Zee, D. S. (1980). Treatment of periodic alternating nystagmus. *Annals of Neurology, 8*(6), 609–611.

Henn, V., Young, L. R., & Finley, C. (1974). Vestibular nucleus units in alert monkeys are also influenced by moving visual fields. *Brain Research, 71*(1), 144–149.

Jeong, H. S., Oh, J. Y., Kim, J. S., Kim, J., Lee, A. Y., & Oh, S. Y. (2007). Periodic alternating nystagmus in isolated nodular infarction. *Neurology, 68*(12), 956–957. https://doi.org/10.1212/01.wnl.0000257111.24769.d2.

Kornhuber, H. H. (1959). Periodic alternating nystagmus (nystagmus alternans) and excitability of the vestibular system. *Archiv für Ohren-, Nasen- und Kehlkopfheilkunde, 174*(3), 182–209.

Laurutis, V. P., & Robinson, D. A. (1986). The vestibulo-ocular reflex during human saccadic eye movements. *The Journal of Physiology, 373*, 209–233.

Leigh, R. J., & Zee, D. S. (2015). *The neurology of eye movements* (5th ed.). New York: Oxford University Press.

Leigh, R. J., Robinson, D. A., & Zee, D. S. (1981). A hypothetical explanation for periodic alternating nystagmus: instability in the optokinetic-vestibular system. *Annals of the New York Academy of Sciences, 374*, 619–635.

Malcolm, R., & Jones, G. M. (1970). A quantitative study of vestibular adaptation in humans. *Acta Oto-Laryngologica, 70*(2), 126–135.

Raphan, T., Matsuo, V., & Cohen, B. (1979). Velocity storage in the vestibulo-ocular reflex arc (VOR). *Experimental Brain Research, 35*(2), 229–248.

Robinson, D. A. (1963). A method of measuring eye movement using a scleral search coil in a magnetic field. *IEEE Transactions on Biomedical Engineering, 10*, 137–145.

Robinson, D. A. (1964). The mechanics of human saccadic eye movement. *The Journal of Physiology, 174*, 245–264.

Robinson, D. A. (1965). The mechanics of human smooth pursuit eye movement. *The Journal of Physiology, 180*(3), 569–591.

Robinson, D. A. (1970). Oculomotor unit behavior in the monkey. *Journal of Neurophysiology, 33*(3), 393–403.

Robinson, D. A. (1975). A quantitative analysis of extraocular muscle cooperation and squint. *Investigative Ophthalmology, 14*(11), 801–825.

Robinson, D. A. (1977a). Linear addition of optokinetic and vestibular signals in the vestibular nucleus. *Experimental Brain Research, 30*(2–3), 447–450.

Robinson, D. A. (1977b). Vestibular and optokinetic symbiosis: An example of explaining by modelling. In *Control of gaze by brain stem neurons* (pp. 49–58). Amsterdam: Elsevier.

Robinson, D. A. (1981). Models of the mechanics of eye movements. In B. L. Zuber (Ed.), *Models of oculomotor behavior and control* (pp. 21–41). Boca Raton: CRC Press.

Robinson, D. A. (1982). The use of matrices in analyzing the three-dimensional behavior of the vestibulo-ocular reflex. *Biological Cybernetics, 46*(1), 53–66.

Robinson, D. A. (1992). Implications of neural networks for how we think about brain function. *Behavioral and Brain Sciences, 15*(4), 644–655. https://doi.org/10.1017/S0140525X00072563.

Robinson, D. A., O'Meara, D. M., Scott, A. B., & Collins, C. C. (1969). Mechanical components of human eye movements. *Journal of Applied Physiology, 26*(5), 548–553.

Shadmehr, R. (2017). Distinct neural circuits for control of movement vs. holding still. *Journal of Neurophysiology, 117*, 1431–1460, 2017.

Shaikh, A. G., Zee, D. S., Crawford, J. D., & Jinnah, H. A. (2016). Cervical dystonia: Disorder of a neural integrator. *Brain, 139*, 2590–2599.

Skavenski, A. A., & Robinson, D. A. (1973). Role of abducens neurons in vestibuloocular reflex. *Journal of Neurophysiology, 36*(4), 724–738. https://doi.org/10.1152/jn.1973.36.4.724.

Waespe, W., & Henn, V. (1977). Neuronal activity in the vestibular nuclei of the alert monkey during vestibular and optokinetic stimulation. *Experimental Brain Research, 27*(5), 523–538.

Waespe, W., Cohen, B., & Raphan, T. (1985). Dynamic modification of the vestibulo-ocular reflex by the nodulus and uvula. *Science, 228*(4696), 199–202. https://doi.org/10.1126/science.3871968.

Westheimer, G. (1954). Mechanism of saccadic eye movements. *A.M.A. Archives of Ophthalmology, 52*(5), 710–724.

Young, L. R., & Oman, C. M. (1969). Model for vestibular adaptation to horizontal rotation. *Aerospace Medicine, 40*(10), 1076–1080.

Part I
Basic Science

Multisensory Integration: Mathematical Solution of Inherent Sensory Ambiguities

Tatyana A. Yakusheva

Abstract The brain integrates information from multiple sensory modalities to generate appropriate motor output and create perceptual experiences of the environment. Multisensory integration is evident at the single-neuron level in the cerebral cortex as well as the subcortical areas or brainstem nuclei. In the last two decades, the cerebellum has received increasing interest as an essential structure for multisensory integration. Studies have shown that the cerebellum integrates vestibular signals with signals from other sensory modalities to generate predictions of our inertial motion, orientation, postural control, and gaze stabilization. Here, we review recent literature on the cerebellar role in the integration of vestibular, visual, and proprioceptive signals for spatial navigation. First, we present evidence that the cerebellum contributes to solving the ambiguity found in vestibular afferent information. Theoretical and behavioral evidence indicates that this vestibular sensory ambiguity is resolved by the central nervous system using a combination of otolith signal and rotational cues from the semicircular canals. Second, in the light of recent findings, we describe the role of the cerebellum in integrating vestibular and visual information. Third, we describe how the cerebellum may integrate vestibular and proprioceptive cues.

Keywords Cerebellum · Purkinje cell · Vestibular · Deep cerebellar nuclei

T. A. Yakusheva (✉)
Department of Otolaryngology, Washington University School of Medicine, St. Louis, MO, USA

© Springer Nature Switzerland AG 2019   21
A. Shaikh, F. Ghasia (eds.), *Advances in Translational Neuroscience of Eye Movement Disorders*, Contemporary Clinical Neuroscience,
https://doi.org/10.1007/978-3-030-31407-1_2

1 Multisensory Integration Necessary for the Computation of Inertial Motion and Spatial Orientation

1.1 A Vestibular Sensory Ambiguity

The vestibular system represents an excellent example of multimodal integration used by the central nervous system to resolve peripheral sensory ambiguity. The well-known "vestibular sensory ambiguity problem" arises from the fact that the otolith afferents cannot distinguish tilt of the body with respect to gravity from the translation in earth horizontal plane because both signals, at the periphery, carry net linear gravito-inertial acceleration information (Fernandez and Goldberg 1976a, b; Angelaki et al. 2004). Figure 1a illustrates an example of an ambiguous otolith afferent response during inertial acceleration (translation) and gravitational acceleration (tilt). Otolith afferent response is identical during tilt and translation. Hence, if our brain would rely only on information provided by otolith afferents, the actual movement would not be correctly perceived, resulting in problems of spatial navigation. Therefore, the central nervous system must use extra-otolith cues to estimate translation and tilt. Behavioral studies suggest that the central nervous system resolves this sensory ambiguity by combining otolith signals with signals that arise from the semicircular canal and visual cues (Angelaki et al. 1999; Merfeld et al. 1999, 2005a, b; Green and Angelaki 2003; Zupan and Merfeld 2003; MacNeilage et al. 2007).

To understand how our brain solves tilt/translation ambiguity, Angelaki and colleagues used a unique combination of tilt and translation stimuli in a subtractive (tilt − translation) or additive (tilt + translation) fashion such that the net linear acceleration experienced by the head is either null or double (Fig. 1a, bottom stimulus traces). Importantly, the tilt-translation stimuli isolate the signal from the semicircular canals (Angelaki et al. 2004; Shaikh et al. 2005; Yakusheva et al. 2007). As expected, otolith afferents did not respond during "tilt − translation" and doubled their response amplitude during "tilt + translation" because they faithfully encode net linear acceleration (Fig. 1a). Therefore, our brain faces a challenging task, to transform the ambiguous vestibular information provided by the otolith afferents into the appropriate signal to estimate gravity and inertial motions correctly.

1.2 NU Purkinje Cell Responses Reflect the Solution to the Tilt/Translation Ambiguity Problem: A Neural Solution to the Vestibular System Ambiguity

According to the multisensory integration hypothesis, to correctly estimate spatial orientation and translational acceleration, the brain must use gravito-inertial acceleration information from the otolith (α) and angular velocity (ω) signals from the

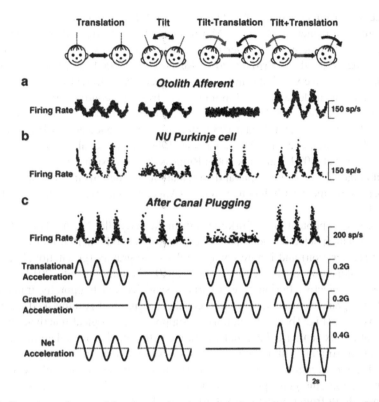

Fig. 1 Experimental protocol (top cartoons) and responses from (**a**) an otolith afferent, (**b**) simple spike responses of an NU Purkinje cell in labyrinthine-intact animals, and (**c**) simple spike responses of an NU Purkinje cell in canal-plugged animals. Responses were recorded during 0.5 Hz "translation," "tilt," and a combination of these stimuli, "tilt − translation" (stimuli out of phase resulted in zero net acceleration) and "tilt + translation" (stimuli in phase resulted in double net acceleration). All stimuli delivered in complete darkness. Note that the translational and tilt stimuli were matched in both amplitude and direction to elicit an identical linear acceleration in the horizontal plane (*bottom traces*). *Straight black* and *curved gray arrows* denote translation and tilt axes of stimulation, respectively. (Replotted from Angelaki et al. 2010 with permission)

semicircular canals (Angelaki et al. 1999; Merfeld and Zupan 2002; Zupan et al. 2002; Green and Angelaki 2004; Green et al. 2005).

Several brain areas such as vestibular nuclei (VN) in the brainstem and rostral fastigial nuclei (rFN) in the cerebellum have been used to search for the neural solution to the peripheral vestibular ambiguity (Angelaki et al. 2004; Green et al. 2005; Shaikh et al. 2005). Recent studies have indicated that the cerebellar nodulus and uvula (NU) may be a place where gravity and inertial motion signals are computed (Yakusheva et al. 2007, 2008; Angelaki and Yakusheva 2009; Angelaki et al. 2010; Laurens et al. 2013a, b). Anatomical studies support the idea that the posterior cerebellar vermis uses multisensory integration to resolve the peripheral vestibular sensory ambiguity problem. These areas of the cerebellar cortex receive ipsilateral projections from more than 70% of vestibular otolith and semicircular canal primary afferents (Carpenter et al. 1972; Marini et al. 1975; Korte and Mugnaini 1979; Kevetter and Perachio 1986; Gerrits et al. 1989; Barmack et al. 1993; Newlands et al. 2002; Maklad and Fritzsch 2003; Kevetter et al. 2004). Furthermore, the cerebellar NU receive vestibular information as secondary afferent mossy fibers from the vestibular nuclei and via climbing fibers from the contralateral inferior olive (Brodal 1976; Bigare and Voogd 1977; Freedman et al. 1977; Groenewegen and Voogd 1977; Brodal and Brodal 1985; Bernard 1987; Kanda et al. 1989; Sato et al. 1989; Thunnissen et al. 1989; Epema et al. 1990; Akaogi et al. 1994a, b; Barmack 1996; Ono et al. 2000; Barmack 2003; Ruigrok 2003). Lesions of the cerebellar NU result in balance problems, spatial disorientation, and deficits in the spatiotemporal properties of the vestibuloocular reflex (Angelaki and Hess 1995a, b; Wearne et al. 1998).

Recently, a population of Purkinje cells in the cerebellar NU has been identified as a translation-selective neurons that exclusively encode inertial motion (translational acceleration) and ignore changes relative to gravity (gravitational acceleration) (Yakusheva et al. 2007, 2008, 2010; Laurens et al. 2013a). These translation-selective Purkinje cells carry a spatially and temporally transformed position signal from the semicircular canals, a signal that is unmasked during tilt − translation (Fig. 1b, third column). In support, after surgical inactivation (plugging) of all six semicircular canals, Purkinje cells failed to modulate during tilt − translation stimulus, and their responses resemble that of the otolith afferent (Fig. 1c). Clearly, in the absence of information from the semicircular canals, the tilt/translation ambiguity problem cannot be solved, and Purkinje cells no longer distinguish tilt from translation (Yakusheva et al. 2007).

The following multisensory computational model depicted in Fig. 2 explains how net linear acceleration information from the otolith (α) and angular velocity signal (ω) from the semicircular canals are transformed and combined in the cerebellum (Yakusheva et al. 2007; Angelaki et al. 2010). How does the spatial and temporal transformation of semicircular canal signal happen? Because the semicircular canal afferents encode angular velocity in 3D space, the first step of canal signal transformation is a spatial transformation that extracts the earth-horizontal component of head rotation (ω_{EH}). This component represents tilt rotation, i.e., changes of the head orientation with respect to gravity. The second step is a temporal transformation of canal-driven signals such as to represent tilt position. Semicircular canal afferents carry angular velocity signal. Therefore, this signal must be integrated ($\int \omega_{EH}$) to obtain an angular position (tilt). Angular position, when combined with otolith information, cancels the gravitational component of the net linear acceleration, and the translation could be computed (Fig. 2) (Green and Angelaki 2004; Green et al. 2005; Yakusheva et al. 2007). In support of this multisensory model, cerebellar NU Purkinje cells modulate during tilt rotation, but not during yaw rotation, suggesting that they carry only the earth-horizontal (ωEH) component, but not the earth-vertical component (ωEV) of angular velocity (Yakusheva et al. 2007, 2008). The signal transformation mentioned above explains why, although anatomically NU receives inputs from horizontal and vertical semi-

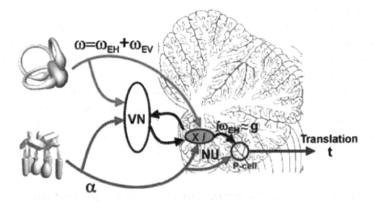

Fig. 2 Schematic illustration of our working hypothesis regarding the relationship between the processing of the semicircular canal and otolith signals within the NU. Semicircular canal afferents carry head-referenced angular velocity (ω), but the NU encodes only the earth-horizontal component (ω_{EH}). This signal is temporally integrated ($\int\omega_{EH}$) and used to cancel the gravitational component (g) of net linear acceleration (α), the signal carried by otolith afferents. The resulting output is the inertial component (t). (Replotted with permission from Angelaki et al. 2010 with permission)

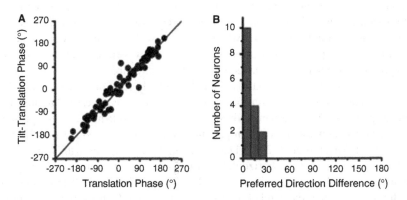

Fig. 3 Spatiotemporal matching of canal-driven and otolith-driven signals. (a) Response phase during the 0.5 Hz "tilt − translation" stimulus (canal-driven component) is plotted as a function of the respective phase during "translation" (otolith-driven component) ($n = 72$; data along the best-responding stimulus direction). Phase has been expressed relative to tilt velocity. (b) Distribution of the difference in preferred directions between the 0.5 Hz "tilt − translation" and "translation" stimulus conditions. (Replotted with permission from Yakusheva et al. 2007)

circular canals, NU Purkinje cells do not respond to yaw rotation but do respond to roll and tilt rotations (Kevetter and Perachio 1986; Fushiki and Barmack 1997; Kevetter et al. 2004; Yakusheva et al. 2007, 2008).

Importantly, a canal-driven signal should match otolith-driven signal not only temporally but also spatially (Fig. 3). Thus, the preferred direction of otolith-driven signal is transformed to align with the orientation of semicircular canals. Indeed, it has been suggested that cerebellar NU Purkinje cells carry during translation a combination of acceleration and velocity rather than linear acceleration signal (Yakusheva et al. 2007, 2008, 2010). The questions remaining are as follows: where does the transformation of the vestibular signal take place? Does the cerebellar NU receive already transformed vestibular afferent information from the vestibular or deep cerebellar nuclei? Does the transformation happen within the cerebellar circuitry of NU? In support of latter, a recent study in the ventral paraflocculus demonstrated that GABAergic inhibition within cerebellar cortex circuitry regulates input-output gain and spatial tuning of the Purkinje cells (Blazquez and Yakusheva 2015). More studies are needed to address these questions.

1.3 Gravity Signal Computed by the Cerebellar Nodulus/Uvula Purkinje Cell

Gravity and inertial motion signals are necessary for motor coordination, postural control, spatial orientation, and navigation (Horstmann and Dietz 1990; Merfeld et al. 1999; Senot et al. 2005; Zago and Lacquaniti 2005; Green and Angelaki 2007; Laurens et al. 2011). Studies in humans indicate that patients with cerebellar ataxia cannot properly estimate the direction of the gravity vector while exposed to visually rotating scenes (Dakin and Rosenberg 2018). Moving visual scenes cause the perception of tilt in the observer, which increases with stimulus velocity (Dichgans et al. 1972). This perceptual illusion is caused by the way the central nervous system integrates vestibular and visual signals, wherein vestibular signals estimate the gravity direction and the visual cue signals the body rotation. Dakin et al. (2018) reported that, in contrast to controls, visual cue prevails over vestibular cues in cerebellar patients. Other human studies have shown that self-motion perception during visual rotation was reduced in patients with cerebellar dysfunction (Bronstein et al. 2008; Bertolini et al. 2012; Dahlem et al. 2016). Patients with vestibulocerebellar lesions experienced horizontal positional nystagmus that could be explained by an erroneously biased estimation of the gravity (Choi et al. 2018). Thus, these findings suggest that the cerebellum uses an internal model to compute and process the vestibular signal in 3D space (Wolpert et al. 1998; Green and Angelaki 2007; Cullen et al. 2011).

Recently, a group of cerebellar NU Purkinje cells was identified that selectively encoded gravity signal, tilt-selective Purkinje cells (Laurens et al. 2013b). It was proposed that gravity information is computed by an internal model (Merfeld 1995; Merfeld et al. 1999; Laurens et al. 2011). In these experiments, the response of NU Purkinje cells was characterized during off-vertical axis rotation (OVAR). Using OVAR, it was demonstrated that not only tilt-selective NU Purkinje cells encode the head orientation of the gravity during arbitrary rotations in space but that translation-selective Purkinje cells responded to the illusion of translation produced by the OVAR (Vingerhoets et al. 2007; Wood et al. 2007).

During tilt/translation paradigm, tilt-selective Purkinje cells responded similarly to tilt, tilt − translation, and tilt + translation but would not modulate to translation stimuli only (Laurens et al. 2013b). This study suggested that the tilt-selective and translation-selective NU Purkinje cells are complementary to each other, such that their net sum represents the net linear acceleration encoded by the otolith afferents. It was reported that overall responses of tilt-selective Purkinje cells were smaller compared to translation-selective neurons. The tilt-selective Purkinje cells were also anatomically distinct; they were found close to the midline and in the anterior part of the cerebellar nodulus. Together, these studies demonstrated that two neuronal populations in the posterior cerebellar vermis encode translation and tilt signals that are computed using internal model (Yakusheva et al. 2007, Laurens et al. 2013a, b). These computed signals are essential for self-motion perception and could be used by hippocampal place cells for spatial mapping (Burguiere et al. 2005).

A puzzle that remains to be solved is whether the earth-vertical component (ω_{EV}) of angular velocity is processed by the cerebellum as well. As discussed above, despite anatomical evidence of horizontal canal input, the cerebellar NU Purkinje cells do not modulate during yaw rotations (Barmack and Yakhnitsa 2003; Yakusheva et al. 2007, 2008, 2010). Perhaps, the yaw-selective Purkinje cells similarly to tilt-selective Purkinje cells are localized in unexplored areas within the vestibulocere-bellum − their existence remains to be discovered.

2 Visual-Vestibular Integration in the Cerebellum for Self-Motion Perception and Spatial Navigation

During self-motion perception, the central nervous system integrates visual (optic flow), vestibular, somatosensory, and proprioceptive cues (Brandt et al. 1974; Dichgans et al. 1974; Hlavacka et al. 1992, 1996; Gu et al. 2008; Fetsch et al. 2009; Hensbroek et al. 2015). Self-motion in light conditions generates a type of visual stimulation called flow, which represents the motion of the visual image on the retina due to movements of the body through space. Optic flow is processed by the terminal nuclei of the accessory optic system (AOS) and the pretectum (Simpson et al. 1988; Gamlin 2006; Giolli et al. 2006). The AOS is linked to the vestibular system, and its efferents target the vestibular nuclei. Thus, they contribute to the vestibular system in gaze stabilization during rotations (Giolli et al. 2006). Behavioral studies have shown that large-field optic flow stimulation generates self-motion perception (Brandt et al. 1973; Berthoz et al. 1975). Head motion information detected by the vestibular system may assist the visual system in separating optic flow originated from motion through a stationary background (self-motion) vs. optic flow and optic flow induced by moving objects. For instance, during sports, athlete should track the motion of the object while moving through the field to accurately estimate the time point of meeting with the object.

Visual-vestibular integration has been found in cortical and subcortical as well as brainstem and cerebellar regions such as the dorsal medial superior temporal (MSTd) area (Gu et al. 2006, 2008), the ventral intraparietal (VIP) area (Chen et al. 2011a), the visual posterior Sylvian (VPS) area (Chen et al. 2011b), the vestibular nuclei (Henn et al. 1974; Waespe and Henn 1977; Daunton and Thomsen 1979), and the cerebellar flocculus (Waespe et al. 1981; Waespe and Henn 1981). However, Bryan and Angelaki in 2009 showed that the vestibular nuclei and deep cerebellar nuclei failed to respond to optokinetic stimulation when animals fixated at a stationary target, suggesting that the response of vestibular and deep cerebellar nuclei during an optokinetic response is the result of activation of optokinetic nystagmus and gaze stabilization system (Bryan and Angelaki 2009).

The vestibulocerebellum could integrate visual and vestibular information. Anatomical studies suggest that the cerebellar NU receives both vestibular and visual information via mossy fibers and climbing fibers (Korte and Mugnaini 1979; Brodal and Brodal 1985; Gerrits et al. 1989; Epema et al. 1990; Barmack 2003; Ruigrok 2003). Several studies in different species (frog, rat, rabbit, pigeon) have shown that Purkinje cells in the vestibulocerebellum show complex spike responses to whole-field visual motion (Ansorge and Grusser-Cornehls 1977; Blanks and Precht 1983; Kano et al. 1990a, b; Kusunoki et al. 1990; Wylie and Frost 1991, 1993, 1996; Wylie et al. 1993).

In macaque cerebellar dorsal uvula, Purkinje cells respond to optokinetic stimuli as well (Heinen and Keller 1992; Heinen and Keller 1996). Purkinje cell complex spikes in the cerebellar flocculus and nodulus of anesthetized rabbits respond to optokinetic stimuli suggesting that this response reflected the activation of visual pathways and is independent of eye movements (Kano et al. 1991a, b). In the alert cat, Purkinje cell simple spike activity in the cerebellar flocculus responds to vertical head rotation and vertical optokinetic stimuli (Fukushima et al. 1996).

Experiments in pigeons provide evidence that the complex spike activity of Purkinje cells modulated differently to the visual optic within the vestibulocerebellar complex (Wylie and Frost 1991, 1993; Wylie et al. 1993). Purkinje cells in the flocculus exhibit the best response to rotational visual flowfields, while cerebellar nodulus and uvula preferred translational optic flow. Human studies revealed that visually induced illusion of self-motion activated the cerebellar NU (Dieterich et al. 2000; Kleinschmidt et al. 2002). In the macaque, the cerebellar NU translation-selective Purkinje cells carry visual signal in addition to vestibular signal (Yakusheva et al. 2013). Yakusheva et al. (2013) characterized visual (optic flow) and vestibular responses of macaque cerebellar NU Purkinje cells to translations in three dimensions. The visual-vestibular stimuli with a Gaussian velocity profile were previously used to characterized cortical neurons, thus allowing comparison of the visual-vestibular signal carried by the cerebellar NU with the cortical areas such as MSTd, VIP, VPS, and PIVC (Gu et al. 2006, 2008; Fetsch et al. 2009; Chen et al. 2011a, b). Figure 4 shows the response of NU Purkinje cell simple spike responses to 3D vestibular translation and optic flow. This example neuron had a single-peak response in their tuning at different directions. Similar to cortical areas, cerebellar NU Purkinje cells showed excitatory double-peak responses as well as, albeit less frequently, inhibitory single-peak responses (Yakusheva et al. 2013).

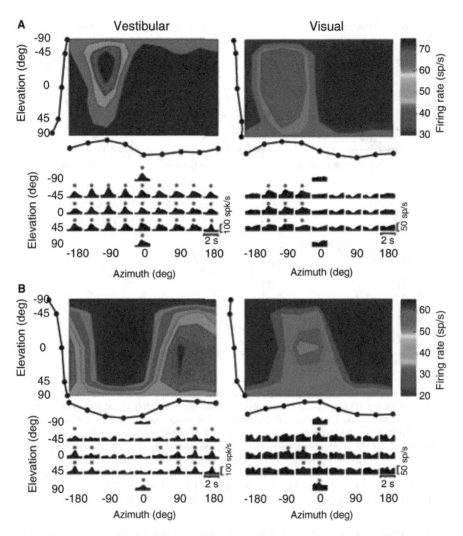

Fig. 4 Two single-peaked examples of NU Purkinje cell responses during 3D vestibular and visual (optic flow) translations. (**a**) Top, congruent example cell. Color-contour maps, showing 3D direction tuning profiles (Lambert cylindrical projection) at peak time for vestibular (1.12 s) and visual (0.9 s) responses with preferred directions: [azimuth, elevation] = [−86°, −18°] and [−94°, −12°], respectively. Tuning curves along the margins illustrate mean firing rates plotted versus elevation or azimuth (averaged across azimuth or elevation, respectively): bottom, response PSTHs. Red stars indicate significant responses. (**b**) Opposite example cell. Vestibular, [azimuth, elevation] = [131°, 22°] and peak time, 1.18 s; visual, [azimuth, elevation] = [−48°, 25°] and peak time, 0.98 s. (Replotted with permission from Yakusheva et al. 2013)

As it has been demonstrated in cortical areas, two types of multisensory visual-vestibular Purkinje cells were found in the cerebellar nodulus and uvula, "congruent" and "opposite" cells. In "congruent" neurons, the directional tuning of the visual and vestibular components was approximately matched (Fig. 4a), whereas

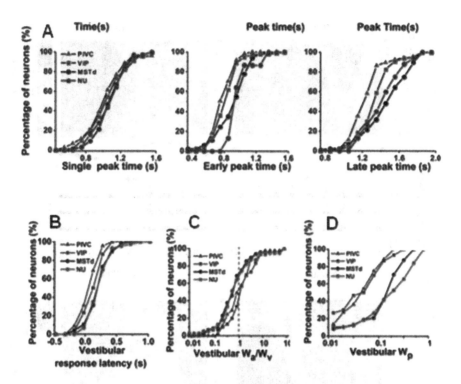

Fig. 5 Comparison of cerebellar NU and cortical areas' (PIVC, VIP, and MSTd) neuronal responses to the 3D translation. Cumulative distributions were plotted: (**a**) Distribution of peak times for single-peaked neurons and double-peaked neurons (early and late peak times) during vestibular motion; (**b**) response latency; (**c**) the ratio of acceleration to velocity weights (w_a/w_v) from model AccVel (accelration+velocity); (**d**) the position weights (w_p) from the model AccVelPos (acceleration + velocity + position). *PIVC*, parietoinsular vestibular cortex; *VIP*, ventral intraparietal; *MSTd*, dorsal medial superior temporal. The NU Purkinje cells have striking similarities in neuronal responses to the vestibular stimuli with MSTd, even though PIVC is considering the primarily vestibular cerebral cortical area. (Replotted with permission from Yakusheva et al. 2013)

for "opposite" neurons, the directional tuning of visual and vestibular components was separated by about 180 deg phase (Fig. 4b). Approximately 90% of recorded Purkinje cells showed significant directional tuning to vestibular, wherein 27% have directional tuning to both visual and vestibular stimuli. Visually tuned Purkinje cells were localized near the midline, in the anterior part of the cerebellar nodulus and ventral uvula (Yakusheva et al. 2013). Evidence suggests that the medial part of the cerebellar NU participates in controlling the time constant for roll, while the lateral part participates in horizontal optokinetic and vestibular time constants (Wearne et al. 1998). Overall, the response of multisensory NU Purkinje cells showed more robust responses to vestibular cues than to optic flow. One of the most intriguing results of this study was that the spatiotemporal characteristics of the vestibular signal, but not visual, carried by NU Purkinje cells were similar to those found in MSTd neurons (Fig. 5). For instance, the vestibular response peak times, latency, the contribution of the velocity, acceleration, and position were remarkably similar between cerebellar NU and MSTd. These findings suggest the existence of multi-synaptic interconnectivity between cerebellar NU and MSTd that could represent cerebro-cerebellar interactions.

Further studies are necessary to better understand visual signal processing by the cerebellar NU and how it is integrated with information from other sensory modalities. For example, it would be necessary to study Purkinje cell complex spike activity during the same visual-vestibular stimuli. Future studies should also address the question of whether self-motion perception is improved after combining vestibular and optic flow stimuli. Lastly, to fully understand visual and vestibular signal processing by the cerebellum, a computational model could facilitate the interpretation of existent and future experimental results.

3 Role of Vestibular and Proprioceptive Signal Integration in the Representation of Different Reference Frame

During self-motion, our peripheral vestibular sensors, the semicircular canals, and the otolith organs provide information about the head motion because they are physically fixed in the head. However, our daily life activities including locomotion, postural control, and navigation require for our brain to process information about the motion and orientation of our body in space that is sensed by the proprioceptive system. Thus, information from both sensory systems should be integrated such that our brain could accurately compute self-motion during navigation. Such computation most likely requires from both sensory systems reference frame transformation to be appropriately combined.

Human studies have demonstrated the role of vestibular and proprioceptive cue integration in motor and posture controls, as well as in self-motion perception (Mergner et al. 1983; Israel and Berthoz 1989; Mergner et al. 1991; Berthoz et al. 1995; Mittelstaedt and Mittelstaedt 2001; Sun et al. 2003; Campos et al. 2009; Siegle et al. 2009; Frissen et al. 2011). In these studies, vestibular and proprioceptive stimuli were presented either separately or combined. For example, during a walk in place, such as a treadmill, only proprioceptive system is activated without input from the vestibular system. During passive motion through space, the central nervous system receives only the vestibular signal. Finally, during active walk-through space, vestibular and proprioceptive inputs are combined. Evidence suggests that both proprioceptive and vestibular cues contribute to the estimation of the traveled distance walking. During curvilinear locomotion, Frissen et al. (2011) showed that humans estimate their velocity of motion through the space using integrated vestibular-proprioceptive signals. They argue that vestibular-proprioceptive sensory integration could be explained by the maximum-likelihood estimate (MLE) model (Ernst and Banks 2002). Thus, finding the site of the brain areas that integrate vestibular and proprioceptive cues is fundamental for understanding how the brain computes an estimate of self-motion. The neural correlates of vestibular and proprioceptive integrations have been shown on the level of the cervical spinal cord neurons (Ezure and Wilson 1984; Wilson et al. 1984, 1986) and the vestibular nuclei (Brink et al. 1980; Boyle and Pompeiano 1981; Kasper et al. 1988; Wilson et al. 1990; Gdowski et al. 2000; Gdowski and McCrea 2000; Medrea and Cullen 2013). It has been shown that non-eye movement vestibular neuronal responses during rotation with the head free correlated better with body velocity (Gdowski and McCrea 1999). When vestibular neurons were tested during active and passive motions, it was shown that most of the neurons were most sensitive to a passive head-on-trunk rotation as well as to whole-body rotation but was significantly reduced during freely moving head (McCrea et al. 1999; Roy and Cullen 2001). It was suggested that vestibular signals related to active head movements were canceled primarily by subtraction of an efference copy signal of head motion. Thus, several studies in primates concluded that vestibular nuclei similar to the vestibular afferents carry the vestibular signal in head-centered coordinates (Roy and Cullen 2001, 2004; Cullen et al. 2011; Carriot et al. 2013). Accurate estimation of self-motion during navigation requires from our brain to differentiate head from body movement (Roy and Cullen 2001; Brooks and Cullen 2009; Carriot et al. 2013). Thus, the CNS must perform the reference frame transformation of vestibular information from a head-centered to a body-centered to be properly combined with extra-vestibular cues such as proprioceptive and visual.

The cerebellum has been suggested as a brain area where head-to-body reference frame transformation for a vestibular signal might take place. Furthermore, this transformed vestibular signal could be combined with proprioceptive and visual information. Proprioceptive information from the muscle and joints are conveyed to the cerebellum through spinocerebellar pathways by mossy and climbing fibers (Murphy et al. 1973a, b; Swenson and Castro 1983a, b; Quy et al. 2011). It has been demonstrated that some vestibular neurons in the rostral medial portion of the deep cerebellar nuclei, the rostral fastigial nuclei (rFN), encode the vestibular signal in the body-centered reference frame (Kleine et al. 2004; Shaikh et al. 2004; Brooks and Cullen 2009). The rFN receive input from the cerebellar cortex, anterior and posterior vermis (Manto and Pandolfo 2002; Ito 2006). The anterior vermis has been suggested to integrate the neck proprioceptive signal, while posterior vermis provides strong vestibular input (Andre et al. 1998; Manzoni et al. 1998a, b; Yakusheva et al. 2007, 2008, 2010; Zhang et al. 2016).

Shaikh et al. (2004) showed that the responses of rFN neuron during 2D translation were systematically changing with head position suggesting the transformation of the vestibular signal from head to a body-centered reference frame at the cerebellum (Fig. 6). These experiments revealed that otolith signal activated during translational stimuli is combined with the signal from the neck proprioceptors at the level of cerebellar rFN to achieve the head-to-body reference frame transformation. At least half of the population of cerebellar rFN reported carrying signal in the body-centered reference frame (Fig. 6a) and a half in the head-centered reference frame (Fig. 6b). Similar results were demonstrated by Brooks and Cullen (2009) during rotational stimuli. A recent study suggested that this property of cerebellar rFN to integrate vestibular signal and proprioceptive signal extend to 3D motion (Martin et al. 2018). This study demonstrated that like the previous study in a 2D plane (Shaikh et al. 2004) for some rFN neurons spatiotemporal tunings varied with

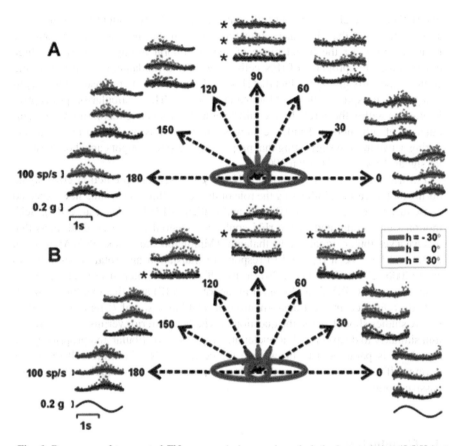

Fig. 6 Responses of two rostral FN neurons during passive whole-body translation (0.5 Hz) at different directions in the horizontal plane. Data were recorded for the three head-on-body positions: straight-ahead ($h = 0°$, blue), 30° to the right ($h = -30°$, red) or to the left ($h = 30°$) head-on-trunk positions. Superimposed solid lines represent best-fit sine functions. (**a**) Example of FN neuron encoding motion in a body reference frame, where the firing rate of the cell is independent of head-on-trunk position. (**b**) Example of FN neuron encoding motion in a head reference frame. The firing rate of the cell changes for the different head-on-trunk positions. In both panels, the minimum responses are marked with asterisks. The black traces represent the linear acceleration stimuli. Motion stimuli are defined relative to the body and, thus, change direction relative to the head (head drawing). (Replotted with permission from Shaikh et al. 2004)

changes in head position in vertical and horizontal planes indicating that these neurons carry translational signal in the head-centered reference frame. Some rFN (about 16%) were reported carrying 3D fully transformed translational signal in the body-centered reference frame, and their spatiotemporal tunings were independent of the head orientation. Many rFN cells carry 3D translational signal in the reference frame that was not purely head-centered or body-centered indicating mixed representation of vestibular responses in rFN. Perhaps, it could also suggest the existence of a partially transformed population of rFN cells. It has been suggested that rFN cells are close to being fully transformed to body-centered coordinates as a population and that the partial transformation on the individual level could be necessary to cover the broad range of spatiotemporal tuning responses in three dimensions. This property of rFN cells could play a fundamental role in postural control especially giving the fact that rFN sends projections to brainstem areas regulating vestibulospinal reflexes (Homma et al. 1995). The vestibular-proprioceptive interaction within the cerebellum could be essential for active locomotion and computation of self-motion. Evidence suggests that neurons in rFN do not modulate during active motion suggesting that proprioceptive signal might cancel out vestibular signal (Brooks and Cullen 2009).

Furthermore, it has been shown rFN cell carry sensory prediction error signal during motor learning indicating the role of the cerebellum in learning new expected consequences of self-motion (Brooks and Cullen 2013; Brooks et al. 2015). Self-motion signal computed in the rFN could be essential for the cortical areas that receive a vestibular signal via the thalamus (Middleton and Strick 1997; Meng et al. 2007) and have been reported to respond to translation and rotation stimuli (Gu et al. 2006, 2008; Chen et al. 2011a, b). Of particular interest are parietoinsular vestibular cortex (PIVC) and ventral intraparietal (VIP) area that exhibited a head-to-body reference frame transformation of vestibular signal (Chen et al. 2013). Future studies should tackle the question of whether posterior vermis of the cerebellum similarly to deep cerebellar nuclei integrates both vestibular and proprioceptive signals. It is paramount to investigate the cerebellar NU Purkinje cell responses during self-generated motions since its properties have been studied only during passive motion.

4 Conclusion

Accumulated evidence indicates that the cerebellum processes multisensory information from vestibular, visual, and proprioceptive systems to correctly estimate self-motion during spatial orientation and navigation. Single-unit recordings in behaving animals from the cerebellar NU Purkinje cells suggest about its key role in the detection of inertial motion and gravity signals using internal model. These studies also indicate that vestibular afferent signal received by the CNS from otolith and semicircular canals undergoes several steps of transformation and that this transformation might happen within the cerebellar NU. Cerebellar NU Purkinje cell responses to visual optic flow suggest that visual signal processed by the cerebellum might contribute to the self-motion perception. We need further studies to understand the visual-vestibular signal integration by the cerebellum fully. The results obtained from recordings in the deep cerebellar nuclei revealed the transformation from head-centered to body-centered reference frame on the level of cerebellar rFN neurons and the role of the proprioceptive signal during active motion. These reference frame transformations allow the brain to distinguish body motion from head motion and are fundamental for active locomotion, posture control, and spatial navigation. Thus, these studies could be an important foundation for future studies addressing multisensory integration in the cerebellum.

Acknowledgment The author thanks Aasef Shaikh and Pablo Blazquez for reading and providing helpful comments on the manuscript. The manuscript was supported by NIH grant R01DC014276. The author declares no other competing financial interests.

References

Akaogi, K., Sato, Y., Ikarashi, K., & Kawasaki, T. (1994a). Mossy fiber projections from the brain stem to the nodulus in the cat. An experimental study comparing the nodulus, the uvula and the flocculus. *Brain Research, 638*(1–2), 12–20.

Akaogi, K., Sato, Y., Ikarashi, K., & Kawasaki, T. (1994b). Zonal organization of climbing fiber projections to the nodulus in the cat. *Brain Research, 638*(1–2), 1–11.

Andre, P., Manzoni, D., & Pompeiano, O. (1998). Spatiotemporal response properties of cerebellar Purkinje cells to neck displacement. *Neuroscience, 84*(4), 1041–1058.

Angelaki, D. E., & Hess, B. J. (1995a). Inertial representation of angular motion in the vestibular system of rhesus monkeys. II. Otolith-controlled transformation that depends on an intact cerebellar nodulus. *Journal of Neurophysiology, 73*(5), 1729–1751.

Angelaki, D. E., & Hess, B. J. (1995b). Lesion of the nodulus and ventral uvula abolish steady-state off-vertical axis otolith response. *Journal of Neurophysiology, 73*(4), 1716–1720.

Angelaki, D. E., & Yakusheva, T. A. (2009). How vestibular neurons solve the tilt/translation ambiguity. Comparison of brainstem, cerebellum, and thalamus. *Annals of the New York Academy of Sciences, 1164*, 19–28.

Angelaki, D. E., McHenry, M. Q., Dickman, J. D., Newlands, S. D., & Hess, B. J. (1999). Computation of inertial motion: Neural strategies to resolve ambiguous otolith information. *The Journal of Neuroscience, 19*(1), 316–327.

Angelaki, D. E., Shaikh, A. G., Green, A. M., & Dickman, J. D. (2004). Neurons compute internal models of the physical laws of motion. *Nature, 430*(6999), 560–564.

Angelaki, D. E., Yakusheva, T. A., Green, A. M., Dickman, J. D., & Blazquez, P. M. (2010). Computation of egomotion in the macaque cerebellar vermis. *Cerebellum, 9*(2), 174–182.

Ansorge, K., & Grusser-Cornehls, U. (1977). Visual and visual-vestibular responses of frog cerebellar neurons. *Experimental Brain Research, 29*(3–4), 445–465.

Barmack, N. H. (1996). GABAergic pathways convey vestibular information to the beta nucleus and dorsomedial cell column of the inferior olive. *Annals of the New York Academy of Sciences, 781*, 541–552.

Barmack, N. H. (2003). Central vestibular system: Vestibular nuclei and posterior cerebellum. *Brain Research Bulletin, 60*(5–6), 511–541.

Barmack, N. H., & Yakhnitsa, V. (2003). Cerebellar climbing fibers modulate simple spikes in Purkinje cells. *The Journal of Neuroscience, 23*(21), 7904–7916.

Barmack, N. H., Baughman, R. W., Errico, P., & Shojaku, H. (1993). Vestibular primary afferent projection to the cerebellum of the rabbit. *Journal of Comparative Neurology, 327*(4), 521–534.

Bernard, J. F. (1987). Topographical organization of olivocerebellar and corticonuclear connections in the rat--an WGA-HRP study: I. Lobules IX, X, and the flocculus. *Journal of Comparative Neurology, 263*(2), 241–258.

Berthoz, A., Pavard, B., & Young, L. R. (1975). Perception of linear horizontal self-motion induced by peripheral vision (linearvection) basic characteristics and visual-vestibular interactions. *Experimental Brain Research, 23*(5), 471–489.

Berthoz, A., Israel, I., Georges-Francois, P., Grasso, R., & Tsuzuku, T. (1995). Spatial memory of body linear displacement: What is being stored? *Science, 269*(5220), 95–98.

Bertolini, G., Ramat, S., Bockisch, C. J., Marti, S., Straumann, D., & Palla, A. (2012). Is vestibular self-motion perception controlled by the velocity storage? Insights from patients with chronic degeneration of the vestibulo-cerebellum. *PLoS One, 7*(6), e36763.

Bigare, F., & Voogd, J. (1977). Cerebello-vestibular projections in the cat [proceedings]. *Acta Morphologica Neerlando-Scandinavica, 15*(4), 323–325.

Blanks, R. H., & Precht, W. (1983). Responses of units in the rat cerebellar flocculus during opto-kinetic and vestibular stimulation. *Experimental Brain Research, 53*(1), 1–15.

Blazquez, P. M., & Yakusheva, T. A. (2015). GABA-A inhibition shapes the spatial and temporal response properties of Purkinje cells in the macaque cerebellum. *Cell Reports, 11*(7), 1043–1053.

Boyle, R., & Pompeiano, O. (1981). Convergence and interaction of neck and macular vestibular inputs on vestibulospinal neurons. *Journal of Neurophysiology, 45*(5), 852–868.

Brandt, T., Dichgans, J., & Koenig, E. (1973). Differential effects of central verses peripheral vision on egocentric and exocentric motion perception. *Experimental Brain Research, 16*(5), 476–491.

Brandt, U., Fluur, E., & Zylberstein, M. Z. (1974). Relationship between flight experience and vestibular function in pilots and nonpilots. *Aerospace Medicine, 45*(11), 1232–1236.

Brink, E. E., Hirai, N., & Wilson, V. J. (1980). Influence of neck afferents on vestibulospinal neurons. *Experimental Brain Research, 38*(3), 285–292.

Brodal, A. (1976). The olivocerebellar projection in the cat as studied with the method of retrograde axonal transport of horseradish peroxidase. II. The projection to the uvula. *Journal of Comparative Neurology, 166*(4), 417–426.

Brodal, A., & Brodal, P. (1985). Observations on the secondary vestibulocerebellar projections in the macaque monkey. *Experimental Brain Research, 58*(1), 62–74.

Bronstein, A. M., Grunfeld, E. A., Faldon, M., & Okada, T. (2008). Reduced self-motion perception in patients with midline cerebellar lesions. *Neuroreport, 19*(6), 691–693.

Brooks, J. X., & Cullen, K. E. (2009). Multimodal integration in rostral fastigial nucleus provides an estimate of body movement. *The Journal of Neuroscience, 29*(34), 10499–10511.

Brooks, J. X., & Cullen, K. E. (2013). The primate cerebellum selectively encodes unexpected self-motion. *Current Biology, 23*(11), 947–955.

Brooks, J. X., Carriot, J., & Cullen, K. E. (2015). Learning to expect the unexpected: Rapid updating in primate cerebellum during voluntary self-motion. *Nature Neuroscience, 18*(9), 1310–1317.

Bryan, A. S., & Angelaki, D. E. (2009). Optokinetic and vestibular responsiveness in the macaque rostral vestibular and fastigial nuclei. *Journal of Neurophysiology, 101*(2), 714–720.

Burguiere, E., Arleo, A., Hojjati, M., Elgersma, Y., De Zeeuw, C. I., Berthoz, A., & Rondi-Reig, L. (2005). Spatial navigation impairment in mice lacking cerebellar LTD: A motor adaptation deficit? *Nature Neuroscience, 8*(10), 1292–1294.

Campos, J. L., Siegle, J. H., Mohler, B. J., Bulthoff, H. H., & Loomis, J. M. (2009). Imagined self-motion differs from perceived self-motion: Evidence from a novel continuous pointing method. *PLoS One, 4*(11), e7793.

Carpenter, M. B., Stein, B. M., & Peter, P. (1972). Primary vestibulocerebellar fibers in the monkey: Distribution of fibers arising from distinctive cell groups of the vestibular ganglia. *The American Journal of Anatomy, 135*(2), 221–249.

Carriot, J., Brooks, J. X., & Cullen, K. E. (2013). Multimodal integration of self-motion cues in the vestibular system: Active versus passive translations. *The Journal of Neuroscience, 33*(50), 19555–19566.

Chen, A., DeAngelis, G. C., & Angelaki, D. E. (2011a). Representation of vestibular and visual cues to self-motion in ventral intraparietal cortex. *The Journal of Neuroscience, 31*(33), 12036–12052.

Chen, A., DeAngelis, G. C., & Angelaki, D. E. (2011b). Convergence of vestibular and visual self-motion signals in an area of the posterior sylvian fissure. *The Journal of Neuroscience, 31*(32), 1617–1627.

Chen, X., Deangelis, G. C., & Angelaki, D. E. (2013). Diverse spatial reference frames of vestibular signals in parietal cortex. *Neuron, 80*(5), 1310–1321.

Choi, S. Y., Jang, J. Y., Oh, E. H., Choi, J. H., Park, J. Y., Lee, S. H., & Choi, K. D. (2018). Persistent geotropic positional nystagmus in unilateral cerebellar lesions. *Neurology, 91*(11), e1053–e1057.

Cullen, K. E., Brooks, J. X., Jamali, M., Carriot, J., & Massot, C. (2011). Internal models of self-motion: Computations that suppress vestibular reafference in early vestibular processing. *Experimental Brain Research, 210*(3–4), 377–388.

Dahlem, K., Valko, Y., Schmahmann, J. D., & Lewis, R. F. (2016). Cerebellar contributions to self-motion perception: Evidence from patients with congenital cerebellar agenesis. *Journal of Neurophysiology, 115*(5), 2280–2285.

Dakin, C. J., & Rosenberg, A. (2018). Gravity estimation and verticality perception. *Handbook of Clinical Neurology, 159*, 43–59.

Dakin, C. J., Peters, A., Giunti, P., Day, B. L. (2018). Cerebellar Degeneration Increases Visual Influence on Dynamic Estimates of Verticality. Current Biology 28(22):3589–3598.e3

Daunton, N., & Thomsen, D. (1979). Visual modulation of otolith-dependent units in cat vestibular nuclei. *Experimental Brain Research, 37*(1), 173–176.

Dichgans, J., Held, R., Young, L. R., & Brandt, T. (1972). Moving visual scenes influence the apparent direction of gravity. *Science, 178*(4066), 1217–1219.

Dichgans, J., Diener, H. C., & Brandt, T. (1974). Optokinetic-graviceptic interaction in different head positions. *Acta Oto-Laryngologica, 78*(5–6), 391–398.

Dieterich, M., Bucher, S. F., Seelos, K. C., & Brandt, T. (2000). Cerebellar activation during optokinetic stimulation and saccades. *Neurology, 54*(1), 148–155.

Epema, A. H., Gerrits, N. M., & Voogd, J. (1990). Secondary vestibulocerebellar projections to the flocculus and uvulo-nodular lobule of the rabbit: A study using HRP and double fluorescent tracer techniques. *Experimental Brain Research, 80*(1), 72–82.

Ernst, M. O., & Banks, M. S. (2002). Humans integrate visual and haptic information in a statistically optimal fashion. *Nature, 415*(6870), 429–433.

Ezure, K., & Wilson, V. J. (1984). Interaction of tonic neck and vestibular reflexes in the forelimb of the decerebrate cat. *Experimental Brain Research, 54*(2), 289–292.

Fernandez, C., & Goldberg, J. M. (1976a). Physiology of peripheral neurons innervating otolith organs of the squirrel monkey. I. Response to static tilts and to long-duration centrifugal force. *Journal of Neurophysiology, 39*(5), 970–984.

Fernandez, C., & Goldberg, J. M. (1976b). Physiology of peripheral neurons innervating otolith organs of the squirrel monkey. III. Response dynamics. *Journal of Neurophysiology, 39*(5), 996–1008.

Fetsch, C. R., Turner, A. H., DeAngelis, G. C., & Angelaki, D. E. (2009). Dynamic reweighting of visual and vestibular cues during self-motion perception. *The Journal of Neuroscience, 29*(49), 15601–15612.

Freedman, S. L., Voogd, J., & Vielvoye, G. J. (1977). Experimental evidence for climbing fibers in the avian cerebellum. *Journal of Comparative Neurology, 175*(2), 243–252.

Frissen, I., Campos, J. L., Souman, J. L., & Ernst, M. O. (2011). Integration of vestibular and proprioceptive signals for spatial updating. *Experimental Brain Research, 212*(2), 163–176.

Fukushima, K., Chin, S., Fukushima, J., & Tanaka, M. (1996). Simple-spike activity of floccular Purkinje cells responding to sinusoidal vertical rotation and optokinetic stimuli in alert cats. *Neuroscience Research, 24*(3), 275–289.

Fushiki, H., & Barmack, N. H. (1997). Topography and reciprocal activity of cerebellar Purkinje cells in the uvula-nodulus modulated by vestibular stimulation. *Journal of Neurophysiology, 78*(6), 3083–3094.

Gamlin, P. D. (2006). The pretectum: Connections and oculomotor-related roles. *Progress in Brain Research, 151*, 379–405.

Gdowski, G. T., & McCrea, R. A. (1999). Integration of vestibular and head movement signals in the vestibular nuclei during whole-body rotation. *Journal of Neurophysiology, 82*(1), 436–449.

Gdowski, G. T., & McCrea, R. A. (2000). Neck proprioceptive inputs to primate vestibular nucleus neurons. *Experimental Brain Research, 135*(4), 511–526.

Gdowski, G. T., Boyle, R., & McCrea, R. A. (2000). Sensory processing in the vestibular nuclei during active head movements. *Archives Italiennes de Biologie, 138*(1), 15–28.

Gerrits, N. M., Epema, A. H., van Linge, A., & Dalm, E. (1989). The primary vestibulocerebellar projection in the rabbit: Absence of primary afferents in the flocculus. *Neuroscience Letters, 105*(1–2), 27–33.

Giolli, R. A., Blanks, R. H., & Lui, F. (2006). The accessory optic system: Basic organization with an update on connectivity, neurochemistry, and function. *Progress in Brain Research, 151*, 407–440.

Green, A. M., & Angelaki, D. E. (2003). Resolution of sensory ambiguities for gaze stabilization requires a second neural integrator. *The Journal of Neuroscience, 23*(28), 9265–9275.

Green, A. M., & Angelaki, D. E. (2004). An integrative neural network for detecting inertial motion and head orientation. *Journal of Neurophysiology, 92*(2), 905–925.

Green, A. M., & Angelaki, D. E. (2007). Coordinate transformations and sensory integration in the detection of spatial orientation and self-motion: From models to experiments. *Progress in Brain Research, 165*, 155–180.

Green, A. M., Shaikh, A. G., & Angelaki, D. E. (2005). Sensory vestibular contributions to constructing internal models of self-motion. *Journal of Neural Engineering, 2*(3), S164–S179.

Groenewegen, H. J., & Voogd, J. (1977). The parasagittal zonation within the olivocerebellar projection. I. Climbing fiber distribution in the vermis of cat cerebellum. *Journal of Comparative Neurology, 174*(3), 417–488.

Gu, Y., Watkins, P. V., Angelaki, D. E., & DeAngelis, G. C. (2006). Visual and nonvisual contributions to three-dimensional heading selectivity in the medial superior temporal area. *The Journal of Neuroscience, 26*(1), 73–85.

Gu, Y., Angelaki, D. E., & Deangelis, G. C. (2008). Neural correlates of multisensory cue integration in macaque MSTd. *Nature Neuroscience, 11*(10), 1201–1210.

Heinen, S. J., & Keller, E. L. (1992). Cerebellar uvula involvement in visual motion processing and smooth pursuit control in monkey. *Annals of the New York Academy of Sciences, 656*, 775–782.

Heinen, S. J., & Keller, E. L. (1996). The function of the cerebellar uvula in monkey during optokinetic and pursuit eye movements: Single-unit responses and lesion effects. *Experimental Brain Research, 110*(1), 1–14.

Henn, V., Young, L. R., & Finley, C. (1974). Vestibular nucleus units in alert monkeys are also influenced by moving visual fields. *Brain Research, 71*(1), 144–149.

Hensbroek, R. A., Ruigrok, T. J., van Beugen, B. J., Maruta, J., & Simpson, J. I. (2015). Visuovestibular information processing by unipolar brush cells in the rabbit flocculus. *Cerebellum, 14*(5), 578–583.

Hlavacka, F., Mergner, T., & Schweigart, G. (1992). Interaction of vestibular and proprioceptive inputs for human self-motion perception. *Neuroscience Letters, 138*(1), 161–164.

Hlavacka, F., Mergner, T., & Bolha, B. (1996). Human self-motion perception during translatory vestibular and proprioceptive stimulation. *Neuroscience Letters, 210*(2), 83–86.

Homma, Y., Nonaka, S., Matsuyama, K., & Mori, S. (1995). Fastigiofugal projection to the brainstem nuclei in the cat: An anterograde PHA-L tracing study. *Neuroscience Research, 23*(1), 89–102.

Horstmann, G. A., & Dietz, V. (1990). A basic posture control mechanism: The stabilization of the centre of gravity. *Electroencephalography and Clinical Neurophysiology, 76*(2), 165–176.

Israel, I., & Berthoz, A. (1989). Contribution of the otoliths to the calculation of linear displacement. *Journal of Neurophysiology, 62*(1), 247–263.

Ito, M. (2006). Cerebellar circuitry as a neuronal machine. *Progress in Neurobiology, 78*(3–5), 272–303.

Kanda, K., Sato, Y., Ikarashi, K., & Kawasaki, T. (1989). Zonal organization of climbing fiber projections to the uvula in the cat. *Journal of Comparative Neurology, 279*(1), 138–148.

Kano, M., Kano, M. S., Kusunoki, M., & Maekawa, K. (1990a). Nature of optokinetic response and zonal organization of climbing fiber afferents in the vestibulocerebellum of the pigmented rabbit. II. The nodulus. *Experimental Brain Research, 80*(2), 238–251.

Kano, M. S., Kano, M., & Maekawa, K. (1990b). Receptive field organization of climbing fiber afferents responding to optokinetic stimulation in the cerebellar nodulus and flocculus of the pigmented rabbit. *Experimental Brain Research, 82*(3), 499–512.

Kano, M., Kano, M. S., & Maekawa, K. (1991a). Optokinetic response of simple spikes of Purkinje cells in the cerebellar flocculus and nodulus of the pigmented rabbit. *Experimental Brain Research, 87*(3), 484–496.

Kano, M., Kano, M. S., & Maekawa, K. (1991b). Simple spike modulation of Purkinje cells in the cerebellar nodulus of the pigmented rabbit to optokinetic stimulation. *Neuroscience Letters, 128*(1), 101–104.

Kasper, J., Schor, R. H., & Wilson, V. J. (1988). Response of vestibular neurons to head rotations in vertical planes. II. Response to neck stimulation and vestibular-neck interaction. *Journal of Neurophysiology, 60*(5), 1765–1778.

Kevetter, G. A., & Perachio, A. A. (1986). Distribution of vestibular afferents that innervate the sacculus and posterior canal in the gerbil. *Journal of Comparative Neurology, 254*(3), 410–424.

Kevetter, G. A., Leonard, R. B., Newlands, S. D., & Perachio, A. A. (2004). Central distribution of vestibular afferents that innervate the anterior or lateral semicircular canal in the mongolian gerbil. *Journal of Vestibular Research, 14*(1), 1–15.

Kleine, J. F., Guan, Y., Kipiani, E., Glonti, L., Hoshi, M., & Buttner, U. (2004). Trunk position influences vestibular responses of fastigial nucleus neurons in the alert monkey. *Journal of Neurophysiology, 91*(5), 2090–2100.

Kleinschmidt, A., Thilo, K. V., Buchel, C., Gresty, M. A., Bronstein, A. M., & Frackowiak, R. S. (2002). Neural correlates of visual-motion perception as object- or self-motion. *NeuroImage, 16*(4), 873–882.

Korte, G. E., & Mugnaini, E. (1979). The cerebellar projection of the vestibular nerve in the cat. *Journal of Comparative Neurology, 184*(2), 265–277.

Kusunoki, M., Kano, M., Kano, M. S., & Maekawa, K. (1990). Nature of optokinetic response and zonal organization of climbing fiber afferents in the vestibulocerebellum of the pigmented rabbit. I. The flocculus. *Experimental Brain Research, 80*(2), 225–237.

Laurens, J., Valko, Y., & Straumann, D. (2011). Experimental parameter estimation of a visuo-vestibular interaction model in humans. *Journal of Vestibular Research, 21*(5), 251–266.

Laurens, J., Meng, H., & Angelaki, D. E. (2013a). Computation of linear acceleration through an internal model in the macaque cerebellum. *Nature Neuroscience, 16*(11), 1701–1708.

Laurens, J., Meng, H., & Angelaki, D. E. (2013b). Neural representation of orientation relative to gravity in the macaque cerebellum. *Neuron, 80*(6), 1508–1518.

MacNeilage, P. R., Banks, M. S., Berger, D. R., & Bulthoff, H. H. (2007). A Bayesian model of the disambiguation of gravitoinertial force by visual cues. *Experimental Brain Research, 179*(2), 263–290.

Maklad, A., & Fritzsch, B. (2003). Partial segregation of posterior crista and saccular fibers to the nodulus and uvula of the cerebellum in mice, and its development. *Brain Research. Developmental Brain Research, 140*(2), 223–236.

Manto, M., & Pandolfo, M. (2002). *The cerebellum and its disorders*. Cambridge/New York: Cambridge University Press.

Manzoni, D., Pompeiano, O., & Andre, P. (1998a). Convergence of directional vestibular and neck signals on cerebellar purkinje cells. *Pflügers Archiv, 435*(5), 617–630.

Manzoni, D., Pompeiano, O., & Andre, P. (1998b). Neck influences on the spatial properties of vestibulospinal reflexes in decerebrate cats: Role of the cerebellar anterior vermis. *Journal of Vestibular Research, 8*(4), 283–297.

Marini, G., Provini, L., & Rosina, A. (1975). Macular input to the cerebellar nodulus. *Brain Research, 99*(2), 367–371.

Martin, C. Z., Brooks, J. X., & Green, A. M. (2018). Role of rostral fastigial neurons in encoding a body-centered representation of translation in three dimensions. *The Journal of Neuroscience, 38*(14), 3584–3602.

McCrea, R. A., Gdowski, G. T., Boyle, R., & Belton, T. (1999). Firing behavior of vestibular neurons during active and passive head movements: Vestibulo-spinal and other non-eye-movement related neurons. *Journal of Neurophysiology, 82*(1), 416–428.

Medrea, I., & Cullen, K. E. (2013). Multisensory integration in early vestibular processing in mice: The encoding of passive vs. active motion. *Journal of Neurophysiology, 110*(12), 2704–2717.

Meng, H., May, P. J., Dickman, J. D., & Angelaki, D. E. (2007). Vestibular signals in primate thalamus: Properties and origins. *The Journal of Neuroscience, 27*(50), 13590–13602.

Merfeld, D. M. (1995). Modeling the vestibulo-ocular reflex of the squirrel monkey during eccentric rotation and roll tilt. *Experimental Brain Research, 106*(1), 123–134.

Merfeld, D. M., & Zupan, L. H. (2002). Neural processing of gravitoinertial cues in humans. III. Modeling tilt and translation responses. *Journal of Neurophysiology, 87*(2), 819–833.

Merfeld, D. M., Zupan, L., & Peterka, R. J. (1999). Humans use internal models to estimate gravity and linear acceleration. *Nature, 398*(6728), 615–618.

Merfeld, D. M., Park, S., Gianna-Poulin, C., Black, F. O., & Wood, S. (2005a). Vestibular perception and action employ qualitatively different mechanisms. I. Frequency response of VOR and perceptual responses during Translation and Tilt. *Journal of Neurophysiology, 94*(1), 186–198.

Merfeld, D. M., Park, S., Gianna-Poulin, C., Black, F. O., & Wood, S. (2005b). Vestibular perception and action employ qualitatively different mechanisms. II. VOR and perceptual responses during combined Tilt & Translation. *Journal of Neurophysiology, 94*(1), 199–205.

Mergner, T., Nardi, G. L., Becker, W., & Deecke, L. (1983). The role of canal-neck interaction for the perception of horizontal trunk and head rotation. *Experimental Brain Research, 49*(2), 198–208.

Mergner, T., Siebold, C., Schweigart, G., & Becker, W. (1991). Human perception of horizontal trunk and head rotation in space during vestibular and neck stimulation. *Experimental Brain Research, 85*(2), 389–404.

Middleton, F. A., & Strick, P. L. (1997). Cerebellar output channels. *International Review of Neurobiology, 41*, 61–82.

Mittelstaedt, M. L., & Mittelstaedt, H. (2001). Idiothetic navigation in humans: Estimation of path length. *Experimental Brain Research, 139*(3), 318–332.

Murphy, J. T., MacKay, W. A., & Johnson, F. (1973a). Differences between cerebellar mossy and climbing fibre responses to natural stimulation of forelimb muscle proprioceptors. *Brain Research, 55*(2), 263–289.

Murphy, J. T., MacKay, W. A., & Johnson, F. (1973b). Responses of cerebellar cortical neurons to dynamic proprioceptive inputs from forelimb muscles. *Journal of Neurophysiology, 36*(4), 711–723.

Newlands, S. D., Purcell, I. M., Kevetter, G. A., & Perachio, A. A. (2002). Central projections of the utricular nerve in the gerbil. *Journal of Comparative Neurology, 452*(1), 11–23.

Ono, S., Kushiro, K., Zakir, M., Meng, H., Sato, H., & Uchino, Y. (2000). Properties of utricular and saccular nerve-activated vestibulocerebellar neurons in cats. *Experimental Brain Research, 134*(1), 1–8.

Quy, P. N., Fujita, H., Sakamoto, Y., Na, J., & Sugihara, I. (2011). Projection patterns of single mossy fiber axons originating from the dorsal column nuclei mapped on the aldolase C compartments in the rat cerebellar cortex. *Journal of Comparative Neurology, 519*(5), 874–899.

Roy, J. E., & Cullen, K. E. (2001). Selective processing of vestibular reafference during self-generated head motion. *The Journal of Neuroscience, 21*(6), 2131–2142.

Roy, J. E., & Cullen, K. E. (2004). Dissociating self-generated from passively applied head motion: Neural mechanisms in the vestibular nuclei. *The Journal of Neuroscience, 24*(9), 2102–2111.

Ruigrok, T. J. (2003). Collateralization of climbing and mossy fibers projecting to the nodulus and flocculus of the rat cerebellum. *Journal of Comparative Neurology, 466*(2), 278–298.

Sato, Y., Kanda, K., Ikarashi, K., & Kawasaki, T. (1989). Differential mossy fiber projections to the dorsal and ventral uvula in the cat. *Journal of Comparative Neurology, 279*(1), 149–164.

Senot, P., Zago, M., Lacquaniti, F., & McIntyre, J. (2005). Anticipating the effects of gravity when intercepting moving objects: Differentiating up and down based on nonvisual cues. *Journal of Neurophysiology, 94*(6), 4471–4480.

Shaikh, A. G., Meng, H., & Angelaki, D. E. (2004). Multiple reference frames for motion in the primate cerebellum. *The Journal of Neuroscience, 24*(19), 4491–4497.

Shaikh, A. G., Green, A. M., Ghasia, F. F., Newlands, S. D., Dickman, J. D., & Angelaki, D. E. (2005). Sensory convergence solves a motion ambiguity problem. *Current Biology, 15*(18), 1657–1662.

Siegle, J. H., Campos, J. L., Mohler, B. J., Loomis, J. M., & Bulthoff, H. H. (2009). Measurement of instantaneous perceived self-motion using continuous pointing. *Experimental Brain Research, 195*(3), 429–444.

Simpson, J. I., Leonard, C. S., & Soodak, R. E. (1988). The accessory optic system. Analyzer of self-motion. *Annals of the New York Academy of Sciences, 545,* 170–179.

Sun, H. J., Lee, A. J., Campos, J. L., Chan, G. S., & Zhang, D. H. (2003). Multisensory integration in speed estimation during self-motion. *CyberPsychology and Behaviour, 6*(5), 509–518.

Swenson, R. S., & Castro, A. J. (1983a). The afferent connections of the inferior olivary complex in rats. An anterograde study using autoradiographic and axonal degeneration techniques. *Neuroscience, 8*(2), 259–275.

Swenson, R. S., & Castro, A. J. (1983b). The afferent connections of the inferior olivary complex in rats: A study using the retrograde transport of horseradish peroxidase. *The American Journal of Anatomy, 166*(3), 329–341.

Thunnissen, I. E., Epema, A. H., & Gerrits, N. M. (1989). Secondary vestibulocerebellar mossy fiber projection to the caudal vermis in the rabbit. *Journal of Comparative Neurology, 290*(2), 262–277.

Vingerhoets, R. A., Van Gisbergen, J. A., & Medendorp, W. P. (2007). Verticality perception during off-vertical axis rotation. *Journal of Neurophysiology, 97*(5), 3256–3268.

Waespe, W., & Henn, V. (1977). Neuronal activity in the vestibular nuclei of the alert monkey during vestibular and optokinetic stimulation. *Experimental Brain Research, 27*(5), 523–538.

Waespe, W., & Henn, V. (1981). Visual-vestibular interaction in the flocculus of the alert monkey. II. Purkinje cell activity. *Experimental Brain Research, 43*(3–4), 349–360.

Waespe, W., Buttner, U., & Henn, V. (1981). Input-output activity of the primate flocculus during visual-vestibular interaction. *Annals of the New York Academy of Sciences, 374,* 491–503.

Wearne, S., Raphan, T., & Cohen, B. (1998). Control of spatial orientation of the angular vestibuloocular reflex by the nodulus and uvula. *Journal of Neurophysiology, 79*(5), 2690–2715.

Wilson, V. J., Ezure, K., & Timerick, S. J. (1984). Tonic neck reflex of the decerebrate cat: Response of spinal interneurons to natural stimulation of neck and vestibular receptors. *Journal of Neurophysiology, 51*(3), 567–577.

Wilson, V. J., Schor, R. H., Suzuki, I., & Park, B. R. (1986). Spatial organization of neck and vestibular reflexes acting on the forelimbs of the decerebrate cat. *Journal of Neurophysiology, 55*(3), 514–526.

Wilson, V. J., Yamagata, Y., Yates, B. J., Schor, R. H., & Nonaka, S. (1990). Response of vestibular neurons to head rotations in vertical planes. III. Response of vestibulocollic neurons to vestibular and neck stimulation. *Journal of Neurophysiology, 64*(6), 1695–1703.

Wolpert, D. M., Miall, R. C., & Kawato, M. (1998). Internal models in the cerebellum. *Trends in Cognitive Sciences, 2*(9), 338–347.

Wood, S. J., Reschke, M. F., Sarmiento, L. A., & Clement, G. (2007). Tilt and translation motion perception during off-vertical axis rotation. *Experimental Brain Research, 182*(3), 365–377.

Wylie, D. R., & Frost, B. J. (1991). Purkinje cells in the vestibulocerebellum of the pigeon respond best to either translational or rotational wholefield visual motion. *Experimental Brain Research, 86*(1), 229–232.

Wylie, D. R., & Frost, B. J. (1993). Responses of pigeon vestibulocerebellar neurons to optokinetic stimulation. II. The 3-dimensional reference frame of rotation neurons in the flocculus. *Journal of Neurophysiology, 70*(6), 2647–2659.

Wylie, D. R., & Frost, B. J. (1996). The pigeon optokinetic system: Visual input in extraocular muscle coordinates. *Visual Neuroscience, 13*(5), 945–953.

Wylie, D. R., Kripalani, T., & Frost, B. J. (1993). Responses of pigeon vestibulocerebellar neurons to optokinetic stimulation. I. Functional organization of neurons discriminating between translational and rotational visual flow. *Journal of Neurophysiology, 70*(6), 2632–2646.

Yakusheva, T. A., Shaikh, A. G., Green, A. M., Blazquez, P. M., Dickman, J. D., & Angelaki, D. E. (2007). Purkinje cells in posterior cerebellar vermis encode motion in an inertial reference frame. *Neuron, 54*(6), 973–985.

Yakusheva, T., Blazquez, P. M., & Angelaki, D. E. (2008). Frequency-selective coding of translation and tilt in macaque cerebellar nodulus and uvula. *The Journal of Neuroscience, 28*(40), 9997–10009.

Yakusheva, T., Blazquez, P. M., & Angelaki, D. E. (2010). Relationship between complex and simple spike activity in macaque caudal vermis during three-dimensional vestibular stimulation. *The Journal of Neuroscience, 30*(24), 8111–8126.

Yakusheva, T. A., Blazquez, P. M., Chen, A., & Angelaki, D. E. (2013). Spatiotemporal properties of optic flow and vestibular tuning in the cerebellar nodulus and uvula. *The Journal of Neuroscience, 33*(38), 15145–15160.

Zago, M., & Lacquaniti, F. (2005). Visual perception and interception of falling objects: A review of evidence for an internal model of gravity. *Journal of Neural Engineering, 2*(3), S198–S208.

Zhang, X. Y., Wang, J. J., & Zhu, J. N. (2016). Cerebellar fastigial nucleus: From anatomic construction to physiological functions. *Cerebellum & Ataxias, 3*, 9.

Zupan, L. H., & Merfeld, D. M. (2003). Neural processing of gravito-inertial cues in humans. IV. Influence of visual rotational cues during roll optokinetic stimuli. *Journal of Neurophysiology, 89*(1), 390–400.

Zupan, L. H., Merfeld, D. M., & Darlot, C. (2002). Using sensory weighting to model the influence of canal, otolith and visual cues on spatial orientation and eye movements. *Biological Cybernetics, 86*(3), 209–230.

Vestibular Perception: From Bench to Bedside

Heiko M. Rust, Barry M. Seemungal, and Amir Kheradmand

Abstract Classic experiments over several decades examined the physiology and pathophysiology of a critical brainstem function called vestibulo-ocular reflex. These studies provided a wealth of information on how the brain, particularly the cerebellum and brainstem, computes the representation of our own motion in order to generate compensatory movements. Contemporary literature over last two decades started focusing on an equally important aspects of vestibular function – the motion perception and spatial orientation. From both physiological and computational standpoints, these studies further extended the application of cerebellar principles (for the control of vestibulo-ocular reflex) to thalamic and cortical function, emphasising on cerebello-cerebral connections. This chapter provides a concise review of the physiology and pathophysiology of vestibular perception and discusses seminal work from our laboratories.

Keywords Motion perception · Tilt · Vestibular cortex · Transcranial magnetic stimulation

1 Introduction

Our ability to walk and navigate in the environment requires that the brain estimate not only our postural orientation with respect to gravity (to stop us from falling over) but also our position relative to environmental landmarks (to stop us from getting lost). To perform both functions – i.e. maintaining a stable posture and navigating in our environment – the brain combines signals of tilt (relative to gravity) and

H. M. Rust · B. M. Seemungal
Department of Neurology, Imperial College, London, UK

A. Kheradmand (✉)
Department of Neurology, The Johns Hopkins University, Baltimore, MD, USA
e-mail: akherad@jhu.edu; http://vorlab.jhu.edu/

© Springer Nature Switzerland AG 2019
A. Shaikh, F. Ghasia (eds.), *Advances in Translational Neuroscience of Eye Movement Disorders*, Contemporary Clinical Neuroscience,
https://doi.org/10.1007/978-3-030-31407-1_3

signals of angular motion, from a variety of internal sensory signals, including the vestibular apparatus, as well as external (environmental) cues. Thus, for determining how upright we are relative to gravity, the brain utilises the peripheral vestibular end-organ signals – mainly otolithic – as well as those from proprioceptive (e.g. joint angle signals) as well as environmental cues gleaned from visual inputs (and sound inputs).

A different type of spatial orientation to that of our perceived position with respect to gravity is that required for navigating in our environment. Ecologically, such orientation is performed while walking. A key concept is that the brain can convert signals of our body motion to our body position in space, either relative to some starting point in space – called egocentric orientation – or relative to an environmental landmark – called allocentric orientation. This type of spatial orientation differs from our perception of our body tilt (relative to gravity) since the vestibular organ provides a direct indication of our tilt with respect to gravity from the otolith organs. In contrast, walking in the environment requires updating our position using both linear movements and angular rotations (i.e. rotations about an earth axis, otherwise known as 'yaw-plane' rotations). Although, as described below, the vestibular system virtually always works in concert with other sensory systems, it appears however that, at least for yaw-plane orientation, the vestibular semicircular canal signals are critical. This is exemplified by the performance of patients who have complete loss of peripheral function in whom, when walking in the dark, updating their orientation when walking in a straight line is preserved (indicating somatosensory-related spatial updating) but who get lost when making a turn (Glasauer et al. 2002). This thus indicates the necessity of semicircular canal signals for spatial updating during locomotor turns in the dark and requires that the brain convert the vestibular nerve signal of head angular velocity to angular position. The ability for the brain to perform this signal transformation (angular velocity to angular position) is *equivalent* (i.e. we do not assume that the brain actually works like this) to an artificial circuit performing a mathematical temporal integration of the head velocity signal to position.

It follows that, if the brain *derives* our spatial orientation – at least for yaw-plane orientation – from self-motion signals, then how the brain estimates our self-motion will affect our angular spatial orientation (note, as described before, vestibular cues of head tilt are directly obtained from the otolith organ and are not a derived measure). Thus, at least for semicircular canal signals of head angular motion, there is a convergence of vestibular and large-field visual motion stimuli as early as the brainstem primary vestibular neurones (Allum et al. 1976; Henn et al. 1974; Reisine and Raphan 1992; Waespe and Henn 1977). Given this low-order convergence, it is also unsurprising that vestibular-responsive cerebral cortical neurons (as shown in primate studies) are also ubiquitously multimodal (Bremmer et al. 2002; Grüsser et al. 1990a, b; Klam and Graf 2003; Schlack et al. 2002). This visual–vestibular convergence manifests in the 'train illusion', i.e. our sensation of self-motion engendered by a train moving past us (a phenomenon called 'vection'). The effect of sensory inputs – e.g. the train illusion – is also modulated by non-sensory cognitive cues such as environmental context, since the train illusion is more easily provoked when sitting in a stationary train, which could move at any moment, compared to sitting on the train platform, when we are very unlikely to move (Guedry Jr. 1974).

The otoliths – which are biological linear accelerometers – measure our tilt, since gravity ('g') is equivalent to us accelerating upwards into space while standing on an elevator where such an acceleration would enable our feet to be firmly planted onto the elevator floor. Equally, the otoliths will respond to any acceleration, and this does not have to be g. It is unlikely that humans evolved to transduce non-g high-acceleration situations, but artificial environments of the modern world push our brain's capacity to the limits; e.g. an aviator being catapulted off the deck of an aircraft carrier (the steam catapults of the US Navy accelerate a typical aircraft from zero to 200 knots within 4 seconds) will perceive a strong sense of upwards tilt or have an illusion of visual tilt when looking at the instrument panels within the cockpit, which are known as somatogravic and oculogravic illusions (Guedry Jr. 1974). These illusions result from a horizontal linear acceleration of the aircraft combined with g giving a resultant acceleration (via vector addition) of a tilt. Likewise, in microgravity, astronauts often report difficulty distinguishing the spacecraft floors, walls and ceiling surfaces from one another and have tilt perception errors. Fortunately, such extreme artificial environments are uncommon, and normally, the brain will disambiguate tilt and linear acceleration by combining canal and otolith signals. In this process, the posterior cerebellum carries out the computations necessary to distinguish between tilt and translation (Yakusheva et al. 2007), although this mechanism is not fail-safe in extreme environments as in the aviator on an aircraft carrier.

Thus, it is convenient to split vestibular perception into two main (but partially overlapping) components: (i) an awareness of our position in space: 'where am I?' and (ii) a sensation of self-motion: 'am I moving?' Our internal estimates of self-motion may be used by the brain for balance control. Theoretically, this is most likely for movements that introduce a change with respect to gravity – e.g. roll and pitch – since, when standing, excessive sway (i.e. roll and pitch movements) may result in a fall. Indeed, balance control in the elderly may be linked to elevated self-motion perceptual thresholds in the roll plane (Karmali et al. 2017; Seemungal 2005; Seemungal et al. 2004). In contrast, ecological yaw-plane rotations (i.e. that of normal body and head movements) are, by definition, orthogonal to gravity and hence, in isolation, will not result in a body attitude that will exceed the limits of one's stance.

Although ecologically vestibular signals are virtually always activated together with other sensory modalities, in the laboratory, it is possible to isolate the contribution of vestibular signals, to investigate brain processes involved in spatial orientation and self-motion perception. Such experiments typically involve passive whole-body movement (or change in orientation) of human subjects in the dark. In this chapter, we discuss the results of these studies and focus on the contribution of vestibular inputs and their interaction with other sensory modalities that mediate our coherent perception of self-motion and spatial orientation, and in addition, when available, we discuss their neural correlates.

2   The Peripheral Vestibular Apparatus and Vestibular Reflex Functioning

Stimulation of the vestibular organs results in a cascade of sensory and motor consequences, some as part of low-order reflex mechanisms and others as higher-order outputs, e.g. orientation with respect to the surrounding environment or motion perception. Any understanding of higher-order sensory functioning requires a clear understanding of the sensory apparatus that transduces external stimuli, how these signals are processed, and how they are funnelled towards the high-order brain structures.

The peripheral vestibular apparatuses – which transduce head acceleration stimuli – are sited within the bony labyrinths. These peripheral organs comprise the semicircular canals, which transduce angular acceleration, and the otoliths, which transduce linear acceleration or tilt with respect to gravity. The most well-known vestibular reflex is the vestibulo-ocular reflex (VOR), which is vital for maintaining visual stability with changes in the head position (Fig. 1). In humans, in addition to the horizontal and vertical VOR, lateral head tilts (i.e. with respect to gravity) lead to changes in the torsional eye position in the opposite direction of the head tilt. This ocular counter-roll (OCR) is a constrained, phylogenetically old vestibular reflex and, in contrast to horizontal or vertical VOR, does not match the magnitude of the head tilt. Most available clinical tests of vestibular function are focused on evaluation of VOR function, such as those test used to interrogate semicircular canal function (e.g. via bithermal caloric responses), or those that are used to interrogate otolith function (e.g. vestibular-evoked myogenic potentials or VEMP). In this context, the lack of routine vestibular perceptual testing in the clinic is problematic since only a (relatively) small part of the vestibular system can be interrogated using VOR testing.

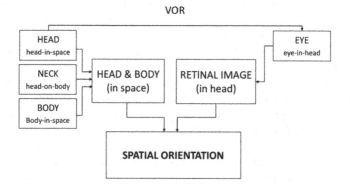

Fig. 1 Perception of spatial orientation involves integration of vestibular inputs that encode the head position in space with sensory inputs that encode the eye and body positions. In this process, through the vestibulo-ocular reflex (VOR), the position of the head determines the eye position

3   Vestibular Perception

3.1   Multimodal Contributions

The brain uses all available information, including vestibular, visual, proprioceptive, and somatosensory cues – such as neck and trunk position – as well as cognitive information from previous experience in estimating the orientation of our body
and motion with respect to our environment (Cullen 2018; Guedry Jr. 1974;
Seemungal 2014; Kheradmand and Winnick 2017; Alberts et al. 2016b; Barbieri
et al. 2008; Barra et al. 2012; Böhmer and Mast 1999; Carriot et al. 2011; Ceyte
et al. 2009; Day and Wade 1969; McKenna et al. 2004; Riccio et al. 1992; Trousselard
et al. 2004; Wade 1968; Clemens et al. 2011; Bray et al. 2004; Wright and Horak
2007). Indeed, multimodal sensory integration is the basis for our stable perception
of the world in upright orientation, despite frequent changes in our eye, head and
body positions (a feature known as 'orientation constancy'). An example of a disrupted orientation constancy is seen in the room tilt illusion (Akdal et al. 2017) –
which most commonly occurs with acute migraine or acute stroke and less
commonly during otolith crisis of Tumarkin in Meniere's disease – in which the
patient complains of a perception of the world being either upside down or tilted,
and is presumably mediated by central otolith pathways.

   In addition to vestibular inputs, sensory modalities that encode the neck and
trunk positions modulate the brain's estimate of our body or environmental tilt with
respect to gravity (Alberts et al. 2016b; Barbieri et al. 2008; Barra et al. 2012;
Böhmer and Mast 1999; Carriot et al. 2011; Ceyte et al. 2009; Day and Wade 1969;
McKenna et al. 2004; Riccio et al. 1992; Trousselard et al. 2004; Wade 1968). The
relative strengths of sensory inputs' influence upon our perceived spatial orientation
are modulated by context and over time, e.g. while upright, vestibular inputs have a
greater influence on our spatial orientation, whereas somatosensory inputs dominate
when we are recumbent (Alberts et al. 2016b; Lechner-Steinleitner 1978).
Somatosensory neck inputs dominate during a maintained head tilt, when there is a
gradual drift in estimating upright orientation, typically in the direction of the head
tilt – and of a magnitude greater than that provoked by an equal body tilt – which is
followed by a post-adaptive perceptual bias which unsurprisingly is not related to
adaption in the VOR (Fig. 3) (Lechner-Steinleitner 1978; Tarnutzer et al. 2013;
Wade 1968, 1970; Otero-Millan and Kheradmand 2016).

   Our daily environment is rich with visual cues that can also affect spatial orientation through integration with vestibular inputs. Strong effects of visual cues can be
seen in various settings, ranging from an entire tilted furnished room to simple cues
such as a square visual frame, in which a perpetual bias can be induced by the orientation of the frame (known as the rod-and-frame effect) (Asch and Witkin 1948;
Cian et al. 2001; Dyde and Milner 2002; Fiori et al. 2014; Groen et al. 2002; Haji-
Khamneh and Harris 2010; Howard 1982; Howard and Childerson 1994; Jenkin
et al. 2003; Kupferberg et al. 2009; Mittelstaedt 1986; Tomassini et al. 2014;
Vingerhoets et al. 2009; Zoccolotti et al. 1992). Similarly, background visual

motion, such as large-field visual motion, can bias our spatial orientation, and these visual effects are more dominant when our body is tilted away from the vertical, consistent with context-dependent modulation of sensory cues in affecting our spatial perception (Alberts et al. 2016a; Corbett and Enns 2006; Dichgans et al. 1972; Dyde et al. 2006; Goto et al. 2003; Guerraz et al. 2001; Lopez et al. 2007; Vingerhoets et al. 2008).

Self-motion perception can be considered a primary vestibular sensation, being prominently linked to activation of the semicircular canals and, to a lesser extent, the otoliths (Guedry Jr. 1974; Fetsch et al. 2011; Karmali et al. 2014). Regarding non-vestibular cues engendering self-motion sensation, large-field visual motion cues easily provoke self-motion sensation, whereas somatosensory cues only weakly elicit self-motion sensation. During real whole-body angular rotations in the dark, preserved semicircular canal function is critical in maintaining self-motion perception, as evidenced by a vestibular patient's inability to detect and/or spatially orientate during either passive or active whole-body turns (Glasauer et al. 2002; Fetsch et al. 2011). In contrast, somatosensory inputs assume a greater importance for detecting linear motion as evidenced by performance of vestibular patinets in passive linear motion threshold and tilt perception studies (Gianna et al. 1995, 1996; Wiest et al. 2001; Yates et al. 2000). Having said that, although somatosensory cues support the perception of self-motion when engendered by other cues, rarely do somatosensory cues elicit vivid sensations of self-motion on their own.

3.2   Ambiguity of Vestibular Sensations

In the process of integrating vestibular inputs into vestibular-mediated perceptions, the compensatory movement of the eyes through the VOR response is critical for maintaining visual stability with changes in the head position (Fig. 1). In humans, a lateral head tilt (i.e. changes in the head position in the roll plane), as opposed to head movement in the horizontal (yaw) or vertical (pitch) plane, leads to a partial VOR response that does not match the magnitude of the head movement (i.e. OCR response) (Kheradmand and Winnick 2017; Otero-Millan and Kheradmand 2016). Because of this partial OCR response, lateral head tilt with respect to gravity results in a mismatch between the head position in space and eye position within the orbit, which requires neural integration of sensory inputs that are encoded in different reference frames in order to maintain spatial orientation (Fig. 2a). Such discrepancies between vestibular inputs and other sensory modalities highlight the importance of higher-order processing that maintains perceptual stability while we interact with the surrounding environment. Such processing demand can be measured by a psychophysical task known as the subjective visual vertical (SVV), in which a visual line is used to measure perceived earth-vertical orientation in the absence of orientation cues (Fig. 2b) (Kheradmand et al. 2016; Kheradmand and Winnick 2017). In the upright position, where the visual and vestibular reference frames are all aligned with the axis of gravity, perceived spatial orientation typically remains within two

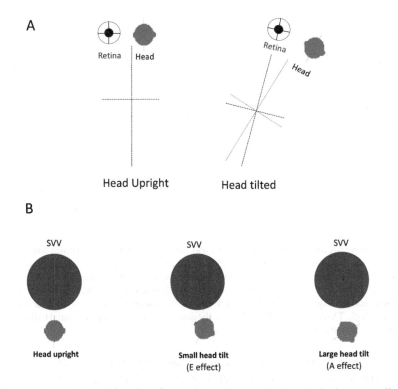

Fig. 2 (**a**) In the upright position, the reference frames for visual and vestibular inputs are aligned with the axis of gravity. During a lateral head tilt, however, the ocular counter-roll (OCR) only partially compensates for the amount of the head tilt, and these reference frames become separated. (**b**) Systematic biases in spatial orientation are shown as the subjective visual vertical (SVV) errors during lateral head tilts. Typically, SVV errors are within two degrees of earth-vertical in the upright position. The E-effect occurs at small tilt angles with an SVV error deviation in the opposite direction of the head tilt. The A-effect occurs at large tilt angles with an SVV error deviation in the direction of the head tilt

degrees of earth-vertical (Kheradmand and Winnick 2017). During whole-body or head tilts, however, SVV errors increase as the visual and vestibular reference frames become separated. Thus, SVV errors during head tilt reflect challenges for the brain in maintaining a common reference frame for spatial orientation based on vestibular and visual inputs (Fig. 3) (Van Beuzekom and Van Gisbergen 2000; Howard 1982; Tarnutzer et al. 2009; Mittelstaedt 1983).

Usually, SVV errors are biased in the direction of the tilt at angles greater than 60° (known as Aubert or A-effect), and they are often biased in the opposite direction of the tilt at smaller tilt angles (known as Entgegengesetzt or E-effect) (Fig. 2b) (Aubert 1861; Howard 1982; Mittelstaedt 1983; Müller 1916; Van Beuzekom and Van Gisbergen 2000). These systematic errors do not correspond with the perceived head tilt position, which generally remains more accurate than spatial orientation (Bronstein 1999; Kaptein and Van Gisbergen 2004; Mittelstaedt 1983; Van

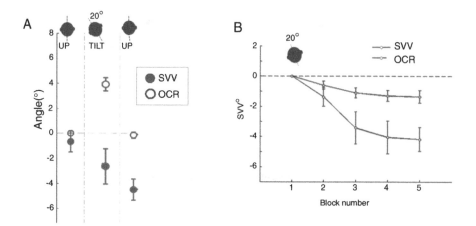

Fig. 3 (**a**) Subjective visual vertical (SVV) and ocular counter-roll (OCR) measured simultaneously at three head positions: upright, 20° head tilt and back to upright (12 subjects). The SVV errors are dissociated from the OCR during and after head tilt. (**b**) Continuous SVV recording for 500 trials (5 blocks of 100 trials ~10 mins) along with simultaneous OCR measurements during 20° left head tilt (eight subjects). Both SVV and OCR are normalised to their baseline values from the first block. There is a larger SVV drift for the duration of the head tilt compared to the OCR. The error bars show the standard error of the mean

Beuzekom et al. 2001; Van Beuzekom and Van Gisbergen 2000). The errors of spatial orientation are also dissociated from the ocular counter-roll (i.e. the VOR response during head tilt) (Fig. 3a) (Otero-Millan and Kheradmand 2016). Such dissociations show that the spatial orientation is not simply determined by the vestibular inputs that encode head position and, thus, further sensory processing and integration must take place within the neural networks involved in spatial orientation.

The ambiguity of self-motion perception for yaw-plane rotations or lateral motion is well described as experienced by us all when looking at a moving train which engenders a sense of self-motion. The brain processes that aim to reduce the ambiguity of self versus visual motion have been studied extensively (Bense et al. 2001; Bottini et al. 2001; Dieterich and Brandt 2008; Probst et al. 1985, 1986; Probst and Wist 1990; Stephan et al. 2005; Wenzel et al. 1996). One theory considers that there is a reciprocal visual–vestibular inhibition for motion perception such that the brain generates a percept either of self-motion (i.e. 'our train is moving') or of object motion ('the other train is moving'). The support for such a hypothesis runs from psychophysics (Probst et al. 1985, 1986), neurophysiology (Probst and Wist 1990), neuroimaging (Bense et al. 2001; Bottini et al. 2001; Dieterich and Brandt 2008; Stephan et al. 2005; Wenzel et al. 1996) and visual cortical excitability studies in humans (Seemungal et al. 2013).

The strong sensation of self-motion and imbalance one perceives when stepping off a fast spinning 'merry-go-round' or the complaint of a continuous vertigo in a patient with a vestibular neuritis (with a third-degree vestibular nystagmus) clearly reflects a coupling of the VOR and vestibular-mediated motion perception. However, an uncoupling of vestibular reflexes and perception can occur naturally in subjects who are repeatedly exposed to strong vestibular and visual stimuli (e.g. pirouetting in dancers). Such uncoupling is also seen in the ballet dancers during whole-body step-rotations (Nigmatullina et al. 2015). Ballet dancers are known to have attenuated vestibular responses (Osterhammel et al. 1968; Okada et al. 1999). The control group, on the other hand, showed the usual congruency between high-order perceptual response and low-order VOR responses.

3.3   The Neural Correlates of Vestibular Sensations

As mentioned previously, vestibular-guided spatial orientation – e.g. updating our directional heading during locomotor turns in the dark – requires the brain to convert the vestibular signals of head angular velocity to travelled angle via a process called path integration (Mittelstaedt and Mittelstaedt 1980; Seemungal et al. 2007). The neural correlates that convert the inertial signals of self-motion to travelled distance (i.e. the neural correlates of path integration) have been a focus of animal and human research. One potential parallel is eye movement control that similarly requires a velocity-to-position conversion which is mediated by known neural substrates in the brainstem and cerebellar circuits, the so-called ocular motor neural integrator. For vestibular-guided spatial orientation, the velocity-to-position conversion could emanate from a bottom-up process, such with perceived travelled distance being an output of the ocular motor neural integrator. Current evidence, however, does not support a bottom-up model for human vestibular-guided spatial orientation. In primates, vestibular thalamic relay circuits do not possess head-in-space signals but show vestibular-related head velocity signals (Meng et al. 2007). Indeed, Kaski et al. (2016) showed that focal cortical stroke in patients does not affect vestibular-derived motion perception. In contrast, they found that temporoparietal lesions resulted in a spatial deficit when navigating passively and rotating under vestibular guidance in a contralesional direction.

  Functional magnetic resonance imaging (fMRI) and positron emission tomography (PET) have enabled the exploration of the cerebral cortical regions that are activated by vestibular stimulation (Kheradmand and Winnick 2017; Lopez et al. 2012). Given that functional neuroimaging requires subjects to remain still, yet the ecological vestibular stimulus is head motion, a variety of surrogate means for stimulating the vestibular apparatus during human neuroimaging studies have been used – from bithermal caloric irrigation or galvanic stimulation. This means that the cortical brain activity related to such stimulation may not be related to vestibular perception simply on the account that the brain activity being located is in the cerebral cortex. Nevertheless, studies in healthy subjects undergoing vestibular stimulation in the neuroimaging scanners have shown signal in the insula, superior temporal gyrus, inferior parietal lobule, parietal operculum, somatosensory cortex, cingulate gyrus, frontal cortex and hippocampus (Bense et al. 2001; Bottini et al. 2001;

Dieterich et al. 2003; Eickhoff et al. 2006; Emri et al. 2003; Indovina et al. 2005; Lobel et al. 1998; Stephan et al. 2005; Suzuki et al. 2001; zu Eulenburg et al. 2012). A meta-analysis suggested that the core region showing activation across these studies involves the retroinsular cortex and parietal operculum (Lopez et al. 2012).

Several studies, including neuroimaging studies, support the notion that the 'higher-order' vestibular processing that contributes to spatial orientation is localised to the temporoparietal junction (TPJ), and it shares the same neural substrate with perception of body orientation, heading perception and visual gravitational motion (Blanke et al. 2015; Bosco et al. 2008; Cazzato et al. 2015; Donaldson et al. 2015; Hansen et al. 2015; Igelström and Graziano 2017; Indovina et al. 2005; Jáuregui Renaud 2015; Kaski et al. 2016; Lacquaniti et al. 2013; Lopez et al. 2008a; Roberts et al. 2016; Saj et al. 2014; Silani et al. 2013; Ventre-Dominey 2014; Lester and Dassonville 2014; Karnath and Dieterich 2006; Kerkhoff 1999; Utz et al. 2011). The role of TPJ in this aspect of vestibular perception has been studied in patients with TPJ lesions and with noninvasive brain stimulation in healthy subjects (Kheradmand et al. 2015; Fiori et al. 2015; Santos-Pontelli et al. 2016; Lester and Dassonville 2014; Karnath and Dieterich 2006; Kerkhoff 1999; Utz et al. 2011). Similar to TPJ lesions, the disruptive effect of transcranial magnetic stimulation (TMS) at the right TPJ in healthy controls can alter spatial orientation with respect to gravity (Fig. 4a) (Kheradmand et al. 2015; Otero-Millan et al. 2018). This effect is measured as a TMS-induced SVV shift and is dissociated from the OCR response during head tilt. The dissociation of TMS-induced SVV shift from OCR shows that

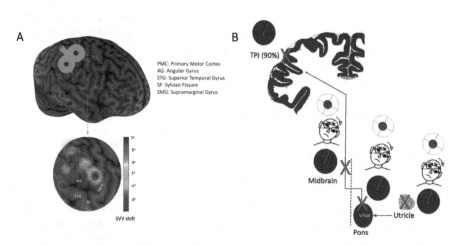

Fig. 4 (a) An example of TMS effect within the right TPJ in one subject. The inhibitory effect of TMS at the posterior aspect of supramarginal gyrus (SMG) results in larger SVV shift compared to other locations. (b) The SVV deviations with peripheral vestibular and brainstem lesions (x) are mainly related to pathological changes in the vestibulo-ocular pathways, which is often associated with commensurate deviations in the eye and head positions in the roll plane. These deviations are ipsilesional with caudal lesions and contralesional with rostral lesions. With TPJ lesions, the SVV deviations are mostly contralesional (90% of patients) and are not associated with pathological vestibulo-ocular changes

the TPJ cortical mechanisms are primarily involved in processing vestibular inputs, in contrast to subcortical regions that can influence spatial orientation directly through vestibulo-ocular responses (Fig. 4b).

Another method of localising vestibular perceptual functioning is to obtain psychophysical measurements in human subjects in the laboratory and then to obtain structural neuroimaging in a separate session. By correlating between subject variability in behaviour and in brain imaging parameters, one can make structural–functional comparisons across the group of individuals (Nigmatullina et al. 2015). To address neuroanatomical correlates of vestibular motion perception, Nigmatullina et al. (Nigmatullina et al. 2015) collected behavioural and structural neuroimaging data from 49 young, healthy, right-handed females who were either ballerinas or rowers (Fig. 5). Ballerinas' long-term pirouetting training is a model of extreme vestibular adaptation, while rowers acted as controls matched for physical activity. One of their main findings using a voxel-based morphometry (VBM) was that ballet dancers show reduced grey matter (GM) density in the vestibular cerebellum compared to controls (in a whole brain analysis). They then found that the duration of vertigo following a rotational stop (constant angular 90 °/s velocity) in dancers was correlated with the density of vestibular cerebellar grey matter (shorter responses were associated with lower grey matter density), but this relationship was positively correlated in controls. In both controls and dancers, the duration of the evoked reflexive VOR response (following a stopping response) was positively correlated. These data suggest that the vestibular cerebellum is key in mediating the uncoupling for vestibular reflex and perceptual functions. Indeed, isolated cerebellar strokes are associated with acute vertigo, even without a nystagmus, indicating the importance of the vestibular cerebellum in processing the signals of self-motion that ascend to perceptual regions (Amarenco et al. 1990; Disher et al. 1991; Horii et al. 2006; Huang and Yu 1985; Lee and Cho 2004; Lee et al. 2006; Lee et al. 2003; Seemungal 2007). In addition, the same study – via diffusion tensor imaging (DTI) – demonstrated the extensive cerebral cortical white matter network – which interestingly was bilateral – that correlated with vestibular motion perception. The DTI parameters did not reach significance in the dancers' group, perhaps indicating an attenuation of vestibular motion signals at cortical level in dancers. Comparing groups, however, a core white matter network was identified, and this was focussed in the TPJ region bilaterally. These data support the notion that vestibular self-motion perception is mediated by an extensive and bilateral white matter circuit in the cerebral cortex. Indeed, a network concept of self-motion perception predicts that single focal brain lesions will not lead to significant lateralised impairments of vestibular self-motion perception (Seemungal 2014). Evidence for this was provided by Kaski et al. who showed that focal brain lesions in the TPJ affected vestibular spatial but not vestibular motion perception (Kaski et al. 2016). Indeed, the clinical observation of the loss of vestibular motion perception in elderly patients (Seemungal 2005), hypothesised to be due to the cortical network disruption from white matter disease, was instrumental in the development of a specific technique to probe self-motion perception as a factor in falls in the elderly (Seemungal 2005).

In the first study to investigate cortical aspects of vestibular function using brain stimulation, it was shown that repetitive transcranial magnetic stimulation (rTMS)

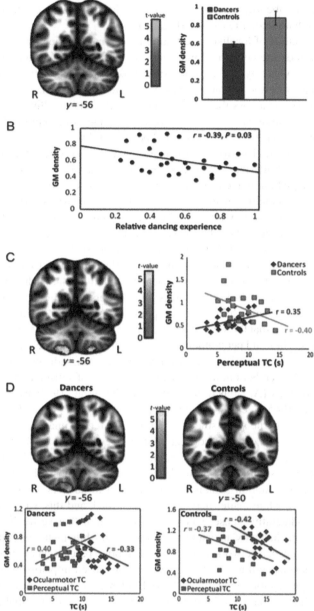

Fig. 5 Results of grey matter (GM) analysis in ballet dancers versus controls. (**a**) Whole-brain GM comparison showed a reduction in GM density in the posterior cerebellum in dancers compared with controls (dancers<controls). This region was then used as a mask in subsequent GM analyses (**c, d**). The inset graph on the right shows the mean GM density of the posterior cerebellum region in the two groups. (**b**) The mean GM density of the cerebellar regions from dancers and

to the posterior parietal cortex disrupted vestibular spatial percepts of angular position for contralateral rotations, but not perceived angular velocity of rotation (Seemungal et al. 2008). These findings were also confirmed in a study assessing patients with acute infarcts involving the right TPJ (Fig. 6) (Kaski et al. 2016). In this study involving patients with less than 2 weeks of their infarct (typically lesion studies combine patients of highly variable lesion ages) and using voxel-based lesion–symptom mapping (VLSM) techniques, the effect of acute right hemisphere lesions on perceived angular position, velocity, and motion duration during whole-body rotations in the dark were explored. First, compared to healthy controls, out of 18 acute stroke patients tested, only four patients with damage to the TPJ showed impaired spatial orientation performance for leftward (contralesional) compared to rightward (ipsilesional) rotations. Second, only patients with TPJ damage showed a congruent underestimation in both their travelled distance (perceived as shorter) and motion duration (perceived as briefer) for leftward compared to rightward rotations. All 18 patients tested showed normal self-motion perception. These data suggest that the TPJ may encode vestibular-guided movement in a form that reflects the relationship between travelled distance (s), velocity of motion (v), and duration of motion (t), i.e. $s = \int v.dt$, a suggestion which was previously made by Berthoz and colleagues in a pioneering study (Berthoz et al. 1995) but also subsequently supported by rTMS studies (Seemungal et al. 2007) (Kaski et al. 2016). Thus, an impaired cortical integration of vestibular head motion inputs could underpin other types of egocentric topographical disorientation syndromes associated with focal posterior right hemispheric lesions (Aguirre and D'Esposito 1999).

4 Vestibular Perception in a Clinical Context

Vestibular dysfunction can affect perceptions of spatial orientation or self-motion. Patients with peripheral vestibular loss may develop pronounced visual dependence which is implicated in the development of a chronic maladaptive syndrome called PPPD or 'persistent postural–perceptual dizziness' which manifests with an excessive sensitivity to visual motion stimuli and busy visual environments (Lopez et al. 2006, 2007; Dichgans et al. 1972; Goto et al. 2003; Guerraz et al. 2001;

Fig. 5 (continued) control analysis (**a**) negatively correlates with the relative dancing experience (1 = maximum dancing experience). (**c**) Intergroup interaction revealed a cluster, where GM density shows opposing correlations for perceptual time constants (TCs) in dancers and controls. (**d**) In dancers, a significant cluster (left panel) was found in which GM density correlates negatively with ocular motor time constants (TCs) and positively with perceptual TC. In controls, significant cluster (right panel) was found where GM density negatively correlates with both ocular motor and perceptual TCs (**c**, **d**). The inset graphs, with corresponding r values, illustrate the correlations by plotting the mean GM value of the significant cluster for each subject with the corresponding psychophysical measures (**a**, **c**, **d**). The significant clusters are shown at $P < 0.05$ (corrected) superimposed on a structural T1 image (grey) in the coronal view with a Talairach coordinate (mm). Color bars indicates t values. R, right; L, left

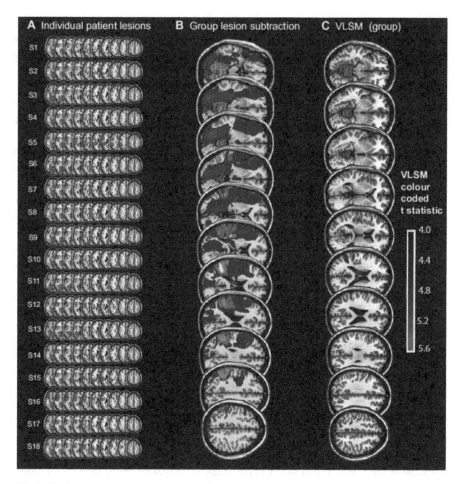

Fig. 6 Right hemispheric lesions in 18 stroke patients taken from Kaski et al. 2016. Brain lesion maps and analysis. (**a**) Lesion map of all stroke patients (patients S1–S18). (**b**) Lesion subtraction analysis for patients S1–S18 localised the position task deficit to the temporoparietal junction (TPJ) shown in yellow. (**c**) Voxel-based lesion–symptom mapping (VLSM) analysis. The bar on the far right gives the colour coding for the significance level for the VLSM analysis. The most significant regions are in the angular gyrus and just reaching the superior temporal gyrus, with further less significant voxels in the middle temporal gyrus

Popkirov et al. 2018). Patients with peripheral vestibular loss usually have more pronounced A-effect during head tilt consistent with reduced *weight* of vestibular inputs in the process of sensory integration for spatial orientation (Bronstein 1999; Bronstein et al. 1996; Dai et al. 1989; Graybiel et al. 1968; Lopez et al. 2008b; Müller et al. 2016; Toupet et al. 2014). The effect of visual cues is altered in these patients, with a pronounced visual dependence towards the side of vestibular loss (Lopez et al. 2006) (Dichgans et al. 1972; Goto et al. 2003; Guerraz et al. 2001; Lopez et al. 2007). Vestibular-mediated spatial orientation is also altered with lesions involving the cerebral hemispheres or brainstem. The effect of these 'central' lesions are mainly described in the context of ischaemic infarcts (Brandt et al. 1994; Baier et al. 2012; Barra et al. 2010; Kerkhoff and Zoelch 1998; Rousseaux et al. 2015). With hemispheric lesions, spatial deviations with respect to gravity are mainly contralesional (i.e. away from the side of the lesion), whereas only about 10% of patients may have ipsilesional deviations (i.e. towards the side of the lesion) (Baier et al. 2013; Barra et al. 2010; Brandt et al. 1994; Kerkhoff and Zoelch 1998; Pérennou et al. 2008; Saj et al. 2005; Yelnik et al. 2002). With brainstem lesions, however, spatial deviations are always ipsilesional with caudal brainstem involvement and contralesional with rostral brainstem involvement (Dieterich and Brandt 1993). In addition, the extent of deviations with hemispheric lesions is usually less than brainstem or peripheral vestibular lesions (Dieterich and Brandt 1993; Hafström et al. 2004). These anatomical differences could be linked to pathological vestibulo-ocular changes with brainstem lesions that can directly affect the orientation of the images on the retina, whereas at the level of TPJ, spatial orientation is primarily linked to vestibular processing and integration with other sensory modalities (Fig. 4b) (Brodsky et al. 2006).

The hemispheric lesions that cause spatial deviations converge largely around the inferior parietal lobule within the TPJ (Kheradmand and Winnick 2017). A subset of these patients also have robust postural deviations and actively resist correcting their false postural orientation back to upright position (Danells et al. 2004; Dieterich and Brandt 1992; Johannsen et al. 2006; Paci et al. 2009; Pérennou et al. 2008). Patients with this phenomenon, which is known as 'pusher syndrome', are unable to learn to walk again even with proper assistance and have longer-lasting spatial deviations (Mansfield et al. 2015). Pushing behaviour is highly correlated with spatial deviation, suggesting that these patients actively align their body with their erroneous spatial perception.

Another disorder associated with spatial disorientation is vestibular migraine. Patients with vestibular migraine typically complain of dizziness, disorientation or sense of disequilibrium, often triggered or aggravated with changes in the head or body positions. This type of migraine presentation accounts for the most common cause of episodic dizziness (Balaban 2016; Dieterich et al. 2016; Furman et al. 2013; Lempert et al. 2012). Patients with vestibular migraine were found to have larger errors of spatial orientation compared with healthy controls (Winnick et al. 2018). These larger errors were in the opposite direction of the head tilt, consistent with overestimation of the head position in the process of sensory integration for spatial orientation (i.e. larger E effect). In keeping with this result, patients mostly reported dizziness in the same head direction that induced larger errors of spatial perception. Other studies have shown high variability in spatial orientation with the head in upright position and more pronounced postural sway in these patients compared to those with other types of migraine or healthy controls (Asai et al. 2009; Furman et al. 2005; Kandemir et al. 2014; Teggi et al. 2009). Altogether, these findings suggest that visuospatial symptoms and dizziness in vestibular migraine could be related to sensory processes involved in integration of vestibular inputs into perception of spatial orientation. Patients with vestibular migraine are also found to have reduced motion detection thresholds in the roll plane (Lewis et al. 2011).

Perceptuo-reflex uncoupling, whereby there is a loss of correspondence between vestibular reflex functioning and perception (i.e. symptoms), can be observed in acute and chronic vestibular patients. Although patients with a vestibular neuritis show strong coupling between symptoms and signs (i.e. as the vestibular nystagmus subsides so does their vertigo), acute patients with isolated cerebellar strokes not involving the brainstem may have strong sensations of vertigo without nystagmus (Amarenco et al. 1990; Disher et al. 1991; Horii et al. 2006; Huang and Yu 1985; Lee and Cho 2004; Lee et al. 2006; Lee et al. 2003; Seemungal 2007). Indeed, vertigo is the commonest – and occasionally the only – symptom in patients with cerebellar strokes (Amarenco et al. 1990, Disher et al. 1991; Horii et al. 2006; Huang and Yu 1985; Lee 2014; Lee and Cho 2004; Lee et al. 2006; Lee et al. 2003; Seemungal 2007). In contrast to acute patients, patients with a previous vestibular neuritis may show VOR recovery in around half of patients; however, a key finding is that recovery of VOR functioning is not correlated with symptomatic burden (Bergenius and Perols 1999; Jacobson and McCaslin 2003; Kammerlind et al. 2005; Okinaka et al. 1993; Palla et al. 2008). Whether the vestibular cerebellum mediates symptomatic recovery following a peripheral vestibular neuritis remains speculative but is supported by its involvement in experience-dependent, perceptuo-reflex uncoupling between nystagmus and perception in dancers (Nigmatullina et al. 2015).

References

Abe, H., Kondo, T., Oouchida, Y., Suzukamo, Y., Fujiwara, S., & Izumi, S.-I. (2012). Prevalence and length of recovery of pusher syndrome based on cerebral hemispheric lesion side in patients with acute stroke. *Stroke, 43,* 1654–1656. https://doi.org/10.1161/STROKEAHA.111.638379.

Aguirre, G. K., & D'Esposito, M. (1999). Topographical disorientation: A synthesis and taxonomy. *Brain, 122*(Pt 9), 1613–1628.

Akdal, G., Toydemir, H. E., Tanrıverdizade, T., & Halmagyi, G. M. (2017). Room tilt illusion: A symptom of both peripheral and central vestibular disorders. *Acta Neurologica Belgica, 117*(1), 363–365.

Alberts, B. B. G. T., de Brouwer, A. J., Selen, L. P. J., & Medendorp, W. P. (2016a). A Bayesian account of visuo-vestibular interactions in the rod-and-frame task. *eneuro,* ENEURO.0093-16.2016. https://doi.org/10.1523/ENEURO.0093-16.2016.

Alberts, B. B. G. T., Selen, L. P. J., Bertolini, G., Straumann, D., Medendorp, W. P., & Tarnutzer, A. A. (2016b). Dissociating vestibular and somatosensory contributions to spatial orientation. *Journal of Neurophysiology, 116,* 30–40. https://doi.org/10.1152/jn.00056.2016.

Allum, J. H., Graf, W., Dichgans, J., & Schmidt, C. L. (1976). Visual-vestibular interactions in the vestibular nuclei of the goldfish. *Experimental Brain Research, 26*(5), 463–485.

Amarenco, P., Roullet, E., Hommel, M., et al. (1990). Infarction in the territory of the medial branch of the posterior inferior cerebellar artery. *Journal of Neurology, Neurosurgery, and Psychiatry, 53,* 731–735.

Asai, M., Aoki, M., Hayashi, H., Yamada, N., Mizuta, K., & Ito, Y. (2009). Subclinical deviation of the subjective visual vertical in patients affected by a primary headache. *Acta Oto-Laryngologica, 129,* 30–35. https://doi.org/10.1080/00016480802032785.

Asch, S. E., & Witkin, H. A. (1948). Studies in space orientation; perception of the upright with displaced visual fields. *Journal of Experimental Psychology, 38*, 325–337.

Aubert, H. (1861). Eine scheinbare bedeutende Drehung von Objecten bei Neigung des Kopfes nach rechts oder links. *Archiv für pathologische Anatomie und Physiologie und für klinische Medicin, 20*, 381–393.

Baggio, J. A. O., Mazin, S. S. C., Alessio-Alves, F. F., Barros, C. G. C., Carneiro, A. A. O., Leite, J. P., et al. (2016). Verticality perceptions associate with postural control and functionality in stroke patients. *PLoS One, 11*, e0150754. https://doi.org/10.1371/journal.pone.0150754.

Baier, B., Eulenburg, P. z., Best, C., Geber, C., Müller-Forell, W., Birklein, F., et al. (2013). Posterior insular cortex – A site of vestibular–somatosensory interaction? *Brain and Behavior, 3*, 519–524. https://doi.org/10.1002/brb3.155.

Baier, B., Suchan, J., Karnath, H.-O., & Dieterich, M. (2012). Neural correlates of disturbed perception of verticality. *Neurology, 78*, 728–735. https://doi.org/10.1212/WNL.0b013e318248e544.

Balaban, C. D. (2016). Chapter 3 – Neurotransmitters in the vestibular system. In J. M. Furman & T. Lempert (Eds.), *Handbook of clinical neurology neuro-otology* (pp. 41–55). Cambridge, MA: Elsevier. https://doi.org/10.1016/B978-0-444-63437-5.00003-0.

Baloh, R. W. (2003). Clinical practice. Vestibular neuritis. *The New England Journal of Medicine, 348*(11), 1027–1032.

Barbieri, G., Gissot, A.-S., Fouque, F., Casillas, J.-M., Pozzo, T., & Pérennou, D. (2008). Does proprioception contribute to the sense of verticality? *Experimental Brain Research, 185*, 545–552. https://doi.org/10.1007/s00221-007-1177-8.

Barra, J., Benaim, C., Chauvineau, V., Ohlmann, T., Gresty, M., & Perennou, D. (2008). Are rotations in perceived visual vertical and body axis after stroke caused by the same mechanism? *Stroke, 39*, 3099–3101. https://doi.org/10.1161/STROKEAHA.108.515247.

Barra, J., Marquer, A., Joassin, R., Reymond, C., Metge, L., Chauvineau, V., et al. (2010). Humans use internal models to construct and update a sense of verticality. *Brain, 133*, 3552–3563. https://doi.org/10.1093/brain/awq311.

Barra, J., Pérennou, D., Thilo, K. V., Gresty, M. A., & Bronstein, A. M. (2012). The awareness of body orientation modulates the perception of visual vertical. *Neuropsychologia, 50*, 2492–2498. https://doi.org/10.1016/j.neuropsychologia.2012.06.021.

Beevor, C. E. (1909). Remarks on paralysis of the movements of the trunk in hemiplegia, and the muscles which are affected. *British Medical Journal, 1*, 881.

Bender, M., & Jung, R. (1948). Abweichungen der subjektiven optischen Vertikalen und Horizontalen bei Gesunden und Hirnverletzten. *European Archives of Psychiatry and Clinical Neuroscience, 181*, 193–212.

Bense, S., Stephan, T., Yousry, T. A., Brandt, T., & Dieterich, M. (2001). Multisensory cortical signal increases and decreases during vestibular galvanic stimulation (fMRI). *Journal of Neurophysiology, 85*(2), 886–899.

Bergenius, J., & Perols, O. (1999). Vestibular neuritis: A follow-up study. *Acta Otolaryngology, 119*(8), 895–899.

Bermúdez Rey, M. C., Clark, T. K., Wang, W., Leeder, T., Bian, Y., & Merfeld, D. M. (2016). Vestibular perceptual thresholds increase above the age of 40. *Frontiers in Neurology, 7*, 162. eCollection.

Berthoz, A. (1997). Parietal and hippocampal contribution to topokinetic and topographic memory. *Philosophical Transactions of the Royal Society of London. Series B, Biological Sciences, 352*(1360), 1437–1448.

Berthoz, A., Israel, I., Georges-Francois, P., Grasso, R., & Tsuzuku, T. (1995). Spatial memory of body linear displacement: What is being stored? *Science, 269*, 95–98.

Bjerver, K., & Silfverskiöld, B. P. (1968). Lateropulsion and imbalance in Wallenberg's syndrome. *Acta Neurologica Scandinavica, 44*, 91–100.

Blanke, O., Landis, T., Spinelli, L., & Seeck, M. (2004). Out-of-body experience and autoscopy of neurological origin. *Brain, 127*, 243–258. https://doi.org/10.1093/brain/awh040.

Blanke, O., Ortigue, S., Landis, T., & Seeck, M. (2002). Neuropsychology: Stimulating illusory own-body perceptions. *Nature, 419*, 269–270.

Blanke, O., Perrig, S., Thut, G., Landis, T., & Seeck, M. (2000). Simple and complex vestibular responses induced by electrical cortical stimulation of the parietal cortex in humans. *Journal of Neurology, Neurosurgery, and Psychiatry, 69*(4), 553–556.

Blanke, O., Slater, M., & Serino, A. (2015). Behavioral, neural, and computational principles of bodily self-consciousness. *Neuron, 88*, 145–166. https://doi.org/10.1016/j.neuron.2015.09.029.

Bohannon, R. W. (1996). Ipsilateral pushing in stroke. *Archives of Physical Medicine and Rehabilitation, 77*, 524.

Böhmer, A., & Mast, F. (1999). Assessing otolith function by the subjective visual vertical. *Annals of the New York Academy of Sciences, 871*, 221–231.

Bonan, I. V., Hubeaux, K., Gellez-Leman, M. C., Guichard, J. P., Vicaut, E., & Yelnik, A. P. (2007). Influence of subjective visual vertical misperception on balance recovery after stroke. *Journal of Neurology, Neurosurgery, and Psychiatry, 78*, 49–55. https://doi.org/10.1136/jnnp.2006.087791.

Bonan, I. V., Leman, M. C., Legargasson, J. F., Guichard, J. P., & Yelnik, A. P. (2006). Evolution of subjective visual vertical perturbation after stroke. *Neurorehabilitation and Neural Repair, 20*, 484–491. https://doi.org/10.1177/1545968306289295.

Bosco, G., Carrozzo, M., & Lacquaniti, F. (2008). Contributions of the human temporoparietal junction and MT/V5+ to the timing of interception revealed by transcranial magnetic stimulation. *The Journal of Neuroscience, 28*, 12071–12084. https://doi.org/10.1523/JNEUROSCI.2869-08.2008.

Bottini, G., Karnath, H. O., Vallar, G., Sterzi, R., Frith, C. D., Frackowiak, R. S., & Paulesu, E. (2001). Cerebral representations for egocentric space: Functional-anatomical evidence from caloric vestibular stimulation and neck vibration. *Brain, 124*, 1182–1196.

Braem, B., Honoré, J., Rousseaux, M., Saj, A., & Coello, Y. (2014). Integration of visual and haptic informations in the perception of the vertical in young and old healthy adults and right braindamaged patients. *Neurophysiologie Clinique/Clinical Neurophysiology, 44*, 41–48. https://doi.org/10.1016/j.neucli.2013.10.137.

Brandt, T., & Dieterich, M. (1994). Vestibular syndromes in the roll plane: Topographic diagnosis from brainstem to cortex. *Annals of Neurology, 36*, 337–347. https://doi.org/10.1002/ana.410360304.

Brandt, T., & Dieterich, M. (2000). Perceived vertical and lateropulsion: Clinical syndromes, localization, and prognosis. *Neurorehabilitation and Neural Repair, 14*, 1–12. https://doi.org/10.1177/154596830001400101.

Brandt, T., Dieterich, M., & Danek, A. (1994). Vestibular cortex lesions affect the perception of verticality. *Annals of Neurology, 35*, 403–412. https://doi.org/10.1002/ana.410350406.

Brandt, T., Strupp, M., & Dieterich, M. (2014). Towards a concept of disorders of "higher vestibular function". *Frontiers in Integrative Neuroscience, 8*, 47. https://doi.org/10.3389/fnint.2014.00047.

Bray, A., Subanandan, A., Isableu, B., Ohlmann, T., Golding, J. F., & Gresty, M. A. (2004). We are most aware of our place in the world when about to fall. *Current Biology, 14*, R609–R610. https://doi.org/10.1016/j.cub.2004.07.040.

Bremmer, F., Klam, F., Duhamel, J. R., Ben Hamed, S., & Graf, W. (2002). Visual-vestibular interactive responses in the macaque ventral intraparietal area (VIP). *The European Journal of Neuroscience, 16*(8), 1569–1586.

Brodsky, M. C., Donahue, S. P., Vaphiades, M., & Brandt, T. (2006). Skew deviation revisited. *Survey of Ophthalmology, 51*, 105–128. https://doi.org/10.1016/j.survophthal.2005.12.008.

Bronstein, A. M. (1999). The interaction of otolith and proprioceptive information in the perception of verticality. The effects of labyrinthine and CNS disease. *Annals of the New York Academy of Sciences, 871*, 324–333.

Bronstein, A. M., Yardley, L., Moore, A. P., & Cleeves, L. (1996). Visually and posturally mediated tilt illusion in Parkinson's disease and in labyrinthine defective subjects. *Neurology, 47*, 651–656.

Brookes, G. B., Gresty, M. A., Nakamura, T., & Metcalfe, T. (1993). Sensing and controlling rotational orientation in normal subjects and patients with loss of labyrinthine function. *The American Journal of Otology, 14*, 349–351.

Brooks, J. X., & Cullen, K. E. (2013). The primate cerebellum selectively encodes unexpected self-motion. *Current Biology, 23*(11), 947–955. https://doi.org/10.1016/j.cub.2013.04.029.

Bury, N., & Bock, O. (2016). Role of gravitational versus egocentric cues for human spatial orientation. *Experimental Brain Research, 234*, 1013–1018. https://doi.org/10.1007/s00221-015-4526-z.

Carriot, J., Cian, C., Paillard, A., Denise, P., & Lackner, J. R. (2011). Influence of multisensory graviceptive information on the apparent zenith. *Experimental Brain Research, 208*, 569–579. https://doi.org/10.1007/s00221-010-2505-y.

Cazzato, V., Mian, E., Serino, A., Mele, S., & Urgesi, C. (2015). Distinct contributions of extrastriate body area and temporoparietal junction in perceiving one's own and others' body. *Cognitive, Affective, & Behavioral Neuroscience, 15*, 211–228. https://doi.org/10.3758/s13415-014-0312-9.

Ceyte, H., Cian, C., Trousselard, M., & Barraud, P.-A. (2009). Influence of perceived egocentric coordinates on the subjective visual vertical. *Neuroscience Letters, 462*, 85–88. https://doi.org/10.1016/j.neulet.2009.06.048.

Cian, C., Raphel, C., & Barraud, P. A. (2001). The role of cognitive factors in the rod-and-frame effect. *Perception, 30*, 1427–1438. https://doi.org/10.1068/p3270.

Clark, T. K., Newman, M. C., Oman, C. M., Merfeld, D. M., & Young, L. R. (2015). Modeling human perception of orientation in altered gravity. *Frontiers in Systems Neuroscience, 9*, 68. https://doi.org/10.3389/fnsys.2015.00068.

Clemens, I. A. H., Vrijer, M. D., Selen, L. P. J., Gisbergen, J. A. M. V., & Medendorp, W. P. (2011). Multisensory processing in spatial orientation: An inverse probabilistic approach. *The Journal of Neuroscience, 31*, 5365–5377. https://doi.org/10.1523/JNEUROSCI.6472-10.2011.

Clément, G., & Eckardt, J. (2005). Influence of the gravitational vertical on geometric visual illusions. *Acta Astronautica, 56*, 911–917.

Clément, G., Tilikete, C., & Courjon, J. H. (2008). Retention of habituation of vestibulo-ocular reflex and sensation of rotation in humans. *Experimental Brain Research, 190*, 307–315.

Colnat-Coulbois, S., Gauchard, G. C., Maillard, L., Barroche, G., Vespignani, H., Auque, J., et al. (2011). Management of postural sensory conflict and dynamic balance control in late-stage Parkinson's disease. *Neuroscience, 193*, 363–369. https://doi.org/10.1016/j.neuroscience.2011.04.043.

Corbett, J. E., & Enns, J. T. (2006). Observer pitch and roll influence: The rod and frame illusion. *Psychonomic Bulletin & Review, 13*, 160–165.

Cullen, K. E. (2018). Multisensory integration and the perception of self motion. *Oxford Research Wncyclopedia of Neuroscience, 19*. https://doi.org/10.1093/acrefore/9780190264086.013.91.

Dai, M. J., Curthoys, I. S., & Halmagyi, G. M. (1989). Linear acceleration perception in the roll plane before and after unilateral vestibular neurectomy. *Experimental Brain Research, 77*, 315–328.

Danells, C. J., Black, S. E., Gladstone, D. J., & McIlroy, W. E. (2004). Poststroke "Pushing". *Stroke, 35*, 2873–2878. https://doi.org/10.1161/01.STR.0000147724.83468.18.

Day, R. H., & Wade, N. J. (1968). Involvement of neck proprioceptive system in visual after-effect from prolonged head tilt. *The Quarterly Journal of Experimental Psychology, 20*, 290–293. https://doi.org/10.1080/14640746808400163.

Day, R. H., & Wade, N. J. (1969). Mechanisms involved in visual orientation constancy. *Psychological Bulletin, 71*, 33.

De Renzi, E., Faglioni, P., & Scotti, G. (1971). Judgment of spatial orientation in patients with focal brain damage. *Journal of Neurology, Neurosurgery, and Psychiatry, 34*, 489–495.

De Ridder, D., Van Laere, K., Dupont, P., Menovsky, T., & Van de Heyning, P. (2007). Visualizing out-of-body experience in the brain. *The New England Journal of Medicine, 357*, 1829–1833. https://doi.org/10.1056/NEJMoa070010.

Dichgans, J., & Brandt, T. (1978). Visual-vestibular interaction: Effects on self-motion perception and postural control. In R. Held, H. W. Leibowitz, & H. L. Teuber (Eds.), *Handbook of sensory physiology* (Vol. VIII: *Perception.*, pp. 755–804). Berlin: Springer.

Dichgans, J., Diener, H. C., & Brandt, T. (1974). Optokinetic-graviceptive interaction in different head positions. *Acta Oto-Laryngologica, 78*, 391–398.

Dichgans, J., Held, R., Young, L. R., & Brandt, T. (1972). Moving visual scenes influence the apparent direction of gravity. *Science, 178*, 1217–1219.

Dieterich, M., Bense, S., Lutz, S., Drzezga, A., Stephan, T., Bartenstein, P., et al. (2003). Dominance for vestibular cortical function in the nondominant hemisphere. *Cerebral Cortex, 13*, 994–1007.

Dieterich, M., & Brandt, T. (1992). Wallenberg's syndrome: Lateropulsion, cyclorotation, and subjective visual vertical in thirty-six patients. *Annals of Neurology, 31*, 399–408. https://doi.org/10.1002/ana.410310409.

Dieterich, M., & Brandt, T. (1993). Ocular torsion and tilt of subjective visual vertical are sensitive brainstem signs. *Annals of Neurology, 33*, 292–299. https://doi.org/10.1002/ana.410330311.

Dieterich, M., & Brandt, T. (2008). Functional brain imaging of peripheral and central vestibular disorders. *Brain, 131*, 2538–2552.

Dieterich, M., Obermann, M., & Celebisoy, N. (2016). Vestibular migraine: The most frequent entity of episodic vertigo. *Journal of Neurology, 263*, 82–89. https://doi.org/10.1007/s00415-015-7905-2.

Disher, M. J., Telian, S. A., & Kemink, J. L. (1991). Evaluation of acute vertigo: Unusual lesions imitating vestibular neuritis. *The American Journal of Otology, 12*, 227–231.

Dolowitz, D. A., Henriksson, N. G., & Forssman, B. (1963). Laterotorsion: A vestibular spinal reflex. *Laryngoscope, 73*, 893–905.

Donaldson, P. H., Rinehart, N. J., & Enticott, P. G. (2015). Noninvasive stimulation of the temporoparietal junction: A systematic review. *Neuroscience and Biobehavioral Reviews, 55*, 547–572. https://doi.org/10.1016/j.neubiorev.2015.05.017.

Dyde, R. T., Jenkin, M. R., & Harris, L. R. (2006). The subjective visual vertical and the perceptual upright. *Experimental Brain Research, 173*, 612–622. https://doi.org/10.1007/s00221-006-0405-y.

Dyde, R. T., & Milner, A. D. (2002). Two illusions of perceived orientation: One fools all of the people some of the time; the other fools all of the people all of the time. *Experimental Brain Research, 144*, 518–527. https://doi.org/10.1007/s00221-002-1065-1.

Eickhoff, S. B., Weiss, P. H., Amunts, K., Fink, G. R., Zilles, K. (2006). Identifying human parieto-insular vestibular cortex using fMRI and cytoarchitectonic mapping. *Human Brain Mapping, 27*, 611–621. https://doi.org/10.1002/hbm.20205

Emri M. (2003). Cortical Projection of Peripheral Vestibular Signaling. *Journal of Neurophysiology, 89*, 2639–2646. https://doi.org/10.1152/jn.00599.2002

Fernandez, C., & Goldberg, J. M. (1971). Physiology of peripheral neurons innervating semicircular canals of the squirrel monkey. II. Response to sinusoidal stimulation and dynamics of peripheral vestibular system. *Journal of Neurophysiology, 34*(4), 661–675.

Fetsch, C. R., Pouget, A., DeAngelis, G. C., & Angelaki, D. E. (2011). Neural correlates of reliability-based cue weighting during multisensory integration. *Nature Neuroscience, 15*(1), 146–154.

Finkelstein, A., Ulanovsky, N., Tsodyks, M., & Aljadeff, J. (2018). Optimal dynamic coding by mixed-dimensionality neurons in the head-direction system of bats. *Nature Communications, 9*(1), 3590.

Fiori, F., Candidi, M., Acciarino, A., David, N., & Aglioti, S. M. (2015). The right temporoparietal junction plays a causal role in maintaining the internal representation of verticality. *Journal of Neurophysiology, 114*, 2983–2990. https://doi.org/10.1152/jn.00289.2015.

Fiori, F., David, N., & Aglioti, S. M. (2014). Processing of proprioceptive and vestibular body signals and self-transcendence in Ashtanga yoga practitioners. *Frontiers in Human Neuroscience, 8*, 734. https://doi.org/10.3389/fnhum.2014.00734.

Friedmann, G. (1970). The judgement of the visual vertical and horizontal with peripheral and central vestibular lesions. *Brain, 93*, 313–328.

Funk, J., Finke, K., Müller, H. J., Utz, K. S., & Kerkhoff, G. (2011). Visual context modulates the subjective vertical in neglect: Evidence for an increased rod-and-frame-effect. *Neuroscience, 173*, 124–134. https://doi.org/10.1016/j.neuroscience.2010.10.067.

Furman, J. M., Marcus, D. A., & Balaban, C. D. (2013). Vestibular migraine: Clinical aspects and pathophysiology. *The Lancet Neurology, 12*, 706–715. https://doi.org/10.1016/S1474-4422(13)70107-8.

Furman, J. M., Sparto, P. J., Soso, M., & Marcus, D. (2005). Vestibular function in migraine-related dizziness: A pilot study. *Journal of Vestibular Research, 15*, 327–332.

Gandor, F., Basta, D., Gruber, D., Poewe, W., & Ebersbach, G. (2016). Subjective visual vertical in PD patients with lateral trunk flexion. *Parkinsons Disease, 2016*, 7489105. https://doi.org/10.1155/2016/7489105.

Gentaz, E., Badan, M., Luyat, M., & Touil, N. (2002). The manual haptic perception of orientations and the oblique effect in patients with left visuo-spatial neglect. *Neuroreport, 13*, 327–331.

Gianna, C., Heimbrand, S., & Gresty, M. (1996). Thresholds for detection of motion direction during passive lateral whole-body acceleration in normal subjects and patients with bilateral loss of labyrinthine function. *Brain Research Bulletin, 40*(5–6), 443–447; discussion 448–9.

Gianna, C. C., Heimbrand, S., Nakamura, T., & Gresty, M. A. (1995). Thresholds for perception of lateral motion in normal subjects and patients with bilateral loss of vestibular function. *Acta Oto-Laryngologica. Supplementum, 520*(Pt 2), 343–346.

Glasauer, S., Amorim, M. A., Viaud-Delmon, I., & Berthoz, A. (2002). Differential effects of labyrinthine dysfunction on distance and direction during blindfolded walking of a triangular path. *Experimental Brain Research, 145*, 489–497.

Glasauer, S., Amorim, M. A., Vitte, E., & Berthoz, A. (1994). Goal-directed linear locomotion in normal and labyrinthine-defective subjects. *Experimental Brain Research, 98*(2), 323–335.

Goto, F., Kobayashi, H., Saito, A., Hayashi, Y., Higashino, K., Kunihiro, T., et al. (2003). Compensatory changes in static and dynamic subjective visual vertical in patients following vestibular schwanoma surgery. *Auris Nasus Larynx, 30*, 29–33.

Graybiel, A., Miller, E. F., Newsom, B. D., & Kennedy, R. S. (1968). The effect of water immersion on perception of the oculogravic illusion in normal and labyrinthine-defective subjects. *Acta Oto-Laryngologica, 65*, 599–610.

Green, A. M., Shaikh, A. G., & Angelaki, D. E. (2005). Sensory vestibular contributions to constructing internal models of self-motion. *Journal of Neural Engineering, 2*, S164–S179.

Groen, E. L., Jenkin, H. L., & Howard, I. P. (2002). Perception of self-tilt in a true and illusory vertical plane. *Perception, 31*, 1477–1490. https://doi.org/10.1068/p3330.

Guedry, F. E., Jr. (1974). Psychophysics of vestibular sensation. In H. H. Kornhuber (Ed.), *Handbook of sensory physiology* (Vol. VI, Pt 2., pp. 3–154). Berlin: Springer.

Guerraz, M., Poquin, D., & Ohlmann, T. (1998). The role of head-centric spatial reference with a static and kinetic visual disturbance. *Perception & Psychophysics, 60*, 287–295.

Guerraz, M., Yardley, L., Bertholon, P., Pollak, L., Rudge, P., Gresty, M. A., et al. (2001). Visual vertigo: Symptom assessment, spatial orientation and postural control. *Brain, 124*, 1646–1656.

Guldin, W. O., & Grüsser, O. J. (1998). Is there a vestibular cortex? *Trends in Neurosciences, 21*, 254–259.

Grüsser, O. J., Pause, M., & Schreiter, U. (1990a). Localization and responses of neurones in the parieto-insular vestibular cortex of awake mon-keys (Macaca fascicularis). *The Journal of Physiology, 430*, 537–557.

Grüsser, O. J., Pause, M., & Schreiter, U. (1990b). Vestibular neurones in the parieto-insular cortex of monkeys (Macaca fascicularis): Visual and neck receptor responses. *The Journal of Physiology, 430*, 559–583.

Hafström, A., Fransson, P.-A., Karlberg, M., & Magnusson, M. (2004). Idiosyncratic compensation of the subjective visual horizontal and vertical in 60 patients after unilateral vestibular deafferentation. *Acta Oto-Laryngologica, 124*, 165–171.

Hagström, L., Hörnsten, G., & Silfverskiöld, B. P. (1969). Oculostatic and visual phenomena occurring in association with Wallenberg's syndrome. *Acta Neurologica Scandinavica, 45*(5), 568–582.

Haji-Khamneh, B., & Harris, L. R. (2010). How different types of scenes affect the Subjective Visual Vertical (SVV) and the Perceptual Upright (PU). *Vision Research, 50*, 1720–1727. https://doi.org/10.1016/j.visres.2010.05.027.

Hansen, K. A., Chu, C., Dickinson, A., Pye, B., Weller, J. P., & Ungerleider, L. G. (2015). Spatial selectivity in the temporoparietal junction, inferior frontal sulcus, and inferior parietal lobule. *Journal of Vision, 15*, 15–15. https://doi.org/10.1167/15.13.15.

Henn, V., Young, L. R., & Finley, C. (1974). Vestibular nucleus units in alert monkeys are also influenced by moving visual fields. *Brain Research, 71*(1), 144–149.

Hörnsten, G. (1974). Wallenberg's syndrome. I. General symptomatology, with special reference to visual disturbances and imbalance. *Acta Neurologica Scandinavica, 50*(4), 434–446.

Horak, F. B., Shupert, C. L., & Mirka, A. (1989). Components of postural dyscontrol in the elderly: A review. *Neurobiology of Aging, 10*, 727–738.

Horii, A., Okumura, K., Kitahara, T., & Kubo, T. (2006). Intracranial vertebral artery dissection mimicking acute peripheral vertigo. *Acta Oto-Laryngologica, 126*, 170–173.

Howard, I. P. (1982). *Human visual orientation.* New York: Wiley.

Howard, I. P., & Childerson, L. (1994). The contribution of motion, the visual frame, and visual polarity to sensations of body tilt. *Perception, 23*, 753–762. https://doi.org/10.1068/p230753.

Huang, C. Y., & Yu, Y. L. (1985). Small cerebellar strokes may mimic labyrinthine lesions. *Journal of Neurology, Neurosurgery, and Psychiatry, 48*, 263–265.

Igelström, K. M., & Graziano, M. S. A. (2017). The inferior parietal lobule and temporoparietal junction: A network perspective. *Neuropsychologia, 105*, 70–83. https://doi.org/10.1016/j.neuropsychologia.2017.01.001.

Indovina, I., Maffei, V., Bosco, G., Zago, M., Macaluso, E., & Lacquaniti, F. (2005). Representation of visual gravitational motion in the human vestibular cortex. *Science, 308*, 416–419. https://doi.org/10.1126/science.1107961.

Ionta, S., Heydrich, L., Lenggenhager, B., Mouthon, M., Fornari, E., Chapuis, D., et al. (2011). Multisensory mechanisms in temporo-parietal cortex support self-location and first-person perspective. *Neuron, 70*, 363–374. https://doi.org/10.1016/j.neuron.2011.03.009.

Isableu, B., Gueguen, M., Fourré, B., Giraudet, G., & Amorim, M.-A. (2008). Assessment of visual field dependence: Comparison between the mechanical 3D rod-and-frame test developed by Oltman in 1968 with a 2D computer-based version. *Journal of Vestibular Research, 18*, 239–247.

Isableu, B., Ohlmann, T., Cremieux, J., & Amblard, B. (1997). Selection of spatial frame of reference and postural control variability. *Experimental Brain Research, 114*, 584–589.

Jacobson, G. P., & McCaslin, D. L. (2003). Agreement between functional and electrophysiologic measures in patients with unilateral peripheral vestibular system impairment. *Journal of the American Academy of Audiology, 14*(5), 231–238.

Jáuregui Renaud, K. (2015). Vestibular function and depersonalization/derealization symptoms. *Multisensory Research, 28*, 637–651. https://doi.org/10.1163/22134808-00002480.

Jenkin, H. L., Dyde, R. T., Jenkin, M. R., Howard, I. P., & Harris, L. R. (2003). Relative role of visual and non-visual cues in determining the direction of "up": Experiments in the York tilted room facility. *Journal of Vestibular Research, 13*, 287–293.

Jenkin, H. L., Jenkin, M. R., Dyde, R. T., & Harris, L. R. (2004). Shape-from-shading depends on visual, gravitational, and body-orientation cues. *Perception, 33*, 1453–1461. https://doi.org/10.1068/p5285.

Johannsen, L., Fruhmann Berger, M., & Karnath, H.-O. (2006). Subjective visual vertical (SVV) determined in a representative sample of 15 patients with pusher syndrome. *Journal of Neurology, 253*, 1367–1369. https://doi.org/10.1007/s00415-006-0216-x.

Kahane, P., Hoffmann, D., Minotti, L., & Berthoz, A. (2003). Reappraisal of the human vestibular cortex by cortical electrical stimulation study. *Annals of Neurology, 54*(5), 615–624.

Kammerlind, A. S., Ledin, T. E., Skargren, E. I., & Odkvist, L. M. (2005). Long-term follow-up after acute unilateral vestibular loss and comparison between subjects with and without remaining symptoms. *Acta Oto-Laryngologica, 125*(9), 946–953.

Kandemir, A., Çelebisoy, N., & Köse, T. (2014). Perception of verticality in patients with primary headache disorders. *The Journal of International Advanced Otology, 10*, 138–143. https://doi.org/10.5152/iao.2014.25.

Kaptein, R. G., & Van Gisbergen, J. A. M. (2004). Interpretation of a discontinuity in the sense of verticality at large body tilt. *Journal of Neurophysiology, 91*, 2205–2214. https://doi.org/10.1152/jn.00804.2003.

Karmali, F., Bermúdez Rey, M. C., Clark, T. K., Wang, W., & Merfeld, D. M. (2017). Multivariate analyses of balance test performance, vestibular thresholds, and age. *Frontiers in Neurology, 8*, 578. https://doi.org/10.3389/fneur.2017.00578. eCollection 2017.

Karmali, F., Lim, K., & Merfeld, D. M. (2014). Visual and vestibular perceptual thresholds each demonstrate better precision at specific frequencies and also exhibit optimal integration. *Journal of Neurophysiology, 111*(12), 2393–2403. https://doi.org/10.1152/jn.00332.2013.

Karnath, H.-O., & Broetz, D. (2003). Understanding and treating "pusher syndrome". *Physical Therapy, 83*, 1119–1125.

Karnath, H.-O., & Dieterich, M. (2006). Spatial neglect--a vestibular disorder? *Brain, 129*, 293–305. https://doi.org/10.1093/brain/awh698.

Karnath, H.-O., Ferber, S., & Dichgans, J. (2000). The origin of contraversive pushing Evidence for a second graviceptive system in humans. *Neurology, 55*, 1298–1304. https://doi.org/10.1212/WNL.55.9.1298.

Karnath, H.-O., & Rorden, C. (2012). The anatomy of spatial neglect. *Neuropsychologia, 50*, 1010–1017. https://doi.org/10.1016/j.neuropsychologia.2011.06.027.

Kaski, D., Quadir, S., Nigmatullina, Y., Malhotra, P. A., Bronstein, A. M., & Seemungal, B. M. (2016). Temporoparietal encoding of space and time during vestibular-guided orientation. *Brain, 139*, 392–403. https://doi.org/10.1093/brain/awv370.

Kerkhoff, G. (1999). Multimodal spatial orientation deficits in left-sided visual neglect. *Neuropsychologia, 37*, 1387–1405. https://doi.org/10.1016/S0028-3932(99)00031-7.

Kerkhoff, G., & Zoelch, C. (1998). Disorders of visuospatial orientation in the frontal plane in patients with visual neglect following right or left parietal lesions. *Experimental Brain Research, 122*, 108–120. https://doi.org/10.1007/s002210050497.

Kheradmand, A., Gonzalez, G., Otero-Millan, J., & Lasker, A. (2016). Visual perception of upright: Head tilt, visual errors and viewing eye. *Journal of Vestibular Research, 25*, 201–209. https://doi.org/10.3233/VES-160565.

Kheradmand, A., Lasker, A., & Zee, D. S. (2015). Transcranial magnetic stimulation (TMS) of the supramarginal gyrus: A window to perception of upright. *Cerebral Cortex, 25*, 765–771. https://doi.org/10.1093/cercor/bht267.

Kheradmand, A., & Winnick, A. (2017). Perception of upright: Multisensory convergence and the role of temporo-parietal cortex. *Frontiers in Neurology, 8*. https://doi.org/10.3389/fneur.2017.00552.

Klam, F., & Graf, W. (2003). Vestibular response kinematics in posterior parietal cortex neurons of macaque monkeys. *The European Journal of Neuroscience, 18*(4), 995–1010.

Kupferberg, A., Glasauer, S., Stein, A., & Brandt, T. (2009). Influence of uninformative visual cues on gravity perception. *Annals of the New York Academy of Sciences, 1164*, 403–405. https://doi.org/10.1111/j.1749-6632.2009.03851.x.

Lacquaniti, F., Bosco, G., Indovina, I., La Scaleia, B., Maffei, V., Moscatelli, A., et al. (2013). Visual gravitational motion and the vestibular system in humans. *Frontiers in Integrative Neuroscience, 7*. https://doi.org/10.3389/fnint.2013.00101.

Lechner-Steinleitner, S. (1978). Interaction of labyrinthine and somatoreceptor inputs as determinants of the subjective vertical. *Psychological Research, 40*, 65–76.

Lee, H., & Cho, Y. W. (2004). A case of isolated nodulus infarction presenting as a vestibular neuritis. *Journal of the Neurological Sciences, 221*, 117–119.

Lee, H., Sohn, S. I., Cho, Y. W., Lee, S. R., Ahn, B. H., Park, B. R., & Baloh, R. W. (2006). Cerebellar infarction presenting isolated vertigo: Frequency and vascular topographical patterns. *Neurology, 67*(7), 1178–1183.

Lee, H., Yi, H. A., Cho, Y. W., Sohn, C. H., Whitman, G. T., Ying, S., & Baloh, R. W. (2003). Nodulus infarction mimicking acute peripheral vestibulopathy. *Neurology, 60*, 1700–1702.

Lee, H. (2014). Isolated vascular vertigo. *Journal of Stroke, 16*(3), 124–130.

Leigh, R. J., & Zee, D. S. (2015). *The neurology of eye movements.* New York: Oxford Univ Press.

Lempert, T., Olesen, J., Furman, J., Waterston, J., Seemungal, B., Carey, J., et al. (2012). Vestibular migraine: diagnostic criteria. *Journal of Vestibular Research, 22*, 167.

Lester, B. D., & Dassonville, P. (2014). The role of the right superior parietal lobule in processing visual context for the establishment of the egocentric reference frame. *Journal of Cognitive Neuroscience, 26*, 2201–2209. https://doi.org/10.1162/jocn_a_00636.

Lewis, R. F., Priesol, A. J., Nicoucar, K., Lim, K., & Merfeld, D. M. (2011). Dynamic tilt thresholds are reduced in vestibular migraine. *Journal of Vestibular Research: Equilibrium & Orientation, 21*, 323.

Lipshits, M., Bengoetxea, A., Cheron, G., & McIntyre, J. (2005). Two reference frames for visual perception in two gravity conditions. *Perception, 34*, 545–555. https://doi.org/10.1068/p5358.

Lopez, C., Bachofner, C., Mercier, M., & Blanke, O. (2009). Gravity and observer's body orientation influence the visual perception of human body postures. *Journal of Vision, 9*, 1–1. https://doi.org/10.1167/9.5.1.

Lopez, C., & Blanke, O. (2011). The thalamocortical vestibular system in animals and humans. *Brain Research Reviews, 67*, 119–146.

Lopez, C., Blanke, O., & Mast, F. W. (2012). The human vestibular cortex revealed by coordinate-based activation likelihood estimation meta-analysis. *Neuroscience, 212*, 159–179.

Lopez, C., Halje, P., & Blanke, O. (2008a). Body ownership and embodiment: Vestibular and multisensory mechanisms. *Neurophysiologie Clinique/Clinical Neurophysiology, 38*, 149–161. https://doi.org/10.1016/j.neucli.2007.12.006.

Lopez, C., Lacour, M., Ahmadi, A. E., Magnan, J., & Borel, L. (2007). Changes of visual vertical perception: A long-term sign of unilateral and bilateral vestibular loss. *Neuropsychologia, 45*, 2025–2037. https://doi.org/10.1016/j.neuropsychologia.2007.02.004.

Lopez, C., Lacour, M., Léonard, J., Magnan, J., & Borel, L. (2008b). How body position changes visual vertical perception after unilateral vestibular loss. *Neuropsychologia, 46*, 2435–2440. https://doi.org/10.1016/j.neuropsychologia.2008.03.017.

Lopez, C., Lacour, M., Magnan, J., & Borel, L. (2006). Visual field dependence-independence before and after unilateral vestibular loss. *Neuroreport, 17*, 797–803. https://doi.org/10.1097/01.wnr.0000221843.58373.c8.

Lobel, E., Kleine, J. F., Le Bihan D., Leroy-Willig, A., Berthoz A. (1998). Functional MRI of galvanic vestibular stimulation. *Journal of Neurophysiology, 80*, 2699–2709.

Mansfield, A., Fraser, L., Rajachandrakumar, R., Danells, C. J., Knorr, S., & Campos, J. (2015). Is perception of vertical impaired in individuals with chronic stroke with a history of "pushing"? *Neuroscience Letters, 590*, 172–177. https://doi.org/10.1016/j.neulet.2015.02.007.

Mast, F., & Jarchow, T. (1996). Perceived body position and the visual horizontal. *Brain Research Bulletin, 40*, 393–397; discussion 397–398.

Mast, F. W. (2000). Human perception of verticality: Psychophysical experiments on the centrifuge and their neuronal implications. *Japanese Psychological Research, 42*, 194–206. https://doi.org/10.1111/1468-5884.00146.

McKenna, G. J., Peng, G. C. Y., & Zee, D. S. (2004). Neck muscle vibration alters visually perceived roll in normals. *Journal of the Association for Research in Otolaryngology, 5*, 25–31. https://doi.org/10.1007/s10162-003-4005-2.

Meng, H., May, P. J., Dickman, J. D., & Angelaki, D. E. (2007). Vestibular signals in primate thalamus: Properties and origins. *The Journal of Neuroscience, 27*, 13590–13602.

Mikellidou, K., Cicchini, G. M., Thompson, P. G., & Burr, D. C. (2015). The oblique effect is both allocentric and egocentric. *Journal of Vision, 15*, 24. https://doi.org/10.1167/15.8.24.

Mittelstaedt, H. (1983). A new solution to the problem of the subjective vertical. *Naturwissenschaften, 70*, 272–281. https://doi.org/10.1007/BF00404833.

Mittelstaedt, H. (1986). The subjective vertical as a function of visual and extraretinal cues. *Acta Psychologica, 63*, 63–85.

Mittelstaedt, M. L., & Mittelstaedt, H. (1980). Homing by path integration in a mammal. *Naturwissenschaften, 67*, 566–567.

Müller, G. (1916). Über das Aubertsche Phänomen. *Z Sinnesphysiol, 49*, 109–246.

Müller, J. A., Bockisch, C. J., & Tarnutzer, A. A. (2016). Spatial orientation in patients with chronic unilateral vestibular hypofunction is ipsilesionally distorted. *Clinical Neurophysiology, 127*, 3243–3251. https://doi.org/10.1016/j.clinph.2016.07.010.

Nigmatullina, Y., Hellyer, P. J., Nachev, P., Sharp, D. J., & Seemungal, B. M. (2015). The neuroanatomical correlates of training-related perceptuo-reflex uncoupling in dancers. *Cerebral Cortex, 25*(2), 554–562.

Okada, T., Grunfeld, E., Shallo-Hoffmann, J., & Bronstein, A. M. (1999). Vestibular perception of angular velocity in normal subjects and in patients with congenital nystagmus. *Brain, 122*, 1293–1303.

Okinaka, Y., Sekitani, T., Okazaki, H., Miura, M., & Tahara, T. (1993). Progress of caloric response of vestibular neuronitis. *Acta Oto-Laryngologica. Supplementum, 503*, 18–22.

Osterhammel, P., Terkildsen, K., & Zilstorff, K. (1968). Vestibular habituation in ballet dancers. *Acta Oto-Laryngologica, 66*(3), 221–228.

Otero-Millan, J., & Kheradmand, A. (2016). Upright perception and ocular torsion change independently during head tilt. *Frontiers in Human Neuroscience, 10*. https://doi.org/10.3389/fnhum.2016.00573.

Otero-Millan, J., Winnick, A., & Kheradmand, A. (2018). Exploring the role of temporoparietal cortex in upright perception and the link with torsional eye position. *Frontiers in Neurology, 9*. https://doi.org/10.3389/fneur.2018.00192.

Paci, M., Baccini, M., & Rinaldi, L. A. (2009). Pusher behaviour: A critical review of controversial issues. *Disability and Rehabilitation, 31*, 249–258. https://doi.org/10.1080/09638280801928002.

Paci, M., Matulli, G., Megna, N., Baccini, M., & Baldassi, S. (2011). The subjective visual vertical in patients with pusher behaviour: A pilot study with a psychophysical approach. *Neuropsychological Rehabilitation, 21*, 539–551. https://doi.org/10.1080/09602011.2011.583777.

Palla, A., Straumann, D., & Bronstein, A. M. (2008). Vestibular neuritis: Vertigo and the high-acceleration vestibulo-ocular reflex. *Journal of Neurology, 255*(10), 1479–1482.

Pastor, A. M., De la Cruz, R. R., & Baker, R. (1994). Eye position and eye velocity integrators reside in separate brainstem nuclei. *Proceedings of the National Academy of Sciences of the United States of America, 91*, 807–811.

Pedersen, P. M., Wandel, A., Jørgensen, H. S., Nakayama, H., Raaschou, H. O., & Olsen, T. S. (1996). Ipsilateral pushing in stroke: Incidence, relation to neuropsychological symptoms, and impact on rehabilitation. The Copenhagen stroke study. *Archives of Physical Medicine and Rehabilitation, 77*, 25–28. https://doi.org/10.1016/S0003-9993(96)90215-4.

Pérennou, D. A., Amblard, B., Laassel, E. M., Benaim, C., Hérisson, C., & Pélissier, J. (2002). Understanding the pusher behavior of some stroke patients with spatial deficits: A pilot study. *Archives of Physical Medicine and Rehabilitation, 83*, 570–575.

Pérennou, D. A., Amblard, B., Leblond, C., & Pélissier, J. (1998). Biased postural vertical in humans with hemispheric cerebral lesions. *Neuroscience Letters, 252*, 75–78.

Pérennou, D. A., Mazibrada, G., Chauvineau, V., Greenwood, R., Rothwell, J., Gresty, M. A., et al. (2008). Lateropulsion, pushing and verticality perception in hemisphere stroke: A causal relationship? *Brain, 131*, 2401–2413. https://doi.org/10.1093/brain/awn170.

Pérennou, D., Piscicelli, C., Barbieri, G., Jaeger, M., Marquer, A., & Barra, J. (2014). Measuring verticality perception after stroke: Why and how? *Neurophysiologie Clinique/Clinical Neurophysiology, 44*, 25–32. https://doi.org/10.1016/j.neucli.2013.10.131.

Piscicelli, C., Barra, J., Davoine, P., Chrispin, A., Nadeau, S., & Pérennou, D. (2015a). Inter- and intra-rater reliability of the visual vertical in subacute stroke. *Stroke, 46*, 1979–1983. https://doi.org/10.1161/STROKEAHA.115.009610.

Piscicelli, C., Nadeau, S., Barra, J., & Pérennou, D. (2015b). Assessing the visual vertical: How many trials are required? *BMC Neurology, 15*, 215. https://doi.org/10.1186/s12883-015-0462-6.

Popkirov et al. (2018). https://www.ncbi.nlm.nih.gov/pubmed/29208729

Probst, T., Brandt, T., & Degner, D. (1986). Object-motion detection affected by concurrent self-motion perception: Psychophysics of a new phenomenon. *Behavioural Brain Research, 22*(1), 1–11.

Probst, T., Straube, A., & Bles, W. (1985). Differential effects of ambivalent visual-vestibular somatosensory stimulation on the perception of self-motion. *Behavioural Brain Research, 16*(1), 71–79.

Probst, T., & Wist, E. R. (1990). Electrophysiological evidence for visual-vestibular interaction in man. *Neuroscience Letters, 108*(3), 255–260.

Punt, T. D., & Riddoch, M. J. (2002). Towards a theoretical understanding of pushing behaviour in stroke patients. *Neuropsychological Rehabilitation, 12*, 455–472. https://doi.org/10.1080/09602010244000246.

Radtke, A., Popov, K., Bronstein, A. M., & Gresty, M. A. (2000). Evidence for a vestibulo-cardiac reflex in man. *Lancet, 356*(9231), 736–737.

Rancz, E. A., Moya, J., Drawitsch, F., Brichta, A. M., Canals, S., & Margrie, T. W. (2015). Widespread vestibular activation of the rodent cortex. *The Journal of Neuroscience, 35*(15), 5926–5934. https://doi.org/10.1523/JNEUROSCI.1869-14.2015.

Rees, G. (2007). Neural correlates of the contents of visual awareness in humans. *Philosophical Transactions of the Royal Society of London. Series B, Biological Sciences, 362*, 877–886.

Reisine, H., & Raphan, T. (1992). Unit activity in the vestibular nuclei of monkeys during off-vertical axis rotation. *Annals of the New York Academy of Sciences, 656*, 954–956.

Reynolds, R. F., & Bronstein, A. M. (2003). The broken escalator phenomenon. Aftereffect of walking onto a moving platform. *Experimental Brain Research, 151*(3), 301–308. Epub 2003 Jun 12.

Riccio, G. E., Martin, E. J., & Stoffregen, T. A. (1992). The role of balance dynamics in the active perception of orientation. *Journal of Experimental Psychology. Human Perception and Performance, 18*, 624–644.

Rizk, S., Ptak, R., Nyffeler, T., Schnider, A., & Guggisberg, A. G. (2013). Network mechanisms of responsiveness to continuous theta-burst stimulation. *The European Journal of Neuroscience, 38*(8), 3230–3238.

Roberts, R. E., Ahmad, H., Arshad, Q., Patel, M., Dima, D., Leech, R., et al. (2016). Functional neuroimaging of visuo-vestibular interaction. *Brain Structure and Function, 222*(5), 2329–2343. https://doi.org/10.1007/s00429-016-1344-4.

Rossi, M., Soto, A., Santos, S., Sesar, A., & Labella, T. (2009). A prospective study of alterations in balance among patients with Parkinson's disease. *ENE, 61*, 171–176. https://doi.org/10.1159/000189270.

Rousseaux, M., Braem, B., Honoré, J., & Saj, A. (2015). An anatomical and psychophysical comparison of subjective verticals in patients with right brain damage. *Cortex, 69*, 60–67. https://doi.org/10.1016/j.cortex.2015.04.004.

Rousseaux, M., Honoré, J., Vuilleumier, P., & Saj, A. (2013). Neuroanatomy of space, body, and posture perception in patients with right hemisphere stroke. *Neurology, 81*, 1291–1297.

Rubenstein, R. L., Norman, D. M., Schindler, R. A., & Kaseff, L. (1980). Cerebellar infarction--a presentation of vertigo. *Laryngoscope, 90*(3), 505–514.

Sack, A. T., Camprodon, J. A., Pascual-Leone, A., & Goebel, R. (2005). The dynamics of inter-hemispheric compensatory processes in mental imagery. *Science, 308*, 702–704.

Saj, A., Cojan, Y., Musel, B., Honoré, J., Borel, L., & Vuilleumier, P. (2014). Functional neuroanatomy of egocentric versus allocentric space representation. *Neurophysiologie Clinique, 44*, 33–40. https://doi.org/10.1016/j.neucli.2013.10.135.

Saj, A., Honoré, J., Coello, Y., & Rousseaux, M. (2005). The visual vertical in the pusher syndrome: Influence of hemispace and body position. *Journal of Neurology, 252*, 885–891. https://doi.org/10.1007/s00415-005-0716-0.

Santos-Pontelli, T. E. G., Pontes-Neto, O. M., de Araujo, D. B., Santos, A. C., & Leite, J. P. (2011). Persistent pusher behavior after a stroke. *Clinics (São Paulo, Brazil), 66*, 2169–2171.

Santos-Pontelli, T. E. G., Rimoli, B. P., Favoretto, D. B., Mazin, S. C., Truong, D. Q., Leite, J. P., et al. (2016). Polarity-dependent misperception of subjective visual vertical during and after transcranial direct current stimulation (tDCS). *PLoS One, 11*, e0152331. https://doi.org/10.1371/journal.pone.0152331.

Schlack, A., Hoffmann, K. P., & Bremmer, F. (2002). Interaction of linear vestibular and visual stimulation in the macaque ventral intraparietal area (VIP). *The European Journal of Neuroscience, 16*(10), 1877–1886.

Scocco, D. H., Wagner, J. N., Racosta, J., Chade, A., & Gershanik, O. S. (2014). Subjective visual vertical in Pisa syndrome. *Parkinsonism & Related Disorders, 20*, 878–883. https://doi.org/10.1016/j.parkreldis.2014.04.030.

Seemungal, B. M. (2005). *The mechanisms and loci of human vestibular perception*. Doctoral thesis, University of London.

Seemungal, B. M. (2007). Neuro-otological emergencies. *Current Opinion in Neurology, 20*(1), 32–39.

Seemungal et al. (2013). https://www.ncbi.nlm.nih.gov/pubmed/22291031

Seemungal, B. M. (2014). The cognitive neurology of the vestibular system. *Current Opinion in Neurology, 27*(1), 125–132.

Seemungal, B. M. (2015). The components of vestibular cognition--motion versus spatial perception. *Multisensory Research, 28*(5–6), 507–524.

Seemungal, B. M., Glasauer, S., Gresty, M. A., & Bronstein, A. M. (2007). Vestibular perception and navigation in the congenitally blind. *Journal of Neurophysiology, 97*(6), 4341–4356.

Seemungal, B., Gunaratne, I., Fleming, I., Gresty, M., & Bronstein, A. (2004). Perceptual and nystagmic thresholds of vestibular function in yaw. *Journal of Vestibular Research, 14*, 461–466.

Seemungal, B. M., Masaoutis, P., Green, D. A., Plant, G. T., & Bronstein, A. M. (2011). Symptomatic recovery in Miller Fisher Syndrome parallels vestibular-perceptual and not vestibular-ocular reflex function. *Frontiers in Neurology, 2*, 2.

Seemungal, B. M., Rizzo, V., Gresty, M. A., Rothwell, J. C., & Bronstein, A. M. (2008). Posterior parietal rTMS disrupts human Path Integration during a vestibular navigation task. *Neuroscience Letters, 437*, 88–92.

Seemungal, B. M., Rizzo, V., Gresty, M. A., Rothwell, J. C., & Bronstein, A. M. (2009). Perceptual encoding of self-motion duration in human posterior parietal cortex. *Annals of the New York Academy of Sciences, 1164*, 236–238.

Shinder, M. E., & Taube, J. S. (2019). Three-dimensional tuning of head direction cells in rats. *Journal of Neurophysiology, 121*(1), 4–37. https://doi.org/10.1152/jn.00880.2017.

Silani, G., Lamm, C., Ruff, C. C., & Singer, T. (2013). Right supramarginal gyrus is crucial to overcome emotional egocentricity bias in social judgments. *The Journal of Neuroscience, 33*, 15466–15476. https://doi.org/10.1523/JNEUROSCI.1488-13.2013.

Stephan, T., Deutschländer, A., Nolte, A., Schneider, E., Wiesmann, M., Brandt, T., & Dieterich, M. (2005). Functional MRI of galvanic vestibular stimulation with alternating currents at different frequencies. *NeuroImage, 26*(3), 721–732.

Suzuki, M., Kitano, H., Ito, R., Kitanishi, T., Yazawa, Y, Ogawa T., Shiino, A., Kitajima K. (2001). Cortical and subcortical vestibular response to caloric stimulation detected by functional magnetic resonance imaging. *Brain Res Cogn Brain Res, 12*, 441–449.

Tarnutzer, A. A., Bertolini, G., Bockisch, C. J., Straumann, D., & Marti, S. (2013). Modulation of internal estimates of gravity during and after prolonged roll-tilts. *PLoS One, 8*, e78079. https://doi.org/10.1371/journal.pone.0078079.

Tarnutzer, A. A., Bockisch, C., Straumann, D., & Olasagasti, I. (2009). Gravity dependence of subjective visual vertical variability. *Journal of Neurophysiology, 102*, 1657–1671. https://doi.org/10.1152/jn.00007.2008.

Teggi, R., Colombo, B., Bernasconi, L., Bellini, C., Comi, G., & Bussi, M. (2009). Migrainous vertigo: Results of caloric testing and stabilometric findings. *Headache: The Journal of Head and Face Pain, 49*, 435–444. https://doi.org/10.1111/j.1526-4610.2009.01338.x.

Thömke, F., Marx, J. J., Iannetti, G. D., Cruccu, G., Fitzek, S., Urban, P. P., et al. (2005). A topodiagnostic investigation on body lateropulsion in medullary infarcts. *Neurology, 64*, 716–718. https://doi.org/10.1212/01.WNL.0000152040.27264.1A.

Tomassini, A., Solomon, J. A., & Morgan, M. J. (2014). Which way is down? Positional distortion in the tilt illusion. *PLoS One, 9*, e110729. https://doi.org/10.1371/journal.pone.0110729.

Toupet, M., Van Nechel, C., & Bozorg Grayeli, A. (2014). Influence of body laterality on recovery from subjective visual vertical tilt after vestibular neuritis. *Audiology & Neuro-Otology, 19*, 248–255. https://doi.org/10.1159/000360266.

Trousselard, M., Barraud, P., Nougier, V., Raphel, C., & Cian, C. (2004). Contribution of tactile and interoceptive cues to the perception of the direction of gravity. *Brain Research. Cognitive Brain Research, 20*, 355–362. https://doi.org/10.1016/j.cogbrainres.2004.03.008.

Tyrrell, R. A., & Owens, D. A. (1988). A rapid technique to assess the resting states of the eyes and other threshold phenomena: The modified binary search (MOBS). *Behavior Research Methods, Instruments, & Computers, 20*, 137–141.

Utz, K. S., Keller, I., Artinger, F., Stumpf, O., Funk, J., & Kerkhoff, G. (2011). Multimodal and multispatial deficits of verticality perception in hemispatial neglect. *Neuroscience, 188*, 68–79. https://doi.org/10.1016/j.neuroscience.2011.04.068.

Van Beuzekom, A. D., Medendorp, W. P., & Van Gisbergen, J. A. M. (2001). The subjective vertical and the sense of self orientation during active body tilt. *Vision Research, 41*, 3229–3242. https://doi.org/10.1016/S0042-6989(01)00144-4.

Van Beuzekom, A. D., & Van Gisbergen, J. A. (2000). Properties of the internal representation of gravity inferred from spatial-direction and body-tilt estimates. *Journal of Neurophysiology, 84*, 11–27. https://doi.org/10.1152/jn.2000.84.1.11/F.

Ventre-Dominey, J. (2014). Vestibular function in the temporal and parietal cortex: Distinct velocity and inertial processing pathways. *Frontiers in Integrative Neuroscience, 8*. https://doi.org/10.3389/fnint.2014.00053.

Vimal, V. P., Lackner, J. R., & DiZio, P. (2018). Learning dynamic control of body yaw orientation. *Experimental Brain Research, 236*(5), 1321–1330.

Vingerhoets, R. A. A., De Vrijer, M., Van Gisbergen, J. A. M., & Medendorp, W. P. (2009). Fusion of visual and vestibular tilt cues in the perception of visual vertical. *Journal of Neurophysiology, 101*, 1321–1333. https://doi.org/10.1152/jn.90725.2008.

Vingerhoets, R. A. A., Medendorp, W. P., & Van Gisbergen, J. A. M. (2008). Body-tilt and visual verticality perception during multiple cycles of roll rotation. *Journal of Neurophysiology, 99*, 2264–2280. https://doi.org/10.1152/jn.00704.2007.

Wade, N. J. (1968). Visual orientation during and after lateral head, body, and trunk tilt. *Perception & Psychophysics, 3*, 215–219. https://doi.org/10.3758/BF03212730.

Wade, N. J. (1970). Effect of prolonged tilt on visual orientation. *The Quarterly Journal of Experimental Psychology, 22*, 423–439. https://doi.org/10.1080/14640747008401916.

Wade, N. J., & Day, R. H. (1968). Development and dissipation of a visual spatial aftereffect from prolonged head tilt. *Journal of Experimental Psychology, 76*, 439–443.

Waespe, W., & Henn, V. (1977). Neuronal activity in the vestibular nuclei of the alert monkey during vestibular and optokinetic stimulation. *Experimental Brain Research, 27*(5), 523–538.

Wenzel, R., Bartenstein, P., Dieterich, M., Danek, A., Weindl, A., Minoshima, S., Ziegler, S., Schwaiger, M., & Brandt, T. (1996). Deactivation of human visual cortex during involuntary ocular oscillations. A PET activation study. *Brain, 119*, 101–110.

Wiest, G., Demer, J. L., Tian, J., Crane, B. T., & Baloh, R. W. (2001). Vestibular function in severe bilateral vestibulopathy. *Journal of Neurology, Neurosurgery, and Psychiatry, 71*(1), 53–57.

Winnick, A., Sadeghpour, S., Otero-Millan, J., Chang, T.-P., & Kheradmand, A. (2018). Errors of upright perception in patients with vestibular migraine. *Frontiers in Neurology, 9*. https://doi.org/10.3389/fneur.2018.00892.

Wright, W. G., & Horak, F. B. (2007). Interaction of posture and conscious perception of gravitational vertical and surface horizontal. *Experimental Brain Research, 182*, 321–332. https://doi.org/10.1007/s00221-007-0990-4.

Yakusheva, T. A., Shaikh, A. G., Green, A. M., Blazquez, P. M., Dickman, J. D., & Angelaki, D. E. (2007). Purkinje cells in posterior cerebellar vermis encode motion in an inertial reference frame. *Neuron, 54*, 973–985.

Yardley, L. (1990). Contribution of somatosensory information to perception of the visual vertical with body tilt and rotating visual field. *Perception & Psychophysics, 48*, 131–134.

Yates, B. J., Jian, B. J., & Cotter, L. A. (2000). Responses of vestibular nucleus neurons to tilt following chronic bilateral removal of vestibular inputs. *Experimental Brain Research, 130*, 151–158.

Yelnik, A. P., Lebreton, F. O., Bonan, I. V., Colle, F. M. C., Meurin, F. A., Guichard, J. P., et al. (2002). Perception of verticality after recent cerebral hemispheric stroke. *Stroke, 33*, 2247–2253. https://doi.org/10.1161/01.STR.0000027212.26686.48.

zu Eulenburg, P., Caspers, S., Roski, C., Eickhoff, S. B., (2012). Meta-analytical definition and functional connectivity of the human vestibular cortex. *NeuroImage, 60*, 162–169. https://doi.org/10.1016/j.neuroimage.2011.12.032

Zoccolotti, P., Antonucci, G., Goodenough, D. R., Pizzamiglio, L., & Spinelli, D. (1992). The role of frame size on vertical and horizontal observers in the rod-and-frame illusion. *Acta Psychologica, 79*, 171–187.

Distorted Gravity and Distorted Eyes: Who Is at Fault – The Cerebellum or Brainstem?

Alexander A. Tarnutzer

Abstract Sensory input from the otolith organs and the semicircular canals is combined in the vestibular nuclei and forwarded to other brainstem, cerebellar, and higher cortical areas by the central graviceptive pathways. This network forms the basis for internal estimates of the direction of gravity. Lateralized lesions along these pathways will result in vestibular tone imbalance in the roll plane. This can be observed both at the level of brainstem reflexes as the ocular tilt reaction (OTR; consisting of ocular torsion, head tilt, and skew deviation) and at the higher level of cortical behavioral paradigms (such as the subjective visual vertical, SVV). Recent research has demonstrated a wide network of cerebellar and brainstem areas contributing to these internal estimates. While cerebellar lesions including the dentate nucleus, nodulus, or flocculus result in contraversive OTR and shifts of perceived vertical, lesions affecting the biventer lobule, the middle cerebellar peduncle, the tonsil, and the inferior semilunar lobule are associated with ipsiversive shifts of OTR and perceived vertical. Patients with brainstem lesions below the crossing of the graviceptive pathways at the level of the pons present with ipsilesional OTR and tilts of perceived vertical, while lesions above the crossing demonstrate contraversive shifts. Specifically, ipsiversive shifts were noted for lesions affecting the medial longitudinal fasciculus (MLF) and the medial vestibular nucleus, whereas contraversive shifts were associated with lesions of the rostral interstitial nucleus of the MLF and the interstitial nucleus of Cajal. Distortions in verticality perception and OTR therefore can be mediated both by brainstem and cerebellar lesions.

Keywords Ocular torsion · Subjective visual vertical · Internal estimates of direction of gravity · Ocular tilt reaction · Perception

A. A. Tarnutzer (✉)
Department of Neurology, University Hospital Zurich and University of Zurich, Zurich, Switzerland

Neurology, Cantonal Hospital of Baden, Baden, Switzerland
e-mail: alexander.tarnutzer@access.uzh.ch

© Springer Nature Switzerland AG 2019   73
A. Shaikh, F. Ghasia (eds.), *Advances in Translational Neuroscience of Eye Movement Disorders*, Contemporary Clinical Neuroscience,
https://doi.org/10.1007/978-3-030-31407-1_4

1 Introduction

The central vestibular pathways (CVP) interconnect a network of brainstem, cerebellar, thalamic, and cortical areas involved in the processing, integration, and perception of "graviceptive" input (Angelaki et al. 2009; Brandt and Dieterich 2017). Otolith (utricular and saccular) and semicircular canal (SCC) inputs converge at the brainstem vestibular nuclei; project via bilateral ascending pathways through the medial lateral fasciculus (MLF) to the ocular motor nuclei (N. III, N. IV, and N. VI), the interstitial nucleus of Cajal (INC), the rostral interstitial nucleus of the MLF (riMLF), and various thalamic subnuclei; and end in the temporo-peri-Sylvian vestibular cortex (TPSVC) (Dieterich and Brandt 2015). The TPSVC is considered the human analogue of the parieto-insular vestibular cortex (PIVC) in nonhuman primates (Kahane et al. 2003), where vestibular input converges with other sensory signals and subserves perception of spatial orientation and navigation (Dieterich and Brandt 2008). In the upright position, the internal representation of gravity and spontaneous head orientation aligns with gravity, while the torsional orientation of the eyes is symmetric (binocular fundoscopic torsion = $0°$). Depending on the exact topography of the lesion, vestibular tone imbalances lead to roll-tilts of perception, head, and body as well as to misalignments of the visual axes (skew deviation) and binocular torsion (Dieterich and Brandt 2015).

In the 1990s, the role of the brainstem in distorting the eyes and the perception of gravity was demonstrated by a series of landmark studies (Brandt and Dieterich 1993, 1994; Dieterich and Brandt 1993a), while only anecdotal evidence was available at that time on the role of the cerebellum in generating internal estimates of the direction of gravity (Mossman and Halmagyi 1997). So during these days, researchers would have clearly identified the brainstem as a source of torsional deviations of the eyes and the perceived direction of gravity. However, since then, the role of the cerebellum in estimating the direction of gravity has been investigated with much greater detail, now requiring a more differentiated answer regarding failure of which neural networks are linked to vestibular tone imbalances in the roll plane.

In this chapter, I will mainly focus on the contribution of the cerebellum and the brainstem to internal estimates of the direction of gravity and summarize research published within the last 10–15 years. Special emphasis will be put on the anatomical/functional correlation of specific cerebellar/brainstem lesions, underlining the localizing value of these findings. However, the pattern in thalamic and cerebral lesions will also be discussed briefly.

There is a rich literature on neuronal networks responsible for discriminating tilt and translation in nonhuman primates (see, for example, Shaikh et al. 2005; Merfeld et al. 1999). This included the identification of Purkinje cells in the caudal cerebellar vermis with responses that reflect estimates of head tilt (Laurens et al. 2013) and provide frequency-selective coding of tilt and translation (Yakusheva et al. 2008). However, these studies are beyond the scope of this chapter.

2 Measuring Internal Estimates of the Direction of Gravity

2.1 Perceptual Measurements Using Higher Cortical Behavioral Paradigms

Internal estimates of the direction of gravity can be assessed in different domains (see Fig. 1) – with vision-based paradigms being most frequently used. In the subjective visual vertical (SVV) task, an illuminated line is aligned with perceived direction of vertical in darkness ("subjective visual vertical" or SVV). Discarding retinal input, paradigms assessed verticality perception in the haptic domain ("subjective haptic vertical" or SHV (Schuler et al. 2010)), based on verbal reports of current whole-body roll position (Kaptein and Van Gisbergen 2004) or based on self-positioning along the principle axes ("subjective postural vertical/horizontal") (Perennou et al. 2008). All these paradigms have in common that they provide information both about the accuracy (i.e., the degree of veracity) and the precision (i.e., the degree of reproducibility) of perceived vertical.

Under static conditions and in darkness, the SVV is mainly influenced by otolith signals and proprioceptive input (Barra et al. 2010), since darkness excludes visual references and the absence of rotation excludes a contribution by semicircular canal stimulation. The SVV in healthy human subjects when upright is accurate (i.e., within a range of ±2.5° of earth vertical (Perennou et al. 2008; Dieterich and Brandt 1993b)), whereas while roll-tilted, systematic errors toward roll overestimation (E-effect, observed at moderate roll angles up to 60° and for roll angles beyond 120–135°) and roll underestimation (A-effect, for roll angles between 60 and 120°) have been extensively studied (Kaptein and Van Gisbergen 2004; Tarnutzer et al. 2009a; Clemens et al. 2011; De Vrijer et al. 2008). It was suggested that A- and E-effects are a consequence of how various sensory signals are centrally integrated into a unified percept of vertical (Mittelstaedt 1983). While Mittelstaedt puts a focus

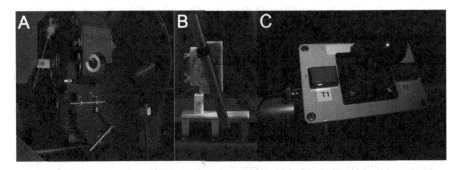

Fig. 1 Illustration of different experimental paradigms used to assess verticality perception. Whereas for the subjective visual vertical (SVV, panel **a**) a luminous line is aligned with perceived vertical, a rod is used instead in the subjective haptic vertical (SHV, panel **b**). Self-adjustments in darkness can be controlled by a joystick (subjective postural vertical, SPV, panel **c**) that allows active positioning in the roll plane using a motorized chair as shown in panel **a**

on an imbalanced otolith input to cause the A- and E-effects (Mittelstaedt 1983), more recent modeling of perceived vertical underlined the role of a noisy but accurate otolith signal (Tarnutzer et al. 2009a; Clemens et al. 2011; De Vrijer et al. 2008). To maximize the precision of verticality estimates, this noisy otolith estimate is combined with a bias that represents prior knowledge about where vertical is located. This bias refers to the body's longitudinal axis and is based on the assumption that small roll angles are more likely than large roll angles. Using this modeling approach, verticality estimates at small roll angles are accurate and precise; however, at larger roll angles, both A- and E-effects arise, reflecting the experimental data well (Tarnutzer et al. 2009a; De Vrijer et al. 2008). Likewise, SVV precision is dependent on roll angle. Compared to upright, it decreases with increasing head-roll angle and is minimal around 120–150° (Tarnutzer et al. 2009a; De Vrijer et al. 2008). This pattern of SVV variability was successfully reproduced using an otolith model estimating the gravitational vertical based on the characteristics (e.g., firing rate, orientation of polarization vectors) of a group of utricular and saccular otolith afferents and a prior biasing errors toward the body's longitudinal axis (Tarnutzer et al. 2009a; Clemens et al. 2011; De Vrijer et al. 2008; MacNeilage et al. 2007; Tarnutzer et al. 2012a).

2.2 Brainstem Reflexive Measurements of Internal Estimates of Direction of Gravity

Internal estimates of the direction of gravity not only provide a sense of vertical but also contribute to reflexive responses at the level of the brainstem and the cerebellum controlling eye and head position in the roll plane. The goal of these reflexive responses is to stabilize posture by keeping torsional eye position and head roll orientation aligned with gravity. Likewise, vertical eye position is kept aligned and perpendicular to the gravitational vertical. A biased internal estimate of the direction of gravity, e.g., due to unilateral damage of the central vestibular pathways, will result in distortion of these brainstem reflexes – a clinical finding referred to as (partial or complete) ocular tilt reaction (OTR; for details, see below in sect. 3).

Internal estimates of the direction of gravity can be assessed by quantifying ocular torsion using fundus photography, by measuring head roll and vertical ocular divergence (skew) (for details, see (Dieterich and Brandt 1993b)). Specifically, head roll will result in a modification of torsional eye position. This reflex is mediated by the otolith organs and has been termed ocular counter roll (OCR) (Collewijn et al. 1985). OCR compensates for approximately 10% of head roll under static conditions (Bockisch and Haslwanter 2001). Considering that both verticality perception and static OCR (Fernandez et al. 1972; Schor et al. 1984) rely on otolith input, it is not surprising that OCR shows a similar pattern of roll-dependent modulation in precision as the SVV (Haustein 1992; Tarnutzer et al. 2009b) and that errors in the perceived direction of gravity are correlated to torsional eye position (Pavlou et al.

2003; Wade and Curthoys 1997). This was demonstrated on an individual subject basis by Tarnutzer and colleagues: they reported a high correlation between the trial-to-trial variability of perceptual (SVV) and brainstem reflexive torsional eye movements (ocular torsion as measured by modified scleral search coils) in different whole-body roll positions (Tarnutzer et al. 2009b).

3   Lesions Along the Graviceptive Pathways

Unilateral lesions along the graviceptive pathways result in vestibular tone imbalances. Specifically, lesions along the central vestibular pathways shift verticality estimates and result in perceptual signs (e.g., SVV roll-tilt) and ocular motor signs (deviations in ocular torsion, head tilt, and skew deviation (i.e., a dissociation of vertical eye position). Whereas a complete OTR is found only in 20% of patients with lateralized brainstem lesions, partial OTR, i.e., isolated ocular torsion (83%) or skew deviation (31%), are more frequently noted (Brandt and Dieterich 1994). The most sensitive sign indicating damage along the central vestibular pathways in patients with brainstem lesions, however, is perceptual change as assessed by the SVV (94%) (Brandt and Dieterich 1994).

3.1   Peripheral Vestibular Lesions and Lesions at the Level of the Brainstem

Unilateral peripheral vestibular lesions are associated with ipsilesional roll-tilts of the perceived direction of gravity and an ipsilesional OTR (Bohmer and Rickenmann 1995; Curthoys et al. 1991; Halmagyi et al. 1979; Anastasopoulos et al. 1997). Compared to central lesions, signs of OTR seem to be less frequent and probably disappear quickly, as noted for skew deviation (Kattah et al. 2009). For brainstem lesions affecting the ascending central vestibular pathways, the location of the lesion predicts the direction of roll-tilt of both perceptual and ocular motor responses. Note, however, that likely several distinct and independently ascending pathways contain graviceptive input and these pathways may cross or not (Kirsch et al. 2016).

Whereas earlier studies have shown that brainstem lesions caudal to the pons involving the vestibular nuclei or the root entry zone are associated with ipsiversive tilts of the SVV and an ipsiversive OTR, more cranial (i.e., pontomesencephalic) brainstem lesions yielded contraversive deviations of the SVV and a contraversive OTR (Brandt and Dieterich 1994; Dieterich and Brandt 1993b; Halmagyi et al. 1990; Lee et al. 2005). Applying modern statistical lesion-behavior mapping analysis, Baier and colleagues recently provided more detailed information on the brainstem networks involved in processing otolith input. These authors studied perceptual biases in the estimated direction of gravity in 79 patients with unilateral ischemic brainstem lesions (Baier et al. 2012a). Overall, pathological SVV roll-tilts (i.e., deviations larger than 2.5°) were found in 53% of patients. Pathological binocular torsion was noted in 7/29 patients (24%) with contraversive SVV roll-tilt and in 4/32 patients (13%) with ipsiversive SVV roll-tilt. A complete OTR was identified in seven of the 61 patients (11%) tested for ocular torsion (Baier et al. 2012a). This study confirmed roll-tilt of the SVV to be a brainstem sign, and statistics linked ipsiversive shifts of the SVV with lesions of the medial longitudinal fasciculus (MLF) and the medial vestibular nucleus, whereas contraversive shifts were associated with lesions of the rostral interstitial nucleus of the MLF, the superior cerebellar peduncle, the oculomotor nucleus, and the interstitial nucleus of Cajal. Compared to the OTR, SVV roll-tilts were the more sensitive brainstem sign (53% vs. 18%). The lower rate in patients with abnormal SVV roll-tilts in the study by Baier and colleagues (Baier et al. 2012a) compared to a previous study (Dieterich and Brandt 1993b) reporting on verticality perception in patients with brainstem lesions (53% vs. 94%) is potentially related to the improvements in neuroimaging, detecting much smaller brainstem lesions than in the early 1990s or differences in recording delays (Baier et al. 2012a) (Table 1).

In patients with unilateral internuclear ophthalmoplegia, accompanying contraversive roll-tilts of SVV have been reported, suggesting that the central vestibular pathways run in parallel to the MLF (Zwergal et al. 2008a). An ipsilateral graviceptive pathway connecting the vestibular nuclei and the posterolateral thalamus was proposed, and pontomesencephalic stroke including the medial part of the medial lemniscus was linked to ipsiversive roll-tilt of the SVV in one study (Zwergal et al. 2008b). However, using statistical lesion-behavior mapping analysis, this was not confirmed by Baier and colleagues (Baier et al. 2012a), leaving the role of the medial lemniscus in the brainstem graviceptive network unclear.

3.2   Cerebellar Lesions and Their Effect on Graviception

Vestibular input is forwarded from the vestibular nuclei to the cerebellum, which plays an essential role in the initiation, coordination, and adaptation of movements. Involvement of the cerebellum in the processing of graviceptive input was therefore hypothesized. Whereas this was supported by anecdotal reports since the late 1990s (Mossman and Halmagyi 1997; Lee et al. 2005; Min et al. 1999; Park et al. 2013), larger case series confirmed cerebellar involvement in graviception in 2008 and 2009 (Baier et al. 2008; Baier and Dieterich 2009). In a group of 43 patients with unilateral cerebellar infarction, contraversive SVV roll-tilt (58%), ocular torsion (35%), and skew deviation (14%) were noted more frequently than ipsiversive SVV roll-tilt (26%), ocular torsion (14%), and skew deviation (9%) (Baier and Dieterich 2009). In another study, Baier and colleagues applied lesion-mapping techniques in 31 patients with acute lateralized cerebellar lesions presenting with signs of ocular tilt reaction and correlated clinical signs of vestibular tone imbalance with structural changes on MR imaging. As previously shown for brainstem lesions (Dieterich and

Table 1 Lesion locations (peripheral and central) and the resulting direction of both brainstem (OTR) and perceptual biases

Area	Specific location	Pattern of shifts	OTR bias	Perceptual bias
Peripheral-vestibular	Vestibular nerve (Halmagyi et al. 1979)	Ipsilesional	Yes	Yes
Brainstem	Medial vestibular nucleus (Baier et al. 2012a)	Ipsilesional	Yes	Yes
	Root entry zone (Lee et al. 2005)	Ipsilesional	Yes	Yes
	Medial longitudinal fasciculus (Baier et al. 2012a)	Ipsilesional	Yes	Yes
	Medial part of the medial lemniscus (Zwergal et al. 2008b)	Ipsilesional		
	Rostral interstitial nucleus of the MLF (Baier et al. 2012a)	Contralesional	Yes	Yes
	Superior cerebellar peduncle (Baier et al. 2012a)	Contralesional	Yes	Yes
	Oculomotor nucleus (Baier et al. 2012a)	Contralesional	Yes	Yes
	Interstitial nucleus of Cajal (Baier et al. 2012a)	Contralesional	Yes	Yes
Cerebellum	Dentate nucleus (Min et al. 1999; Baier et al. 2008; Tarnutzer et al. 2012b)	Contralesional	Yes	Yes
	Biventer lobule (Baier et al. 2008)	Ipsilesional	Yes	Yes
	Inferior semilunar lobule (Baier et al. 2008)	Ipsilesional	Yes	Yes
	Middle cerebellar peduncle (Baier et al. 2008)	Ipsilesional	Yes	Yes
	Pyramis (Baier et al. 2008)	Ipsilesional	Yes	Yes
	Uvula (Baier et al. 2008)	Ipsilesional	Yes	Yes
	Tonsil (Baier et al. 2008)	Ipsilesional	Yes	Yes
	Flocculus (Park et al. 2013; Baier et al. 2008)	Ipsilesional	Yes	Yes
	Nodulus (Mossman and Halmagyi 1997; Min et al. 1999; Kim et al. 2009; Tarnutzer et al. 2015)	Contralesional	N/A	N/A
Thalamus	Dorsolateral and dorsomedial thalamic subnuclei (Baier et al. 2016)	Contralesional	No	Yes
	Inferior and medial thalamic areas (Baier et al. 2016)	Ipsilesional	No	Yes
Cortex	Insular cortex (Brandt et al. 1994, Baier et al. 2012b)[a]	Contralesional > ipsilesional	No	Yes
	Insular gyrus (Brandt et al. 1994)	Contralesional > ipsilesional	No	Yes
	Middle and superior temporal gyrus (Brandt et al. 1994; Baier et al. 2012b)	Contralesional > ipsilesional	No	Yes

(continued)

Table 1 (continued)

Area	Specific location	Pattern of shifts	OTR bias	Perceptual bias
	Inferior frontal gyrus (Baier et al. 2012b)	Contralesional > ipsilesional	No	Yes
	Pre- and postcentral gyrus (Baier et al. 2012b)	Contralesional > ipsilesional	No	Yes
	Rolandic operculum (Baier et al. 2012b)	Contralesional > ipsilesional	No	Yes
	Inferior parietal lobe (Baier et al. 2012b)	Contralesional > ipsilesional	No	Yes
	Superior longitudinal fascicle (Baier et al. 2012b)	Contralesional > ipsilesional	No	Yes
	Inferior/superior occipitofrontal fascicle (Baier et al. 2012b)	Contralesional > ipsilesional	No	Yes

[a]Lesions restricted to the posterior insular cortex were not associated with significant shifts of perceived vertical or OTR shifts (Baier et al. 2013)

Brandt 1993b), perceptual measurements of the estimated direction of gravity (such as the SVV) were the most sensitive signs of vestibular tone imbalance, being present in all 31 patients (100%) in the study by Baier and colleagues (Baier et al. 2008). Signs of OTR were noted less frequently: ocular torsion was found in 17/31 (55%) patients, skew deviation in 8/31 (26%), and head roll in 12/31 (39%). A complete OTR was present in 8/31 (26%) patients. In this study, the dentate nucleus was identified as a critical anatomical cerebellar structure within the central vestibular pathways. Specifically, contraversive SVV tilts and contraversive signs of the OTR were linked to dentate nucleus lesions (Baier et al. 2008), whereas in patients with ipsiversive SVV tilts and ipsiversive OTR, the dentate nucleus was spared, and lesions were located in the biventer lobule and the inferior semilunar lobule, the middle cerebellar peduncle, the pyramis and the uvula, the tonsil, and the flocculus (Baier et al. 2008). These findings confirmed previous observations in single cases or small case series, reporting contraversive OTR and SVV roll-tilt in patients with lateralized cerebellar lesions including the nodulus and the dentate nucleus or the uvula (Min et al. 1999), the nodulus, and the uvula (Mossman and Halmagyi 1997) and in lesions restricted to the flocculus (Park et al. 2013) or the nodulus (Kim et al. 2009). Several hypotheses have been proposed to explain contraversive tilts of SVV and OTR in patients with lesions affecting the nodulus: whereas some proposed loss of inhibition due to nodular damage and resulting increased resting firing rate in ipsilesional secondary otolith neurons in the ipsilesional vestibular nuclei (Mossman and Halmagyi 1997), others have suggested a role of the vestibulocerebellum in modulating the activity in otolith-ocular connections in order to control for torsional and vertical eye position. Damage would then result in skew deviation (Zee 1996). A dissociation between perceptual and ocular motor responses was recently described in a single patient with an isolated heminodular stroke. Whereas otolith-perceptual integration was abolished in this patient (see Fig. 2), otolith-ocular

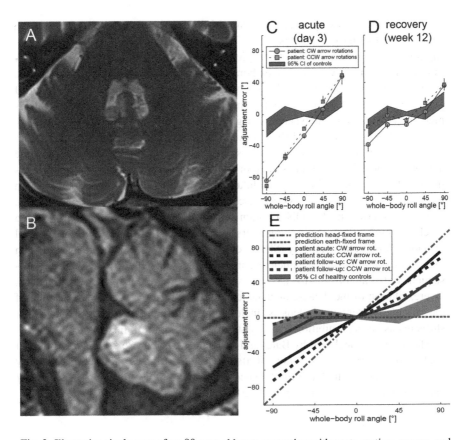

Fig. 2 Illustrative single case of an 80-year-old man presenting with acute vertigo, nausea, and gait imbalance. Acute MR imaging demonstrated a right heminodulus with possible minor involvement of the upper pole of the right cerebellar tonsil on axial (panel **a**, axial T2-weighted sequence) and sagittal (panel **b**, fluid-attenuated inversion recovery (FLAIR) sequence) sequences. (MR images: courtesy of the Institute of Neuroradiology, University Hospital Zurich, Switzerland). Perceptual estimates of the direction of gravity in this patient are shown in panels **c** (on day 3) and **d** (after 12 weeks) and compared to 11 healthy human subjects (95% confidence interval (CI) in gray). Average adjustment errors (±1SD) in the patient are presented separately for trials with CW (gray-filled circles interconnected by a solid black line) and CCW (gray-filled squares interconnected by a dashed black line) arrow rotations. Predictions for adjustments when the nodulus is damaged are shown in panel **e**. Based on the assumption that the laterality of the lesion induces an additional bias of fixed size in all positions, the offset as determined in upright position (panel **c**) was subtracted from all positions. Acutely after right-sided heminodular stroke, perceived vertical matched the prediction for a head-fixed frame closely. On follow-up, perceived vertical was in-between a head-fixed frame and an earth-fixed frame, suggesting partial recovery of otolith-input integration after heminodular loss. (Modified after (Tarnutzer et al. 2015), reused with permission)

reflexes remained preserved, underlining the importance of the nodulus in spatial orientation (Tarnutzer et al. 2015).

The dentate nucleus, however, not only receives graviceptive input from the vestibular nuclei but also is interconnected with the inferior olivary nucleus (ION) by the rubro-dentato-olivary tract ("Guillain-Mollaret triangle") (Pearce 2008). In a single patient with right-sided pontomesencephalic hemorrhage and subsequent disruption of the Guillain-Mollaret triangle, the role of the dentate nucleus in processing vestibular input was further elaborated (Tarnutzer et al. 2012b). In the acute stage, this patient demonstrated a slight ipsilesional (i.e., rightward) deviation of perceived vertical. On follow-up 4 months later, clinical worsening was accompanied by vertical-torsional nystagmus, right-sided hypertrophy of the inferior olivary nucleus (ION), and a marked shift of the SVV toward the left side (i.e., away from the side of original hemorrhage and subsequent hypertrophy of the ION). The authors hypothesized that this switch in SVV deviation is a consequence of removing GABA-mediated inhibitory modulation of the ION neurons provided through the central tegmental tract. As a consequence, synchronized oscillations of large groups of neurons in the ION are then sent to the cerebellar cortex and the deep cerebellar nuclei through climbing fibers and result in a loss of the inhibitory control provided through this pathway (Shaikh et al. 2010). This mechanism may lead to an overexcitation of the dentate nucleus contralateral to the hypertrophic ION and consequently to a roll-tilt toward the side of the overactive dentate nucleus. Overexcitation of the dentate nucleus leading to a shift in SVV toward the side of the stimulated dentate nucleus is analogous to the reported contralesional SVV shifts in case of an inhibition (e.g., by stroke or bleeding) of the dentate nucleus.

However, as concluded by Baier and colleagues, the cerebellar mechanisms that lead to contraversive or ipsiversive shifts in internal estimates of direction of gravity are far from clear, emphasizing the need for further studies focusing on cerebellar anatomy and physiology (Baier et al. 2008).

3.3   Dissociation of Perceptual and Ocular Motor Shifts in the Thalamus and the Cortex

Previously, predominantly contraversive SVV roll-tilts without OTR have been reported for acute cerebral lesions in the temporo-parietal cortex including the posterior insula, the insular gyrus, and the middle/superior temporal gyrus (Brandt et al. 1994). Recently, statistical voxel-wise lesion-behavior mapping confirmed an association between SVV roll-tilt and the insular cortex and the (right) inferior frontal gyrus and also identified the right superior temporal gyrus, the pre- and postcentral gyrus, the rolandic operculum, and the inferior parietal lobe as cortical structures likely involved in the vestibulo-cortical network for verticality perception in acute stroke patients (Baier et al. 2012b, 2013). Subcortical structures involved in this network included the superior longitudinal fasciculus and the inferior/superior occipitofrontal fasciculus (Baier et al. 2012b). In accordance with previous studies, shifts in SVV were more often contralesional than ipsilesional (59% vs. 37%) and of larger size for right-sided lesions. As pointed out by these authors, none of their patients presented with isolated lesions of one of the identified associated areas (as the insular cortex or the inferior frontal gyrus), leaving it open whether isolated lesions of these regions are sufficient to result in a bias in verticality perception (Baier et al. 2012b). The importance of this limitation was more recently confirmed. Of note, acute lesions restricted to the posterior insular cortex – a region considered core to the human vestibular cortical network – did not result in vertigo, deficits of verticality perception, or other vestibular otolith deficits such as ocular torsion or skew deviation (Baier et al. 2013). This led the authors to the conclusion that "lesions of the posterior insular cortex have to be combined with lesions of adjacent regions of the cortical and subcortical vestibular network to cause vestibular otolith deficits" (Baier et al. 2013).

Among different modalities, patients with hemispheric stroke showed contralesional errors in verticality perception of similar size for the visual vertical and the haptic vertical, whereas postural vertical offsets were more pronounced in right hemispheric lesions (Perennou et al. 2008).

In the thalamus, dissociations between verticality perception and otolith-mediated brainstem reflexes have been observed as well. This included lesions of the posterolateral thalamus (Karnath et al. 2000). The authors observed a significant body tilt, while the SVV remained accurate, postulating separate pathways for sensing body orientation besides the pathways projecting to the vestibular cortex.

In a series of 35 patients with acute thalamic infarctions, Dieterich and Brandt reported SVV roll-tilts in 69% of patients, with ipsilesional shifts noted more frequently than contralesional shifts (44 vs. 25%) (Dieterich and Brandt 1993a). More recently, voxel-based morphometric lesion behavior mapping in 37 patients with thalamic infarctions has demonstrated significant shifts in verticality perception in 59% of patients with contralesional roll-tilts being more frequent than ipsilesional roll-tilts (41% vs. 19%). Noteworthy, two distinct anatomical sites were involved in the directional processing of graviceptive signals in the thalamus: Whereas contralesional roll-tilts of the SVV were linked to lesions of the dorsolateral and dorsomedial thalamic subnuclei, ipsilesional roll-tilts were associated with lesions in more inferior and medial areas (Baier et al. 2016).

4   Conclusions

Based on current knowledge, it becomes obvious that the networks for internally estimating direction of gravity include brainstem and cerebellar structures. Therefore, both brainstem and cerebellar lesions may result in a lateralized disruption of the central vestibular pathways and subsequent deviations of eye torsion and verticality perception. Additional clinical brainstem or cerebellar signs and imaging will further elucidate to the lesion location.

Acknowledgments   I thank Dr. Christopher J. Bockisch for carefully reading and commenting on the manuscript.

Funding Details　No funding was received for this manuscript.

Disclosure Statement　Dr. Tarnutzer has nothing to disclose.

References

Anastasopoulos, D., Haslwanter, T., Bronstein, A., Fetter, M., & Dichgans, J. (1997). Dissociation between the perception of body verticality and the visual vertical in acute peripheral vestibular disorder in humans. *Neuroscience Letters, 233*(2–3), 151–153.

Angelaki, D. E., Gu, Y., & DeAngelis, G. C. (2009). Multisensory integration: Psychophysics, neurophysiology, and computation. *Current Opinion in Neurobiology, 19*(4), 452–458.

Baier, B., & Dieterich, M. (2009). Ocular tilt reaction: A clinical sign of cerebellar infarctions? *Neurology, 72*(6), 572–573. https://doi.org/10.1212/01.wnl.0000342123.39308.32.

Baier, B., Bense, S., & Dieterich, M. (2008). Are signs of ocular tilt reaction in patients with cerebellar lesions mediated by the dentate nucleus? *Brain, 131*(Pt 6), 1445–1454.

Baier, B., Thomke, F., Wilting, J., Heinze, C., Geber, C., & Dieterich, M. (2012a). A pathway in the brainstem for roll-tilt of the subjective visual vertical: Evidence from a lesion-behavior mapping study. *The Journal of Neuroscience, 32*(43), 14854–14858. https://doi.org/10.1523/JNEUROSCI.0770-12.2012. 32/43/14854 [pii].

Baier, B., Suchan, J., Karnath, H. O., & Dieterich, M. (2012b). Neural correlates of disturbed perception of verticality. *Neurology, 78*(10), 728–735. https://doi.org/10.1212/WNL.0b013e318248e544. WNL.0b013e318248e544 [pii].

Baier, B., Zu Eulenburg, P., Best, C., Geber, C., Muller-Forell, W., Birklein, F., & Dieterich, M. (2013). Posterior insular cortex – a site of vestibular-somatosensory interaction? *Brain and Behavior, 3*(5), 519–524. https://doi.org/10.1002/brb3.155.

Baier, B., Conrad, J., Stephan, T., Kirsch, V., Vogt, T., Wilting, J., Muller-Forell, W., & Dieterich, M. (2016). Vestibular thalamus: Two distinct graviceptive pathways. *Neurology, 86*(2), 134–140. https://doi.org/10.1212/WNL.0000000000002238.

Barra, J., Marquer, A., Joassin, R., Reymond, C., Metge, L., Chauvineau, V., & Perennou, D. (2010). Humans use internal models to construct and update a sense of verticality. *Brain, 133*(Pt 12), 3552–3563. doi:awq311 [pii]. https://doi.org/10.1093/brain/awq311.

Bockisch, C. J., & Haslwanter, T. (2001). Three-dimensional eye position during static roll and pitch in humans. *Vision Research, 41*(16), 2127–2137.

Bohmer, A., & Rickenmann, J. (1995). The subjective visual vertical as a clinical parameter of vestibular function in peripheral vestibular diseases. *Journal of Vestibular Research, 5*(1), 35–45.

Brandt, T., & Dieterich, M. (1993). Skew deviation with ocular torsion: A vestibular brainstem sign of topographic diagnostic value. *Annals of Neurology, 33*(5), 528–534.

Brandt, T., & Dieterich, M. (1994). Vestibular syndromes in the roll plane: Topographic diagnosis from brainstem to cortex. *Annals of Neurology, 36*(3), 337–347.

Brandt, T., & Dieterich, M. (2017). The dizzy patient: Don't forget disorders of the central vestibular system. *Nature Reviews. Neurology*. https://doi.org/10.1038/nrneurol.2017.58.

Brandt, T., Dieterich, M., & Danek, A. (1994). Vestibular cortex lesions affect the perception of verticality. *Annals of Neurology, 35*(4), 403–412.

Clemens, I. A., De Vrijer, M., Selen, L. P., Van Gisbergen, J. A., & Medendorp, W. P. (2011). Multisensory processing in spatial orientation: An inverse probabilistic approach. *The Journal of Neuroscience, 31*(14), 5365–5377. https://doi.org/10.1523/JNEUROSCI.6472-10.2011. 31/14/5365 [pii].

Collewijn, H., van der, S. J., Ferman, L., & Jansen, T. C. (1985). Human ocular counterroll: Assessment of static and dynamic properties from electromagnetic scleral coil recordings. *Experimental Brain Research, 59*(1), 185–196.

Curthoys, I. S., Halmagyi, G. M., & Dai, M. J. (1991). The acute effects of unilateral vestibular neurectomy on sensory and motor tests of human otolithic function. *Acta Oto-Laryngologica. Supplementum, 481*, 5–10.

De Vrijer, M., Medendorp, W. P., & Van Gisbergen, J. A. (2008). Shared computational mechanism for tilt compensation accounts for biased verticality percepts in motion and pattern vision. *Journal of Neurophysiology, 99*(2), 915–930.

Dieterich, M., & Brandt, T. (1993a). Thalamic infarctions: Differential effects on vestibular function in the roll plane (35 patients). *Neurology, 43*(9), 1732–1740.

Dieterich, M., & Brandt, T. (1993b). Ocular torsion and tilt of subjective visual vertical are sensitive brainstem signs. *Annals of Neurology, 33*(3), 292–299. https://doi.org/10.1002/ana.410330311.

Dieterich, M., & Brandt, T. (2008). Functional brain imaging of peripheral and central vestibular disorders. *Brain, 131*(Pt 10), 2538–2552. doi: awn042 [pii]. https://doi.org/10.1093/brain/awn042.

Dieterich, M., & Brandt, T. (2015). The bilateral central vestibular system: Its pathways, functions, and disorders. *Annals of the New York Academy of Sciences, 1343*, 10–26. https://doi.org/10.1111/nyas.12585.

Fernandez, C., Goldberg, J. M., & Abend, W. K. (1972). Response to static tilts of peripheral neurons innervating otolith organs of the squirrel monkey. *Journal of Neurophysiology, 35*(6), 978–987.

Halmagyi, G. M., Gresty, M. A., & Gibson, W. P. (1979). Ocular tilt reaction with peripheral vestibular lesion. *Annals of Neurology, 6*(1), 80–83.

Halmagyi, G. M., Brandt, T., Dieterich, M., Curthoys, I. S., Stark, R. J., & Hoyt, W. F. (1990). Tonic contraversive ocular tilt reaction due to unilateral meso-diencephalic lesion. *Neurology, 40*(10), 1503–1509.

Haustein, W. (1992). Head-centric visual localization with lateral body tilt. *Vision Research, 32*(4), 669–673.

Kahane, P., Hoffmann, D., Minotti, L., & Berthoz, A. (2003). Reappraisal of the human vestibular cortex by cortical electrical stimulation study. *Annals of Neurology, 54*(5), 615–624.

Kaptein, R. G., & Van Gisbergen, J. A. (2004). Interpretation of a discontinuity in the sense of verticality at large body tilt. *Journal of Neurophysiology, 91*(5), 2205–2214.

Karnath, H. O., Ferber, S., & Dichgans, J. (2000). The neural representation of postural control in humans. *Proceedings of the National Academy of Sciences of the United States of America, 97*(25), 13931–13936. https://doi.org/10.1073/pnas.240279997.

Kattah, J. C., Talkad, A. V., Wang, D. Z., Hsieh, Y. H., & Newman-Toker, D. E. (2009). HINTS to diagnose stroke in the acute vestibular syndrome: Three-step bedside oculomotor examination more sensitive than early MRI diffusion-weighted imaging. *Stroke, 40*(11), 3504–3510.

Kim, H. A., Lee, H., Yi, H. A., Lee, S. R., Lee, S. Y., & Baloh, R. W. (2009). Pattern of otolith dysfunction in posterior inferior cerebellar artery territory cerebellar infarction. *Journal of the Neurological Sciences, 280*(1–2), 65–70. doi:S0022-510X(09)00051-3 [pii]. https://doi.org/10.1016/j.jns.2009.02.002.

Kirsch, V., Keeser, D., Hergenroeder, T., Erat, O., Ertl-Wagner, B., Brandt, T., & Dieterich, M. (2016). Structural and functional connectivity mapping of the vestibular circuitry from human brainstem to cortex. *Brain Structure & Function, 221*(3), 1291–1308. https://doi.org/10.1007/s00429-014-0971-x.

Laurens, J., Meng, H., & Angelaki, D. E. (2013). Neural representation of orientation relative to gravity in the macaque cerebellum. *Neuron, 80*(6), 1508–1518. https://doi.org/10.1016/j.neuron.2013.09.029.

Lee, H., Lee, S. Y., Lee, S. R., Park, B. R., & Baloh, R. W. (2005). Ocular tilt reaction and anterior inferior cerebellar artery syndrome. *Journal of Neurology, Neurosurgery, and Psychiatry, 76*(12), 1742–1743. https://doi.org/10.1136/jnnp.2005.069104.

MacNeilage, P. R., Banks, M. S., Berger, D. R., & Bulthoff, H. H. (2007). A Bayesian model of the disambiguation of gravitoinertial force by visual cues. *Experimental Brain Research, 179*(2), 263–290.

Merfeld, D. M., Zupan, L., & Peterka, R. J. (1999). Humans use internal models to estimate gravity and linear acceleration. *Nature, 398*(6728), 615–618. https://doi.org/10.1038/19303.

Min, W., Kim, J., Park, S., & Suh, C. (1999). Ocular tilt reaction due to unilateral cerebellar lesion. *Neuro-Ophthalmology, 22*(2), 81–85.

Mittelstaedt, H. (1983). A new solution to the problem of the subjective vertical. *Naturwissenschaften, 70*(6), 272–281.

Mossman, S., & Halmagyi, G. M. (1997). Partial ocular tilt reaction due to unilateral cerebellar lesion. *Neurology, 49*(2), 491–493.

Park, H. K., Kim, J. S., Strupp, M., & Zee, D. S. (2013). Isolated floccular infarction: Impaired vestibular responses to horizontal head impulse. *Journal of Neurology, 260*(6), 1576–1582. https://doi.org/10.1007/s00415-013-6837-y.

Pavlou, M., Wijnberg, N., Faldon, M. E., & Bronstein, A. M. (2003). Effect of semicircular canal stimulation on the perception of the visual vertical. *Journal of Neurophysiology, 90*(2), 622–630.

Pearce, J. M. (2008). Palatal Myoclonus (syn. Palatal Tremor). *European Neurology, 60*(6), 312–315. doi:000159929 [pii]. https://doi.org/10.1159/000159929.

Perennou, D. A., Mazibrada, G., Chauvineau, V., Greenwood, R., Rothwell, J., Gresty, M. A., & Bronstein, A. M. (2008). Lateropulsion, pushing and verticality perception in hemisphere stroke: A causal relationship? *Brain, 131*(Pt 9), 2401–2413. https://doi.org/10.1093/brain/awn170.

Schor, R. H., Miller, A. D., & Tomko, D. L. (1984). Responses to head tilt in cat central vestibular neurons. I. Direction of maximum sensitivity. *Journal of Neurophysiology, 51*(1), 136–146.

Schuler, J. R., Bockisch, C. J., Straumann, D., & Tarnutzer, A. A. (2010). Precision and accuracy of the subjective haptic vertical in the roll plane. *BMC Neuroscience, 11*, 83. https://doi.org/10.1186/1471-2202-11-83.

Shaikh, A. G., Green, A. M., Ghasia, F. F., Newlands, S. D., Dickman, J. D., & Angelaki, D. E. (2005). Sensory convergence solves a motion ambiguity problem. *Current Biology, 15*(18), 1657–1662. https://doi.org/10.1016/j.cub.2005.08.009.

Shaikh, A. G., Hong, S., Liao, K., Tian, J., Solomon, D., Zee, D. S., Leigh, R. J., & Optican, L. M. (2010). Oculopalatal tremor explained by a model of inferior olivary hypertrophy and cerebellar plasticity. *Brain, 133*(Pt 3), 923–940. doi:awp323 [pii]. https://doi.org/10.1093/brain/awp323.

Tarnutzer, A. A., Bockisch, C., Straumann, D., & Olasagasti, I. (2009a). Gravity dependence of subjective visual vertical variability. *Journal of Neurophysiology, 102*(3), 1657–1671.

Tarnutzer, A. A., Bockisch, C. J., & Straumann, D. (2009b). Head roll dependent variability of subjective visual vertical and ocular counterroll. *Experimental Brain Research, 195*(4), 621–626.

Tarnutzer, A. A., Bockisch, C. J., Olasagasti, I., & Straumann, D. (2012a). Egocentric and allocentric alignment tasks are affected by otolith input. *Journal of Neurophysiology, 107*(11), 3095–3106. https://doi.org/10.1152/jn.00724.2010. jn.00724.2010 [pii].

Tarnutzer, A. A., Palla, A., Marti, S., Schuknecht, B., & Straumann, D. (2012b). Hypertrophy of the inferior olivary nucleus impacts perception of gravity. *Frontiers in Neurology, 3*, 79. https://doi.org/10.3389/fneur.2012.00079.

Tarnutzer, A. A., Wichmann, W., Straumann, D., & Bockisch, C. J. (2015). The cerebellar nodulus: Perceptual and ocular processing of graviceptive input. *Annals of Neurology, 77*(2), 343–347. https://doi.org/10.1002/ana.24329.

Wade, S. W., & Curthoys, I. S. (1997). The effect of ocular torsional position on perception of the roll-tilt of visual stimuli. *Vision Research, 37*(8), 1071–1078.

Yakusheva, T., Blazquez, P. M., & Angelaki, D. E. (2008). Frequency-selective coding of translation and tilt in macaque cerebellar nodulus and uvula. *The Journal of Neuroscience, 28*(40), 9997–10009. https://doi.org/10.1523/JNEUROSCI.2232-08.2008.

Zee, D. S. (1996). Considerations on the mechanisms of alternating skew deviation in patients with cerebellar lesions. *Journal of Vestibular Research, 6*(6), 395–401.

Zwergal, A., Cnyrim, C., Arbusow, V., Glaser, M., Fesl, G., Brandt, T., & Strupp, M. (2008a). Unilateral INO is associated with ocular tilt reaction in pontomesencephalic lesions: INO plus. *Neurology, 71*(8), 590–593. https://doi.org/10.1212/01.wnl.0000323814.72216.48.

Zwergal, A., Buttner-Ennever, J., Brandt, T., & Strupp, M. (2008b). An ipsilateral vestibulothalamic tract adjacent to the medial lemniscus in humans. *Brain, 131*(Pt 11), 2928–2935. https://doi.org/10.1093/brain/awn201.

Magnetic Vestibular Stimulation

Bryan K. Ward

Abstract Strong static magnetic fields such as those in MRI machines can induce vertigo and nystagmus. The mechanism is a Lorentz force, generated in the inner ear fluids of the labyrinth by the interaction between normal ionic currents entering into labyrinthine hair cells and the strong static magnetic field of the MRI machine. The Lorentz force is constant and displaces the cupulae of the semicircular canals to a deviated position, simulating a response to a constant acceleration of the head with its consequent sustained nystagmus. The Lorentz effect scales with the strength of the magnetic field, but is not harmful, except for inducing transient dizziness and nausea. Magnetic vestibular stimulation (MVS) has implications for studies of vestibular physiology and adaptation, the interpretation of functional MRI studies, and human safety when undergoing diagnostic imaging studies.

Keywords Magnetic field · MRI · Vertigo · Lorentz force · Nystagmus

It was recently discovered that strong magnetic fields like those of magnetic resonance imaging (MRI) machines can cause nystagmus and sensations of vertigo in normal healthy adults. The effect is thought to be the result of a Lorentz force, generated by the interactions of the normal ionic currents of the inner ear and the strong static magnetic field of an MRI machine, leading to a pressure on the cupulae of the semicircular canals. This chapter will review the evidence supporting a Lorentz force mechanism that explains the vertigo and nystagmus experienced by humans in strong static magnetic fields. The chapter will also discuss implications of this phenomenon for understanding vestibular physiology, as well as for interpretations of current functional MRI research and human safety in strong magnetic fields.

B. K. Ward (✉)
Johns Hopkins University School of Medicine, Johns Hopkins Outpatient Center, Department of Otolaryngology-Head and Neck Surgery, Baltimore, MD, USA
e-mail: bward15@jhmi.edu

© Springer Nature Switzerland AG 2019   89
A. Shaikh, F. Ghasia (eds.), *Advances in Translational Neuroscience of Eye Movement Disorders*, Contemporary Clinical Neuroscience,
https://doi.org/10.1007/978-3-030-31407-1_5

1   History

MRI uses a strong static magnetic field to arrange the spins of hydrogen atoms. The amount of signal obtained from an MRI scan is proportional to and dependent upon the strength of the static magnetic field. Greater magnetic field strength contributes to decreased imaging times and higher image resolution. These benefits have there-fore driven the development of technology that generates stronger static magnetic fields. Since the first MRI scans in humans in 1977, there has been steady progress in the development of stronger static magnetic fields. Currently, most clinical scan-ners use 1.5 tesla (T) or 3 T magnetic field strengths, but scanners for research purposes are available at strengths of 7 T, 9.4 T, and 11.5 T, and clinical 7 T scanners are in development.

Patients and technicians working around these strong MRI machines often report transient sensations of vertigo (Schenck et al. 1992; Kangarlu et al. 1999; de Vocht et al. 2006). The frequency of these reports of dizziness around MRI machines has increased as more powerful magnetic fields are used for both clinical and research MRI studies (Schaap et al. 2014). Building on these reports, Houpt et al. began studying the behavioral responses of rats and mice exposed to strong static magnetic fields. They observed circling behaviors in the animals exposed to high-magnetic-field MRI machines and that the behaviors could be curtailed by performing a laby-rinthectomy (Houpt et al. 2003, 2007). These studies suggested that the labyrinth was involved in generating circling behavior in the animals after being in a strong magnetic field; however, the evidence had not sufficiently accumulated to propose a mechanistic interpretation.

The first observation of eye movement responses to a static magnetic field was serendipitous; Vincenzo Marcelli and colleagues were using functional MRI (fMRI) to study the central processing of a caloric peripheral vestibular stimulus (Marcelli et al. 2009). A caloric test is performed by recording eye movements after stimulat-ing the labyrinth by irrigating the ear canal with warm or cool water. Marcelli noticed that when study subjects were in darkness with infrared illumination in a 1.5 T MRI, some had a slow, persistent horizontal nystagmus that was present prior to irrigating the ear canal with water. Marcelli et al. speculated this nystagmus was a result of the static magnetic field (Marcelli et al. 2009).

A challenge encountered by early investigations into the causes of dizziness around MRI machines was that the perception of rotation that technologists and patients experience around these scanners is transient. Since the perception of rota-tion was the principal evidence for magnetic vestibular stimulation, these reports of a transient experience of rotation prompted investigations into transient stimuli causing magnetic vestibular stimulation (Glover et al. 2007). An example of a tran-sient magnetic stimulus is electromagnetic induction in which movement through a magnetic field induces a current in a wire. Electromagnetic induction would occur only during movement through a magnetic field and would therefore be transient, occurring only around the time of movement. Establishing a link between eye movements and the static magnetic field was important for sorting candidate mechanisms of magnetic vestibular stimulation. By providing a clear signal that could be observed under different experimental conditions, eye movements could be used to understand what parts of the labyrinth might be affected.

2 Observations in Humans

The perception people experience when entering a 7 T magnetic field supine in darkness can vary by head position but is often a whole-body (or in some cases just head) roll where the axis is roughly perpendicular to the navel when supine. The sensation of vertigo is transient, adapting completely in about a minute in a 7 T magnetic field (Mian et al. 2013). The nystagmus in humans, however, can persist long beyond the time after the sense of rotation has ceased (up to 90 minutes thus far) (Ward and Zee 2016).

The key to observing the nystagmus evoked by the magnetic field is removing visual fixation. Visual fixation can suppress an unwanted nystagmus that is due to asymmetry in the peripheral vestibular system. Removing visual fixation is used in the clinic when diagnosing patients with vestibular disorders in order to bring out a nystagmus. Marcelli's observation of a *persistent* nystagmus without visual fixation was important to understanding the mechanism of magnetic vestibular stimulation.

A predominantly horizontal nystagmus can be observed in all normal humans without visual fixation the entire time they are lying supine in a magnetic field of at least 1.5 T (Roberts et al. 2011). The nystagmus in an MRI machine is a jerk nystagmus typical of a vestibular stimulus, with both slow-phase and quick-phase components (Fig. 1). The slow-phase component is the part of a vestibular nystagmus attributable to excitation of the peripheral vestibular system.

The static magnetic field of an MRI machine is always "on." When images are acquired in an MRI, pulses of radio waves are used to excite hydrogen nuclei and magnetic field gradients are used to locate the signals in space. The nystagmus in the magnetic field is seen regardless of whether any MR images are being acquired, however, implying that the static magnetic field creates the effect and not the radio-frequency pulses or any transient magnetic fields that are used when obtaining MR images. Furthermore, the velocity of the slow-phase component of nystagmus scales roughly linearly with the strength of the *static* magnetic field, being barely noticeable in 1.5 T MRI scanners and easily observed in 7 T MRI scanners.

Prior speculation suggested a transient stimulus for the origin of magnetic vestibular stimulation; therefore, Roberts et al. compared the effect of rates of motion into and out of the MRI bore on the eye movement responses. Although different velocities of entry into the magnetic field change the stimulus intensity of electromagnetic induction, the velocity of entry into the MRI does not affect the nystagmus. In fact, the nystagmus persists even though the subject is lying still in the magnetic field. These observations indicate the nystagmus is not caused by motion through the magnetic field. The nystagmus in an MRI is absent in patients with bilateral vestibular hypofunction, suggesting that similar to the results from earlier

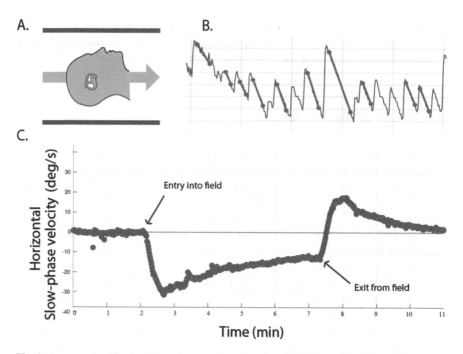

Fig. 1 An example of horizontal nystagmus of a subject in a 7 T MRI machine. The subject enters the magnetic field head first with the magnetic field directed toward the feet (**a**). Part **b** shows eye position versus time with slope of slow-phase eye velocity highlighted in red. The slow-phase eye velocity (SPV) of each beat of nystagmus (red dot) is shown as a function of time in Part **c**. The SPV increases to a peak after entry before partially adapting. Once the subject exits the magnet, an aftereffect is observed in which the nystagmus beats in the opposite direction. The decay of the aftereffect reflects dissipation of the adaptation that occurred when in the MRI

rodent studies by Houpt et al., the effect requires intact peripheral vestibular function on at least one side.

The direction of nystagmus provides additional clues about the mechanism. Both the direction and the intensity of the observed nystagmus depend on the position of the head with respect to the magnetic field. In normal humans, the nystagmus is primarily horizontal when the person is lying supine in the magnet (i.e., usual position for a diagnostic MRI scan, Fig. 1). The direction of nystagmus can reverse with extreme head pitch and with feet-first entry into the magnetic field (e.g., the direction changes from left-beating to right-beating). All human subjects also have a null position where there is no nystagmus observed, and if they pitch their head beyond this null position, the direction of nystagmus reverses (Roberts et al. 2011). The static magnetic fields of MRI machines have a polarity, with the north pole of the MRI bore at one end and the south pole at the other. Inside the magnetic field bore of an MRI machine, the magnetic field is homogeneous and static, which means all the magnetic field vectors point in the same direction at the center of the bore. The varying effects on nystagmus with head pitch imply a relationship between the position of the head and the orientation of the static magnetic field. The nystagmus also reverses direction when entering the magnetic field feet first and when entering head first into the *back* of the magnet, suggesting that the direction of nystagmus also depends on the polarity (north to south direction) of the magnetic field.

Although the nystagmus persists throughout the time a subject is lying in the MRI bore, there is some decay in the velocity of the slow-phase component of nystagmus (Fig. 1). Immediately upon exiting the magnetic field, an aftereffect occurs in which the subject perceives rotation in the direction opposite that experienced when entering the magnet but of similar duration (Mian et al. 2013). The direction of the nystagmus also reverses, and the duration of this nystagmus aftereffect depends upon the duration of exposure to the MRI (Jareonsettasin et al. 2016). For short duration exposures in which no adaptation has occurred, there is no aftereffect upon leaving the magnetic field (Roberts et al. 2011).

3 Mechanism

Roberts et al. synthesized these observations and hypothesized that a Lorentz force generated by the interactions between normal ionic currents in the inner ear and the static magnetic field could induce a constant displacement of the cupula resulting in a persistent horizontal nystagmus (Roberts et al. 2011). Although contributions from alternative mechanisms such as electromagnetic induction (mentioned above) may contribute to the perceptions of dizziness around MRI machines, these alternatives are increasingly felt to be unlikely as data accumulates in support of the Lorentz force hypothesis. It is important to keep in mind that this effect happens regardless of whether any MR images are being acquired and therefore is a result of the static magnetic field and not the use of time-variant magnetic fields or the radiofrequency pulses when obtaining MR images.

3.1 What Is a Lorentz Force?

A Lorentz force is a force imposed on a charged particle moving through a magnetic field. If the charged particles are flowing in a uniform direction such as through a wire, there will be a force imposed on the wire when placed in a magnetic field. A similar effect occurs in magnetohydrodynamics (study of magnetic properties of electrically conducting liquids), in which charges flowing through a conducting fluid experience a force when in a magnetic field (Fig. 2). In magnetohydrodynamics, the formula for this is $F = B \times hj$. The magnitude of the Lorentz force (F) is proportional to the strength of the magnetic field (B), the current density (j), and the length (h) over which the current is flowing. The Lorentz force, the magnetic field, and the current are vectors and, therefore, have a direction. The Lorentz force is the cross product of the magnetic field and current vectors. To determine the direction of the Lorentz force, the "right-hand rule" is applied, where the thumb is placed in

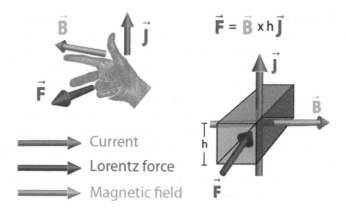

Fig. 2 Cartoon of a Lorentz force. A Lorentz force is represented by the eq. $F = B \times hj$, where F is the Lorentz force, B is the static magnetic field, j is the current density, and h is the height over which the current flows. Note that the Lorentz force is proportional to the current and intensity of the magnetic field. The direction of the Lorentz force is determined by the right-hand rule as shown

the direction of the current, the index finger is in the direction of the magnetic field, and the resulting Lorentz force is in the direction of the middle finger (or coming out the palm). Recall that whenever there is a difference between the magnetic field and current vectors, a cross product will result in a Lorentz force. When the vectors align, there is no cross product and no force.

3.2   How Does a Lorentz Force Occur in the Inner Ear?

The inner ear is an ideal environment in which a magnetohydrodynamic force can occur, where a conductive fluid must transmit both electric current and pressure. Inner ear hair cells and vestibular afferents have a spontaneous discharge rate. In the vestibular system, a resting neural discharge or firing rate of vestibular afferents is critical to the physiology of the system. This resting discharge rate allows afferents to modulate bidirectionally, either increasing in activity when excited or decreasing in activity when inhibited (Fig. 3) (Lowenstein et al. 1936). Modulating around a resting discharge rate allows the system to maintain linearity across a range of different frequency stimuli and eliminates a stimulus threshold. The spontaneous discharge rate of the vestibular system requires constant recycling of ions through the potassium-rich endolymph. Potassium ions are generated in the dark cells near the semicircular canal ampullae and the vestibule and are secreted into the endolymph (Kimura 1969). The potassium ions enter the apical ends of hair cells via the mechanoelectric transduction channels and sustain their resting discharge. This process sustains a constant current traveling through the endolymph and entering hair cells (Fig. 4). It also fulfills a criterion for a Lorentz force, in which a conductor is needed

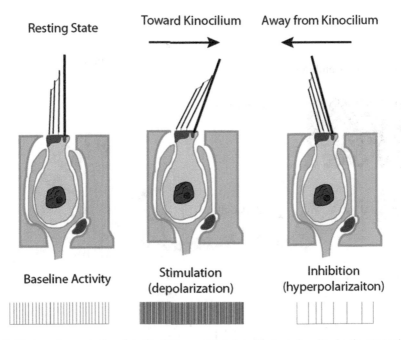

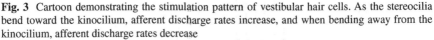

Fig. 3 Cartoon demonstrating the stimulation pattern of vestibular hair cells. As the stereocilia bend toward the kinocilium, afferent discharge rates increase, and when bending away from the kinocilium, afferent discharge rates decrease

(i.e., endolymph) to carry current. When this system is introduced into a magnetic field of sufficient field strength, a Lorentz force is generated in the fluid through which the current is traveling.

The other criterion for the Lorentz force is that the conductive fluid carries a force. The labyrinth consists of three semicircular canals (superior, lateral, and posterior) and two otoconial organs (utricle and saccule). The membranous labyrinth is filled with endolymph and contains the semicircular canal cristae and the otoconial maculae. For the semicircular canals, due to inertia, a head rotation results in a lag of the movement of endolymph relative to the head. This lag creates relative movement of the cupula, causing hair cells to become deflected, either exciting or inhibiting the associated vestibular afferents (Fig. 5). The force in the endolymph fluid due to a head acceleration signals the brain that the head is moving. Endolymph is therefore the conductive fluid needed to carry both electric current and pressure.

Finally, for a Lorentz force to cause magnetic vestibular stimulation, the brain must be able to detect the force. While there may be other environments in the body capable of generating a Lorentz force (e.g., the cochlea), in the labyrinth, the semicircular canal cupulae are shear sensors that can detect the force, leading to the observed nystagmus and vertigo. The presence of cupulae permits the generated Lorentz force to be sensed. The Lorentz force, however, must be strong enough and in the appropriate direction in order to stimulate these sensors.

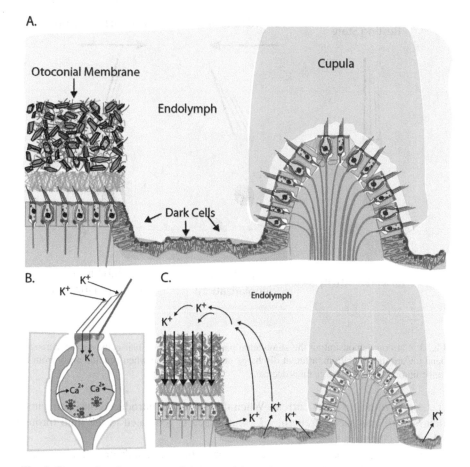

Fig. 4 Cartoon showing anatomy of the otoconial membrane, cupula, and nearby dark cells (**a**). (**b**) Potassium (K+) enters the apical ends of the hair cell stereocilia, leading to depolarization of the hair cell, calcium (Ca 2+) entry into the cell, and release of excitatory neurotransmitter at the synaptic cleft. (**c**) Potassium is secreted into the endolymph by the dark cells and circulates. We hypothesize that there is relative uniform current (potassium ion movement) at the utricular macula

3.3 What Structures of the Inner Ear Are Involved in Generating Magnetic Vestibular Stimulation?

There is a close relationship between reflexive eye movements and stimulation of primary vestibular afferents. Relying on the work of early vestibular physiologists like Ewald, Flourens, and Breuer, we know that we can study eye movements and infer which parts of the inner ear might be stimulated (John Leigh and Zee 2015). When lying supine in the 7 T magnetic field, humans with a normal vestibular system have nystagmus that is a mixture of horizontal and torsional slow-phase components (Otero-Millan et al. 2017), but the pattern is unusual. Clinicians caring for

Fig. 5 (**a**) In a Lorentz force, the force is thought to create a static displacement of the canal cupulae, providing a force to the cupulae akin to a constant head acceleration (**b**)

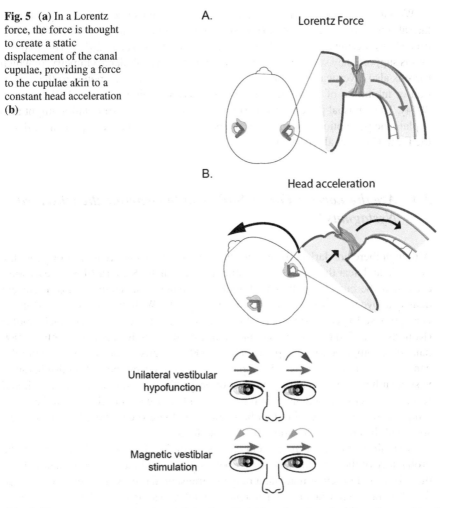

A.

Lorentz Force

B.

Head acceleration

Unilateral vestibular hypofunction

Magnetic vestiblar stimulation

Fig. 6 The eye movements seen in a unilateral vestibular hypofunction where there is a combined horizontal and torsional nystagmus, with slow phases in which the top poles of the eyes rotate in the same direction as the horizontal component. This is due to the summed components of the remaining three semicircular canals on the contralateral side. In magnetic vestibular stimulation, the top poles of the eyes rotate opposite the direction of the horizontal component. This is due to the combination of excitation and inhibition of the superior semicircular canals

patients with vertigo commonly encounter mixed horizontal-torsional nystagmus in patients with unilateral vestibular hypofunction. This pattern of mixed horizontal and torsional nystagmus after a unilateral vestibular lesion is one in which the top poles of the eyes rotate in a slow phase toward the same ear as the horizontal slow-phase component is directed, as if the eyes are a rolling wheel (Fig. 6). In magnetic vestibular stimulation, however, the torsional component is one in which the top poles of the eyes rotate in a slow-phase pattern of nystagmus *away* from the ear toward which the horizontal component of slow phase is directed.

Working from the pattern of eye movements, we presume that the superior and lateral semicircular canals are primarily stimulated in magnetic vestibular stimulation when entering the magnet lying supine (Otero-Millan et al. 2017). This hypothesis is supported by studying patients with unilateral vestibular asymmetry, in which there is also a vertical component to the nystagmus that corresponds to the excitation or inhibition of the intact remaining superior semicircular canal (Ward et al. 2014). This unusual pattern of superior and horizontal canal stimulation might also explain the perception of rotation of the body rolling about an axis perpendicular to the Earth (Mian et al. 2015, 2016).

3.4   Are the Lorentz Forces Sufficient to Generate the Observed Nystagmus?

Although there are dark cells that produce potassium in endolymph throughout the vestibule and near the cristae of the semicircular canals (Kimura 1969), we hypothesize that the current in the endolymph is channeled in relative uniform current density above the sensory structures of the labyrinth. With this assumption, Roberts et al. proposed a geometric model of pressures across the semicircular canal cupulae (Roberts et al. 2011). The human utricular macula contains approximately 30,000 hair cells, compared to the approximately 7000 hair cells of each of the semicircular canal cristae (Lopez et al. 2005; Gopen et al. 2003). Given the resting discharge of vestibular hair cells, the higher number of hair cells in the utricular maculae should result in higher current density in the endolymph above the utricle than in the endolymph above the cristae. Utricular hair cells were therefore proposed as the primary source of the current generating the Lorentz force.

Assuming a range for the current entering individual hair cells depending on the probability of the ion channels being open or closed and the number of hair cells in the cristae and utricular macula, a range of transcupular pressure can be calculated for a Lorentz force when in a 7 T magnetic field. As a simple calculation, these pressures were within a range capable of both displacing the cupula and generating nystagmus when compared to values previously determined by Oman and Young (Oman and Young 1972). This model was supported by a more complex simulation using fluid dynamics by Antunes et al. (Antunes et al. 2012). It appears therefore that in strong magnetic fields like those of a 7 T MRI, the forces generated near the utricle and canal cristae are sufficient to displace cupulae and produce nystagmus.

3.5   The Current Model of Magnetic Vestibular Stimulation

Building on the evidence gathered above, magnetic vestibular stimulation is the result of interactions between strong static magnetic fields and the constant flow of ions through inner ear endolymph that sustains the resting discharge of vestibular

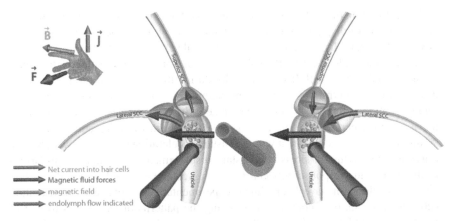

Fig. 7 Simple model for the Lorentz force mechanism of magnetic vestibular stimulation for a subject in an MRI lying supine. The net current is hypothesized to result from the utricular current. When a difference occurs between the magnetic field and current vectors in a strong static magnetic field, a Lorentz force causes a constant pressure on the cupulae of the lateral and superior semicircular canals. The direction of the Lorentz force is determined by the right-hand rule. Note the direction of the forces on the horizontal semicircular canals sums as a horizontal head rotation, but the forces on the superior semicircular canals are opposite, canceling the vertical component but summing the torsional component of nystagmus

afferents. The interaction creates a constant force called a Lorentz force that presses on the cupulae of the lateral and superior semicircular canals (Fig. 7). Due to the presumed high current density above the utricle, the Lorentz force is thought to be primarily generated in the endolymph fluid above the utricle, where the forces are sufficient to cause a constant displacement of the cupulae of both lateral and superior semicircular canals.

The direction of the nystagmus seen in humans depends on the relative orientations of the head (within which the current vector is contained) and the magnetic field vectors. The polarity of magnetic fields can vary by manufacturer. In the 7 T MRI used in our studies, the magnetic field is oriented from the head to toe. In this environment, most normal human subjects develop a horizontal and torsional nystagmus with slow phases to the left. Given the combination of excitation and inhibition of both superior semicircular canals, the vertical components of nystagmus cancel, and the torsional components sum. This results in the unusual pattern of horizontal-torsional nystagmus (Fig. 6) and likely the unusual perception of rotation as well. If the subject pitches their head such that the current and magnetic field vectors are aligned, there is no Lorentz force and no observed nystagmus. This "null" position is present in all subjects, but there is variability ranging up to 50 degrees across subjects (Roberts et al. 2011). The cause may be related to anatomic variations in the orientation of both the bony and membranous labyrinth. If a subject pitches their head forward beyond the null position, the direction of nystagmus reverses. This reversal corresponds to the "right-hand rule" of a Lorentz force. When the vectors cross, the direction of the resultant Lorentz force points in the opposite direction.

If the direction of the Lorentz force aligns with one of the cupulae near the utricle, this leads to displacement of the cupula so long as the head remains in that position. Constant displacement of a canal cupula creates an effect similar to constant acceleration. Recall that if a subject is rotated at a constant velocity, the semicircular canal cupula will displace initially during the step of acceleration but will drift back to its neutral position. The nystagmus in response to a constant velocity rotation will therefore eventually decay to zero. In magnetic vestibular stimulation, the nystagmus persists the entire time a subject is in the magnetic field (up to 90 minutes thus far), consistent with a constant acceleration stimulus in which there is a constant transcupular pressure and static displacement of the cupula. Similarities in the nystagmus response between magnetic vestibular stimulation and constant angular acceleration have been shown experimentally by constantly accelerating subjects in a rotational chair and comparing the patterns of nystagmus to that of magnetic vestibular stimulation in the same subjects (Jareonsettasin et al. 2016).

4 Applications of Magnetic Vestibular Stimulation

Magnetic vestibular stimulation is a novel method of activating the vestibular system that delivers a stimulus akin to a constant acceleration while using only a static magnetic field. The stimulus can induce dizziness as described above, but it is painless and can be delivered for long-duration experiments. Magnetic vestibular stimulation has implications for studies of sensorimotor adaptation, human safety, and functional MRI studies.

4.1 Set-Point Adaptation

In biology, a set point is a level at which a fluctuating physiological state tends to stabilize. A stable platform is important in order to perform accurate movements or to react quickly to sudden environmental changes (Zee et al. 2017). Adapting to new set points is important for organisms that must learn to live in a changing environment, yet the processes by which this learning occurs are unclear. Some environmental changes are transient, while others are more enduring. The brain must learn to handle both short- and long-lasting environmental changes. In the vestibular system, semicircular canals are paired and operate about a resting discharge rate as described above. This arrangement allows rapid corrective responses to changes in head position. For longer-lasting asymmetry, such as when one labyrinth is impaired, the brain must adapt to a new set point in order to eliminate spontaneous nystagmus.

Magnetic vestibular stimulation has many features that make it an ideal stimulus to explore "set-point" adaptation, the mechanism by which the levels of tonic activity within the vestibular system are balanced so as not to cause an unwanted spontaneous nystagmus when the head is not moving. Magnetic vestibular stimulation is easy to perform over long periods of time with minimal discomfort. There is a slow decay in the nystagmus velocity that corresponds to adaptation. An aftereffect upon leaving the static magnetic field appears immediately upon exiting the magnet and reflects learning of a new set point that happened while in the static magnetic field. Magnetic vestibular stimulation has already proved helpful in modeling how the vestibular system adapts to a constant asymmetry by revealing a cascade of adaptation parameters of increasing-duration time constants (Jareonsettasin et al. 2016). The amount of adaptation that occurs in the magnet is associated with the duration of aftereffect. These results imply that for longer-duration changes in set point, the brain adapts to the new set point with increasing certainty.

4.2 Human Safety

As long as people adhere to safety guidelines with respect to ferromagnetic objects near the strong magnetic fields of MRI machines and proper use of intravenous contrast agents, MRI scans are safe. Current clinical MRI scanners commonly use 1.5 T and 3 T magnetic field strengths; however, the FDA recently approved a 7 T MRI machine for human clinical studies. While dizziness and vertigo are uncommon near or in MRI machines with weaker magnetic fields, these symptoms are common near 7 T MRI machines (Schaap et al. 2014; Heilmaier et al. 2011). The promise of improved image resolution with decreased imaging time likely will make these scanners more common in clinical use.

The Lorentz force described above scales linearly with the strength of the magnetic field, and this scaling likely accounts for the increase in nystagmus velocity observed in subjects in stronger magnetic fields, as well as the increased sensation of vertigo. Although the Lorentz force mechanism is not expected to have long-term consequences for vestibular function, a more powerful force on the cupulae in strong MRI machines would be expected to induce vertigo in patients, research subjects, and other medical personnel working around MRI machines. Vertigo can trigger autonomic responses and induce nausea, which could lead to harm in some rare instances (Ward et al. 2015). Some researchers have suggested prophylactic use of anti-nausea medications like diphenhydramine (Thormann et al. 2013), which could be particularly helpful in patients at risk of complications from nausea and vomiting.

4.3 Functional MRI

Studies using functional MRI (fMRI) assess parts of the brain that are more metabolically active when a participant performs a task. More metabolically active areas of the brain correspond to increased neuronal activity. These studies using fMRI look for changes in the blood-oxygenation-level-dependent (BOLD) signal on the images in order to identify regions of the brain that might be activated as a result of a performed task. Magnetic vestibular stimulation may have implications for the design and interpretation of fMRI studies, particularly for a type of fMRI study called resting-state fMRI. In resting-state fMRI, participants are not asked to do any particular task. By averaging data over a longer-duration imaging study, synchronous patterns of BOLD signal can be observed among different regions of the brain, suggesting functional connectivity. Resting-state fMRI has led to consistent patterns of activation that are now identified as functional networks in the brain (Rosazza and Rosazza and Minati 2011).

Researchers using these techniques should be aware that magnetic vestibular stimulation can confound their data. When a sustained nystagmus develops, vision is degraded and fixation impaired. This disturbance is detected by the visual system, which then enlists both immediate and long-term adaptive mechanisms to nullify the unwanted eye drifts. Consequently, simply lying in the MRI bore with the eyes open or closed induces a behavioral challenge to which many parts of the central nervous system respond, including sensory systems that detect image motion on the retina (if the eyes are open), and motor systems that develop counterbalancing behavior to nullify unwanted biases and ensure steady fixation. In fact, evidence of this chain of neural activity having an impact on resting-state fMRI studies has been reported (Boegle et al. 2016, 2017). Magnetic vestibular stimulation may therefore affect data of many fMRI studies. On the other hand, the vestibular stimulation of the magnetic field may prove useful to researchers exploring functional connectivity of parts of the cortex that respond to vestibular stimulation.

4.4 Magnetoreception

The discovery of magnetic vestibular stimulation has prompted speculation that magnetosensation is a new human sense. Studies have looked into the role of a magnetic sense in humans, with mixed results (Kirschvink et al. 1992; Robin Baker 1985). The Earth has a magnetic dipole generated by convection currents occurring within its molten iron core. It is important to note that the strength of the magnetic fields in the studies discussed in this chapter far exceeds, by several orders of magnitude, the magnetic field of the Earth, which varies between 30 and 70 microtesla. Several species are known to detect static magnetic fields of strengths similar to the Earth and appear capable of using this information for navigation. The organ that senses these magnetic fields is unknown (Nordmann et al. 2017), although there is some evidence supporting the vestibular system as important in pigeons (Wu and Dickman 2011, 2012). If humans were capable of detecting magnetic fields, the mechanism, if present, may be entirely different, as the strengths of the fields required to generate nystagmus in a Lorentz force must be far greater than that of the Earth's magnetic field (Roberts et al. 2011; Antunes et al. 2012).

5 Conclusion

Evidence supports a Lorentz force mechanism for the nystagmus observed in human subjects and the dizziness they experience when in strong MRI machines. The observed nystagmus suggests magnetic vestibular stimulation delivers a constant force on semicircular canal cupulae of the superior and lateral semicircular canals akin to a constant acceleration and is generated in the endolymph above the utricle. Magnetic vestibular stimulation is a novel way of stimulating the vestibular labyrinth that may aid understanding of neurophysiology while also having implications for diagnostic imaging and animal navigation.

References

Antunes, A., Glover, P. M., Li, Y., Mian, O. S., & Day, B. L. (2012). Magnetic field effects on the vestibular system: Calculation of the pressure on the cupula due to ionic current-induced Lorentz force. *Physics in Medicine and Biology, 57*(14), 4477–4487.

Boegle, R., Stephan, T., Ertl, M., Glasauer, S., & Dieterich, M. (2016). Magnetic vestibular stimulation modulates default mode network fluctuations. *NeuroImage, 127*(February), 409–421.

Boegle, R., Ertl, M., Stephan, T., & Dieterich, M. (2017). Magnetic vestibular stimulation influences resting-state fluctuations and induces visual-vestibular biases. *Journal of Neurology, 264*(5), 999–1001.

de Vocht, F., van Drooge, H., Engels, H., & Kromhout, H. (2006). Exposure, health complaints and cognitive performance among employees of an MRI scanners manufacturing department. *Journal of Magnetic Resonance Imaging: JMRI, 23*(2), 197–204.

Glover, P. M., Cavin, I., Qian, W., Bowtell, R., & Gowland, P. A. (2007). Magnetic-field-induced Vertigo: A theoretical and experimental investigation. *Bioelectromagnetics, 28*(5), 349–361.

Gopen, Q., Lopez, I., Ishiyama, G., Baloh, R. W., & Ishiyama, A. (2003). Unbiased stereologic type I and type II hair cell counts in human utricular macula. *The Laryngoscope, 113*(7), 1132–1138.

Heilmaier, C., Theysohn, J. M., Maderwald, S., Kraff, O., Ladd, M. E., & Ladd, S. C. (2011). A large-scale study on subjective perception of discomfort during 7 and 1.5 T MRI examinations. *Bioelectromagnetics, 32*(8). Wiley Online Library), 610–619.

Houpt, T. A., Pittman, D. W., Barranco, J. M., Brooks, E. H., & Smith, J. C. (2003). Behavioral effects of high-strength static magnetic fields on rats. *The Journal of Neuroscience: The Official Journal of the Society for Neuroscience, 23*(4), 1498–1505.

Houpt, T. A., Cassell, J. A., Riccardi, C., DenBleyker, M. D., Hood, A., & Smith, J. C. (2007). Rats avoid high magnetic fields: Dependence on an intact vestibular system. *Physiology & Behavior, 92*(4), 741–747.

Jareonsettasin, P., Otero-Millan, J., Ward, B. K., Roberts, D. C., Schubert, M. C., & Zee, D. S. (2016). Multiple time courses of vestibular set-point adaptation revealed by sustained magnetic field stimulation of the labyrinth. *Current Biology: CB, 26*(10), 1359–1366.

John Leigh, R., & Zee, D. S. (2015). *The neurology of eye movements.* 5th edition, New York, NY: Oxford University Press.

Kangarlu, A., Burgess, R. E., Zhu, H., Nakayama, T., Hamlin, R. L., Abduljalil, A. M., & Robitaille, P. M. (1999). Cognitive, cardiac, and physiological safety studies in ultra high field magnetic resonance imaging. *Magnetic Resonance Imaging, 17*(10), 1407–1416.

Kimura, R. S. (1969). Distribution, structure, and function of dark cells in the vestibular labyrinth. *The Annals of Otology, Rhinology, and Laryngology, 78*(3), 542–561.

Kirschvink, J. L., Kobayashi-Kirschvink, A., & Woodford, B. J. (1992). Magnetite biomineraliza-
     tion in the human brain. *Proceedings of the National Academy of Sciences of the United States
     of America, 89*(16), 7683–7687.
Lopez, I., Ishiyama, G., Tang, Y., Tokita, J., Baloh, R. W., & Ishiyama, A. (2005). Regional esti-
     mates of hair cells and supporting cells in the human crista ampullaris. *Journal of Neuroscience
     Research, 82*(3), 421–431.
Lowenstein, B., Otto, Y., & Sand, A. (1936) The activity of the horizontal semi-circular canal of the
     dogfish Scyllium Canicula. http://jeb.biologists.org/content/jexbio/13/4/416.full.pdf.
Marcelli, V., Esposito, F., Aragri, A., Furia, T., Riccardi, P., Tosetti, M., Biagi, L., Marciano, E.,
     & Di Salle, F. (2009). Spatio-temporal pattern of vestibular information processing after brief
     caloric stimulation. *European Journal of Radiology, 70*(2), 312–316.
Mian, O. S., Li, Y., Antunes, A., Glover, P. M., & Day, B. L. (2013). On the Vertigo due to static
     magnetic fields. *PLoS One, 8*(10), e78748.
Mian, O. S., Glover, P. M., & Day, B. L. (2015). Reconciling magnetically induced vertigo and
     nystagmus. *Frontiers in Neurology, 6*(September), 201.
Mian, O. S., Li, Y., Antunes, A., Glover, P. M., & Day, B. L. (2016). Effect of head pitch and roll
     orientations on magnetically induced vertigo. *The Journal of Physiology, 594*(4), 1051–1067.
Nordmann, G. C., Hochstoeger, T., & Keays, D. A. (2017). Magnetoreception-a sense without a
     receptor. *PLoS Biology, 15*(10), e2003234.
Oman, C. M., & Young, L. R. (1972). The physiological range of pressure difference and
     cupula deflections in the human semicircular canal. Theoretical considerations. *Acta Oto-
     Laryngologica, 74*(5), 324–331.
Otero-Millan, J., Zee, D. S., Schubert, M. C., Roberts, D. C., & Ward, B. K. (2017). Three-
     dimensional eye movement recordings during magnetic vestibular stimulation. *Journal of
     Neurology, March.* https://doi.org/10.1007/s00415-017-8420-4.
Roberts, D. C., Marcelli, V., Gillen, J. S., Carey, J. P., Santina, C. C. D., & Zee, D. S. (2011).
     MRI magnetic field stimulates rotational sensors of the brain. *Current Biology: CB, 21*(19),
     1635–1640.
Robin Baker, R. (1985). Magnetoreception by man and other primates. In *Magnetite biominer-
     alization and magnetoreception in organisms* (Topics in geobiology) (pp. 537–561). Boston:
     Springer.
Rosazza, C., & Minati, L. (2011). Resting-state brain networks: Literature review and clinical
     applications. *Neurological Sciences: Official Journal of the Italian Neurological Society and of
     the Italian Society of Clinical Neurophysiology, 32*(5), 773–785.
Schaap, K., Christopher-de Vries, Y., Mason, C. K., de Vocht, F., Portengen, L., & Kromhout, H.
     (2014). Occupational exposure of healthcare and research staff to static magnetic stray fields
     from 1.5-7 tesla MRI scanners is associated with reporting of transient symptoms. *Occupational
     and Environmental Medicine, 71*(6), 423–429.
Schenck, J. F., Dumoulin, C. L., Redington, R. W., Kressel, H. Y., Elliott, R. T., & McDougall,
     I. L. (1992). Human exposure to 4.0-tesla magnetic fields in a whole-body scanner. *Medical
     Physics, 19*(4), 1089–1098.
Thormann, M., Amthauer, H., Adolf, D., Wollrab, A., Ricke, J., & Speck, O. (2013). Efficacy of
     diphenhydramine in the prevention of vertigo and nausea at 7T MRI. *European Journal of
     Radiology, 82*(5). Elsevier), 768–772.
Ward, B., & Zee, D. (2016). Dizziness and vertigo during MRI. *The New England Journal of
     Medicine, 375*(21), e44.
Ward, B. K., Roberts, D. C., Santina, C. C. D., Carey, J. P., & Zee, D. S. (2014). Magnetic ves-
     tibular stimulation in subjects with unilateral labyrinthine disorders. *Frontiers in Neurology,
     5*(March), 28.
Ward, B. K., Zee, D. S., Solomon, D., Gallia, G. L., & Reh, D. D. (2015). CSF leak: A complica-
     tion from vomiting after magnetic vestibular stimulation. *Neurology, 85*(6), 551–552.
Wu, L.-Q., & Dickman, J. D. (2011). Magnetoreception in an avian brain in part mediated by inner
     ear Lagena. *Current Biology: CB, 21*(5), 418–423.

Wu, L.-Q., & Dickman, J. D. (2012). Neural correlates of a magnetic sense. *Science, 336*(6084), 1054–1057.

Zee, D. S., Jareonsettasin, P., & Leigh, R. J. (2017). Ocular stability and set-point adaptation. *Philosophical Transactions of the Royal Society of London. Series B, Biological Sciences, 372*(1718). https://doi.org/10.1098/rstb.2016.0199.

Fixational Eye Movements in Visual, Cognitive, and Movement Disorders

Jorge Otero-Millan

Abstract The eyes never remain completely still. Even when attempting to maintain our gaze stable looking at a small target, fixational eye movements keep the eyes in constant motion. Given the simplicity of the task and the rich dataset that can be obtained, the interest in fixational eye movements has grown in multiple research and clinical fields. First, this chapter reviews the general classes of eye movements and how they may contribute to the main two components of fixational eye movements: drifts and microsaccades. While microsaccades are considered to be part of a continuum with saccades, drift results from a combination of all other smooth eye movements, such as vestibular ocular reflex, smooth pursuit, and vergence. Then, it discusses the methods used to analyze fixational eye movements and what are the typical parameters of interest and some considerations on how they should be measured. Finally, it describes how fixational eye movements are altered in patients affected by different degrees of vision loss, by movement disorders, or by cognitive disorders.

Keywords Microsaccades · Drift · Nystagmus · Amblyopia · Parkinsonism

1   What Are Fixational Eye Movements and Why Study Them?

Fixational eye movements are the eye movements that occur when subjects attempt to look at a small visual target without letting their gaze move away from it. This task may be referred to as maintained fixation, sustained fixation, visual fixation, or just fixation. During fixation, although with some variability, eye movements are finely controlled to optimize information acquisition (Rolfs 2009; Martinez-Conde et al. 2013; Rucci and Poletti 2015). Fixational eye movements compensate for the spatial and temporal limitations of the retina and the visual system in capturing and

J. Otero-Millan (✉)
Department of Neurology, Johns Hopkins University, Baltimore, MD, USA
e-mail: jotero@jhu.com

© Springer Nature Switzerland AG 2019
A. Shaikh, F. Ghasia (eds.), *Advances in Translational Neuroscience of Eye Movement Disorders*, Contemporary Clinical Neuroscience,
https://doi.org/10.1007/978-3-030-31407-1_6

processing the visual scene. First, the retina has the spatial limitation of not having a homogeneous resolution. Only the fovea has the highest concentration of photoreceptions, and even within the fovea, this concentration varies. Thus, in tasks that require seeing very fine spatial detail, fixational eye movements can align the highest-resolution area of the fovea with the object of interest (Poletti et al. 2013). Second, the visual system has the temporal limitation of neural responses adapting to constant stimuli. Objects may appear to disappear or fade from vision if retinal motion is removed or reduced (Coppola and Purves 1996; Martinez-Conde et al. 2006). This combined action of adaptation of photoreceptors and visual neurons and fixational eye movements can be interpreted in general as a whitening filter that eliminates redundancy in the visual stream and thus optimizes transmission of information (Rucci et al. 2018). Third, the retina also has the temporal limitation of decreased responses to stimuli that change too fast, which presents as perceived smear of rapidly moving objects or reduced visual acuity in the presence of retinal motion (Westheimer and McKee 1975). Thus, eye movements must frequently move the direction of gaze to keep it aligned with the object of interest and to prevent fading and eliminate redundancy in the visual stream while at the same time providing enough periods of relative stability when visual acuity can be optimal. The compromise that the oculomotor system has adopted can be observed not only during fixation but also more generally during visual exploration. The eyes move with an alternation of very quick movements and saccades, interleaved with short periods of stability and fixations.

Visual fixation, even within its simplicity as a task, can provide rich data about multiple brain systems. Performing visual fixation accurately involves vision, cognition, and motor control. Subjects need a healthy visual system, so they can locate and perceive the visual target properly. They need their cognitive abilities to understand the instructions and to continuously maintain fixation on the target. Finally, they need to be able to accurately control their eye muscles to precisely control the direction of gaze. Thus, by measuring fixational eye movements, we can simultaneous obtain information about those three major systems in the brain (Alexander et al. 2018) (Fig. 1).

1.1 Classes of Fixational Eye Movements

Fixational eye movements are usually classified based on descriptive criteria. By looking at the waveforms present in eye movement records, two types of eye movements become obviously distinct: microsaccades as quick steplike changes of eye position and drift as slow smooth eye position changes occurring in between microsaccades. If the recording is made with special devices with high enough resolution (Bolger et al. 1992; McCamy et al. 2013a), a third type of eye movement may be appreciated, tremor, which consists of small fast oscillations that occur simultaneously with drift.

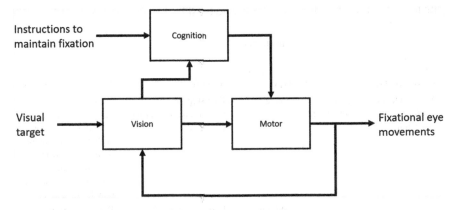

Fig. 1 Schematic representation of the main brain systems involved in fixational eye movements. The visual system is necessary to see the visual stimuli in which to fixate. Cognition is necessary to understand the fixation task and perform it correctly. The oculomotor system should be accurately calibrated to produce the correct eye movements to reach the target and provide a stable position once there

This descriptive classification, however, hides some of the complexity of the oculomotor system. Eye movements can be classified by their specific function, and there is a set of segregated neural structures that correspond with each function. This segregation is of great use in diagnosis and localization of neurological disease, because specific eye movements may be severely deteriorated while others remain intact. The oculomotor system has two main functions, stabilize gaze to reduce retinal image slip and shift gaze to realign the fovea with different parts of the image. These two functions are typically performed in a sequence of alternate states. During the gaze-shifting state, a saccade changes eye position quickly with a minimal role of sensory feedback during the movement. During the image stabilization state, eye movements are controlled by sensory-motor loops in which the appropriate eye movement is continuously monitored using multiple sensory signals to maintain a stable retinal image. The vestibular ocular reflex (VOR) and the optokinetic reflex (OKR) work in concert to generate eye movement that compensates for head movements using vestibular and full-field visual inputs, respectively. Smooth pursuit eye movements are produced to selectively stabilize in the fovea an object moving over a background. The neural integrator generates the necessary drive to the eye muscles to hold stable eye position in the absence of other eye movements. Vergence eye movements occur in combination with saccades, VOR, OKR, and smooth pursuit to generate the appropriate differential eye movements between the two eyes to look at the corresponding position in depth.

While microsaccades can be considered a particular case of saccades that occur during fixation, drift results from the combination of all the other stabilizing movements or even passive movements caused by the elastic properties of the eye. Drift will occur due to noise in each of those systems, miscalibration of their sensory-motor loops or their interrelationships, or abnormal biases or oscillations. Table 1 summarizes all the normal types of eye movements and how they relate to fixational eye movements.

Table 1 Classification of fixational eye movements. NPH: nucleus prepositus hypoglossi, INC: interstitial nucleus of Cajal

Descriptive fixational eye movement type	Description	Functional eye movement type	Main function	Main neural correlate
Microsaccades and/or saccadic intrusions	Quick movements	Saccade	Fast steplike movements to change eye position	Burst of activity in paramedian pontine reticular formation (PPRF) or (horizontal) rostral interstitial nucleus of the medial longitudinal fasciculus (riMLF) (vertical)
		Dynamic overshoot	Fast "bouncing" movement of the eye after a saccade	Burst of activity in PPRF or (horizontal) riMLF (vertical)
Drift	Slow movements	Vestibulo-ocular reflex	Smooth movements to compensate for head movements	Canal-driven activity in the vestibular nuclei
		Optokinetic reflex	Smooth movements to compensate for visual surround movements	Visually driven activity in the vestibular nuclei
		Smooth pursuit	Smooth movements to stabilize a target moving over a surround movement	Activity in the pontine nuclei and flocculus
		Gaze holding	Stable gaze holding at a new position after another eye movement	Sustained activity at NPH (horizontal) and INC (vertical)
		Gaze-evoked drift	An imperfect integrator will tend to drift toward a null position with an exponential decay	Exponential decay of activity in NPH (horizontal) and INC (vertical)
		Glissades	If the gain of the integrator is not well match to its inputs, the eye will drift quickly from the position that ended the previous eye movement to the new	None (passive movement of the eye due to its elastic properties)
		Vergence	Disconjugate movement of the eyes to move and realign the eyes in depth	Vergence neurons in the mesencephalic reticular formation
Tremor	Small fast oscillation			Unknown

2   Methods to Measure and Analyze Fixational Eye Movements

Given the small size of fixational eye movements, it becomes necessary to use eye-tracking devices that have enough temporal and spatial resolution in order to record them and analyze them properly. It is common to use at least 250 Hz and recording systems that can achieve noise levels of 0.1 deg. Currently, the most commonly used family of devices are the infrared video eye trackers which can record up to 2000 Hz. The scleral search coil method (Robinson 1963) is still considered the gold standard providing the best accuracy and precision. Some recent studies have also analyzed fixational eye movements using video recordings of retinal images obtained with an ophthalmoscope (Chung et al. 2015). Some parameters that average data over large windows of time and do not require microsaccade detection may be estimated also in recordings with lower frame rates of 60 Hz, for example, and lower resolution.

2.1   Microsaccade Measurements

Analysis of microsaccades starts with a method to identify them and separate them from other movements such as drift or noise that may be present in the recording. Since microsaccades are so small and noise is always present, there is always a compromise between sensitivity and specificity. Typically, a threshold needs to be set, for example, on eye movement velocity to decide if a movement is fast enough to be considered a microsaccade. A too low threshold will result in many false positives and a too high threshold on many false negative or misses. A popular method in the microsaccade field has been developed by Engbert and Kliegl (Engbert and Kliegl 2003) where the threshold is calculated for each recording based on an estimate of the noise level on the data. Still, the method requires a parameter that adjusts the sensitivity (λ), and different studies have used different values ranging from 4 to 8. Recently, a plethora of new methods have been published incorporating novel machine-learning techniques, but none of them have yet become a new standard in the field (Daye and Optican 2014; Otero-Millan et al. 2014; Andersson et al. 2017; Mihali et al. 2017; Bellet et al. 2018; Zemblys et al. 2018). These methods may eliminate the need for the setting of a sensitive parameter and adapt optimally to each patient. They may further provide metrics to evaluate the quality of the recording (Otero-Millan et al. 2014). Further testing is required to compare those methods with different datasets obtained with different recording systems while subjects perform different tasks.

Microsaccade Rate

Microsaccade rate measures the number of microsaccades that occur per unit of time. To measure, one must divide the total number of microsaccades by the total amount of time recorded. When measuring the total amount of time, there are some considerations that may result in slightly different results, for example, whether periods of blinks or missing data are counted toward the total time or not. This can have a large effect on reported rates for subjects with frequent blinks or large amounts of missing data. Many studies have relied in a binocular criterion (only accept microsaccades detected in both eyes) to reduce the number of false positives. Therefore, if two studies use exactly the same detection method with the same parameters but one has only monocular data, it is expected that this study will report higher rates.

Microsaccade Amplitude

Microsaccade amplitude (or magnitude) refers to the size of the movement and the amount of degrees traveled from beginning to end. Two main issues affect microsaccade amplitude measurements. The first is how each study defines amplitude. Some studies will consider the entire excursion of the eye during the movement, while other studies may consider only the distance between the first and the last point of the microsaccade. These two methods will produce different results if dynamic overshoots are common and large. The second is what is the largest amplitude considered. Many studies have defined any saccade smaller than one degree as a microsaccade, while other studies have used more stringent thresholds of half a degree.

Microsaccade Velocity

To measure how fast microsaccades are, one may measure the maximum or peak velocity during the movement. Velocities of saccades and microsaccades are known to follow a parametric relationship with their magnitude called "main sequence" (Bahill et al. 1975). Larger amplitudes tend to be associated with larger velocities, so it is important to dissociate differences in velocities from differences in magnitudes. To fairly compare velocities, one should correct for amplitude first. One possible approach is to fit a regression to the relationship between amplitudes and velocities and report the parameters of such fit. For small saccades including microsaccades, this relationship tends to be linear, so reporting the slope of a linear fit is a good option. For datasets including larger saccades, it is advisable to use some other fit that includes a saturation term and not only a slope, such as the exponential equation Velocity = $V_{max} [1-\exp(-Amplitude/C)]$, where V_{max} is the saturation velocity and C determines the shape of the relationship (Rosini et al. 2013).

Microsaccade Direction

Microsaccade direction is usually reported with a graphical representation of their distribution, for example, with a polar plot (Engbert 2006; Otero-Millan et al. 2011). When given summary statistics, one must be careful. Given the circularity of the distribution, it is not appropriate to calculate an overall average direction. For example, in a case where most microsaccades are perfectly horizontal with half to the right and half to the left, an average calculation would result with a vertical direction. Other parameters describing this distribution can still be measured. For example, the ratio between the average absolute value of the horizontal and vertical components will measure if the microsaccade tend to be more horizontal or more vertical.

Microsaccade Intervals

The distribution of intervals of time in between microsaccades can also provide useful information. The average microsaccade rate will correspond mathematically with the inverse of the average time interval between microsaccades. However, there may be more information in the shape of the distribution of the intervals. For example, Otero-Millan and colleagues showed that fitting an ex-Gaussian model (Otero-Millan et al. 2008) to the distributions of saccade intervals, it was possible to observe different parametric relationships with the different parameters of the fit. While different viewing tasks affected the exponential component, the amplitude of the subsequent saccade affected the Gaussian component. Amit and colleagues have recently used a similar analysis to analyze distributions of microsaccade intervals (Amit et al. 2019).

2.2 Drift Parameters

Estimating parameters related to microsaccades is relatively easier than to drift. Microsaccades are short events with high velocities, so during the movement, the signal-to-noise ratio is high. Drift is a much slower and continuous movement which will always have noise added into it and is more likely to be affected by artifacts. For that reason, with many recordings, system may not be possible to asses small differences in drift properties since those may be overwhelmed by the noise. It may still be possible to observe difference in between populations when the effect on drifts is large enough.

The parameters of interest for drift will also be velocity and amplitude, although it is possible also to study its spectral properties to identify oscillations. Because the eye may move back and forth during a given drift period, it is more common to measure variability of both velocity of position.

2.3 Saccadic Intrusions

Saccadic intrusions are saccades that intrude or interrupt stable fixation. Recent studies have suggested that saccadic intrusions for a continuum with microsaccades have been given a different name because they have been typically studied in different fields and different populations. Saccadic intrusions in patients are typically larger than in healthy controls and will tend to group together forming specific patterns. Square-wave jerks are the most common type of saccadic intrusions, and they occur when one fixational saccade moves the eye away followed after a short period of stability by another saccade that brings the eye back. Abadi and colleagues have carefully classified the different patterns and measured their frequency and characteristics in healthy controls (Abadi and Gowen 2004).

2.4 Fixation Stability

Many studies of the relationship between visual deficits and eye movements have measured fixation stability. Fixation stability measures the overall movement of the eyes while trying to fixate without measuring specific properties of drift and microsaccades. Different studies have measured similar parameters, but one of the most common ones is the bivariate contour ellipse area (BCEA) which measures the dispersion of eye positions in two dimensions. Similar results will be obtained if the standard deviation of eye position or eye velocity is used.

3 Abnormalities of Fixational Eye Movements

The next sections review the effects that impaired visual input, motor control, or cognitive function has on fixational eye movements. Here, we will not cover nystagmus, which is characterized by a repetitive eye movement pattern that must include abnormally fast drift. Instead, we will review cases where drift may be abnormally fast but without an obvious repetitive pattern. The presence of nystagmus will pose a challenge when analyzing fixational eye movements. In patients with visual problems, it is common to find latent nystagmus or congenital nystagmus with a wide variety of waveforms. Patients suffering from vestibular disorders or neurological disorders that affect the neural integrator or the vestibular system will typically present with jerk nystagmus (waveform formed by linear or exponential slow phases alternating with quick phases or saccades in the opposite directions). Measurements of drift will be inflated by the nystagmus, and the presence or absence of microsaccades will be confounded with the frequent quick phases of the nystagmus. Thus, when studying populations that may present with nystagmus, it may be advisable to separate those patients into different groups for data analysis (Ghasia et al. 2018).

3.1   Visual Impairments

Eye movements and vision have evolved together. The spatiotemporal characteristics of our visual processing stream depend not only on the properties of the retina but also on the pattern of eye movements that continuously move the image projected onto it (Rucci et al. 2007; Rucci and Poletti 2015). Indeed, there are correlations between the pattern of eye movements that different species make and the anatomy of their visual systems (Samonds et al. 2018). It is not always trivial, however, to establish the causal direction of this correlation. For example, do we continuously move our eyes because the visual system adapts to avoid the fading of images (Martinez-Conde et al. 2006, 2013), or does the visual system adapt because the eyes continuously move to optimally encode visual information? Because of this evolutionary correlation between characteristics of the visual system and characteristics of eye movements, it is expected that problems in the visual system that degrade quality of vision may alter eye movements or, by the same token, that problems in eye movement control may degrade the quality of vision.

In some cases, it is clear that a vision problem is not caused by an eye movement problem such as in macular degeneration. So, we can study how losing vision affects eye movements. In other cases, however, when eye movement abnormalities correlate with vision deficits, it may be more complicated to establish which one is the cause and which one is the effect, such as in amblyopia. This is further complicated by the fact that changes in eye movement output could be directly correlated with the current quality of the visual input, but there could also be other changes that are adaptive and occur over time, thus not relating directly to the current quality of vision but to the history of quality of vision.

Blindness

Visual input is important to maintain accurate and appropriate oculomotor behavior. Blindness, from birth or acquired, causes unstable fixation and possibly nystagmus. Patients with congenital blindness have an impaired vestibulo-ocular reflex and are unable to initiate saccades (Leigh and Zee 1980). Schneider and colleagues (Schneider et al. 2013) studied the effect of monocular and binocular visual loss on the stability of gaze. They showed increased fixation instability with binocular viewing in patients compared to controls, especially in patients with binocular vision loss. They hypothesize that the main effect of vision loss on gaze stability is through the neural integrator based on their results showing more conjugate eye movements during saccades and other eye movements that bypass the neural integrator than during gaze holding. Together, these findings suggest that normal development of some eye movements such as saccades requires vision early on in life. Gaze holding, on the other hand, requires constant calibration by visual input, and acquired visual loss will increase fixation instability.

Amblyopia

Amblyopia refers to a reduction of visual acuity that cannot be explained by refractive errors or by any detectable eye disease (Flom and Neumaier 1966). Amblyopia tends to affect only one eye which is referred to as the amblyopic eye, while the eye with normal vision is called the fellow eye.

When amblyope subjects fixate, looking only with their amblyope eye, they show higher fixation instability measured with BCEA than when viewing with the fellow eye, binocularly, and compared with control (Subramanian et al. 2013; Shaikh et al. 2016). The instability is correlated with the severeness of amblyopia (Shaikh et al. 2016), and it is even larger when amblyopia is related to strabismus (Ghasia et al. 2018).

Increased fixation instability could be caused by increased production of microsaccades, increased drift, or a combination of both or even by microsaccades and drifts that have normal properties but that are coupled in a way that increases the overall instability. Increased drift in amblyopia was first shown by Ciuffreda (Ciuffreda et al. 1980), but recent studies have shown that microsaccades are also altered, having larger amplitude and lower frequency (Shi et al. 2012; Shaikh et al. 2016).

Ghasia and colleagues extended some of the findings related to fixational eye movements to other tasks, such as visual search, and found that the distribution of saccade amplitudes is shifted toward larger amplitudes (Chen et al. 2018). This is consistent with the idea that microsaccades and saccades are part of a continuum (Martinez-Conde et al. 2013; Otero-Millan et al. 2013a) and that the distribution of saccades is adapted to the properties of the visual system (Samonds et al. 2018).

Myopia, Defocus, and Absence of Visual Target

Myopia has also been shown to have an effect on microsaccades. Ghasia and colleagues compared fixational eye movements in subjects with and without correction of their refractive error. They found an increase in microsaccade amplitude with correction of myopia, but microsaccade rate remained unchanged (Ghasia and Shaikh 2015). On the other hand, another recent study found that a simulated refractive error did not cause any change in microsaccade properties (Raveendran et al. 2019). These two findings appear to contradict each other, and it is unclear at the moment why these differences occurred. Moreover, if the fixation spot is completely removed, which could be interpreted as extreme defocus, and subjects are asked to fixate on the center of the screen without a target, microsaccade amplitudes are increased and rates reduced (McCamy et al. 2013b).

3.2 Movement Control Impairments

A distributed network across the brain is involved in the accurate control of eye movements. In a simplified model, the frontal cortex, basal ganglia, and superior colliculus (SC) control the initiation of saccades and the inhibition of reflexive saccades, while the cerebellum ensures accuracy of saccades, smooth pursuit, and gaze holding. Impairment in any of these areas will cause a particular pattern of eye movement abnormalities. This entire circuit is also involved in the control of fixation and the generation of microsaccades. Often, abnormalities observed in other eye movements, i.e., slow saccades, will also correspond with the abnormalities present in microsaccades. However, since microsaccades are produced while trying not to move the eyes, there are unique aspects to them that may not be possible to asses by studying other types of eye movements. This might be the case for microsaccade rates and amplitudes.

Among movement disorders, microsaccades have been most studied in parkinsonian syndromes because of the frequent occurrence of saccadic intrusions in these patients that on occasions are visible with the naked eye. In progressive supranuclear palsy (PSP), square-wave jerks are very frequent and large, and they share with saccades their lack of a vertical component and slow velocity (Otero-Millan et al. 2011). In patients with Parkinson's disease (PD), saccade and microsaccade velocities are normal (Bhidayasiri et al. 2001), but the frequency of microsaccades is higher than that in controls (Pinnock et al. 2010; Otero-Millan et al. 2013b). Comparing microsaccade production in PSP and PD, Otero-Millan and colleagues found that while frequency was higher than normal in both groups, the amplitude of microsaccades was larger than normal only in the PSP group (Otero-Millan et al. 2013b).

Recent modeling (Otero-Millan et al. 2018) has shown how the altered properties of microsaccadic amplitude, rate, and velocity in PD and PSP can be simulated by simply modifying two parameters in the model. In this model, microsaccades are generated when fluctuations of activity in the superior colliculus (SC) change the balance of inhibition between omnipause neurons (OPNs) and burst neurons (BNs) in the brain stem. In between microsaccades, OPNs show sustained firing, inhibiting BNs and making them silent. These two populations of neurons inhibit each other, so when the BNs start firing, the circuit behaves as a positive feedback loop with OPNs inhibited by the BNs, thus reducing the inhibition of BNs which further inhibits the OPNs. Lowering the gain of the BNs has the primary effect of reducing the velocities of microsaccades; however, it may also increase the amplitude of microsaccades, since it would take a larger fluctuation of SC activity to break the OPN-BN balance. Increasing the amount of noise that drives the SC fluctuations will have the main effect of increasing the rate of microsaccades. By tweaking these two parameters, it is possible to simulate simultaneously the deficits in three microsaccade properties (velocity, magnitude, and rate) present in PD and PSP populations (Otero-Millan et al. 2013b) (Fig. 2).

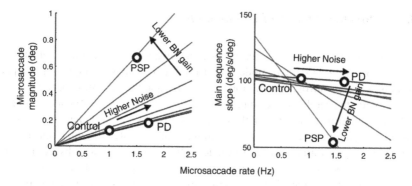

Fig. 2 Modeling of microsaccade production in healthy controls and PSP and PD patients. Simulations of the model varying two parameters (noise level and BN gain) allow to cover the spectrum of microsaccadic rates, velocities, and magnitudes observed on those populations

3.3 Cognitive Impairments

Links between cognition and microsaccade production have been shown in healthy controls. Mental workload induced by asking subjects to perform complicated calculations while fixating on a small target resulted in an increased microsaccade amplitude with reduced frequency (Siegenthaler et al. 2014). Attention also affects microsaccades by biasing their direction sometimes toward and sometimes away from the focus of attention (Hafed and Clark 2002; Engbert and Kliegl 2003; Meyberg et al. 2017). Finally, transient stimulation that can be visual or auditory can temporarily suppress microsaccades (Engbert and Kliegl 2003). It is therefore expected that patients with cognitive impairments could show abnormal microsaccade production.

Fried et al. studied fixational eye movements in attention-deficit hyperactivity disorder (ADHD) patients during a continuous performance test. During the task, subjects can predict when the stimulus is going to appear due to its repetitive timing. Healthy controls tend to suppress their blinks and microsaccades, while ADHD patients don't (Fried et al. 2014). This result, however, may not be generalizable to other tasks like simple visual fixation. Panagiotidi and colleagues studied a non-clinical population of subjects with ADHD traits and found that the score of ADHD traits and microsaccade rate were correlated: subjects with higher scores produced more frequent microsaccades (Panagiotidi et al. 2017).

Alterations in microsaccade in patients with Alzheimer's disease (AD) have also been investigated. While Bylsma and colleagues found an increase in microsaccade rate that correlated with Mini-Mental State Examination (MMSE) score, Kapoula and colleagues did not find a change in microsaccade rate but found that the distribution of microsaccade directions included more oblique microsaccades than healthy controls (Bylsma et al. 1995; Kapoula et al. 2013). In healthy controls, most microsaccades tend to be almost purely horizontal or sometimes vertical, but oblique microsaccades are rare (Otero-Millan et al. 2011).

In autism, the study by Shirama and colleagues found that fixational eye movements were altered in a group of adults with autism spectrum disorder (ASD). They found that instability measured with BCEA was larger in the ASD group than in healthy controls when asked to fixate on the center of the monitor without a fixation target. Under the same conditions, they also found larger and more frequent microsaccades than in controls. When the target was present, their eye movements did not differ between groups (Shirama et al. 2016).

4 Conclusion

The study of fixational eye movements presents opportunities for the development of new diagnostic test and biomarkers that can be used to track progression of patients' efficacy of treatments. There are still some pending questions in the field tough. Future studies need to establish the link between vision deficits and fixation stability and what is the contribution of adaptive mechanism over prolonged periods of low vision. It is also entirely unknown what particular oculomotor system is responsible for the increases in fixation stability. Studies on blind patients appear to subject that the major contribution comes from the neural integrator, but this needs to be confirmed in other deficits such as amblyopia.

References

Abadi, R. V., & Gowen, E. (2004). Characteristics of saccadic intrusions. *Vision Research, 44,* 2675–2690.

Alexander, R. G., Macknik, S. L., & Martinez-Conde, S. (2018). Microsaccade characteristics in neurological and ophthalmic disease. *Frontiers in Neurology, 9.* Available at: https://www.frontiersin.org/articles/10.3389/fneur.2018.00144/full. Accessed 5 Apr 2018.

Amit, R., Abeles, D., & Yuval-Greenberg, S. (2019). Transient and sustained effects of stimulus properties on the generation of microsaccades. *Journal of Vision, 19,* 6–6.

Andersson, R., Larsson, L., Holmqvist, K., Stridh, M., & Nyström, M. (2017). One algorithm to rule them all? An evaluation and discussion of ten eye movement event-detection algorithms. *Behav Res, 49,* 616–637.

Bahill, A. T., Clark, M. R., & Stark, L. (1975). The main sequence, a tool for studying human eye movements. *Mathematical Biosciences, 24,* 191–204.

Bellet, M. E., Bellet, J., Nienborg, H., Hafed, Z. M., Berens, P. (2018). *Human-level saccade detection performance using deep neural networks.* Available at: http://biorxiv.org/lookup/doi/10.1101/359018. Accessed 5 Oct 2018.

Bhidayasiri, R., Riley, D. E., Somers, J. T., Lerner, A. J., Büttner-Ennever, J. A., & Leigh, R. J. (2001). Pathophysiology of slow vertical saccades in progressive supranuclear palsy. *Neurology, 57,* 2070–2077.

Bolger, C., Sheahan, N., Coakley, D., & Malone, J. (1992). High frequency eye tremor: Reliability of measurement. *Clinical Physics and Physiological Measurement, 13,* 151–159.

Bylsma, F. W., Rasmusson, D. X., Rebok, G. W., Keyl, P. M., Tune, L., & Brandt, J. (1995). Changes in visual fixation and saccadic eye movements in Alzheimer's disease. *International Journal of Psychophysiology, 19*, 33–40.

Chen, D., Otero-Millan, J., Kumar, P., Shaikh, A. G., & Ghasia, F. F. (2018). Visual search in amblyopia: Abnormal fixational eye movements and suboptimal sampling strategies. *Investigative Ophthalmology & Visual Science, 59*, 4506–4517.

Chung, S. T. L., Kumar, G., Li, R. W., & Levi, D. M. (2015). Characteristics of fixational eye movements in amblyopia: Limitations on fixation stability and acuity? *Vision Research, 114*, 87–99.

Ciuffreda, K. J., Kenyon, R. V., & Stark, L. (1980). Increased drift in amblyopic eyes. *British Journal of Ophthalmology, 64*, 7–14.

Coppola, D., & Purves, D. (1996). The extraordinarily rapid disappearance of entopic images. *Proceedings of the National Academy of Sciences, 93*, 8001–8004.

Daye, P. M., & Optican, L. M. (2014). Saccade detection using a particle filter. *Journal of Neuroscience Methods, 235*, 157–168.

Engbert, R. (2006). Microsaccades: a microcosm for research on oculomotor control, attention, and visual perception. In *Progress in brain research* (pp. 177–192). Elsevier. Available at: http://linkinghub.elsevier.com/retrieve/pii/S0079612306540099. Accessed 6 July 2015.

Engbert, R., & Kliegl, R. (2003). Microsaccades uncover the orientation of covert attention. *Vision Research, 43*, 1035–1045.

Flom, M. C., & Neumaier, R. W. (1966). Prevalence of amblyopia. *Public Health Reports, 81*, 329–341.

Fried, M., Tsitsiashvili, E., Bonneh, Y. S., Sterkin, A., Wygnanski-Jaffe, T., Epstein, T., & Polat, U. (2014). ADHD subjects fail to suppress eye blinks and microsaccades while anticipating visual stimuli but recover with medication. *Vision Research, 101*, 62–72.

Ghasia, F. F., & Shaikh, A. G. (2015). Uncorrected myopic refractive error increases microsaccade amplitude. *Investigative Ophthalmology & Visual Science* . Available at: http://www.iovs.org/cgi/doi/10.1167/iovs.14-15882. Accessed 2 June 2015, *56*, 2531.

Ghasia, F. F., Otero-Millan, J., & Shaikh, A. G. (2018). Abnormal fixational eye movements in strabismus. *British Journal of Ophthalmology, 102*, 253–259.

Hafed, Z. M., & Clark, J. J. (2002). Microsaccades as an overt measure of covert attention shifts. *Vision Research, 42*, 2533–2545.

Kapoula, Z., Yang, Q., Otero-Millan, J., Xiao, S., Macknik, S. L., Lang, A., Verny, M., & Martinez-Conde, S. (2013). Distinctive features of microsaccades in Alzheimer's disease and in mild cognitive impairment. *Age (Dordr), 36*(2), 535–543.

Leigh, R. J., & Zee, D. S. (1980). Eye movements of the blind. *Investigative Ophthalmology & Visual Science, 19*, 328–331.

Martinez-Conde, S., Macknik, S. L., Troncoso, X. G., & Dyar, T. A. (2006). Microsaccades counteract visual fading during fixation. *Neuron, 49*, 297–305.

Martinez-Conde, S., Otero-Millan, J., & Macknik, S. L. (2013). The impact of microsaccades on vision: Towards a unified theory of saccadic function. *Nature Reviews Neuroscience, 14*, 83–96.

McCamy, M. B., Collins, N., Otero-Millan, J., Al-Kalbani, M., Macknik, S. L., Coakley, D., Troncoso, X. G., Boyle, G., Narayanan, V., Wolf, T. R., & Martinez-Conde, S. (2013a). Simultaneous recordings of ocular microtremor and microsaccades with a piezoelectric sensor and a video-oculography system. *PeerJ, 1*, e14.

McCamy, M. B., Najafian Jazi, A., Otero-Millan, J., Macknik, S. L., & Martinez-Conde, S. (2013b). The effects of fixation target size and luminance on microsaccades and square-wave jerks. *PeerJ, 1*, e9.

Meyberg, S., Sinn, P., Engbert, R., & Sommer, W. (2017). Revising the link between microsaccades and the spatial cueing of voluntary attention. *Vision Research, 133*, 47–60.

Mihali, A., Opheusden, B. V., & Ma, W. J. (2017). Bayesian microsaccade detection. *Journal of Vision, 17*, 13–13.

Otero-Millan, J., Troncoso, X. G., Macknik, S. L., Serrano-Pedraza, I., & Martinez-Conde, S. (2008). Saccades and microsaccades during visual fixation, exploration and search: Foundations for a common saccadic generator. *Journal of Vision, 8*(14), 21.

Otero-Millan, J., Serra, A., Leigh, R. J., Troncoso, X. G., Macknik, S. L., & Martinez-Conde, S. (2011). Distinctive features of saccadic intrusions and microsaccades in progressive supranuclear palsy. *Journal of Neuroscience, 31*, 4379–4387.

Otero-Millan, J., Macknik, S. L., Langston, R. E., & Martinez-Conde, S. (2013a). An oculomotor continuum from exploration to fixation. *PNAS, 110*, 6175–6180.

Otero-Millan, J., Schneider, R., Leigh, R. J., Macknik, S. L., & Martinez-Conde, S. (2013b). Saccades during attempted fixation in parkinsonian disorders and recessive Ataxia: From microsaccades to square-wave jerks Geng JJ, ed. *PLoS One, 8*, e58535.

Otero-Millan, J., Castro, J. L. A., Macknik, S. L., & Martinez-Conde, S. (2014). Unsupervised clustering method to detect microsaccades. *Journal of Vision, 14*, 18–18.

Otero-Millan, J., Optican, L. M., Macknik, S. L., & Martinez-Conde, S. (2018). Modeling the triggering of saccades, microsaccades, and saccadic intrusions. *Frontiers in Neurology, 9*. Available at: https://www.frontiersin.org/articles/10.3389/fneur.2018.00346/full. Accessed 18 July 2018.

Panagiotidi, M., Overton, P. G., & Stafford, T. (2017). Increased microsaccade rate in individuals with ADHD traits. *Journal of Eye Movement Research, 10*. Available at: https://doi.org/10.16910/jemr.10.1.6. Accessed 4 Oct 2017.

Pinnock, R. A., McGivern, R. C., Forbes, R., & Gibson, J. M. (2010). An exploration of ocular fixation in Parkinson's disease, multiple system atrophy and progressive supranuclear palsy. *Journal of Neurology, 257*, 533–539.

Poletti, M., Listorti, C., & Rucci, M. (2013). Microscopic eye movements compensate for nonhomogeneous vision within the fovea. *Current Biology, 23*, 1691–1695.

Raveendran, R. N., Bobier, W., & Thompson, B. (2019). Reduced amblyopic eye fixation stability cannot be simulated using retinal-defocus-induced reductions in visual acuity. *Vision Research, 154*, 14–20.

Robinson, D. A. (1963). A method of measuring eye movements using a scleral coil in a magnetic field. *IEEE Transactions on Biomedical Engineering, 10*, 137–145.

Rolfs, M. (2009). Microsaccades: Small steps on a long way. *Vision Research, 49*, 2415–2441.

Rosini, F., Federighi, P., Pretegiani, E., Piu, P., Leigh, R. J., Serra, A., Federico, A., & Rufa, A. (2013). Ocular-motor profile and effects of memantine in a familial form of adult cerebellar ataxia with slow saccades and square wave saccadic intrusions Martinez-Conde S, ed. *PLoS One, 8*, e69522.

Rucci, M., & Poletti, M. (2015). Control and functions of fixational eye movements. *Annual Review of Vision Science, 1*, 499–518.

Rucci, M., Iovin, R., Poletti, M., & Santini, F. (2007). Miniature eye movements enhance fine spatial detail. *Nature, 447*, 852–855.

Rucci, M., Ahissar, E., & Burr, D. (2018). Temporal coding of visual space. *Trends in Cognitive Sciences, 22*, 883–895.

Samonds, J. M., Geisler, W. S., & Priebe, N. J. (2018). Natural image and receptive field statistics predict saccade sizes. *Nature Neuroscience, 21*, 1591–1599.

Schneider, R. M., Thurtell, M. J., Eisele, S., Lincoff, N., Bala, E., & Leigh, R. J. (2013). Neurological basis for eye movements of the blind. *PLoS One, 8*, e56556.

Shaikh, A. G., Otero-Millan, J., Kumar, P., & Ghasia, F. F. (2016). Abnormal fixational eye movements in amblyopia. *PLoS One, 11*, e0149953.

Shi, X. F., Xu, L., Li, Y., Wang, T., Zhao, K., & Sabel, B. A. (2012). Fixational saccadic eye movements are altered in anisometropic amblyopia. *Restorative Neurology and Neuroscience, 30*, 445–462.

Shirama, A., Kanai, C., Kato, N., & Kashino, M. (2016). Ocular fixation abnormality in patients with autism spectrum disorder. *Journal of Autism and Developmental Disorders, 46*, 1613–1622.

Siegenthaler, E., Costela, F. M., McCamy, M. B., Di Stasi, L. L., Otero-Millan, J., Sonderegger, A., Groner, R., Macknik, S., & Martinez-Conde, S. (2014). Task difficulty in mental arithmetic affects microsaccadic rates and magnitudes. *European Journal of Neuroscience, 39*, 287–294.

Subramanian, V., Jost, R. M., & Birch, E. E. (2013). A quantitative study of fixation stability in amblyopia. *Investigative Ophthalmology & Visual Science, 54*, 1998–2003.

Westheimer, G., & McKee, S. P. (1975). Visual acuity in the presence of retinal-image motion. *JOSA, 65*, 847–850.

Zemblys, R., Niehorster, D. C., Komogortsev, O., & Holmqvist, K. (2018). Using machine learning to detect events in eye-tracking data. *Behavior Research Methods, 50*, 160–181.

An Update on Mathematical Models of the Saccadic Mechanism

Stefano Ramat

Abstract Saccades are the rapid eye movements that we continuously make to shift gaze from one object of interest to another. These voluntary eye movements not only are exemplary to describe various forms of ocular motor and cognitive function but also have a fundamental role as a prototype for understanding general principles in motor neurosciences. Their production requires most of the information processing steps that are needed in all motor control tasks, yet their understanding is far more detailed than that related to other fields in motor control. Indeed, a great deal has been learned about the anatomy and neurophysiology of saccades thanks to a fruitful collaboration between clinical studies, basic science research, and mathematical modeling. In this non-exhaustive review, I will discuss contemporary computational concepts that describe the current understanding of the physiology of saccades in health and disease.

Keywords Computer model · Mathematical simulation · Saccadic system · Ocular motor system

1 Introduction

Saccades are the fastest eye movements produced by the ocular motor system, and they serve to redirect gaze to an object of interest in the visual scene by bringing its image on the fovea, the small area of the retina where visual acuity is highest, for fixation. While looking at a visual scene, we continuously alternate fixation periods, during which our central nervous system (CNS) acquires visual information, with saccades redirecting the fovea to a new target. During natural behavior, we typically make two to three saccades per second (Niemeier et al. 2003). The saccadic mechanism, i.e., the neural circuitry responsible for generating saccades, is active also during the fast phases of nystagmus and REM sleep, which are fast eye movements

S. Ramat (✉)
Department of Electrical, Computer and Biomedical Engineering,
University of Pavia, Pavia, Italy
e-mail: stefano.ramat@unipv.it

© Springer Nature Switzerland AG 2019
A. Shaikh, F. Ghasia (eds.), *Advances in Translational Neuroscience of Eye Movement Disorders*, Contemporary Clinical Neuroscience,
https://doi.org/10.1007/978-3-030-31407-1_7

present also in non-foveate species (Collewijn 1977), while gaze-redirecting saccades are exclusive to foveate ones.

Saccades can be voluntarily triggered after the decision to look at a specific visual, imagined, or remembered target. Alternatively, the generation of saccades can be quasi reflex, evoked by the sudden appearance of a target in the retinal periphery, or reflex, bringing gaze toward an internally computed reference signal, and thus without a predefined target image, as with the fast phases of both vestibular and optokinetic nystagmus.

2   Saccade Kinematics

In spite of how they are generated, saccades are stereotyped, so that their amplitude, duration, and peak velocity follow quite tight relationships, with larger saccades reaching higher peak velocities and having longer durations, which were identified in the 1970s and are commonly called *the saccadic main sequence* (Boghen et al. 1974; Bahill et al. 1975b), exemplified in Fig. 1.

Peak eye velocity increases linearly with saccade size, from microsaccades, i.e. small saccades less than 1° in amplitude intruding fixation (see Martinez-Conde et al. (2013) for a review), up to about 20° saccades; then, the relationship shows a soft saturation with peak velocity reaching, in humans, about 500 deg/s. The main sequence for the amplitude (A)–peak velocity (PV) relationship is typically represented by the equation $PV = V_{Max} \bullet (1 - exp^{-A/C})$ where V_{max} represents the saturation velocity and C is a constant determining how quickly the relationship reaches saturation, having typical values around 500 deg/s and 10 (Rottach et al. 1997), respectively.

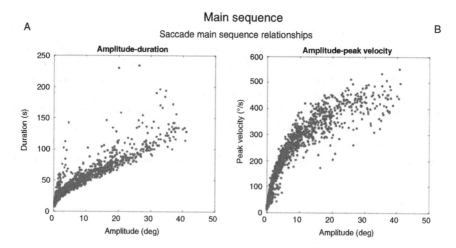

Fig. 1 Saccadic main sequence based on a set of saccades performed by 10 normal subjects and recorded with scleral search coil

Saccade durations (D) range from less than 20 ms for the smallest amplitudes to about 100 ms for the largest ones, and their relationship is linear having the form $D = aA + b$ (Baloh et al. 1975; Inchingolo et al. 1987).

3 Saccades in the Central Nervous System

The production of a typical saccade, i.e., visually driven, less than 15° amplitude and less than 60 ms duration (Bahill et al. 1975a), requires the processing of visual information and the selection of a target and its sensory–motor transformation from retinal coordinates into the appropriate command driving the eye. Numerous areas in the central nervous system are therefore involved in saccade generation, from the brainstem holding motor and premotor circuitry to the cerebellum and the cortex.

The understanding of the processing steps leading to the generation of a saccade has been the object of many modeling efforts, making the saccadic mechanism the most modeled system in neuroscience (for a review, see also Girard and Berthoz 2005; Ramat et al. 2007).

This chapter will attempt to provide a non-exhaustive review of the most significant mathematical models of such system and of the state of the art of our understanding of the saccade generation processes. We will focus our attention on horizontal saccades, which have been studied most, yet will mention some of the relevant differences in the neuronal populations responsible for vertical and torsional saccades.

Following a bottom-up approach, a model of the saccadic generator mechanism needs to take into account the end effector, i.e., the eyeball and the ocular motor plant, which is controlled by the motor neurons of the extraocular muscles, in turn receiving neural commands from the final ocular motor pathway.

4 Early Models of the Saccadic Mechanism

The above observations on saccade duration being generally shorter than 100 ms, i.e., before any visually acquired information may be available to modify them, led to the belief that saccades are ballistic preprogrammed movements that cannot be corrected once triggered and that a refractory period of about 150 ms followed the triggering of a saccade (Westheimer 1954). The first attempt at modeling the saccade generator (Young and Stark 1963), when little was known about the anatomy and the involved physiology of the brainstem, was based on these ideas and considered a retinal error sampling mechanism operating at about 5 Hz followed by a 0.3° dead zone. The error would then be integrated into a step command and fed to Westheimer's second-order plant model (Westheimer 1954), which would produce a saccade of a preprogrammed amplitude, with peak velocity linearly increasing with its size and a fixed duration of 37 ms (Enderle 2002).

It was only with the work of Robinson in 1964 that a more realistic approximation of the plant was introduced, with a significant impact on the understanding of the processing needed to drive it during saccades (Robinson 1964). In fact, Robinson's findings that the dynamics of the eyeball were dominated by the viscosity of extraocular muscles with a time constant of about 200 ms showed that a step of innervation reaching ocular motor neurons, as previously hypothesized for generating a saccade, would cause a relatively slow eye movement reaching the intended destination in about 600 ms (three time constants), with quite different dynamics from those of a saccade. This led to the understanding that the command needed to drive the eyes during a saccade had to produce an intense pulse of force, proportional to eye velocity, to quickly move the eye to a new orbital position, which had to be followed by a force step, to hold it there against the restoring elastic forces. At the same time, knowledge on the electrical activity in muscles and in neurons was growing, and Schaefer's recording in the abducens nucleus of the rabbit (Schaeffer 1965) elucidated how the pulse corresponded to a burst of activity of the same duration of the saccade, modulating the number of recruited muscle motor units and saturating, so that larger saccades needed longer pulse durations (Robinson 1968).

The pulse generator (PG) was then hypothesized in the paramedian pontine reticular formation (PPRF) (Cohen 1971), bursts in the activity of brainstem neurons were recorded in primates (Luschei and Fuchs 1972), and quick phases of nystagmus were shown to be largely comparable to saccades (Ron et al. 1972). In 1973, Skavenski and Robinson (Skavenski and Robinson 1973) working on the frequency response of the vestibulo-ocular reflex (VOR), in the light of the new model of the plant, understood that the experimental results on the phase of the reflex required that the neural command driving it had to be a combination of desired eye velocity and position. They showed that the abducens motor neuron discharge rate was independent of the nature of the eye movement and hypothesized that the eye position component of the command could be obtained by mathematical integration of the desired eye velocity, performed by a neural integrator (NI) shared by all types of eye movements along a "final common path" (FCP). The omnipause neurons (OPN) were recognized as a group of neurons constantly firing during fixation that would pause before saccades in any direction (Luschei and Fuchs 1972).

A first major breakthrough for the modeling of the saccadic mechanism came in 1976, with the finding of two patients with spinocerebellar degeneration producing slow saccades that could modify them in mid-flight (Zee et al. 1976). These results were in agreement with the hypothesis of a local feedback loop for the generation of saccades proposed by Robinson (Robinson 1975), proving that the saccadic mechanism is in fact a closed loop control system driven by an error signal between an intended position of the eye in the orbit and its current position estimate, without a refractory period, therefore dropping the idea of saccades as ballistic preprogrammed movements (Zee et al. 1976) (Fig. 2). The position error was hypothesized as being based on an efference copy of the positional component of the ocular motor command, i.e., taken from the output of the neural integrator in the FCP. The model was further refined in 1979 (Zee and Robinson 1979) with the proposal of a nonlinearity representing the behavior of the pontine bursting neurons as a function of the

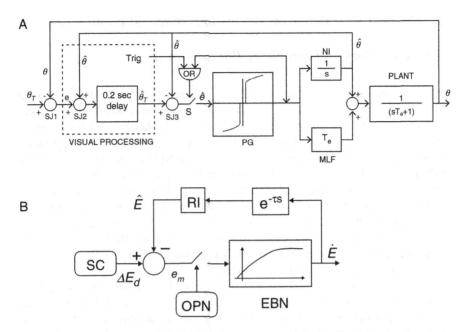

Fig. 2 Panel A: Classical local feedback model (Robinson 1975). An efference copy of eye position as computed by the neural integrator (NI) provides the input to the feedback loop that estimates the spatial position of the target $\hat{\theta}_T$ and retinal error $\hat{e}$. The latter is the input to the pulse generator (PG), a high-gain nonlinear amplifier block producing the saccadic burst that is integrated and sent to a first-order approximation of the ocular motor plant. Panel B: Schematic of the classical local feedback loop model including the later developments up to the 1990s. The delay along the feedback loop (Zee and Robinson 1979), the resettable integrator (Jürgens et al. 1981), and the superior colliculus as the source of a desired eye displacement signal (Ottes et al. 1986)

position error, which is still in use today (Eq. 1) automatically producing the appropriate burst amplitude and duration for the saccade, accounting for the saccadic main sequence (Van Gisbergen et al. 1981):

$$B(e) = \begin{cases} B_m \left(1 - e^{-(e-e_0)/b}\right), & e > e_0 \\ 0, & e \leq e_0 \end{cases} \tag{1}$$

To account for the saccadic oscillations observed in a patient with ocular flutter, the authors proposed that modulating the duration of a delay along the local feedback loop (oscillation frequency is inversely related to the delay (τ) by the relationship $F \approx 1/4\tau$) could be the mechanism explaining them.

A further milestone was the understanding that the saccadic mechanism was driven by desired eye displacement instead than to an absolute position in the orbit. This is compared with the estimate of current displacement computed by a dedicated, resettable integrator (RI) driven by the efference copy of the velocity component of the command (Jürgens et al. 1981).

5 Current Knowledge on Brainstem Neuronal Populations

5.1 Excitatory and Inhibitory Burst Neurons

Excitatory premotor burst neurons (EBN) producing the saccadic pulse fire about 12 ms before the movement of the eye (Van Gisbergen et al. 1981; Henn et al. 1989) and were found to be glutamatergic (Horn 2006). The horizontal EBN lie in the PPRF project directly to the ipsilateral abducens motor neurons, which begin firing 4 ms later, and to the nucleus prepositus hypoglossi (NPH), which contributes to the FCP integrator for horizontal eye movements (Strassman et al. 1986a; Horn et al. 1995). The vertical and torsional EBN lie in the rostral interstitial nuclei of the MLF (riMLF) and project to the corresponding motor neurons, bilaterally for upward saccades and only ipsilaterally for downward, and to the interstitial nucleus of Cajal (INC) (Moschovakis et al. 1991a, b). The latter contributes to the FCP integrator for vertical and torsional eye movements (Horn and Büttner-Ennever 1998) and is the site of vertical inhibitory burst neurons (IBN).

Inhibitory burst neurons (IBN) are glycinergic (Horn 2006), presenting firing patterns that are very similar to those of the EBN (Strassman et al. 1986b); they are considered to implement Sherrington's law of reciprocal innervation for the saccadic system and were hypothesized to contribute stopping saccades once the eye is on target (Leigh and Zee 2015). Horizontal IBN lie in the medullary reticular formation; they receive projections from the contralateral IBN and from the OPN and contralateral SC (Shinoda et al. 2008); they project to the contralateral abducens, EBN, and IBN, as well as the OPN (Shinoda et al. 2011). Vertical and torsional IBN lie in the INC (Izawa et al. 2007; Sugiuchi et al. 2013), and they appear to implement a reciprocal contralateral inhibition similar to that achieved by horizontal IBN, with upward saccades inhibiting downward ones and extorsional ones inhibiting intorsional ones (Sugiuchi et al. 2013).

A horizontal saccade is believed to be driven by the firing of the ipsilateral EBN to the agonist recti, while the ipsilateral IBN inhibit the contralateral antagonist through inhibition of the contralateral EBN. The contralateral IBN briefly fire before saccade end (Van Gisbergen et al. 1981) and were hypothesized to be involved in stopping the saccade (Quaia et al. 1999).

The drive to the motor neurons is a pulse–slide–step (Sylvestre and Cullen 1999), with the slide characterizing the transition between the pulse and the step (Robinson et al. 1990; Straumann et al. 1995; Quaia and Optican 1998).

5.2 Long-Lead Burst Neurons

These neurons are active over 40 ms prior to a saccade and project to the EBN, IBN, OPN, nucleus reticularis tegmenti pontis (NRTP), and cerebellum. They are located in several brainstem areas, the rostral pons, NRTP, the reticular formation, and other

areas and receive input from the SC and the cortical areas involved with the generation of saccades (frontal eye field (FEF), supplementary eye field (SEF), parietal eye
field (PEF)) (Scudder et al. 1996a, b).

5.3 Omnipause Neurons

The OPN lie in the nucleus raphe interpositus (RIP).

These neurons are tonically active and inhibit all premotor burst neurons except
during saccades in all directions (Luschei and Fuchs 1972; Keller 1974; Van
Gisbergen et al. 1981; Strassman et al. 1987), during which they remain silent. They
are also inhibited during blinks (Mays and Morrisse 1995), and their firing is modulated by vergence angle. When a saccade is programmed, the OPN cease firing
about 20 ms before its start and resume their activity at saccade end (Pare and
Guitton 1998).

6 Brainstem Neurons and Saccadic Oscillations: A New Model

The finding of ocular oscillations of saccadic origin in both healthy subjects and
patients (Zee and Robinson 1979; Hain et al. 1986; Ramat et al. 1999; Bhidayasiri
et al. 2001) was an important drive for the development of new models of the saccadic mechanism. The first modeling attempt to explain oscillations was the above-
mentioned delay along the saccadic local feedback loop (Zee and Robinson 1979),
yet the large modulation of delays that would be necessary to reproduce the wide
range of oscillation frequencies observed in experimental data argues against such
possibility (Ramat et al. 2005), as well as the finding that patients could produce
oscillations of different amplitudes with little changes in frequency (Shaikh et al.
2007; Ramat et al. 2008). A new model of the premotor circuitry was then proposed
(Ramat et al. 2005), which differed from previous models on two main aspects: first,
the implementation of an explicit IBN circuitry which took into account the IBN
projection to contralateral IBN and EBN neurons (Strassman et al. 1986b), forming
two positive feedback loops (IBN–EBN–IBN and IBN–IBN) representing a source
of instability that could generate saccadic oscillations, and, second, the implementation of a post-inhibitory rebound mechanism, as first hypothesized for EBN cells by
Enderle for EBN (Enderle and Engelken 1995), in all PBN (Fig. 3a). Post-inhibitory
rebound is a property found in cells having low-threshold t-type Ca^{++} channels that
exhibit bursts of activity and which, when released from inhibition, may fire spontaneously one or more action potentials (Perez-Reyes 2003). Several cell types
showing this properties were reported in the deep cerebellar nuclei (Aizenman and

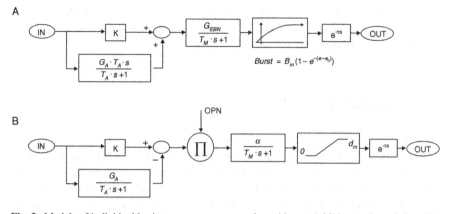

Fig. 3 Models of individual brainstem neurons presenting with post-inhibitory rebound. Panel A: Neuron model by Ramat et al. (2005) in which post-inhibitory rebound is implemented as a high-pass filter governed by an adaptation gain G_A and time constant T_A in parallel to the direct projection to the neuron membrane, i.e., a low-pass filter with gain G_{EBN} and time constant T_M. Panel B: Neuron model by Daye et al. (2013) in which post-inhibitory rebound is due to a low-pass filter (gain G_A and time constant T_A) and the OPN projection has a multiplicative behavior on the neuron activity

Linden 1999; Alviña et al. 2009; Bengtsson et al. 2011), yet it has not been explicitly looked for in PBN. Such mechanism was modeled (Fig. 4) as a high-pass filter showing adaptation, preceding a low-pass filter representing the neuronal membrane, in turn feeding to the output nonlinearity (as in Eq. 1). Thus, when OPN inhibition is lifted, these cells show a rebound of the membrane potential allowing them to fire spontaneously, on both sides. Any imbalance causes the EBN on one side to prevail, briefly exciting the ipsilateral IBN and inhibiting the contralateral EBN, thereby triggering the beginning of oscillations in the model, which are then sustained by the EBN–IBN positive feedback loops. Indeed, as the drive from the initial EBN is extinguished, also the ipsilateral IBN ceases firing, thereby disinhibiting both the contralateral IBN and EBN. The disinhibited IBN then shows rebound firing and also receives an excitation due to the rebound firing of the companion EBN, and such behavior repeats itself alternately on each side, governed by the adaptation properties of the PBN membrane (G_A and T_A in Fig. 4).

The finding of a familial disorder causing saccadic oscillations, microsaccades, and limb tremor in a mother and daughter could also be explained using this model and hypothesizing that oscillations arise in the mentioned feedback loops due to a pathological reduction of glycinergic inhibition (Shaikh et al. 2007, 2008).

Importantly, the model is able to reproduce saccades and their oscillations based on the detailed topology of the brainstem premotor network and on the membrane properties of the modeled cells, instead of proposing new functions specifically tailored to explaining specific experimental findings.

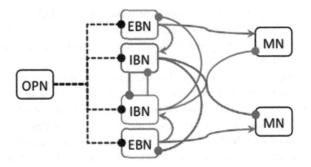

Fig. 4 Brainstem saccade circuitry from the saccadic oscillation model by Ramat et al. (2005) showing the omnipause neurons (OPN), bilateral excitatory burst neurons (EBN), and inhibitory burst neurons (IBN). Inhibitory projections terminate with a round disc; excitatory projections terminate with an arrow. The model presents two positive feedback loops: one between inhibitory burst neurons on the right and left sides (the short IBN–IBN projections) and the other also involving the EBN, i.e., right EBN to right IBN, to left EBN, to left IBN, and back to the right EBN. The positive feedback loops involving neurons presenting with PIR represent the hypothesized mechanism for saccadic oscillations. Suppose that depolarization of the right EBN provides a small output burst. That will drive the right IBN, which in turn inhibits the left EBN. When the burst in the right EBN is over, the inhibition by the right IBN on the left EBN will be lifted, and its membrane potential will show a rebound depolarization. This rebound will cause the left EBN to burst, thereby exciting the left IBN, which, in turn, will lead to inhibition of the original, right EBN. At the end of the firing by the left EBN, the process will be repeated, alternately exciting the right and the left PBN

7   Superior Colliculus

The superior colliculus (SC) is a midbrain structure controlling body orientation in space and extensively involved in saccade generation, sending projections to the long-lead burst generator neurons in the PPRF and RiMLF, the OPN in the RIP, the NRTP, the vestibular nuclei, and the FEF. It is a multilayered structure receiving sensory information from the visual, auditive, and somatosensory systems together with higher-level cognitive signals related to attention and context. The dorsal layers of the SC are said "visual," as they receive direct input from visual cortical areas, and implement a retinotopic map encoding the vector displacement to potential gaze targets (Goldberg and Wurtz 1972; Robinson 1972). Activity at the rostral pole is involved in fixation and the generation of microsaccades near the rostral pole that indicates targets near the fovea, while more caudal neurons encode the eye displacement vector for peripheral targets. A radial receptive field characterizes neurons in such superficial layer, so that their firing behavior depends on the location of the visual target with respect to the center of their receptive field.

The intermediate and deep layers of the SC are instead "motor," or visuomotor, mostly related to the control of eye movements and receiving inputs from visual areas and several cortical areas (PEF, FEF, SEF, dorsolateral prefrontal cortex (DLPFC)), from the NPH and the fastigial nuclei (FN) and from structures in the basal ganglia (the striatum, globus pallidus, subthalamic nucleus). These layers are also topographically organized, and their cells have a motor field, meaning that they fire presaccadic commands before saccades directed to specific areas of the visual field, encoded by the corresponding dorsal layer cells (Girard and Berthoz 2005; Leigh and Zee 2015).

Several groups focused on modeling the superior colliculus and its role in saccade control, proposing a logarithmic mapping between a retinal stimulus and the activated SC cells and from those to the coding of saccade displacement vector based on 2D Gaussian functions (Ottes et al. 1986; Optican 1995). A spatial memory mechanism used in the SC for representing the position of a target and its updating during the execution of a saccade was then modeled as a "moving hill" of activity (Droulez and Berthoz 1991; Lefèvre and Galiana 1992). A central idea of several models was that the SC is part of the local feedback loop of the saccadic mechanism and produces, either directly or indirectly, the signal driving the premotor neurons (Waitzman et al. 1988; Droulez and Berthoz 1991; Lefèvre and Galiana 1992; Optican 1994). Yet, experimental findings have shown that collicular lesions in primates cause only relatively small changes in saccades' dynamics and accuracy (Aizawa and Wurtz 1998; Quaia et al. 1998), arguing against such a fundamental role of the SC in saccade generation.

8   Cerebellum

The involvement of the cerebellum in the generation of saccades was understood as early as the late nineteenth century with the finding that cerebellar electrical stimulation could elicit saccades. Indeed, the cerebellum is extensively involved with the generation of saccades, with the most critical functions being played by the dorsal cerebellar vermis (oculomotor vermis, OMV) and by the caudal fastigial nucleus (cFN) and specifically the fastigial nucleus oculomotor region (FOR), while the function of the cerebellar hemispheres and other nuclei remains less clearly understood (Leigh and Zee 2015).

The cerebellum receives two types of input signals related to saccades: one from the SC through LLBN in the NRTP carried by mossy fibers and one related to motor error through climbing fibers from the inferior olive (IO).

Microstimulation of the OMV while executing a saccade modifies its trajectory (Keller et al. 1983); disruption of the OMV projection to the cFN causes significant ipsilateral hypometria and mild contralateral hypermetria (Sato and Noda 1992). Asymmetrical lesions (Takagi et al. 2000) cause longer latency in ipsilesional saccades, which also become hypometric, and bilateral lesions impair saccade adaptation (Takagi et al. 2000; Kojima et al. 2011). The output of the OMV through its Purkinje cells encodes the stop time of a saccade for landing on target (Thier et al. 2000).

Lesions of the cFN, which are inherently bilateral, cause significant saccade hypermetria (Selhorst et al. 1976), and unilateral pharmacological inactivation of one cFN produces hypermetric ipsilateral and hypometric contralateral saccades

(Robinson et al. 1993; Robinson and Fuchs 2001; Straube et al. 2009; Kojima et al. 2014). The activity of FOR neurons in relation to a saccade is well characterized (Ohtsuka and Noda 1991; Fuchs et al. 1993; Helmchen and Büttner 1995): they fire about 8 ms prior to the onset of contralateral saccades and at the end of ipsilateral saccades.

9 Modeling the Cerebellum and Superior Colliculus in Saccade Generation

Most early models of the saccade-generating mechanism considering the cerebellum were concerned with its role in saccade adaptation (Dean et al. 1994; Schweighofer et al. 1996a, b), which has also been more recently studied in terms of motor learning (Xu-Wilson et al. 2009). A first model assigning an explicit role to the cerebellum in the generation of saccades was that proposed by Dean (1995) focusing on the burst activity of the FOR in accelerating saccades at their onset and choking them at their end when signaled by the local feedback loop. Further progress was represented by the neuromimetic model proposed by Optican and colleagues (Lefèvre et al. 1998; Quaia et al. 1999), which included detailed roles for the SC, the cerebellum, and the brainstem circuitry, incorporating many of the anatomical and neurophysiological findings detailed in the previous paragraph. The overall model architecture (Fig. 5a) revolves around two pathways originating from the cortex: the first, encoding the target in retinotopic coordinates, reaching the SC and the second reaching the cerebellum. The SC determines the onset of the saccade by turning off the OPN and provides the driving signal to the released medium lead burst neuron (MLBN). It also provides the cerebellum with the desired saccade amplitude information through the NRTP. The cerebellum is considered as the site of a displacement integrator (DI) allowing it to monitor the progress of the saccade based on the output of the MLBN and is therefore part of the local feedback loop. Based on this estimate, it provides an additional drive to the ipsilateral MLBN correcting the saccade trajectory and insuring that it gets to the target and projects to the contralateral IBN to stop the saccade when the estimated amplitude matches the desired one provided by the NRTP (Fig. 5b). The model hypothesizes a topographical organization of the FOR, having regions projecting to the horizontal MLBN and others to the vertical ones, which acts as the displacement integrator (i.e., performing a spatial integration) with a locus of activity beginning in the contralateral FOR and spreading to the ipsilateral one with a velocity controlled by the OMV (Quaia et al. 1999).

An evolution of such model, with a control circuitry based on the two pathways involving the SC and the cFN (Lefèvre et al. 1998; Quaia et al. 1999) and a detailed representation of the saccadic premotor circuitry inspired by the saccadic oscillations model (Ramat et al. 2005, 2008) with a new neuron model producing PIR (Fig. 3b), was then proposed by Daye and colleagues (Daye et al. 2013). The model

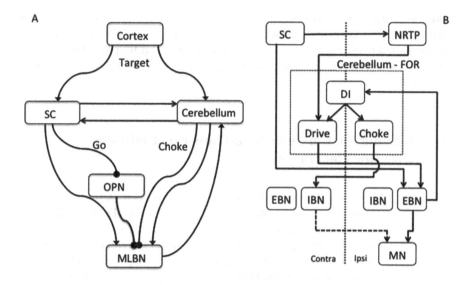

Fig. 5 Conceptual diagrams of the Lefèvre–Optican–Quaia model (Lefèvre et al. 1998; Quaia et al. 1999). Panel A: Overall architecture. The cortex provides the intended target to both the SC and the FOR in the cerebellum. The SC provides both a drive signal to the MLBN in the PPRF and a go signal inhibiting the OPN. The cerebellum also receives the desired displacement information and sends an additional driving signal to the MLBN contributing to the saccade command. The output of the MLBN is fed back to the cerebellum, which monitors the progress of the saccade and stops it when the motor error reaches zero. Panel B: Detail of the projections in the brainstem and cerebellum during a rightward (ipsilateral) saccade. The contralateral superior colliculus (SC) sends a driving signal to the ipsilateral excitatory burst neurons (EBN) and retinotopic target information to the cerebellum through the NRTP, releasing the contralateral FOR, which bursts and provides an additional drive to the ipsilateral EBN. The output of EBN is fed back to the displacement integrator (DI) in the FOR, and its output displacement is compared to the intended one to compute the residual motor error that contributes a drive to the EBN. When the burst of FOR activity being integrated under the velocity control exerted by the cerebellar vermis reaches the ipsilateral FOR, a choke signal excites the contralateral IBN, which in turn inhibits the motor neurons (MN) of the agonist muscle (ipsilateral), stopping the saccade

considers two populations of inhibitory burst neurons, long-lead inhibitory burst neuron (LIBN) and short-lead inhibitory burst neuron (SIBN), the first responsible for turning off the OPN and the second projecting to the group of contralateral PBN; the SC is represented as three populations, one rostral and two caudal receiving the desired eye displacement and an efference copy of eye velocity, and includes the NPH for producing the step of innervation. The model was first fit to reproduce normal saccades; then the change of a few parameters tied to cerebellar dysfunction could account for the deficits of an undiagnosed patient showing asymmetrical amplitude–peak velocity relationships, postsaccadic drift, dynamic overshoot, and ocular flutter.

It was once more the finding of saccadic oscillations and opsoclonus in a peculiar patient that developed the symptom after taking anabolic–androgenic steroids that

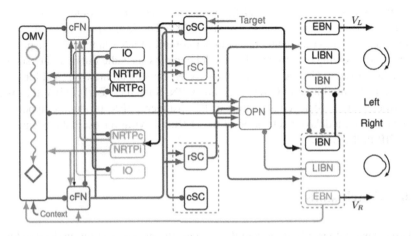

Fig. 6 Architecture of the neuromimetic model of the saccadic system and detail of the areas involved in a rightward saccade (Optican and Pretegiani 2017). The cortex sends target information to the left caudal SC (cSC), which begins firing, inhibiting the rostral SC (rSC), exciting the right group of PBN and the right NRTP (black projections). The left LIBN begin firing and inhibit the OPN, causing the right EBN and IBN to fire and start the saccade. The NRTP produces a locus of inhibition in the OMV, at a location (green circle) depending on the incoming context information. The right EBN activity is fed back to the OMV and controls the speed of the inhibition wave moving toward the right, until it reaches the intended target (red diamond) and it disinhibits the right cFN. The right cFN excites the left group of PBN, with the IBN firing intensely to inhibit the right PBN and choke the saccadic drive, thereby stopping the saccade while disinhibiting the OPN and rSC to prevent further saccadic movements

drove the development of a recent new model of the saccadic mechanism (Optican and Pretegiani 2017; Pretegiani et al. 2017). The model (Fig. 6) shares several features with the previous ones (Lefèvre et al. 1998; Quaia et al. 1999; Ramat et al. 2005; Daye et al. 2013), with PBN being driven by both the SC and the cFN, the two IBN populations with LIBN inhibiting the OPN and IBN reciprocally projecting to each other, forming a positive feedback loop, and to EBN. Differently though, the new model hypothesizes that the OMV (and not the FOR) is part of the saccade feedback loop and implements the displacement integrator, with a pause of activity originating on the side contraversive to the direction of the saccade and releasing the cFN. Such inhibition, under control of an efference copy of EBN activity (eye velocity), spreads to the ipsiversive OMV and releases the corresponding cFN, which excites the contralateral IBN, stops the saccade and reactivates the OPN, thus preventing oscillations. The role of displacement integrator with a wave of inhibition spreading across the OMV is more plausible than the previous hypothesis attributing it to the FOR, since the OMV is continuous across the midline while the FN are not. In this model, the drive to PBN is again provided by both the SC and the cFN, yet the displacement integrator mechanism in the OMV can reconcile with a situation in which the retinotopic target location in the SC and the necessary command are dissociated, as with a moving target. The drive from the SC would be tailored to a saccade of a different size from the needed one, causing a saccade off the main sequence, yet the spreading inhibition in the vermis would stop the saccade on target when it reaches the ipsiversive OMV. The initial location of the vermal inhibition depends on context and is under control of a learning mechanism adjusting it with growing experience. The model explains saccadic flutter and opsoclonus as an alteration of GABA receptors ($GABA_AR$) causing an increase of chloride currents and the inhibition of OMV, cFN, and OPN. Such increased $GABA_AR$ sensitivity would cause a delayed activation of the ipsilateral cFN, producing hypermetric saccades, and a delayed activation of the OPN, allowing for oscillations caused by the PIR in the EBN.

10 Conclusions

As we have witnessed throughout this review, the progress of our understanding of the saccadic mechanism is the outcome of a virtuous synergy between basic science providing new anatomical and physiological knowledge and their interpretations. These become testable when expressed through mathematical models, and the comparison of model predictions with experimental findings sparks new questions and new hypotheses, driving new research and new models. The fascination of the detailed understanding of the roles played by the different CNS areas involved with producing saccades is unique in the domain of motor control research and represents a precious window on understanding the function of our brain.

References

Aizawa, H., & Wurtz, R. H. (1998). Reversible inactivation of monkey superior colliculus. I. Curvature of saccadic trajectory. *Journal of Neurophysiology, 79*, 2082–2096.

Aizenman, C. D., & Linden, D. J. (1999). Regulation of the rebound depolarization and spontaneous firing patterns of deep nuclear neurons in slices of rat cerebellum. *Journal of Neurophysiology, 82*, 1697–1709.

Alviña, K., Ellis-Davies, G., & Khodakhah, K. (2009). T-type calcium channels mediate rebound firing in intact deep cerebellar neurons. *Neuroscience, 158*, 635–641.

Bahill, A. T., Adler, D., & Stark, L. (1975a). Most naturally occurring human saccades have magnitudes of 15 degrees or less. *Investigative Ophthalmology, 14*, 468–469.

Bahill, A. T., Clark, M. R., & Stark, L. (1975b). The main sequence, a tool for studying human eye movements. *Mathematical Biosciences, 24*, 191–204.

Baloh, R. W., Konrad, H. R., Sills, A. W., & Honrubia, V. (1975). The saccade velocity test. *Neurology, 25*, 1071–1076.

Bengtsson, F., Ekerot, C. F., & Jörntell, H. (2011). In vivo analysis of inhibitory synaptic inputs and rebounds in deep cerebellar nuclear neurons. *PLoS One, 6*, e18822.

Bhidayasiri, R., Somers, J. T., Kim, J. I., Ramat, S., Nayak, S., Bokil, H. S., & John Leigh, R. (2001). Ocular oscillations induced by shifts of the direction and depth of visual fixation. *Annals of Neurology, 49*, 24.

Boghen, D., Troost, B. T., Daroff, R. B., Dell'Osso, L. F., & Birkett, J. E. (1974). Velocity characteristics of normal human saccades. *Investigative Ophthalmology, 13*, 619–623.

Cohen, B. (1971). Vestibulo-ocular relations. In P. Bach-y-Rita & C. C. Collins (Eds.), *The control of eye movements* (pp. 105–148). New York: Academic Press.

Collewijn, H. (1977). Eye-and head movements in freely moving rabbits. *The Journal of Physiology, 266,* 471–498.

Daye, P. M., Optican, L. M., Roze, E., Gaymard, B., & Pouget, P. (2013). Neuromimetic model of saccades for localizing deficits in an atypical eye-movement pathology. *Journal of Translational Medicine, 11,* 125.

Dean, P. (1995). Modelling the role of the cerebellar fastigial nuclei in producing accurate saccades: The importance of burst timing. *Neuroscience, 68,* 1059–1077.

Dean, P., Mayhew, J. E. W., & Langdon, P. (1994). Learning and maintaining saccadic accuracy: A model of brainstem–cerebellar interactions. *Journal of Cognitive Neuroscience, 6,* 117–138.

Droulez, J., & Berthoz, A. (1991). A neural network model of sensoritopic maps with predictive short-term memory properties. *Proceedings of the National Academy of Sciences of the United States of America, 88,* 9653–9657.

Enderle, J. D. (2002). Neural control of saccades. *Progress in Brain Research, 140,* 21–49.

Enderle, J. D., & Engelken, E. J. (1995). Simulation of oculomotor post-inhibitory rebound burst firing using a Hodgkin-Huxley model of a neuron. *Biomedical Sciences Instrumentation, 31,* 53–58.

Fuchs, A. F., Robinson, F. R., & Straube, A. (1993). Role of the caudal fastigial nucleus in saccade generation. I. Neuronal discharge pattern. *Journal of Neurophysiology, 70,* 1723–1740.

Girard, B., & Berthoz, A. (2005). From brainstem to cortex: Computational models of saccade generation circuitry. *Progress in Neurobiology, 77,* 215–251.

Goldberg, M. E., & Wurtz, R. H. (1972). Activity of superior colliculus in behaving monkey. I. Visual receptive fields of single neurons. *Journal of Neurophysiology, 35,* 542–559.

Hain, T. C., Zee, D. S., & Mordes, M. (1986). Blink-induced saccadic oscillations. *Annals of Neurology, 19,* 299–301.

Helmchen, C., & Büttner, U. (1995). Saccade-related Purkinje cell activity in the oculomotor vermis during spontaneous eye movements in light and darkness. *Experimental Brain Research, 103,* 198–208.

Henn, V., Hepp, K., & Vilis, T. (1989). Rapid eye movement generation in the primate. Physiology, pathophysiology, and clinical implications. *Revue Neurologique (Paris), 145,* 540–545.

Horn, A. K. E. (2006). The reticular formation. In Horn Neuroanatomy of the Oculomotor System, J.B. Ennever (ed.), Elsevier B.V., Amsterdam. *Progress in brain research* (pp. 127–155).

Horn, A. K., & Büttner-Ennever, J. A. (1998). Premotor neurons for vertical eye movements in the rostral mesencephalon of monkey and human: Histologic identification by parvalbumin immunostaining. *The Journal of Comparative Neurology, 392,* 413–427.

Horn, A. K., Büttner-Ennever, J. A., Suzuki, Y., & Henn, V. (1995). Histological identification of premotor neurons for horizontal saccades in monkey and man by parvalbumin immunostaining. *The Journal of Comparative Neurology, 359,* 350–363.

Inchingolo, P., Spanio, M., & Bianchi, M. (1987). The characteristic peak velocity – Mean velocity of saccadic eye movements in man. In Inchingolo: J.K. O'Regan, A. Lévy-Schoen (Eds.), Elsevier B.V., Amsterdam. *Eye movements from physiology to cognition* (pp. 17–26).

Izawa, Y., Sugiuchi, Y., & Shinoda, Y. (2007). Neural organization of the pathways from the superior colliculus to trochlear motoneurons. *Journal of Neurophysiology, 97,* 3696–3712.

Jürgens, R., Becker, W., & Kornhuber, H. H. (1981). Natural and drug-induced variations of velocity and duration of human saccadic eye movements: Evidence for a control of the neural pulse generator by local feedback. *Biological Cybernetics, 39,* 87–96.

Keller, E. L. (1974). Participation of medial pontine reticular formation in eye movement generation in monkey. *Journal of Neurophysiology, 37,* 316–332.

Keller, E. L., Slakey, D. P., & Crandall, W. F. (1983). Microstimulation of the primate cerebellar vermis during saccadic eye movements. *Brain Research, 288,* 131–143.

Kojima, Y., Soetedjo, R., & Fuchs, A. F. (2011). Effect of inactivation and disinhibition of the oculomotor vermis on saccade adaptation. *Brain Research, 1401,* 30–39.

Kojima, Y., Robinson, F. R., & Soetedjo, R. (2014). Cerebellar fastigial nucleus influence on ipsilateral abducens activity during saccades. *Journal of Neurophysiology, 111*, 1553–1563.

Lefèvre, P., & Galiana, H. L. (1992). Dynamic feedback to the superior colliculus in a neural network model of the gaze control system. *Neural Networks, 5*, 871–890.

Lefèvre, P., Quaia, C., & Optican, L. M. (1998). Distributed model of control of saccades by superior colliculus and cerebellum. *Neural Networks, 11*, 1175–1190.

Leigh, R. J., & Zee, D. S. (2015). *The neurology of eye movements*. Oxford: Oxford University Press.

Luschei, E. S., & Fuchs, A. F. (1972). Activity of brain stem neurons during eye movements of alert monkeys. *Journal of Neurophysiology, 35*, 445–461.

Martinez-Conde, S., Otero-Millan, J., & Macknik, S. L. (2013). The impact of microsaccades on vision: Towards a unified theory of saccadic function. *Nature Reviews. Neuroscience, 14*, 83–96.

Mays, L. E., & Morrisse, D. W. (1995). Electrical stimulation of the pontine omnipause area inhibits eye blink. *Journal of the American Optometric Association, 66*, 419–422.

Moschovakis, A. K., Scudder, C. A., & Highstein, S. M. (1991a). Structure of the primate oculomotor burst generator. I. Medium-lead burst neurons with upward on-directions. *Journal of Neurophysiology, 65*, 203–217.

Moschovakis, A. K., Scudder, C. A., Highstein, S. M., & Warren, J. D. (1991b). Structure of the primate oculomotor burst generator. II. Medium-lead burst neurons with downward on-directions. *Journal of Neurophysiology, 65*, 218–229.

Niemeier, M., Crawford, J. D., & Tweed, D. B. (2003). Optimal transsaccadic integration explains distorted spatial perception. *Nature, 422*, 76–80.

Ohtsuka, K., & Noda, H. (1991). Saccadic burst neurons in the oculomotor region of the fastigial nucleus of macaque monkeys. *Journal of Neurophysiology, 65*, 1422–1434.

Optican, L. M. (1994). Control of saccade trajectory by the superior colliculus. In A. Fuchs, T. Brandt, U. Buettner, & D. S. Zee (Eds.), Optican: A.F. Fuchs, U. Buettner, D. S. Zee (Eds.), Thieme Publishing Group, Stuttgart. *Contemporary ocular motor and vestibular research: A tribute to David A. Robinson* (pp. 98–105).

Optican, L. M. (1995). A field theory of saccade generation: Temporal-to-spatial transform in the superior colliculus. *Vision Research, 35*, 3313–3320.

Optican, L. M., & Pretegiani, E. (2017). A GABAergic dysfunction in the olivary–cerebellar–brainstem network may cause eye oscillations and body tremor. II. Model simulations of saccadic eye oscillations. *Frontiers in Neurology, 8*, 372.

Ottes, F. P., Van Gisbergen, J. A. M., & Eggermont, J. J. (1986). Visuomotor fields of the superior colliculus: A quantitative model. *Vision Research, 26*, 857–873.

Pare, M., & Guitton, D. (1998). Brain stem omnipause neurons and the control of combined eye-head gaze saccades in the alert cat. *Journal of Neurophysiology, 79*, 3060–3076.

Perez-Reyes, E. (2003). Molecular physiology of low-voltage-activated T-type calcium channels. *Physiological Reviews, 83*, 117–161.

Pretegiani, E., Rosini, F., Rocchi, R., Ginanneschi, F., Vinciguerra, C., Optican, L. M., & Rufa, A. (2017). GABAAergic dysfunction in the olivary-cerebellar-brainstem network may cause eye oscillations and body tremor. *Clinical Neurophysiology, 128*, 408–410.

Quaia, C., & Optican, L. M. (1998). Commutative saccadic generator is sufficient to control a 3-D ocular plant with pulleys. *Journal of Neurophysiology, 79*, 3197–3215.

Quaia, C., Aizawa, H., Optican, L. M., & Wurtz, R. H. (1998). Reversible inactivation of monkey superior colliculus. II. Maps of saccadic deficits. *Journal of Neurophysiology, 79*, 2097–2110.

Quaia, C., Lefèvre, P., & Optican, L. M. (1999). Model of the control of saccades by superior colliculus and cerebellum. *Journal of Neurophysiology, 82*, 999–1018.

Ramat, S., Somers, J. T., Vallabh, E. D., & Leigh, R. J. (1999). Conjugate ocular oscillations during shifts of the direction and depth of visual fixation. *Investigative Ophthalmology and Visual Science, 40*, 1681–1686.

Ramat, S., Leigh, R. J., Zee, D. S., & Optican, L. M. (2005). Ocular oscillations generated by coupling of brainstem excitatory and inhibitory saccadic burst neurons. *Experimental Brain Research, 160*, 89–106.

Ramat, S., Leigh, R. J., Zee, D. S., & Optican, L. M. (2007). What clinical disorders tell us about the neural control of saccadic eye movements. *Brain: A Journal of Neurology, 130*, 10–35.

Ramat, S., Leigh, R. J., Zee, D. S., Shaikh, A. G., & Optican, L. M. (2008). Applying saccade models to account for oscillations. *Progress in Brain Research, 171*, 123–130.

Robinson, D. A. (1964). The mechanics of human saccadic eye movement. *The Journal of Physiology, 174*, 245–264.

Robinson, D. A. (1968). Eye movement control in primates. The oculomotor system contains specialized subsystems for acquiring and tracking visual targets. *Science, 161*, 1219–1224.

Robinson, D. A. (1972). Eye movements evoked by collicular stimulation in the alert monkey. *Vision Research, 12*, 1795–1808.

Robinson, D. A. (1975). Oculomotor control signals. In G. Lennerstrand & P. Bach-y-Rita (Eds.), *Basic mechanisms of ocular motility and their clinical implications* (pp. 337–374). Oxford: Pergamon Press.

Robinson, F. R., & Fuchs, A. F. (2001). The role of the cerebellum in voluntary eye movements. *Annual Review of Neuroscience, 24*, 981–1004.

Robinson, D. A., Kapoula, Z., & Goldstein, H. P. (1990). Holding the eye still after a saccade. In *From neuron to action* (pp. 89–96). Berlin, Heidelberg: Springer Berlin Heidelberg.

Robinson, F. R., Straube, A., & Fuchs, A. F. (1993). Role of the caudal fastigial nucleus in saccade generation. II. Effects of muscimol inactivation. *Journal of Neurophysiology, 70*, 1741–1758.

Ron, S., Robinson, D. A., & Skavenski, A. A. (1972). Saccades and the quick phase of nystagmus. *Vision Research, 12*, 2015–2022.

Rottach, K. G., von Maydell, R. D., Das, V. E., Zivotofsky, A. Z., Discenna, A. O., Gordon, J. L., Landis, D. M., & Leigh, R. J. (1997). Evidence for independent feedback control of horizontal and vertical saccades from Niemann-Pick type C disease. *Vision Research, 37*, 3627–3638.

Sato, H., & Noda, H. (1992). Saccadic dysmetria induced by transient functional decortication of the cerebellar vermis. *Experimental Brain Research, 89*, 690.

Schaeffer, K. P. (1965). The excitation pattern of single neurons of the abduccent nerve nucleus in rabbits. *Pflügers Archiv für die Gesamte Physiologie des Menschen und der Tiere, 284*, 31–52.

Schweighofer, N., Arbib, M. A., & Dominey, P. F. (1996a). A model of the cerebellum in adaptive control of saccadic gain. I. The model and its biological substrate. *Biological Cybernetics, 75*, 19–28.

Schweighofer, N., Arbib, M. A., & Dominey, P. F. (1996b). A model of the cerebellum in adaptive control of saccadic gain. II. Simulation results. *Biological Cybernetics, 75*, 29–36.

Scudder, C. A., Moschovakis, A. K., Karabelas, A. B., & Highstein, S. M. (1996a). Anatomy and physiology of saccadic long-lead burst neurons recorded in the alert squirrel monkey. II. Pontine neurons. *Journal of Neurophysiology, 76*, 353–370.

Scudder, C. A., Moschovakis, A. K., Karabelas, A. B., & Highstein, S. M. (1996b). Anatomy and physiology of saccadic long-lead burst neurons recorded in the alert squirrel monkey. I. Descending projections from the mesencephalon. *Journal of Neurophysiology, 76*, 332–352.

Selhorst, J. B., Stark, L., Ochs, A. L., & Hoyt, W. F. (1976). Disorders in cerebellar ocular motor control: I. Saccadic overshoot dysmetria an oculographic, control system and clinico-anatomical analysis. *Brain, 99*, 497–508.

Shaikh, A. G., Miura, K., Optican, L. M., Ramat, S., Leigh, R. J., & Zee, D. S. (2007). A new familial disease of saccadic oscillations and limb tremor provides clues to mechanisms of common tremor disorders. *Brain, 130*, 3020.

Shaikh, A. G., Ramat, S., Optican, L. M., Miura, K., Leigh, R. J., & Zee, D. S. (2008). Saccadic burst cell membrane dysfunction is responsible for saccadic oscillations. *Journal of Neuro-Ophthalmology, 28*, 329–336.

Shinoda, Y., Sugiuchi, Y., Izawa, Y., & Takahashi, M. (2008). Neural circuits for triggering saccades in the brainstem. *Progress in Brain Research, 171*, 79–85.

Shinoda, Y., Sugiuchi, Y., Takahashi, M., & Izawa, Y. (2011). Neural substrate for suppression of omnipause neurons at the onset of saccades. *Annals of the New York Academy of Sciences, 1233*, 100–106.

Skavenski, A. A., & Robinson, D. A. (1973). Role of abducens neurons in vestibuloocular reflex. *Journal of Neurophysiology, 36*, 724–738.

Strassman, A., Highstein, S. M., & McCrea, R. A. (1986a). Anatomy and physiology of saccadic burst neurons in the alert squirrel monkey. I. Excitatory burst neurons. *The Journal of Comparative Neurology, 249*, 337–357.

Strassman, A., Highstein, S. M., & McCrea, R. A. (1986b). Anatomy and physiology of saccadic burst neurons in the alert squirrel monkey. II. Inhibitory burst neurons. *The Journal of Comparative Neurology, 249*, 358–380.

Strassman, A., Evinger, C., McCrea, R. A., Baker, R. G., & Highstein, S. M. (1987). Anatomy and physiology of intracellularly labelled omnipause neurons in the cat and squirrel monkey. *Experimental Brain Research, 67*, 436–440.

Straube, A., Scheuerer, W., Robinson, F. R., & Eggert, T. (2009). Temporary lesions of the caudal deep cerebellar nucleus in nonhuman primates: Gain, offset, and ocular alignment. *Annals of the New York Academy of Sciences, 1164*, 119–126.

Straumann, D., Zee, D. S., Solomon, D., Lasker, A. G., & Roberts, D. C. (1995). Transient torsion during and after saccades. *Vision Research, 35*, 3321–3334.

Sugiuchi, Y., Takahashi, M., & Shinoda, Y. (2013). Input-output organization of inhibitory neurons in the interstitial nucleus of Cajal projecting to the contralateral trochlear and oculomotor nucleus. *Journal of Neurophysiology, 110*, 640–657.

Sylvestre, P. A., & Cullen, K. E. (1999). Quantitative analysis of abducens neuron discharge dynamics during saccadic and slow eye movements. *Journal of Neurophysiology, 82*, 2612–2632.

Takagi, M., Zee, D. S., & Tamargo, R. J. (2000). Effects of lesions of the oculomotor cerebellar vermis on eye movements in primate: Smooth pursuit. *Journal of Neurophysiology, 83*, 2047–2062.

Thier, P., Dicke, P. W., Haas, R., & Barash, S. (2000). Encoding of movement time by populations of cerebellar Purkinje cells. *Nature, 405*, 72–76.

Van Gisbergen, J. A. M., Robinson, D. A., & Gielen, S. (1981). A quantitative analysis of generation of saccadic eye movements by burst neurons. *Journal of Neurophysiology, 45*, 417–442.

Waitzman, D. M., Ma, T. P., Optican, L. M., & Wurtz, R. H. (1988). Superior colliculus neurons provide the saccadic motor error signal. *Experimental Brain Research, 72*, 649–652.

Westheimer, G. (1954). Mechanism of saccadic eye movements. *Archives of Ophthalmology, 52*, 710–724.

Xu-Wilson, M., Chen-Harris, H., Zee, D. S., & Shadmehr, R. (2009). Cerebellar contributions to adaptive control of saccades in humans. *The Journal of Neuroscience, 29*, 12930–12939.

Young, L. R., & Stark, L. (1963). Variable feedback experiments testing a sampled data model for eye tracking movements. *IEEE Transactions on Human Factors in Electronics, HFE-4*, 38–51.

Zee, D. S., & Robinson, D. A. (1979). A hypothetical explanation of saccadic oscillations. *Annals of Neurology, 5*, 405–414.

Zee, D. S., Optican, L. M., Cook, J. D., Robinson, D. A., & Engel, W. K. (1976). Slow saccades in spinocerebellar degeneration. *Archives of Neurology, 33*, 243–251.

The Neural Oculomotor System in Strabismus

Vallabh E. Das

Abstract Binocular alignment and binocular coordination of eye movements are necessary to direct both foveae at targets within 3D space. Unfortunately, individuals suffering from strabismus (ocular misalignment) never develop the necessary alignment and coordination of eye movements for binocular vision. Developmental loss of sensory or motor fusion leads to strabismus in nearly 5% of children, making this disease a significant public health issue. In order to develop a better understanding of this disease, a number of nonhuman primate models have been used successfully. The common feature of existing monkey models is the disruption of binocular vision during the developmental critical period, either by operating on extraocular muscles (EOMs) to redirect the line of sight of one eye or by decorrelating or depriving binocular vision through the use of prisms or occluders. In these models, widespread changes in many visual and oculomotor neural centers have been identified, leading to new insight on the development and maintenance of eye misalignment and other associated strabismus properties.

Keywords Strabismus · Animal model · Nonhuman primate · Eye movements · Neural substrate

1 Introduction

Binocular alignment and binocular coordination of eye movements are features of the normal development of the visual and oculomotor system (von Noorden and Campos 2002). The disruption of binocular vision during the critical period for visual development usually leads to strabismus (ocular misalignment). Developmental sensory strabismus, i.e., strabismus that occurs as a result of disrup-

V. E. Das (✉)
College of Optometry, University of Houston, Houston, TX, USA
e-mail: vdas@central.uh.edu

© Springer Nature Switzerland AG 2019
A. Shaikh, F. Ghasia (eds.), *Advances in Translational Neuroscience of Eye Movement Disorders*, Contemporary Clinical Neuroscience,
https://doi.org/10.1007/978-3-030-31407-1_8

tion in the normal development of binocular vision, affects about 2–4% of children worldwide (Govindan et al. 2005; Greenberg et al. 2007). Other forms of congenital strabismus that are due to misinnervation of motor nerves (e.g., Duane's syndrome), orbital problems (e.g., Marfan's syndrome), and extraocular muscle (EOM) problems (e.g., congenital fibrosis of extraocular muscle) are rare and are not covered in this chapter.

It is likely that the disruption of binocular vision results in a cascade of developmental neural deficits along the visual-oculomotor axis whose behavioral outcome is misaligned eyes and a host of associated oculomotor deficits such as A/V patterns, saccade disconjugacy, and other deficits in binocular coordination of eye movements. Disruption of development of binocular vision also results in many sensory deficits associated with strabismus, including amblyopia, loss of stereopsis, and naso-temporal asymmetry in motion perception when viewing monocularly. This chapter focuses on the discussion of oculomotor deficits and their neural correlates. The reader is directed to other review articles for consideration of various sensory deficits, such as amblyopia, including their neural correlates in visual cortical areas (Crawford et al. 1996; Kiorpes 2016). This chapter also specifically focuses on recent works, including from our own lab, that have used a nonhuman primate model for developmental strabismus to understand disruption in oculomotor neural circuits that leads to eye misalignment and other strabismus properties.

2 Nonhuman Primate Models for Strabismus

Strabismus can be induced in nonhuman primates using either surgical or sensory methods (Das 2016; Walton et al. 2017). Fundamentally, either method can be employed in infant monkeys, whose aim is to disrupt binocular vision during the critical period for development (Boothe et al. 1985; Quick et al. 1989). Surgical methods such as extraocular muscle (EOM) resection, recession, and tenotomy in infant animals are effective in producing a strabismus (Crawford and von Noorden 1979; Economides et al. 2007). These methods affect EOM contractility in addition to disrupting binocular vision during development. However, when attempting to study fundamental oculomotor mechanisms that are driving eye misalignment, it may be preferable to use sensory methods to disrupt binocular vision and leave the periphery intact. In our laboratory, we have successfully used two sensory methods, a daily alternating monocular occlusion (daily AMO) method and an optical prism-viewing method. Both of these are effective in inducing a permanent strabismus along with other oculomotor disruptions that are similar to those observed in humans (Das 2016).

In the daily AMO method, an occluding patch (either goggles or contact lens) is used to block vision in one of the eyes of the infant monkey. The following day, the patch is switched to the other eye and thereafter alternated daily for the first 4–6 months after birth (Tusa et al. 2002). This method prevents binocular vision during the developmental critical period and may be referred to as an example of a binocular vision deprivation rearing paradigm. In the optical prism-viewing method, the infant monkey is fitted with a helmet-like device that houses a 20PD horizontal Fresnel prism in front of one eye and a 20PD vertical Fresnel prism in front of the other eye. Prism-viewing continues for 4–6 months from birth, and during this period the animals are unable to fuse the disparate images that arise from viewing through the prisms. This method prevents binocular vision during the developmental critical period by providing noncorresponding images to the two eyes (Crawford and von Noorden 1980). Both of these sensory methods also do not induce severe amblyopia, and the monkeys can alternately fixate with either eye. The reader is directed to other reviews for discussion of the individual merits and demerits of each of the different rearing paradigms that induce strabismus in nonhuman primates (Das 2016; Walton et al. 2017).

3 Relative Roles of the Brain and the Extraocular Muscles in Strabismus

Before embarking on an investigation through the oculomotor system in search of neural correlates for the strabismic state, it is worthwhile to ask whether motor aspects of eye misalignment should be considered a problem rooted in the brain or in the extraocular muscles (Ghasia and Shaikh 2013; Ghasia et al. 2015). There is generally no argument that developmental strabismus is triggered by disruption of binocular vision and therefore the etiology of the disorder is rooted in the brain. However, it is possible that adaptive changes in muscles can eventually result in a situation in which the brain is no longer driving the state of misalignment on a moment-to-moment basis (Guyton 2006). Evidence for adaptive changes in muscles has been demonstrated in monkey studies by Scott and colleagues in which sarcomeres have been shown to be added or subtracted following eye muscle surgery (Scott 1994). Other evidence for the involvement of the periphery include the identification of hypertrophic or hypotrophic muscles and mislocalization of muscle pulleys in some forms of human strabismus (Oh et al. 2002; Schoeff et al. 2013), changes in satellite cell activation in rabbit EOM following resection which is evidence of muscle remodeling and remodeling of muscle fiber characteristics, and neuromuscular junction density in patients with nystagmus (Antunes-Foschini et al. 2008; Christiansen et al. 2010; Berg et al. 2012).

On the other hand, evidence for an ongoing neural role in setting the state of misalignment and other eye movement abnormalities in strabismus has emerged since the incorporation of a nonhuman primate model for strabismus in oculomotor investigation. The rest of this chapter summarizes this evidence. It is our view that, when the etiology of strabismus is sensory, an important part of the strabismus driving force is neural during development and also in the steady state. However, this conclusion should not be interpreted as rejecting the influence of adaptive changes in extraocular muscle; the final state of misalignment in strabismus is likely due to a combination of brain and muscle involvement that can vary among individual patients.

4 Neural Structures Involved in Eye Misalignment

Figure 1 shows the parts of an anatomical circuit that hypothetically serves vergence eye movements in a normal animal (Noda et al. 1990; Gamlin 1999; Bohlen et al. 2016, 2017; May et al. 2018). Several of the structures shown in the figure have been studied in strabismic monkey models and have provided evidence that structures within this circuit are also involved in strabismus.

4.1 Oculomotor and Abducens Motor Nuclei

The first evidence that oculomotor structures within the brain were involved in maintaining the state of eye misalignment on a moment-to-moment basis came from neural recording studies in the motoneurons of the oculomotor nucleus and

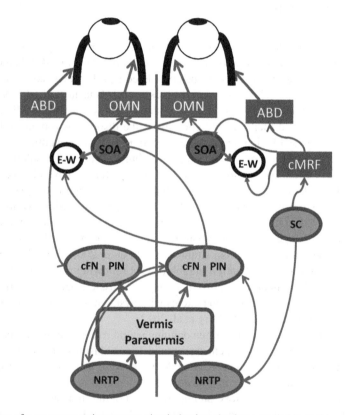

Fig. 1 Part of a neuroanatomic vergence circuit that has also been studied in strabismus monkey models. Anatomical circuit derived from published studies in the literature (Noda et al. 1990; May et al. 1992; Bohlen et al. 2016; May et al. 2018). Legend: OMN oculomotor nucleus, ABD abducens nucleus, SOA supraoculomotor area, EW Edinger-Westphal nucleus, cFN caudal fastigial nucleus, PIN posterior interposed nucleus, SC superior colliculus, cMRF central mesencephalic reticular formation, NRTP nucleus reticularis tegmenti pontis

abducens nuclei in a macaque animal model for strabismus. Das and Mustari (2007) showed that cyclovertical motoneurons within the oculomotor nucleus of strabismic monkeys were driving inappropriate vertical cross-axis movements that resulted in upshoot in abduction and DVD. Joshi and Das (2011) recorded from medial rectus motoneurons within the oculomotor nucleus in two animals with a sensory exotropia and found evidence for a neural drive for horizontal misalignment. Thus, in their study, neuronal responses were fit to a first-order model, and the estimated position and velocity sensitivities and background firing rate (i.e., firing rate when fixating a target at primary position) were found to be similar whether the eye to which the neuron projected was fixating the target or under cover (i.e., the deviated eye). Further, the estimated average values of the neuronal sensitivities were similar to published data in the literature obtained from normal animals. Therefore, in these two animals, the eye deviation or strabismus was fundamentally due to an inappropriate signal from the brain to the extraocular muscles. In another study, Walton et al. (2014a) recorded from abducens nucleus neurons in one monkey with a sensory esotropia and in another monkey with a surgically (bilateral medial rectus tenotomy) induced exotropia. The fundamental finding here was that although the position and velocity sensitivities were similar to that observed in normal monkeys, there was an overall reduction in abducens activity compared to that of the normal animal, providing additional evidence for neural involvement.

Both oculomotor and abducens nuclei responses from two sensory exotropes have been recently recorded (Pullela et al. 2018). The advantage of recording from both motor nuclei in the same animal is that the population neural drive to each of the horizontal muscles can be ascertained providing an overall view of balance of forces being applied to the horizontal recti from the brain. Figure 2 shows the average neural drive of the population of oculomotor and abducens motoneurons pro-

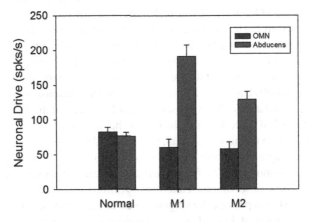

Fig. 2 Total neuronal drive to medial and lateral rectus muscles in normal monkeys during straight-ahead fixation as derived from the literature (Mays and Porter 1984; Fuchs et al. 1988; Sylvestre and Cullen 1999; Miller et al. 2011) and to the medial and lateral rectus muscles of the deviated eye in two exotropic monkeys (M1, M2) with strabismus angle of approximately 30°. Note the relative imbalance in the lateral and medial rectus firing rates in strabismic monkeys compared to the normal animal. Legend: OMN oculomotor nucleus

jecting to the medial and lateral rectus muscles of the deviated eye when the fellow eye is viewing a straight-ahead target. For comparison, the total neuronal drive from the oculomotor and abducens nuclei to the medial and lateral rectus muscles in normal monkeys during straight-ahead fixation as estimated from several published studies is also plotted. The main finding was that the total neuronal drive to the medial rectus of the deviated eye was significantly less than the total neuronal drive to the lateral rectus of the deviated eye in strabismic monkeys. This imbalance in neural drive is significantly contributing to the maintenance of exotropia in these animals.

4.2   Supraoculomotor Area

The presence of neural activity in the motoneurons that appears to drive strabismus does not mean that the neural substrates for strabismus are the motor nuclei themselves. Rather, central structures are likely involved in driving strabismus. The supraoculomotor area (SOA) is the region of the midbrain that is immediately dorsal and dorsolateral to the oculomotor nucleus (May et al. 2018). Neurons in the SOA, also called the midbrain near-response region, have monosynaptic connections to medial rectus motoneurons in the oculomotor nucleus and show responses related exclusively to convergence (near-response) or divergence (far-response) eye movements (Mays 1984; Zhang et al. 1991), i.e., neuronal responses are not modulated during conjugate eye movements such as saccades. Cells with vergence velocity sensitivity have also been identified in this area (Mays et al. 1986).

Cells in the SOA appear to carry a signal related to strabismus angle in animal models for strabismus (Das 2012; Pallus et al. 2018). Figure 3 shows data from a sample cell in the SOA of an animal with exotropia that shows an increase in firing rate when the exotropia angle was less, i.e., reduction in divergent strabismus angle was accompanied by an increase in firing rate similar to what might be expected of a near-response cell in a normal animal. Like the near-response cells in normal animals, these cells also do not modulate during conjugate eye movements. Both near- and far-response cells have been identified in the SOA of strabismic monkeys. When analyzing the population of SOA cells, Das (2012) found that the threshold at which the population of SOA cells began to fire in the strabismic monkeys was shifted toward exotropia (in the normal animal, the threshold is close to 0° of vergence) and the strabismus angle (vergence) sensitivity of the cells was significantly less than that of the normal monkey. This has since been replicated in a recent study by Pallus et al. (2018). These observations led to the hypothesis that the SOA cells which are normally purported to provide the vergence tone to medial rectus muscles necessary for binocular alignment instead provide "reduced tone" in exotropia, leading to maintenance of eye misalignment in these animals (Das 2012). Interestingly, Pallus et al. (2018) found that the strabismus angle sensitivity was reduced in animals with esotropia also. This would suggest that either mechanisms of esotropia are fundamentally different from exotropia or alternatively and in our opinion, more likely,

Fig. 3 Activity of a representative cell in the SOA that shows increased responses when angle of misalignment is small (~20°) compared to when angle of misalignment is large (~30°). Data acquired during alternate cover testing in an exotropic monkey. Angle of misalignment is smaller during right eye viewing than during left eye viewing. Legend: right eye, red; left eye, blue. (Adapted from Das 2012 – copyright *Association for Research in Vision and Ophthalmology*)

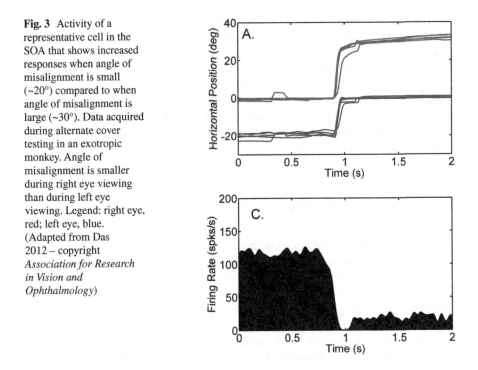

vergence tone signals are applied to both medial and lateral recti via central connections to the oculomotor and abducens nuclei and the balance of "vergence-related" activity to the two motor nuclei produces an outcome of esotropia or exotropia. A structure that provides the abducens motoneurons with vergence signals (like the SOA does to the medial rectus motoneurons) has not yet been identified. One possibility is the central mesencephalic reticular formation (cMRF) or the superior colliculus (SC), as shown in Fig. 1.

4.3 Cerebellum

The SOA receives direct projections from the caudal fastigial nucleus (cFN) and the posterior interposed nucleus (PIN) of the cerebellum (Noda et al. 1990; May et al. 1992). The cFN and PIN have also been shown to contain neurons related to both vergence and accommodation (Zhang and Gamlin 1998). Finally, lesions or degeneration of the midline cerebellum or superior cerebellar peduncle (output of the cerebellum) leads to deficits in alignment and binocular coordination (Ohtsuka et al. 1993; Versino et al. 1996). Therefore, evidence from anatomical, physiological, and lesion studies suggests a role for the cerebellum in binocular control of eye movements (Kheradmand and Zee 2011).

The cerebellum also appears to play a role in strabismus as shown by a study by Joshi and Das (2013). In this study, the authors used muscimol to reversibly inactivate the cFN and PIN in monkeys with strabismus. The main finding was that inactivation of the cFN caused a divergent change in misalignment (reduction of esotropia and increase in exotropia), while inactivation of the PIN caused a convergent change in misalignment (reduction of exotropia). Other aspects of strabismus such as DVD and A/V patterns were unchanged. Therefore, it appears that the reciprocal connections between the cFN/PIN and the SOA are important in the maintenance of the strabismus angle.

4.4 Superior Colliculus

The superior colliculus (SC) is a midbrain structure critical for driving saccadic eye movements (Gandhi and Katnani 2011). The rostral part of the SC is also involved in smooth pursuit (SP), although its role in SP appears to be limited to aspects of target selection/movement initiation and providing a position signal to the smooth-pursuit system during ongoing pursuit (Basso et al. 2000; Krauzlis and Dill 2002; Krauzlis 2004; Gandhi and Katnani 2011).

Although this role of the SC in saccade and smooth-pursuit eye movements has been the primary focus of most studies, there is also evidence supporting the role of the SC in vergence. Convergence-related neurons have been identified in the cat rostral superior colliculus (rSC) (Jiang et al. 1996), and electrical stimulation or pharmacological inactivation produces changes in both accommodation and vergence (Sawa and Ohtsuka 1994; Ohtsuka and Sato 1996; Suzuki et al. 2004). Billitz and Mays (1997) were not able to elicit vergence by electrical stimulation in the SC of the monkey during far viewing, but electrical stimulation during near viewing caused a relaxation of vergence. In another study in nonhuman primates, SC electrical stimulation applied just before or during a vergence-only movement or a combined saccade-vergence movement interfered with the vergence movement (Chaturvedi and van Gisbergen 1999, 2000). Ablations of the rSC in monkeys result in problems with both disparity processing and eye alignment (Lawler and Cowey 1986). Saccade-related neurons in the caudal colliculus show a weak relationship to vergence in that many burst neurons show a reduction in saccade velocity sensitivity when looking at near targets compared to when looking at far targets (Walton and Mays 2003). A recent study by Van Horn and colleagues indeed identified convergence and divergence neurons in the rSC that were modulated during slow vergence, but not conjugate or fast vergence eye movements, thereby postulating that the rSC only contributes to slow vergence (Van Horn et al. 2013). Also, electrical stimulation in this area produced vergence angle changes when looking at near targets.

Considering the evidence for SC involvement in vergence and the potential for disruption in vergence circuits as a neural substrate for strabismus, our lab recently examined the role of the SC in maintaining the state of misalignment in strabismus

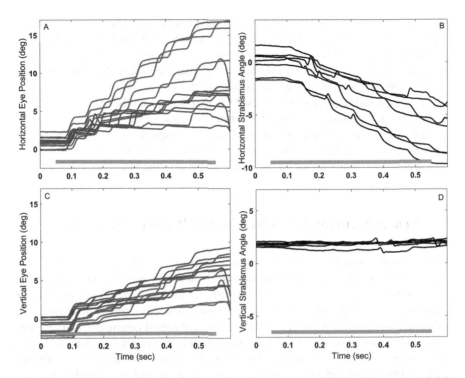

Fig. 4 Raw data showing eye movement responses following electrical stimulation in the left SC of a strabismic monkey during monocular left eye viewing. Panel (**a, c**): Multiple traces of horizontal and vertical eye position of the right (red) and left (blue) eyes. Data are aligned on the start of stimulation. Contralateral (rightward) and upward staircase saccades are evoked during the period of electrical stimulation (shown by green bar). Panel (**b, d**): Horizontal strabismus angle (left eye position – right eye position) shows a significant divergent change in alignment but little vertical change in alignment during stimulation. In all panels, positive values indicate rightward or upward eye positions and negative values indicate leftward or downward eye positions. (Reproduced from Upadhyaya et al. 2016 – copyright *Journal of Neurophysiology*)

monkey models using electrical stimulation techniques (Upadhyaya et al. 2017). Figure 4 shows the effect of electrical stimulation at a rostral site in the SC of an animal with strabismus. In addition to the staircase of saccades that is distinctive of SC stimulation, there was also a significant change in horizontal strabismus angle with little change in vertical strabismus angle. It was found that a variety of either convergent or divergent change in strabismus angle was obtained by electrical stimulation at 51 sites in the SC of three strabismic monkeys. Further analysis showed that the change in strabismus angle was brought about by both disconjugate saccades and disconjugate post-saccadic drift. It was hypothesized that the disconjugate post-saccadic drift could be due to the activation of pools of convergence- or divergence-related neurons in the SC. Recent other studies have also used electrical stimulation and recording techniques to study the SC in strabismic monkeys (Economides et al. 2016, 2018; Fleuriet et al. 2016). Although those studies were primarily directed at examining SC topographic maps, which were determined to be normal, they also showed that electrical stimulation elicited saccades that were disconjugate with significant disconjugate post-saccadic drifts. In follow-up work, the presence of cells related to eye misalignment in the rostral part of the SC was recently reported (Upadhyaya and Das 2018). These are likely the same vergence-related cells reported in normal monkeys (Van Horn et al. 2013).

The main significance of these results is that they place the SC within the vergence circuit that could be involved in setting the angle of misalignment. Additional studies are needed to ascertain the pathway by which the aberrant SC vergence signals reach the abducens and/or oculomotor nuclei. As shown in Fig. 1, the SC has connections to the abducens via the cMRF and also to the SOA and the OMN.

5 Neural Structures Involved in A/V Pattern Strabismus

Pattern strabismus is a fairly common feature associated with strabismus (von Noorden and Campos 2002). Fundamentally, a decrease in exotropia or increase in esotropia in upgaze versus downgaze is defined as an A-pattern strabismus, while an increase in exotropia or decrease in esotropia in upgaze versus downgaze is a V-pattern strabismus. While A and V patterns are the most common, pattern strabismus with X and Y patterns has also been described. This feature has been replicated in monkey models of strabismus and has been a target of investigation. Although pattern strabismus generally refers to changes in static strabismus angles observed during up or down fixation, they are also reproduced during eye movements. Figure 5 shows data from when the animal was performing a horizontal or vertical SP task. Just as might be expected during fixation, during SP also, the angle of strabismus is less during upgaze compared to downgaze. Note that in addition to

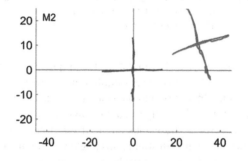

Fig. 5 Strabismus patterns observed during SP in an exotropic strabismic monkeys. Data show the average of several trials of horizontal and vertical SP as the strabismic monkey tracked the moving target with his left eye. The right eye is deviated to the right (exotropia); the angle of misalignment reduces for upward gaze positions compared to downward positions (A-pattern strabismus), and there is a vertical deviation that varies with horizontal gaze position (upshoot in abduction). Positive values indicate rightward and upward positions and negative values indicate leftward and downward positions. Legend: right eye, red; left eye; blue. (Adapted from Agaoglu et al. 2015 – copyright *Association for Research in Vision and Ophthalmology*)

changes in horizontal strabismus angle with vertical gaze, there are also changes in vertical strabismus angle (in this case the monkey showed a dissociated vertical deviation or DVD) with horizontal gaze. The oculomotor manifestation of pattern strabismus is fundamentally an inappropriate cross-axis movement in the deviated eye. Therefore, when the viewing eye makes a purely vertical SP or saccadic eye movement, the deviated eye shows SP with both a vertical and an inappropriate horizontal component (Das et al. 2005).

Das and Mustari (2007) recorded from vertical burst-tonic motoneurons projecting to cyclovertical muscles during horizontal and vertical SP eye movements and found that these neurons were similarly modulated during vertical SP and during the inappropriate cross-axis vertical component that was observed during horizontal SP. Joshi and Das (2011) also reported a similar result from medial rectus motoneurons for the cross-axis horizontal component observed during vertical SP. Fundamentally, these studies provided neurophysiological evidence that the mechanism behind pattern strabismus was neural (Walton et al. 2017).

In an attempt to understand the central mechanisms behind the apparent disconjugacy between the two eyes due to cross-axis movements associated with pattern strabismus, Walton and colleagues have conducted a series of behavioral, electrical stimulation, neural recording, and mathematical modeling studies of cross-axis movements during saccades (Walton et al. 2013; 2014b; Walton and Mustari 2015; Walton and Mustari 2017). In a behavioral study of vertical and oblique saccades, they determined that saccade disconjugacy was due to a disconjugacy in both saccade amplitude and saccade direction. This finding was later corroborated in studies with human patients (Ghasia et al. 2015; Shaikh and Ghasia 2015). Electrical stimulation in the paramedian pontine reticular formation (PPRF – a critical structure for generation of horizontal saccadic eye movements containing the excitatory burst neurons) of strabismic monkeys resulted in disconjugate eye movements with both horizontal and vertical components, suggesting that the PPRF was not organized in the same manner as in a normal monkey (Walton et al. 2013). The electrical stimulation study was followed up by a neural recording study in the PPRF by the same authors, and they found that the PPRF neurons often showed preferred directions that deviated from the horizontal axis (Walton and Mustari 2015). These findings led to the hypothesis that horizontal and vertical saccade control signals are mixed in the PPRF, therefore leading to cross-axis movements and A/V pattern strabismus. Electrical stimulation and recording studies have also been attempted in the SC of animals with strabismus, and although the stimulation leads to disconjugate saccades, the consensus is that the topographic motor maps within the SC are still relatively "normal," i.e., are offset by the magnitude of ocular deviation but are otherwise aligned for the viewing and deviated eyes (Fleuriet et al. 2016; Economides et al. 2018). This suggests that the disconjugacies leading to A/V pattern strabismus are not localized to the SC.

A fundamental assumption necessary to simulate A/V patterns is that there is cross-talk between horizontal and vertical systems (Walton et al. 2017). Recently, Walton and colleagues used control systems modeling to simulate the saccade disconjugacies observed in their monkeys with A/V pattern strabismus (Walton and

Mustari 2017). Among the three models that they tested, they favor a hypothesis wherein the cross-talk between the horizontal and vertical systems is distributed across several nodes in the visual-oculomotor system. An important node where cross-talk may manifest is the neural integrator since cross-axis movements are present in all eye movements and not just saccades. Additional experiments are necessary to establish the role of the neural integrator in A/V pattern strabismus.

6 Visual Suppression and Fixation-Switch Behavior in Strabismus

An interesting behavioral observation in humans and monkeys with strabismus is fixation-switch behavior (van Leeuwen et al. 2001; Das 2009). Thus, during binocular viewing conditions, strabismic subjects will choose to fixate targets with a specific eye depending on the spatial location of the target and may switch fixation depending on the target step. Strabismus with this property is also called alternating strabismus. The spatial patterns of fixation support the idea that fixation-switch behavior follows from visual suppression of a portion of the temporal retina in exotropes and a portion of the nasal retina in esotropes. However, this is likely to be only a partial explanation since targets appearing on the fovea or portions of temporal retina immediately adjacent to the fovea in an exotrope (nasal retina in an esotrope) are still acquired by the "closer" eye (Economides et al. 2012, 2014; Agaoglu et al. 2014). Fundamentally, spatial patterns of fixation in alternating strabismus are such that exotropes will fixate targets on the right with the right eye and targets on the left with their left eye and the boundary at which fixation switch occurs is approximately in between the visual axes of the two eyes, i.e., at approximately half the strabismus angle (Agaoglu et al. 2014). Figure 6 shows an example of multiple saccade trials in which the strabismic animals were viewing binocularly. In the exotrope, targets on the left are acquired by the left eye and targets on the right are acquired by the right eye. Targets that are in between the visual axes of the two eyes may be acquired by either eye. The reverse pattern is observed in the esotrope. Targets on the left are acquired by the right eye and targets on the right are acquired by the left eye. Once again, targets that are in between the visual axes of the two eyes may be acquired by either eye. In the presence of amblyopia, this boundary of right eye versus left eye fixation is shifted toward the amblyopic eye. The ability to switch fixation and the spatial location at which fixation switch occurred may also been used as a marker for the presence of amblyopia (Dickey et al. 1991).

Psychophysical experiments in human exotropes have also confirmed that regions of temporal retina adjacent to the fovea are in fact not suppressed (Economides et al. 2012, 2014). Other experiments in esotropic monkeys have shown that portions of nasal retina adjacent to the fovea are not suppressed (Agaoglu et al. 2014). The human psychophysical experiments that were fundamentally detection-based and the monkey experiments that were fundamentally saccade-

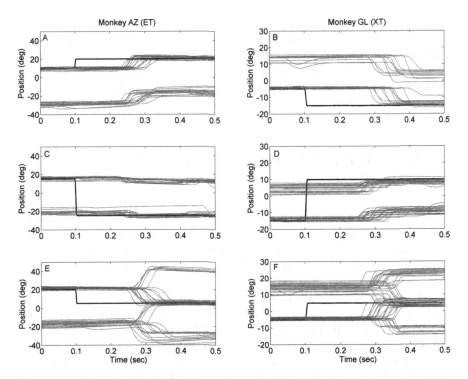

Fig. 6 Saccade behavior under binocular viewing conditions in esotrope (Panels **a, c, e**) and exotrope (Panels **b, d, f**) illustrating spatial patterns of fixation switch. Rightward eye positions are positive, while leftward eye positions are negative. The red traces are right eye position, the blue traces are left eye position, and the black line is the target position. Panels (**a**) and (**b**) illustrate a target step that elicits no fixation switch, i.e., the left eye is fixating the target before and after the target step. Panels (**c**) and (**d**) illustrate a target step for which every trial results in a fixation switch, i.e., the left eye is fixating the target before the target step and the right eye is fixating the target after the target step. Panels (**e**) and (**f**) illustrate a target step for which only some trials elicit a fixation switch. (Figure reproduced from Agaoglu et al. 2014 – copyright *Association for Research in Vision and Ophthalmology*)

based did not provide information on whether depth of visual suppression varies across the retina and whether this might influence target acquisition behavior. Agaoglu and colleagues conducted an experiment in which they showed that motion information provided to the fovea of the deviated eye was indeed processed, albeit weakly, and the motion-evoked optokinetic response could be further modulated by changing the contrast of the visual stimulus (Agaoglu et al. 2015).

The neural circuit by which fixation-switch behavior is driven has not yet been determined. One hypothesis is that target selection areas such as the frontal eye fields, lateral intraparietal area, and the SC are involved in selecting which eye should acquire the target and in generating an appropriately sized and directed saccade. Recently, Economides and colleagues recorded from cells in the SC of monkeys with surgically induced exotropia (Economides et al. 2018). Their primary

finding was that visual receptive fields in the strabismic monkeys have similar retinotopic organization as the normal. There was no evidence for anomalous correspondence, and also both eyes were normally binocularly responsive, i.e., no
evidence for suppression of non-fixating eye. Additional studies of various cell
types in the SC (visuomotor and motor cells) are necessary to determine whether the
target selection circuitry is responsible for acquiring targets with either eye.

7 Neural Plasticity Following Surgical Treatment
of Strabismus

The most common treatment strategy for strabismus is the surgical manipulation of
EOMs to realign the eyes. Surgical approaches have varying levels of success and
permanence, and often patients have to undergo multiple procedures to resolve
residual misalignment (Scheiman and Ciner 1987; Scheiman et al. 1989; Louwagie
et al. 2009; Mohney et al. 2011).

The failure of strabismus correction surgery could be due to adaptive changes
following treatments that occur at the periphery (muscle remodeling) or within central brain areas (central neural adaptation). Adaptive changes at the level of the
EOMs have been reported following surgical manipulation. As described earlier in
this chapter, a study by Scott showed that sarcomere lengths were altered in the
horizontal EOM following surgical manipulation (Scott 1994). Christiansen and
McLoon studied the effects of resection surgery in rabbit EOM and found that there
was an increase in satellite cell activation in both the treated muscle and its antagonist muscle along with the incorporation of new myonuclei, all indicators of muscle
remodeling in response to the surgery (Christiansen and McLoon 2006).

However, changes in inherent EOM contractility may not provide a complete
description of adaptive changes that influence the strabismus angle following treatment as the brain could also adapt innervation patterns following surgery. A recent
series of studies investigated the behavioral and neuronal plasticity following a
typical resect-recess strabismus correction surgery in two adult monkeys with an
exotropia previously induced by prism-rearing (Pullela et al. 2016, 2018). Figure 7a
shows the longitudinal change in alignment starting from day 1 after the surgery of
one of the study animals. In the two animals in the study, misalignment was reduced
by ~70% immediately after surgery but had regressed toward pre-surgical values by
6 months after surgery. Other than misalignment, changes in dynamic saccade and
SP gains were identified, but these were relatively small in magnitude (Pullela
et al. 2016).

Longitudinal recording of population neuronal activity within oculomotor and
abducens nuclei before and after surgical correction showed some interesting patterns of neural adaptation. Figure 7b shows the data from one of the study animals,
in which there was a short-term (within the first month) increase in the population
neural innervation to the recessed lateral rectus muscle. In the other monkey (not

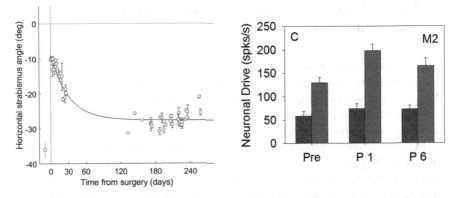

Fig. 7 Longitudinal changes in eye alignment (Panel **a**) and population neuronal drive (Panel **b**) following strabismus correction surgery in an exotropic monkey. Panel (**a**): Each data point in the plot shows the alignment data on days that neuronal data were also acquired. Data point before "0" on the time axis (x-axis) is the average strabismus angle prior to surgery. The plot shows that immediately after surgery, there is a significant reduction in strabismus angle that reverted toward pre-surgical values by about 6 months after surgery. Panel (**b**) shows the average population neuronal activity of neurons within the oculomotor nuclei (purple bars) and the abducens nuclei (green bars) projecting to the muscles of the deviated treated eye when the untreated eye is viewing a straight-ahead target at timepoints "Pre" (before surgical treatment), "P1" (within the first month after treatment), and "P6" (~6 months after treatment). Note the increase in neuronal drive within the abducens nuclei at P1 that effectively acts to negate the intent of surgery. The reversal in alignment observed in Panel A over the first month is likely due to this "negative" adaptive neural effect observed in the abducens neuron responses. (Adapted from Pullela et al. 2018 – copyright *Association for Research in Vision and Ophthalmology*)

shown), there was a short-term (also within one month of treatment) reduction in the population neural innervation to the resected medial rectus muscle. Both of these forms of neural adaptive changes act to drive the eyes toward exotropia, or in other words, these adaptive changes are "fighting" the intent of the treatment, resulting in a relapse to pre-surgical values of strabismus angle (Fig. 7a). The neuronal drive to horizontal recti recovered by six months, but the angle of misalignment was still relapsed, suggesting that there were also adaptive changes in the muscle that occurred over the longer term of ~6 months. Note that the presence of changes in neural drive from motor nuclei does not mean that this area is the site of adaptation. It is likely that central areas of the brain, perhaps vergence circuits, are the site for neural adaptation. In this study, adaptive changes were also observed in the neural drive to the untreated eye, suggesting that premotor areas of the brain that project to both eyes are the site to adaptation.

The counterproductive neural and muscle adaptive changes that were observed only directly explain cases when strabismus surgery fails. Although not proven, it is likely that adaptation within these same mechanisms is responsible for the maintenance of the aligned state when the surgical treatment results in long-term success. When strabismus surgery succeeds, it is likely that residual binocular vision provides the necessary central adaptive signal that changes patterns of innervation from motor nuclei to the EOM to levels appropriate to maintain alignment.

8 Future Area of Investigation: Proprioception and Strabismus

Although great strides have been made in understanding neurophysiological mechanisms underlying strabismus, there are still significant knowledge gaps and one of them is in the role that proprioception might play in the development and maintenance of strabismus. There is some evidence that proprioception plays a role in ocular alignment, but the mechanism is not understood.

EOM stretch or proprioceptive signals are present in a number of regions of the brain, including the oculomotor nucleus, cerebellum, vestibular nuclei, and nucleus prepositus hypoglossi, and most recently have been identified within neurons in area 3a of primary somatosensory cortex (Fuchs and Kornhuber 1969; Tomlinson and Schwarz 1977; Ashton et al. 1988; Kimura et al. 1991; Wang et al. 2007). Lesions of the ophthalmic division of the trigeminal nerve in normal adult monkeys did not produce immediate changes in saccades, pursuit, or the vestibulo-ocular reflex, but in animals with unilateral EOM palsy, trigeminal nerve lesion produced eye misalignment and disconjugate eye movements over a period of weeks (Lewis et al. 1994, 1999, 2001). Also supporting a role of proprioception in eye alignment, patients undergoing multiple strabismus correction surgeries showed changes in visual localization (Steinbach and Smith 1981). Spatial localization deficits were also found in a patient with a congenital trigeminal-oculomotor synkinesis (often associated with strabismus), suggesting aberrant proprioceptive signaling (Lewis and Zee 1993). Support for a proprioceptive role in alignment has also come from studies in a monkey model for superior oblique palsy (SOP) created by sectioning the trochlear nerve transcranially at the level of the cavernous sinus (Shan et al. 2007a, b; Tian et al. 2007). Immediately after the procedure, the animals showed characteristic misalignment of human SOP (Shan et al. 2008). There was also a short-term reduction in misalignment, followed by a gradual increase, despite patching (blocking vision) one or both eyes after surgery, suggesting that proprioceptive mechanisms may be responsible for changes in alignment (Quaia et al. 2008; Shan et al. 2011). A consensus hypothesis has emerged, suggesting that efference copy is primarily necessary for online control of eye movements, while proprioception contributes to the long-term calibration of efference copy and could contribute to the calibration of eye alignment and binocular coordination of eye movements (Steinbach 1986, 1987; Lewis et al. 1994; Donaldson 2000).

Some early studies in the cat have also suggested a role for ocular proprioception in the development of visual cortical properties including binocular vision (see Buisseret (1995) for review). Orientation selectivity in area 17 is impaired in kittens whose EOMs are paralyzed during visual exposure (Buisseret and Gary-Bobo 1979) and can be selectively impaired in dark-reared kittens by detaching specific EOMs, suggesting that active eye movements (possibly stretch of EOMs) are necessary for normal development (Gordon and Gummow 1975; Gary-Bobo et al. 1986). A study in paralyzed and monocularly occluded 4-week-old kittens also suggested that visual stimulation and active eye movements are both needed to produce binocularity in visual cortex (Freeman and Bonds 1979). Loss of cortical binocularity even occurs in adult cats after surgical immobilization of one eye (Fiorentini and Maffei 1974), and this appears to be due to the paralysis itself, rather than abnormal visual input (Maffei and Fiorentini 1976). Studies by Trotter and colleagues showed that sectioning of the ophthalmic division of the trigeminal nerve leads to a significant reduction in the number of area 17 cells with binocular fields, with greatest loss appearing in kittens treated at 5–7 weeks old (near the peak of their critical period) (Trotter et al. 1987). Most cells in area 17 of adult cats that were raised with a trigeminal section performed at 5–12 weeks of age also showed significant deficits in response to interocular phase disparity stimuli, which is a better test of binocular function (Trotter et al. 1993). Although difficult to test precisely in cats, studies of stereoscopic vision and disparity detection show behavioral deficits that mirror the cortical binocularity deficits (Fiorentini et al. 1985; Fiorentini et al. 1986; Graves et al. 1987).

It should be noted that there is controversy on the identity of the proprioceptive receptor in the EOM and the pathway by which proprioceptive signals reach central brain areas. The palisade endings held much promise as the proprioceptive receptor, which incidentally were found to send signals to central brain areas via the oculomotor nerves and not the trigeminal nerve (Lienbacher et al. 2011). If EOM proprioception signals are conveyed by the oculomotor nerve, then the interpretation of studies that employed section of the ophthalmic division of the trigeminal nerve as a way to disrupt proprioception is called into question. However, subsequent findings that palisade endings contained boutons with cholinergic vesicles, which are sometimes associated with postsynaptic acetylcholine receptors, have led to the interpretation that palisades are motor, not sensory (Lukas et al. 2000; Blumer et al. 2009; Zimmermann et al. 2013). In essence, the proprioceptive receptor in the EOM and how these signals are conveyed to central brain areas such as primary somatosensory cortex are still unknown. Therefore, pathways and mechanisms of proprioception in normal oculomotor control will also have to be investigated in parallel in order to interpret its possible role in strabismus.

Acknowledgments The author was supported by NIH grant R01-EY026568 and UHCO core grant P30-EY07551. The author declares no other competing financial interests and thanks Suraj Upadhyaya and Mythri Pullela for their help with figures and for reading and providing helpful suggestions on the manuscript.

References

Agaoglu, M. N., LeSage, S. K., Joshi, A. C., & Das, V. E. (2014). Spatial patterns of fixation-switch behavior in strabismic monkeys. *Investigative Ophthalmology & Visual Science, 55*(3), 1259–1268.

Agaoglu, S., Agaoglu, M. N., & Das, V. E. (2015). Motion information via the nonfixating eye can drive optokinetic nystagmus in strabismus. *Investigative Ophthalmology & Visual Science, 56*(11), 6423–6432.

Antunes-Foschini, R. S., Miyashita, D., Bicas, H. E., & McLoon, L. K. (2008). Activated satellite cells in medial rectus muscles of patients with strabismus. *Investigative Ophthalmology & Visual Science, 49*(1), 215–220.

Ashton, J. A., Boddy, A., Dean, S. R., Milleret, C., & Donaldson, I. M. (1988). Afferent signals from cat extraocular muscles in the medial vestibular nucleus, the nucleus praepositus hypoglossi and adjacent brainstem structures. *Neuroscience, 26*(1), 131–145.

Basso, M. A., Krauzlis, R. J., & Wurtz, R. H. (2000). Activation and inactivation of rostral superior colliculus neurons during smooth-pursuit eye movements in monkeys. *Journal of Neurophysiology, 84*(2), 892–908.

Berg, K. T., Hunter, D. G., Bothun, E. D., Antunes-Foschini, R., & McLoon, L. K. (2012). Extraocular muscles in patients with infantile nystagmus: Adaptations at the effector level. *Archives of Ophthalmology, 130*(3), 343–349.

Billitz, M. S., & Mays, L. E. (1997). Effects of microstimulation of the superior colliculus on vergence and accommodation. *Investigative Ophthalmology & Visual Science, 38*(984).

Blumer, R., Konakci, K. Z., Pomikal, C., Wieczorek, G., Lukas, J. R., & Streicher, J. (2009). Palisade endings: Cholinergic sensory organs or effector organs? *Investigative Ophthalmology & Visual Science, 50*(3), 1176–1186.

Bohlen, M. O., Warren, S., & May, P. J. (2016). A central mesencephalic reticular formation projection to the supraoculomotor area in macaque monkeys. *Brain Structure & Function, 221*(4), 2209–2229.

Bohlen, M. O., Warren, S., & May, P. J. (2017). A central mesencephalic reticular formation projection to medial rectus motoneurons supplying singly and multiply innervated extraocular muscle fibers. *The Journal of Comparative Neurology, 525*(8), 2000–2018.

Boothe, R. G., Dobson, V., & Teller, D. Y. (1985). Postnatal development of vision in human and nonhuman primates. *Annual Review of Neuroscience, 8*, 495–545.

Buisseret, P. (1995). Influence of extraocular muscle proprioception on vision. *Physiological Reviews, 75*(2), 323–338.

Buisseret, P., & Gary-Bobo, E. (1979). Development of visual cortical orientation specificity after dark-rearing: Role of extraocular proprioception. *Neuroscience Letters, 13*(3), 259–263.

Chaturvedi, V., & van Gisbergen, J. A. (1999). Perturbation of combined saccade-vergence movements by microstimulation in monkey superior colliculus. *Journal of Neurophysiology, 81*(5), 2279–2296.

Chaturvedi, V., & Van Gisbergen, J. A. (2000). Stimulation in the rostral pole of monkey superior colliculus: Effects on vergence eye movements. *Experimental Brain Research, 132*(1), 72–78.

Christiansen, S. P., & McLoon, L. K. (2006). The effect of resection on satellite cell activity in rabbit extraocular muscle. *Investigative Ophthalmology & Visual Science, 47*(2), 605–613.

Christiansen, S. P., Antunes-Foschini, R. S., & McLoon, L. K. (2010). Effects of recession versus tenotomy surgery without recession in adult rabbit extraocular muscle. *Investigative Ophthalmology & Visual Science, 51*(11), 5646–5656.

Crawford, M. L., & von Noorden, G. K. (1979). The effects of short-term experimental strabismus on the visual system in Macaca mulatta. *Investigative Ophthalmology & Visual Science, 18*(5), 496–505.

Crawford, M. L., & von Noorden, G. K. (1980). Optically induced concomitant strabismus in monkeys. *Investigative Ophthalmology & Visual Science, 19*(9), 1105–1109.

Crawford, M. L., Harwerth, R. S., Chino, Y. M., & Smith, E. L., 3rd. (1996). Binocularity in prism-reared monkeys. *Eye, 10*(Pt 2), 161–166.

Das, V. E. (2009). Alternating fixation and saccade behavior in nonhuman primates with alternating occlusion-induced exotropia. *Investigative Ophthalmology & Visual Science, 50*(8), 3703–3710.

Das, V. E. (2012). Responses of cells in the midbrain near-response area in monkeys with strabismus. *Investigative Ophthalmology & Visual Science, 53*(7), 3858–3864.

Das, V. E. (2016). Strabismus and oculomotor system: Insights from macaque models. *Annual Review of Vision Science, 2*, 37–59.

Das, V. E., & Mustari, M. J. (2007). Correlation of cross-axis eye movements and motoneuron activity in non-human primates with "A" pattern strabismus. *Investigative Ophthalmology & Visual Science, 48*(2), 665–674.

Das, V. E., Fu, L. N., Mustari, M. J., & Tusa, R. J. (2005). Incomitance in monkeys with strabismus. *Strabismus, 13*(1), 33–41.

Dickey, C. F., Metz, H. S., Stewart, S. A., & Scott, W. E. (1991). The diagnosis of amblyopia in cross-fixation. *Journal of Pediatric Ophthalmology and Strabismus, 28*(3), 171–175.

Donaldson, I. M. (2000). The functions of the proprioceptors of the eye muscles. *Philosophical Transactions of the Royal Society of London. Series B, Biological Sciences, 355*(1404), 1685–1754.

Economides, J. R., Adams, D. L., Jocson, C. M., & Horton, J. C. (2007). Ocular motor behavior in macaques with surgical exotropia. *Journal of Neurophysiology, 98*(6), 3411–3422.

Economides, J. R., Adams, D. L., & Horton, J. C. (2012). Perception via the deviated eye in strabismus. *The Journal of Neuroscience, 32*(30), 10286–10295.

Economides, J. R., Adams, D. L., & Horton, J. C. (2014). Eye choice for acquisition of targets in alternating strabismus. *The Journal of Neuroscience, 34*(44), 14578–14588.

Economides, J. R., Adams, D. L., & Horton, J. C. (2016). Normal correspondence of tectal maps for saccadic eye movements in strabismus. *Journal of Neurophysiology, 116*(6), 2541–2549.

Economides, J. R., Rapone, B. C., Adams, D. L., & Horton, J. C. (2018). Normal topography and binocularity of the superior colliculus in strabismus. *The Journal of Neuroscience, 38*(1), 173–182.

Fiorentini, A., & Maffei, L. (1974). Letter: Change of binocular properties of the simple cells of the cortex in adult cats following immobilization of one eye. *Vision Research, 14*(2), 217–218.

Fiorentini, A., Maffei, L., Cenni, M. C., & Tacchi, A. (1985). Deafferentation of oculomotor proprioception affects depth discrimination in adult cats. *Experimental Brain Research, 59*(2), 296–301.

Fiorentini, A., Cenni, M. C., & Maffei, L. (1986). Impairment of stereoacuity in cats with oculomotor proprioceptive deafferentation. *Experimental Brain Research, 63*(2), 364–368.

Fleuriet, J., Walton, M. M., Ono, S., & Mustari, M. J. (2016). Electrical microstimulation of the superior colliculus in strabismic monkeys. *Investigative Ophthalmology & Visual Science, 57*(7), 3168–3180.

Freeman, R. D., & Bonds, A. B. (1979). Cortical plasticity in monocularly deprived immobilized kittens depends on eye movement. *Science, 206*(4422), 1093–1095.

Fuchs, A. F., & Kornhuber, H. H. (1969). Extraocular muscle afferents to the cerebellum of the cat. *The Journal of Physiology, 200*(3), 713–722.

Fuchs, A. F., Scudder, C. A., & Kaneko, C. R. (1988). Discharge patterns and recruitment order of identified motoneurons and internuclear neurons in the monkey abducens nucleus. *Journal of Neurophysiology, 60*(6), 1874–1895.

Gamlin, P. D. (1999). Subcortical neural circuits for ocular accommodation and vergence in primates. *Ophthalmic & Physiological Optics, 19*(2), 81–89.

Gandhi, N. J., & Katnani, H. A. (2011). Motor functions of the superior colliculus. *Annual Review of Neuroscience, 34*, 205–231.

Gary-Bobo, E., Milleret, C., & Buisseret, P. (1986). Role of eye movements in developmental processes of orientation selectivity in the kitten visual cortex. *Vision Research, 26*(4), 557–567.

Ghasia, F. F., & Shaikh, A. G. (2013). Pattern strabismus: Where does the brain's role end and the muscle's begin? *Journal of Ophthalmology, 2013*, 301256.

Ghasia, F. F., Shaikh, A. G., Jacobs, J., & Walker, M. F. (2015). Cross-coupled eye movement supports neural origin of pattern strabismus. *Investigative Ophthalmology & Visual Science, 56*(5), 2855–2866.

Gordon, B., & Gummow, L. (1975). Effects of extraocular muscle section on receptive fields in cat superior colliculus. *Vision Research, 15*, 1011–1019.

Govindan, M., Mohney, B. G., Diehl, N. N., & Burke, J. P. (2005). Incidence and types of childhood exotropia: A population-based study. *Ophthalmology, 112*(1), 104–108.

Graves, A. L., Trotter, Y., & Fregnac, Y. (1987). Role of extraocular muscle proprioception in the development of depth perception in cats. *Journal of Neurophysiology, 58*(4), 816–831.

Greenberg, A. E., Mohney, B. G., Diehl, N. N., & Burke, J. P. (2007). Incidence and types of childhood esotropia: A population-based study. *Ophthalmology, 114*(1), 170–174.

Guyton, D. L. (2006). The 10th Bielschowsky lecture. Changes in strabismus over time: The roles of vergence tonus and muscle length adaptation. *Binocular Vision & Strabismus Quarterly, 21*(2), 81–92.

Jiang, H., Guitton, D., & Cullen, K. E. (1996). Near-response-related neural activity in the rostral superior colliculus of the cat. *Society for Neuroscience – Abstracts, 22*, 662.

Joshi, A. C., & Das, V. E. (2011). Responses of medial rectus motoneurons in monkeys with strabismus. *Investigative Ophthalmology & Visual Science, 52*(9), 6697–6705.

Joshi, A. C., & Das, V. E. (2013). Muscimol inactivation of caudal fastigial nucleus and posterior interposed nucleus in monkeys with strabismus. *Journal of Neurophysiology, 110*(8), 1882–1891.

Kheradmand, A., & Zee, D. S. (2011). Cerebellum and ocular motor control. *Frontiers in Neurology, 2*, 53.

Kimura, M., Takeda, T., & Maekawa, K. (1991). Contribution of eye muscle proprioception to velocity-response characteristics of eye movements: Involvement of the cerebellar flocculus. *Neuroscience Research, 12*(1), 160–168.

Kiorpes, L. (2016). The puzzle of visual development: Behavior and neural limits. *The Journal of Neuroscience, 36*(45), 11384–11393.

Krauzlis, R. J. (2004). Recasting the smooth pursuit eye movement system. *Journal of Neurophysiology, 91*(2), 591–603.

Krauzlis, R., & Dill, N. (2002). Neural correlates of target choice for pursuit and saccades in the primate superior colliculus. *Neuron, 35*(2), 355–363.

Lawler, K. A., & Cowey, A. (1986). The effects of pretectal and superior collicular lesions on binocular vision. *Experimental Brain Research, 63*(2), 402–408.

Lewis, R. F., & Zee, D. S. (1993). Abnormal spatial localization with trigeminal-oculomotor synkinesis. Evidence for a proprioceptive effect. *Brain, 116*(Pt 5), 1105–1118.

Lewis, R. F., Zee, D. S., Gaymard, B. M., & Guthrie, B. L. (1994). Extraocular muscle proprioception functions in the control of ocular alignment and eye movement conjugacy. *Journal of Neurophysiology, 72*(2), 1028–1031.

Lewis, R. F., Zee, D. S., Goldstein, H. P., & Guthrie, B. L. (1999). Proprioceptive and retinal afference modify postsaccadic ocular drift. *Journal of Neurophysiology, 82*(2), 551–563.

Lewis, R. F., Zee, D. S., Hayman, M. R., & Tamargo, R. J. (2001). Oculomotor function in the rhesus monkey after deafferentation of the extraocular muscles. *Experimental Brain Research, 141*(3), 349–358.

Lienbacher, K., Mustari, M., Ying, H. S., Buttner-Ennever, J. A., & Horn, A. K. (2011). Do palisade endings in extraocular muscles arise from neurons in the motor nuclei? *Investigative Ophthalmology & Visual Science, 52*(5), 2510–2519.

Louwagie, C. R., Diehl, N. N., Greenberg, A. E., & Mohney, B. G. (2009). Long-term follow-up of congenital esotropia in a population-based cohort. *Journal of AAPOS, 13*(1), 8–12.

Lukas, J. R., Blumer, R., Denk, M., Baumgartner, I., Neuhuber, W., & Mayr, R. (2000). Innervated myotendinous cylinders in human extraocular muscles. *Investigative Ophthalmology & Visual Science, 41*(9), 2422–2431.

Maffei, L., & Fiorentini, A. (1976). Asymmetry of motility of the eyes and change of binocular properties of cortical cells in adult cats. *Brain Research, 105*(1), 73–78.

May, P. J., Porter, J. D., & Gamlin, P. D. (1992). Interconnections between the primate cerebellum and midbrain near-response regions. *The Journal of Comparative Neurology, 315*(1), 98–116.

May, P. J., Warren, S., Gamlin, P. D. R., & Billig, I. (2018). An anatomic characterization of the midbrain near response neurons in the macaque monkey. *Investigative Ophthalmology & Visual Science, 59*(3), 1486–1502.

Mays, L. E. (1984). Neural control of vergence eye movements: Convergence and divergence neurons in midbrain. *Journal of Neurophysiology, 51*(5), 1091–1108.

Mays, L. E., & Porter, J. D. (1984). Neural control of vergence eye movements: Activity of abducens and oculomotor neurons. *Journal of Neurophysiology, 52*(4), 743–761.

Mays, L. E., Porter, J. D., Gamlin, P. D., & Tello, C. A. (1986). Neural control of vergence eye movements: Neurons encoding vergence velocity. *Journal of Neurophysiology, 56*(4), 1007–1021.

Miller, J. M., Davison, R. C., & Gamlin, P. D. (2011). Motor nucleus activity fails to predict extraocular muscle forces in ocular convergence. *Journal of Neurophysiology, 105*(6), 2863–2873.

Mohney, B. G., Lilley, C. C., Green-Simms, A. E., & Diehl, N. N. (2011). The long-term follow-up of accommodative esotropia in a population-based cohort of children. *Ophthalmology, 118*(3), 581–585.

Noda, H., Sugita, S., & Ikeda, Y. (1990). Afferent and efferent connections of the oculomotor region of the fastigial nucleus in the macaque monkey. *The Journal of Comparative Neurology, 302*(2), 330–348.

Oh, S. Y., Clark, R. A., Velez, F., Rosenbaum, A. L., & Demer, J. L. (2002). Incomitant strabismus associated with instability of rectus pulleys. *Investigative Ophthalmology & Visual Science, 43*(7), 2169–2178.

Ohtsuka, K., & Sato, A. (1996). Descending projections from the cortical accommodation area in the cat. *Investigative Ophthalmology & Visual Science, 37*(7), 1429–1436.

Ohtsuka, K., Maekawa, H., & Sawa, M. (1993). Convergence paralysis after lesions of the cerebellar peduncles. *Ophthalmologica, 206*(3), 143–148.

Pallus, A. C., Walton, M. M. G., & Mustari, M. J. (2018). Activity of near response cells during disconjugate saccades in strabismic monkeys. *Journal of Neurophysiology, 120*, 2282.

Pullela, M., Degler, B. A., Coats, D. K., & Das, V. E. (2016). Longitudinal evaluation of eye misalignment and eye movements following surgical correction of strabismus in monkeys. *Investigative Ophthalmology & Visual Science, 57*(14), 6040–6047.

Pullela, M., Agaoglu, M. N., Joshi, A. C., Agaoglu, S., Coats, D. K., & Das, V. E. (2018). Neural plasticity following surgical correction of strabismus in monkeys. *Investigative Ophthalmology & Visual Science, 59*(12), 5011–5021.

Quaia, C., Shan, X., Tian, J., Ying, H., Optican, L. M., Walker, M., Tamargo, R., & Zee, D. S. (2008). Acute superior oblique palsy in the monkey: Effects of viewing conditions on ocular alignment and modelling of the ocular motor plant. *Progress in Brain Research, 171*, 47–52.

Quick, M. W., Tigges, M., Gammon, J. A., & Boothe, R. G. (1989). Early abnormal visual experience induces strabismus in infant monkeys. *Investigative Ophthalmology & Visual Science, 30*(5), 1012–1017.

Sawa, M., & Ohtsuka, K. (1994). Lens accommodation evoked by microstimulation of the superior colliculus in the cat. *Vision Research, 34*(8), 975–981.

Scheiman, M., & Ciner, E. (1987). Surgical success rates in acquired, comitant, partially accommodative and nonaccommodative esotropia. *Journal of the American Optometric Association, 58*(7), 556–561.

Scheiman, M., Ciner, E., & Gallaway, M. (1989). Surgical success rates in infantile esotropia. *Journal of the American Optometric Association, 60*(1), 22–31.

Schoeff, K., Chaudhuri, Z., & Demer, J. L. (2013). Functional magnetic resonance imaging of horizontal rectus muscles in esotropia. *Journal of AAPOS, 17*(1), 16–21.

Scott, A. B. (1994). Change of eye muscle sarcomeres according to eye position. *Journal of Pediatric Ophthalmology and Strabismus, 31*(2), 85–88.

Shaikh, A. G., & Ghasia, F. F. (2015). Misdirected horizontal saccades in pan-cerebellar atrophy. *Journal of the Neurological Sciences, 355*(1–2), 125–130.

Shan, X., Tian, J., Ying, H. S., Quaia, C., Optican, L. M., Walker, M. F., Tamargo, R. J., & Zee, D. S. (2007a). Acute superior oblique palsy in monkeys: I. Changes in static eye alignment. *Investigative Ophthalmology & Visual Science, 48*(6), 2602–2611.

Shan, X., Ying, H. S., Tian, J., Quaia, C., Walker, M. F., Optican, L. M., Tamargo, R. J., & Zee, D. S. (2007b). Acute superior oblique palsy in monkeys: II. Changes in dynamic properties during vertical saccades. *Investigative Ophthalmology & Visual Science, 48*(6), 2612–2620.

Shan, X., Tian, J., Ying, H. S., Walker, M. F., Guyton, D., Quaia, C., Optican, L. M., Tamargo, R. J., & Zee, D. S. (2008). The effect of acute superior oblique palsy on torsional optokinetic nystagmus in monkeys. *Investigative Ophthalmology & Visual Science, 49*(4), 1421–1428.

Shan, X., Hamasaki, I., Tian, J., Ying, H. S., Tamargo, R. J., & Zee, D. S. (2011). Vertical alignment in monkeys with unilateral IV section: Effects of prolonged monocular patching and trigeminal deafferentation. *Annals of the New York Academy of Sciences, 1233*, 78–84.

Steinbach, M. J. (1986). Inflow as a long-term calibrator of eye position in humans. *Acta Psychologica, 63*(1–3), 297–306.

Steinbach, M. J. (1987). Proprioceptive knowledge of eye position. *Vision Research, 27*(10), 1737–1744.

Steinbach, M. J., & Smith, D. R. (1981). Spatial localization after strabismus surgery: Evidence for inflow. *Science, 213*(4514), 1407–1409.

Suzuki, S., Suzuki, Y., & Ohtsuka, K. (2004). Convergence eye movements evoked by microstimulation of the rostral superior colliculus in the cat. *Neuroscience Research, 49*(1), 39–45.

Sylvestre, P. A., & Cullen, K. E. (1999). Quantitative analysis of abducens neuron discharge dynamics during saccadic and slow eye movements. *Journal of Neurophysiology, 82*(5), 2612–2632.

Tian, J., Shan, X., Zee, D. S., Ying, H., Tamargo, R. J., Quaia, C., Optican, L. M., & Walker, M. F. (2007). Acute superior oblique palsy in monkeys: III. Relationship to Listing's Law. *Investigative Ophthalmology & Visual Science, 48*(6), 2621–2625.

Tomlinson, R. D., & Schwarz, D. W. (1977). Response of oculomotor neurons to eye muscle stretch. *Canadian Journal of Physiology and Pharmacology, 55*(3), 568–573.

Trotter, Y., Fregnac, Y., & Buisseret, P. (1987). The period of susceptibility of visual cortical binocularity to unilateral proprioceptive deafferentation of extraocular muscles. *Journal of Neurophysiology, 58*(4), 795–815.

Trotter, Y., Celebrini, S., Beaux, J. C., Grandjean, B., & Imbert, M. (1993). Long-term dysfunctions of neural stereoscopic mechanisms after unilateral extraocular muscle proprioceptive deafferentation. *Journal of Neurophysiology, 69*(5), 1513–1529.

Tusa, R. J., Mustari, M. J., Das, V. E., & Boothe, R. G. (2002). Animal models for visual deprivation-induced strabismus and nystagmus. *Annals of the New York Academy of Sciences, 956*, 346–360.

Upadhyaya, S., & Das, V. E. (2018). Properties of cells associated with strabismus angle in the rostral superior colliculus of strabismic monkey. *Investigative Ophthalmology & Visual Science, 59*(9), 1021.

Upadhyaya, S., Meng, H., & Das, V. E. (2017). Electrical stimulation of superior colliculus affects strabismus angle in monkey models for strabismus. *Journal of Neurophysiology, 117*(3), 1281–1292.

Van Horn, M. R., Waitzman, D. M., & Cullen, K. E. (2013). Vergence neurons identified in the rostral superior colliculus code smooth eye movements in 3D space. *The Journal of Neuroscience, 33*(17), 7274–7284.

van Leeuwen, A. F., Collewijn, H., de Faber, J. T., & van der Steen, J. (2001). Saccadic binocular coordination in alternating exotropia. *Vision Research, 41*(25–26), 3425–3435.

Versino, M., Hurko, O., & Zee, D. S. (1996). Disorders of binocular control of eye movements in patients with cerebellar dysfunction. *Brain, 119*(Pt 6), 1933–1950.

von Noorden, G. K., & Campos, E. C. (2002). *Binocular vision and ocular motility: Theory and management of strabismus*. St. Louis: Mosby.

Walton, M. M., & Mays, L. E. (2003). Discharge of saccade-related superior colliculus neurons during saccades accompanied by vergence. *Journal of Neurophysiology, 90*(2), 1124–1139.

Walton, M. M., & Mustari, M. J. (2015). Abnormal tuning of saccade-related cells in pontine reticular formation of strabismic monkeys. *Journal of Neurophysiology, 114*(2), 857–868.

Walton, M. M. G., & Mustari, M. J. (2017). Comparison of three models of saccade disconjugacy in strabismus. *Journal of Neurophysiology, 118*(6), 3175–3193.

Walton, M. M., Ono, S., & Mustari, M. J. (2013). Stimulation of pontine reticular formation in monkeys with strabismus. *Investigative Ophthalmology & Visual Science, 54*(10), 7125–7136.

Walton, M. M., Mustari, M. J., Willoughby, C. L., & McLoon, L. K. (2014a). Abnormal activity of neurons in abducens nucleus of strabismic monkeys. *Investigative Ophthalmology & Visual Science, 56*(1), 10–19.

Walton, M. M., Ono, S., & Mustari, M. (2014b). Vertical and oblique saccade disconjugacy in strabismus. *Investigative Ophthalmology & Visual Science, 55*(1), 275–290.

Walton, M. M. G., Pallus, A., Fleuriet, J., Mustari, M. J., & Tarczy-Hornoch, K. (2017). Neural mechanisms of oculomotor abnormalities in the infantile strabismus syndrome. *Journal of Neurophysiology, 118*(1), 280–299.

Wang, X., Zhang, M., Cohen, I. S., & Goldberg, M. E. (2007). The proprioceptive representation of eye position in monkey primary somatosensory cortex. *Nature Neuroscience, 10*(5), 640–646.

Zhang, H., & Gamlin, P. D. (1998). Neurons in the posterior interposed nucleus of the cerebellum related to vergence and accommodation. I. Steady-state characteristics. *Journal of Neurophysiology, 79*(3), 1255–1269.

Zhang, Y., Gamlin, P. D., & Mays, L. E. (1991). Antidromic identification of midbrain near response cells projecting to the oculomotor nucleus. *Experimental Brain Research, 84*(3), 525–528.

Zimmermann, L., Morado-Diaz, C. J., Davis-Lopez de Carrizosa, M. A., de la Cruz, R. R., May, P. J., Streicher, J., Pastor, A. M., & Blumer, R. (2013). Axons giving rise to the palisade endings of feline extraocular muscles display motor features. *The Journal of Neuroscience, 33*(7), 2784–2793.

Part II
Translational Science

Advanced Vestibular Rehabilitation

Americo A. Migliaccio and Michael C. Schubert

Abstract Vestibular rehabilitation is a critical tool facilitating recovery after peripheral or central vestibular deficits. In this chapter, we will discuss state-of-the-art clinical practice strategies. We will focus on incremental vestibular ocular reflex (VOR) adaptation, with specific focus on the influence of training target contrast on the adaptive process, the influence of position and velocity error signals driving the adaptive process, and the role of active versus passive head rotation. We will then discuss the important effects of aging on VOR adaptation. Finally, we will summarize new developments in vestibular rehabilitation and provide a succinct review of vestibular rehabilitation in atypical medical populations.

Keywords VOR adaptation · Incremental velocity error signal · Head impulse · Active head rotation

A. A. Migliaccio
Balance and Vision Laboratory, Neuroscience Research Australia, Sydney, NSW, Australia

University of New South Wales, Sydney, NSW, Australia

Department of Otolaryngology – Head and Neck Surgery, Johns Hopkins University, Baltimore, MD, USA

M. C. Schubert (✉)
Laboratory of Vestibular NeuroAdaptation, Department of Otolaryngology – Head and Neck Surgery, Johns Hopkins University School of Medicine, Baltimore, MD, USA

Department of Physical Medicine and Rehabilitation, Johns Hopkins University, Baltimore, MD, USA
e-mail: mschube1@jhmi.edu

© Springer Nature Switzerland AG 2019
A. Shaikh, F. Ghasia (eds.), *Advances in Translational Neuroscience of Eye Movement Disorders*, Contemporary Clinical Neuroscience,
https://doi.org/10.1007/978-3-030-31407-1_9

1   Current Best Practice

Vestibular rehabilitation (VPT) is recognized as a critical component addressing the impairments of patients with disorders to the peripheral vestibular labyrinth and the pathways that mediate its afference. Meta-analysis evidence supports that VPT should be the primary intervention when treating vestibular hypofunction and benign paroxysmal positional vertigo (Hillier and McDonnell 2011, 2016; McDonnell and Hillier 2015; Hilton and Pinder 2004). For this chapter, we will focus on new advances in VPT as related to the treatment of vestibular hypofunction. In particular, we will focus on gaze stability training, which represents the uniquely "vestibular" component of VPT. While balance exercises are also critical for rehabilitation in vestibular labyrinth and nerve hypofunction, it is the gaze stability component that distinguishes VPT as most valuable.

The current standard of care in delivering VPT involves the prescription of gaze and gait stability exercises, which evidence shows improves visual acuity during head rotation, reduces fall risk, and improves both static and dynamic balance (Herdman et al. 2003, 2007, Hall et al. 2004; Schubert et al. 2008a). Clinical practice guidelines (CPG) are now available that specify clinicians prescribe a home exercise program of gaze stability exercises for a minimum of three times per day for a total of 12 minutes per day in patients with acute/subacute vestibular hypofunction (Hall et al. 2016). For patients with chronic vestibular hypofunction, the recommendations are a minimum of 20 minutes per day (Hall et al. 2016). Unfortunately, compliance performing these exercises is relatively poor (Meldrum et al. 2012; Huang et al. 2014).

2   Incremental VOR Adaptation

The angular vestibulo-ocular reflex (VOR) maintains images stable on the retina during rapid head rotations by rotating the eyes in the opposite direction to the head. Typically, the gain of the VOR is unity, that is, eye velocity magnitude is equal to head velocity magnitude. In this case, eye velocity divided by head velocity equals 1. However, after an injury to the peripheral vestibular organ or nerve, the VOR becomes weakened and does not generate an eye velocity of equal magnitude to the head velocity, in which case the gain is less than 1 (Halmagyi et al. 1990; Aw et al. 1996). When the VOR is injured, retinal image slip occurs that results in gaze instability during rapid head movements.

Rehabilitation exercises attempt to improve gaze instability during active head movements by increasing or "adapting up" the VOR gain and enlisting other oculomotor systems (Schubert et al. 2006, 2008a). However, the amount of VOR gain adaptation that occurs from these exercises is minimal, with little evidence that the VOR gain to passive head rotation changes at all (Herdman et al. 2007). In 2008, Schubert et al. developed the incremental velocity error method

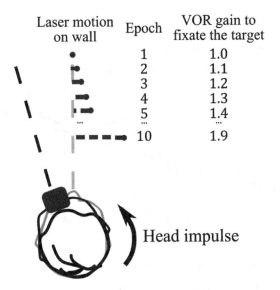

Fig. 1 Motion of the laser target during a training session where the left side is being adapted. The laser target appears stationary on the wall (gain of 1) for all rightward rotations. For leftward rotations, the VOR gain demand for epoch one is set to 1. The gain demand increases every epoch by 0.1. The dashed line shows the path of the laser target as seen by a stationary observer. For instance, during epoch 1, the laser appears to remain stationary. By epoch 10, the laser draws a long line on the wall in the opposite direction to the head – a gain of 1.9. (Modified from Todd et al. 2018)

(Schubert et al. 2008a) for VOR adaptation. This method tasks subjects to make active head rotations while viewing a laser-projected target that moves in the opposite direction of the head at a fraction of the head velocity (Fig. 1; US Patent # 9782068; Todd et al. 2018). The target velocity then gradually increases over a period of 15 minutes until it reaches the head velocity (still oppositely directed).

We recently showed unilateral VOR short-term adaptation is possible in humans (Migliaccio and Schubert 2013). Nine normal subjects underwent the IVE protocol for rotations to one side (adapting side), but for rotations toward the other (non-adapting) side, the visual stimulus was removed. The result was a significant VOR gain increase toward the adapting side for both active (23%) and passive (11%) head rotations, suggesting that VOR short-term adaptation was the main mechanism minimizing retinal image slip (Fig. 2). In a later study, we showed that unilateral VOR adaptation in patients with an isolated peripheral vestibular lesion was possible for rotations toward the side(s) with VOR hypofunction (Migliaccio and Schubert 2014). The increase in VOR gain toward the ipsilesional ear was less consistent in patients compared to controls, but this may have been a compliance issue. Two of the six patients quickly learned the adaptation training exercise, i.e., within 5 minutes, and performed the training without complication. During active head impulses, both these subjects had significant adaptation toward the ipsilesional adapting side (+43.2% and + 39.2%) and minimal adaptation toward the non-adapting side (−14.3% and − 5.6%) (Fig. 3). These increases were greater than those

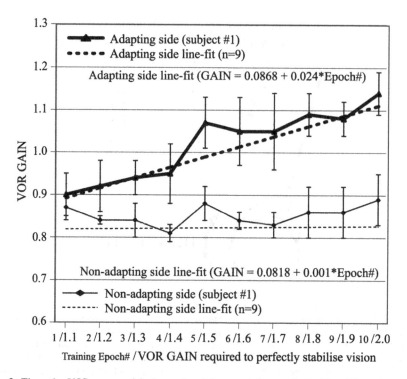

Fig. 2 The active VOR measured during each training epoch for a typical subject. The active VOR gain for head impulses toward the adapting side steadily increased from 0.9 to 1.14 during unilateral incremental training. In contrast, the gain toward the non-adapting side did not vary during the training. Best-line fits (adapting side – thick-dashed black; non-adapting side – thin-dashed black) for the training data across all subjects ($n = 9$) show that there was a significant relationship between gain and training epoch number only for head impulses toward the adapting side. (From Migliaccio and Schubert 2013)

seen in previous studies on bilateral and unilateral incremental adaptation training in unilateral vestibular hypofunction (UVH) subjects (Schubert et al. 2008b) and normal subjects (Migliaccio and Schubert 2013), respectively. Out of the remaining four, two learned the adaptation training exercise reasonably well but were inconsistent in terms of rotating their head with the correct velocity profile and keeping their eyes on the laser target. Notwithstanding, there was evidence of unilateral adaptation in these subjects also. The remaining two patients found the training task too difficult and could not proceed with the experiment. This study determined that the undesirable VOR gain increase (8%) toward the non-adapting side during active head rotations in Migliaccio and Schubert (2013) occurred because the target was extinguished during head rotations toward the non-adapting side. This unwanted adaptation was prevented when a stationary target was presented during head rotations toward the non-adapting side, thereby introducing an asymmetrical training error signal for bilateral head rotation (Migliaccio and Schubert 2014).

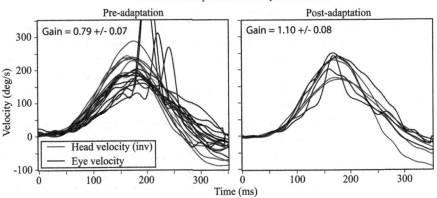

Fig. 3 Comparison of ipsilesional (adapting side) active VOR gain pre- and post-adaptation training. In this UVH patient, the gain increased from 0.79 ± 0.07 to 1.10 ± 0.08, a gain increase of ~40%. In this same patient, the contralesional non-adapting gain went from 0.90 ± 0.5 preadaptation to 0.85 ± 0.8 post-adaptation, a ~5% decrease that was not significant. (From Migliaccio and Schubert 2014)

Gimmon et al. examined whether the right and left VOR gains could be synchronously adapted in opposing directions (Gimmon et al. 2018). To do so, they used three separate VOR adaptation sessions, randomized such that the VOR was adapted up-bilaterally, down-bilaterally, or mixed (one side up, opposite side down). Each subject made active (self-generated) head impulse rotations for 15 min while following the IVE method. VOR training demand changed by 10% every 90 seconds. The results revealed the human VOR can be simultaneously driven in opposite directions suggesting functionally independent VOR circuits for each side of head rotation. It appears, therefore, that the brain relies on the salient error signal to drive bilateral VOR adaptation. This fits with the credit assignment theory, which states that resource assignment toward motor learning occurs when the brain considers the error signal accurate and thus worthy of resource expenditure (Kording et al. 2007). In the case of a bilateral error signal, the VOR from each side can be driven independently; in the case of unilateral adaptation, the brain presumes the error signal to be accurate and therefore makes effort toward bilateral adaptation.

3 The Influence of Training Target Contrast on VOR Adaptation

Like the VOR, smooth pursuit and optokinetic systems seek to stabilize images on the retina using an image slip error signal (e.g., Cohen et al. 1981; Büttner and Büttner-Ennever 2006). Studies have shown that visual contrast affects the perceived velocity of a stimulus and the smooth pursuit following gain (= eye/target velocity) (Thompson 1982, 1983). Visual contrast also affects the slow-phase velocity of optokinetic nystagmus such that it increases with increasing contrast for a given speed (Sumnall et al. 2003). For both smooth pursuit and optokinetic nystagmus, contrast must pass a certain threshold for the detection of image motion. As contrast increases, the internal estimate of stimulus speed reconstructed from prior retinal (movement of the retinal image) and extraretinal signals (efference copy and other proprioceptive cues) becomes more accurate (Waddington and Harris 2015).

In a recent study, we examined the effect of visual target contrast on human unilateral VOR adaptation (Mahfuz et al. 2017). When the visual target during visual-vestibular mismatch training had contrast levels between 1.5 k and 1.4 M, the VOR gain increase due to adaptation training did not change, suggesting that at these levels, the effect of contrast on VOR adaptation had saturated. When target contrast was reduced to 261, the active and passive VOR gain increase toward the adapting side was no longer significant and was similar to the non-adapting side, suggesting that 261 was below the threshold needed for adaptation to occur. Modeling suggested that the contrast threshold for VOR adaptation to occur was 250 and that a contrast level of above 1000 was required for robust VOR adaptation (Fig. 4). Compare these values to the maximum contrast level obtained from a typical LCD monitor (24″ 1920 x 1080 resolution VS243 Asus, Taiwan) of 37.7 in an otherwise dark room.

Low image contrast may explain why classic human studies using normal lighting and magnifying or minifying lenses with full fields of view to drive VOR adaptation required long periods of training to significantly affect the VOR gain (Gauthier and Robinson 1975; Gonshor and Jones 1976). Increasing contrast to 1.5 k was sufficient for robust and rapid VOR adaptation to occur (Mahfuz et al. 2017). However, it is unlikely that an increase in contrast is the only factor. A prior study under similar lighting conditions to the maximum contrast level in Mahfuz et al. (2017) found that incremental adaptation training, rather than "all at once" training (as occurs during classic lens training), resulted in significantly larger gain increases (Schubert et al. 2008b), suggesting that visual contrast is but one of several factors affecting VOR adaptation. Contrast might also explain why many patients with incomplete vestibular hypofunction often have slow and incomplete recovery of VOR gains despite long exposure to the normal visual environment that typically drives the VOR gain to unity. We suggest that visual target contrast should be an important consideration during VOR adaptation/rehabilitation training. These data suggest that optimal VOR gain training can be achieved using a bright target (as emitted by a typical laser pointer) in low ambient lighting, i.e., the typical lighting of a room with curtains closed and lights turned off. *With critical rehabilitation implications, this study also established the incremental VOR gain adaptation method can be achieved in conditions of incomplete dark* (Fig. 5, <8 lux).

Piece-wise linear regressions of VOR gain % increase versus contrast level

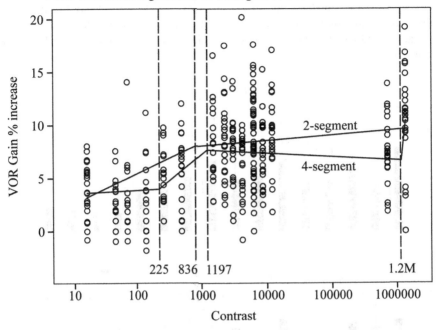

Fig. 4 Scatterplot of VOR gain percentage increase toward the adapting side (after pooling active and passive gains) versus the contrast level plot on a $\log_{10}$ scale. Two-segment piecewise regression ($R^2 = 0.26$) determined a knot at 836, indicating that at this contrast level, there was a significant difference between the two lines, respectively, modeling the data below and above this contrast level. Three-segment analysis revealed a lower knot at 854 and an upper knot at 1.3 M. Four-segment modeling ($R^2 = 0.28$) revealed three knots: 225 (lower), 1197 (middle), and 1.2 M (upper). Overall, our analysis suggests there was a transition period starting at ~225 and ending at ~1000 (i.e., between 836 and 1197) where the VOR gain percentage increase gradually increased. Below 225, there was no significant (but constant) adaptation, and above 1000, there was maximal adaptation. (From Mahfuz et al. 2017)

4 Position and Velocity Error Signal Driving VOR Adaptation

Retinal image movement is the likely feedback signal that drives VOR modification/adaptation for different viewing contexts. However, it was not clear whether a retinal image position or velocity error was used primarily as the feedback signal. In a recent study, we examined the effect of varying the retinal image position error update rates (frequencies) to understand the role of visual feedback in human VOR adaptation (Fadaee and Migliaccio 2016).

Using the human unilateral incremental VOR adaptation technique, the study showed that adaptation declined as the retinal image position error signal used to drive adaptation became less frequent, i.e., as the update rate decreased from 50 (the target position was updated once every 20 ms) to 15 Hz. For 50 and 20 Hz update rates, there was a significant increase in VOR gain toward the adapting side for both the active and passive VOR, whereas at 15 Hz, there was no longer any adaptation (Fig. 6).

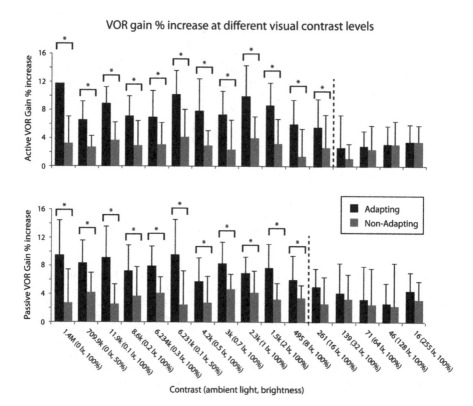

Fig. 5 Mean (± SD) adapting (black bars) and non-adapting (gray bars) side pre- to post-adaptation training VOR gain increase as a percentage for the active (top row) and passive (bottom row) VOR across all subjects (*n* = 12) for each contrast level. In parentheses, beside each contrast level is the ambient light level in lux and the target brightness as a percentage of maximum laser power. Statistical analysis indicates there is a significant difference in percentage gain increase between the adapting and non-adapting sides at all contrast levels above and equal to 1.5 k (left side of vertical dashed line). (* denotes a significant increase (t-test between adapting and non-adapting-side percentage increase, $p < 0.05$)). (From Mahfuz et al. 2017)

Fig. 6 (continued) increased from ~1 to ~1.2 during unilateral incremental training at 50 Hz. Between epochs 2 and 10, the difference in gain between the adapting and non-adapting side (gray traces) became increasingly significant with increasing epoch number. Similarly, adaptation training at 20 Hz initially showed a trend of increasing difference between adapting and non-adapting-side gains, which became significant for epochs 7 to 10. In contrast, adaptation training at 15 Hz results in minimal unilateral adaptation as a group, so that the difference between sides is no longer a function of increasing epoch number (*$p < 0.05$, **$p < 0.01$, and ***$p < 0.001$). (From Fadaee and Migliaccio 2016)

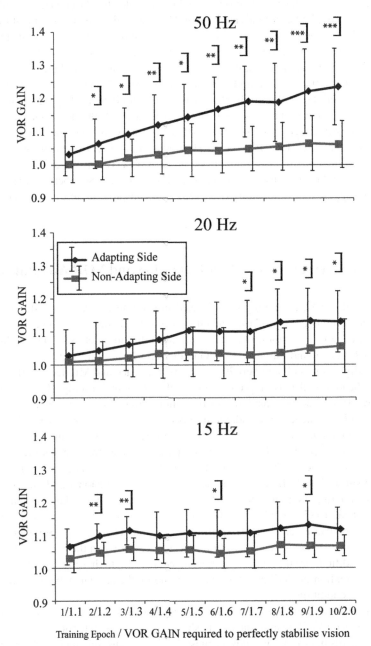

Fig. 6 The active VOR measured during each training epoch across all subjects (pooled data, mean ± SD) at three laser target position update rates (50 Hz [$n = 13$], 20 Hz [$n = 12$], and 15 Hz [$n = 9$]). The active VOR gain for head impulses toward the adapting side (black traces) steadily

Unlike strobe lighting, the visual stimulus during training in this study provided both a position and a velocity error signal because although the laser target position (in-space) update rate was a programmable variable, the laser target itself was always on, providing constant visual input. Consequently, when target position was updated during each head impulse, it drove adaptation to increase the VOR gain (> 1), but when target position was not updated (i.e., laser position did not change during most of the head impulse), it essentially drove the gain toward unity (gain = 1). In other words, the position and velocity errors drove the VOR gain to change in opposite directions. In this study, the active VOR gain increased by ~23% at 50 Hz update rate, which is the same as that observed at the 1000 Hz update rate used previously (Migliaccio and Schubert 2013). Similarly, in this study, the passive VOR gain increased by ~10% at 20 Hz update rate (i.e., two position cues per impulse), which is similar to the ~11% increase observed at the 50 times greater (1000 Hz) update rate used previously (Migliaccio and Schubert 2013). These findings suggest that a position error signal too can provide an effective adaptive drive for the VOR.

5   Active Versus Passive Head Rotation and VOR Adaptation

The majority of VOR training stimuli used in prior VOR adaptation studies have consisted of predictable head motion, either via self-generation or whole-body sinusoidal passive head rotation (Hattori et al. 2000; Solomon et al. 2003; Shelhamer et al. 1994; Paige and Sargent 1991; Gauthier and Robinson 1975; Schubert et al. 2008b; Migliaccio and Schubert 2013 and 2014, Fadaee and Migliaccio 2016). The passive head impulse is unpredictable in direction and timing (Halmagyi and Curthoys 1988) and is considered to be a more physiologically relevant stimulus than a single-frequency sinusoidal stimulus (as would be obtained from chair testing) due to its high-frequency content (up to 6 Hz). Additionally, the VOR must be responsive to unpredictable head motion, particularly at frequencies greater than 1 Hz (and velocities >100°/s) where the VOR becomes the main vision-stabilizing mechanism. Below 1 Hz (and velocities <100°/s), other vision-stabilizing systems such as smooth pursuit and the optokinetic reflex are likely to play the major role (Meyer et al. 1985). Finally, motor learning within the VOR is context specific (human: Shelhamer et al. 1992; primate: Yakushin et al. 2003; Schubert et al. 2008c), which suggests that adaptation will be greatest when the VOR training and testing conditions are the same. The training context might therefore result in differences in retention of VOR adaptation depending on the similarity between training and testing conditions.

We recently examined the effects of passive versus active head movement training contexts on VOR adaptation and its short-term (1 hour) retention (Mahfuz et al. 2018). The results from this study suggest that in humans, the magnitude of unilateral VOR adaptation after active or passive head impulse training is the same for both the active and passive VOR. In other words, the active or passive head rotation context of the training had no effect on the VOR gain increase, suggesting that most of the adaptation observed, regardless of whether the stimulus was active or passive, was occurring in the central vestibular pathways common to both the active and passive VOR, as opposed to the nonoverlapping pathways. Similarly, the training context did not affect retention of VOR adaptation. The only factor that significantly affected retention over the 1 hour immediately after training was the brief exposure to a stationary fixation target before each head impulse during VOR testing (Fig. 7).

Findings from this study suggest that as long as active head impulses have similar velocity profiles to passive head impulses, then optimal VOR training can be performed with active only head rotations, eliminating the need for expensive and bulky equipment (e.g., a rotary chair) or human assistance to deliver passive head impulses. Evidence from this study supports recent Cochrane meta-analysis studies that provide strong recommendations for vestibular rehabilitation providers to prescribe gaze stability exercises using active head rotation (Hillier and McDonnell 2011, 2016; McDonnell and Hillier 2015). Additionally, recent evidence suggests active head rotation training improves postural control (Matsugi et al. 2017). The only factor that affected short-term retention was the duration of exposure to a de-adaptation stimulus, which in this study drove the VOR gain down to unity. Presumably, retention would not be lost in vestibular patients whose ipsilesional VOR gain was increased due to training, but still below unity, because of the lack of a de-adaptation stimulus. In this case, real-world visual conditions would drive the VOR gain up to unity in vestibular patients and reinforce the training, rather than down to unity, and cancel the training as was the case in the healthy subjects used in this study.

6   The Effect of Aging on VOR Adaptation

Aging can have a profound impact on vestibular system function, as can be appreciated from the prevalence of vestibular-related disorders in the elderly. By the age of 70–80 years, a significant percent of our population has vestibular dysfunction – which is linked to increased dizziness and risk of falls (Agrawal et al. 2009). The prevalence of vestibular vertigo in this age group is three times that for young adults (< 30 years) (Neuhauser and Lempert 2009). Some of these symptoms are thought to be, in part at least, a consequence of senescence of the vestibular periphery (Ishiyama 2009). For example, approximately 40% of patients referred for fall risk assessment were found to have VOR deficits including abnormal gain, phase lead, and asymmetries (Jacobson et al. 2008). Aging has been shown to associate with loss of vestibular hair cells, vestibular afferents, and cells in the central vestibular nuclei (Johnsson 1971; Bergstrom 1973; Ross et al. 1976; Rosenhall 1973; Richter 1980; Lopez et al. 1997; Alvarez et al. 1998; Merchant et al. 2000; Velazquez-Villasenor et al. 2000; Park et al. 2001; Rauch et al. 2001).

We recently examined the effects of aging on the VOR and VOR adaptation in 30-month-old mice (equivalent to an 80-year-old human; Yuan et al. 2009)

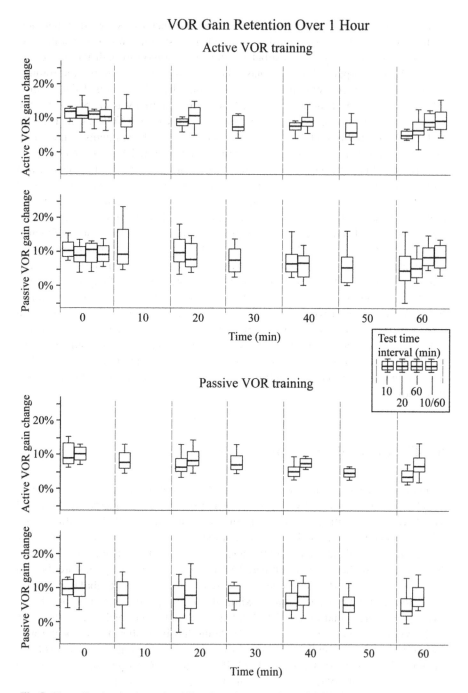

Fig. 7 Normalized active (rows 1 and 3) and passive (rows 2 and 4) VOR gain percentage increases after active (top two rows) and passive (bottom two rows) VOR training toward the adapting side only. VOR gains were measured every 10 (first boxplot between dashed lines), 20 (second boxplot

(Khan et al. 2017). This study showed that the baseline VOR gain for aged mice was significantly lower to controls during high-frequency, high-acceleration, transient head rotations (Khan et al. 2017). The study found that aging minimally affected the static counter-tilt response and sinusoidal VOR gain but significantly affected VOR adaptation, particularly for gain-decrease adaptation training (Fig. 8).

It is possible that vestibular plasticity mechanisms behind VOR adaptation and compensation normalize VOR function as structural losses to the vestibular system occur with aging. Presumably, there is a threshold of vestibular loss below which these plasticity mechanisms can no longer mask the underlying vestibular system deficit. This might explain why human VOR function is steady up to age 70, followed by a steep decline (Matiño-Soler et al. 2015). If VOR plasticity can compensate for potentially large vestibular losses due to aging, one could ask why does the behavioral VOR in patients with partial lesions to their vestibular organs not fully recover? The reason could be due to the sudden onset of the lesion. Our prior studies have shown that significant VOR adaptation can be induced in a short period of time if the VOR-adapting stimulus challenges the VOR gain to change in small increments (Schubert et al. 2008b; Migliaccio and Schubert 2013, 2014; Fadaee and Migliaccio 2016; Mahfuz et al. 2017, 2018). In contrast, when the required adaptive change is large, for example, when the VOR gain must double, then VOR adaptation becomes more difficult and less optimal (Gauthier and Robinson 1975; Schubert et al. 2008b). Unlike most other injuries, structural degradation due to aging is a gradual process, and so small losses in vestibular system function might be more easily accommodated by the VOR.

Some of this loss in VOR adaptation is likely explained by degeneration of the vestibulocerebellum, which plays a major role during VOR-dependent adaptation. Purkinje cells in the cerebellar flocculus are well positioned to encode information used to induce plasticity (Ito 1982; Lisberger and Pavelko 1986; Lisberger and Fuchs 1978). It has been postulated that gain-increase and gain-decrease VOR adaptation occurs through two distinct cerebellar mechanisms, with the former involving long-term depression (LTD) and the latter long-term potentiation (LTP) at Purkinje cells in the cerebellar flocculus, although other pathways and sites in the VOR circuitry are also implicated (Schonewille et al. 2011; Titley and Hansel 2016). Age-related degeneration of Purkinje cells could explain the Khan et al. (2017) results, if Purkinje cell death affected these two mechanisms differently. It is known that Purkinje cells in the cerebellum are particularly affected by aging with their number decreasing with age (Andersen et al. 2003; Rogers et al. 1984; Woodruff-Pak et al. 2010; Zhang et al. 2010). In humans, the decline in Purkinje

---

Fig. 7 (continued) between dashed lines), or 60 minutes (third boxplot between dashed lines) for 60 minutes after adaptation training. In addition, the VOR gain was tested every 10 minutes, but the visual fixation light before the start of each impulse was only provided at testing times 0 and 60 minutes after adaptation training (fourth boxplot between dashed lines). Each box shows the median and goes from the first to the third quartile with whiskers denoting the maximum and minimum values. Taken together, these data suggest that the number of times the VOR is tested after training affects retention more so than the time lapsed after training. Gains measured immediately after active training and 60 minutes later with no other testing between these times (fourth boxplot) showed no significant loss in retention. (From Mahfuz et al. 2018)

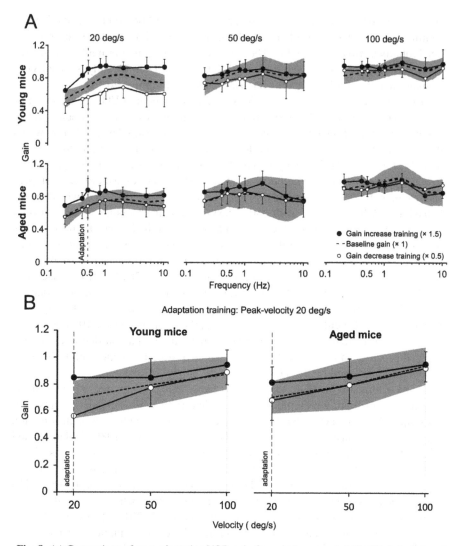

Fig. 8 (a) Comparison of post-adaptation VOR gain for gain-increase (×1.5) (filled circles) and gain-decrease (×0.5) (open circles) adaptation training in young (top panel) and aged mice (bottom panel). The baseline VOR response is shown as dashed lines (mean). For both young and aged mice, the difference between VOR gains post gain increase and gain decrease after adaptation training was maximal when tested at the training velocity (20 o/s; left column). The effect of gain-increase adaptation was significant for both young and aged mice, whereas gain-decrease adaptation was only significant for young mice. The vertical dashed line represents the adaptation training frequency 0.5 Hz. (b) Comparison of gain-increase (×1.5) (filled circles) and gain-decrease (×0.5) (open circles) adaptation pooled across frequencies for each stimulus peak velocity (20, 50, and 100 o/s) in young (left panel) and aged mice (right panel). The mean baseline VOR response is shown by the dashed line. The vertical dashed line denotes the adaptation training velocity. VOR gain-decrease adaptation was ~85% lower in aged mice compared to young mice. VOR gain-increase adaptation was also significantly reduced in aged mice, albeit by a lesser extent of ~30%. (From Khan et al. 2017)

cell population is less marked before age 60 but becomes pronounced afterward (Hall et al. 1975). Intriguingly, we observed a similar asymmetrical effect on VOR adaptation and compensation in alpha-9-knockout mice (these have an impaired cholinergic efferent vestibular system) (Hübner et al. 2015, 2017). Notably, cholinergic neurotransmission in the cerebellar flocculus has also been implicated in impaired gain-decrease VOR adaptation (Prestori et al. 2013). These findings raise the possibility of an involvement of the cholinergic system in the age-related changes in VOR adaptation.

7   New Developments in Vestibular Rehabilitation

7.1   Ipsilesional-Only Rotation

A recent case study used ipsilesional passive head rotation to reduce VOR gain asymmetry in a patient with a unilateral vestibular hypofunction (Binetti et al. 2017). The authors report that until applying the following method, the patient was unable to recover. The authors applied 10 epochs of 15 passive ipsilesional-only head impulses while the patient attempted to keep eyes focused on a stationary target (i.e., times one VOR exercise). The patient rested 30 seconds between each training epoch. This was repeated for five consecutive days. The authors used video HIT equipment to ensure head velocities were greater than 120 d/s. After training, the ipsilesional VOR gain improved 24% (pre 0.57 ± 0.11, post 0.71 ± 0.10). At 6- and 12-month follow-up, the patient reported no further vestibular symptoms.

7.2   Galvanic Vestibular Stimulation

Preliminary studies using galvanic vestibular stimulation suggest it may be able to improve postural stability in standing or while walking (Mulavara Fiedler et al. 2011; Pal et al. 2009). More recently, applying weak, nonlinear electrical signals that are imperceptible to human sensory threshold (stochastic resonance – SR) holds some promise at improving vestibular function and imbalance. Serrador et al. showed an improved gain of ocular counter roll (OCR) in elderly with an initially reduced OCR gain (Fig. 9). In those elderly with healthy OCR gain, the application of SR was negligible for the OCR (Serrador et al. 2018). Wuehr et al. used SR to reduce the variability of stride time, stride length, and base of support in patients with bilateral vestibular loss compared to sham control (Wuehr et al. 2016). Furthermore, the authors report most of the patients noted an improved sensation of walking balance most pronounced during slow gait speed.

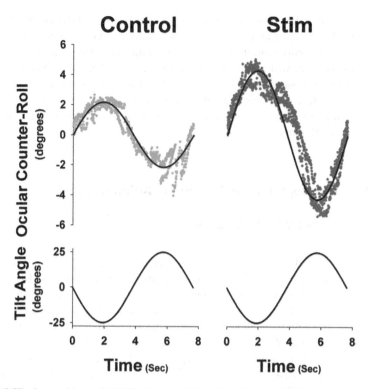

Fig. 9 OCR of one subject to 0.125 Hz during ±25° passive roll tilt during both no stimulation and SN trials. The bottom panel demonstrates the gain calculations from the linear regression. This subject demonstrated a significant increase in OCR from control (left side) to stimulation (right side). (From Serrador et al. 2018)

7.3 Auditory Feedback

Encouraging, though preliminary, evidence suggests auditory feedback may improve both balance and gait. A case series study suggests that patients with bilateral vestibular hypofunction and hearing impairment used hearing to improve both duration standing on foam and gait speed (Shayman et al. 2017).

7.4 Transcranial Cerebellar Direct Current Stimulation

Sixteen patients with chronic dizziness (> 6 months) due to UVH (e.g., vestibular neuritis, Ramsay Hunt syndrome) were treated for five days each performing three 25 min sessions of vestibular rehabilitation exercises every 4 hours. Subjects were randomized to the Sham group (VPT only) or Stim group (Koganemaru et al. 2017). On the first day only, the Stim group received 20 minutes of transcranial cerebellar direct current stimulation (tcDCS, 2 mA of electrical constant direct current). The tcDCS was applied via electrodes placed 2 cm below the inion, medial to the mastoid apophysis. Those subjects receiving the tcDCS showed significantly improved dizziness handicap inventory (DHI) scores for the physical and functional domains, in addition to the total scores compared with the VPT-only group.

7.5 Home Computerized Dynamic Visual Acuity

Crane and Schubert developed a home training device based on computerized dynamic visual acuity testing that involved wearing a rate sensor and interactive online software that subjects accessed using their own computers and internet service. The training asked subjects to move their heads and identify letters that flashed on the monitor only during rapid head rotation (>100 d/s) (Crane and Schubert 2018). Data revealed that subjects who previously tried vestibular rehabilitation but had no appreciable improvement in their dizziness handicap inventory (DHI) showed significantly improved DHI scores after completing 4 weeks of the home therapy regimen. Initial mean DHI was 42 (range, 32–56) and post DHI (4 weeks) was 11.5 ($p < 0.004$; range, 0–16). Three of the subjects continued to do the home treatment and noted a further reduction in DHI (range, 2–6; $p < 0.002$).

7.6 Virtual Reality

There remains an impression that virtual reality (VR) applications hold great promise as an exciting and better method to deliver vestibular rehabilitation. However, recent studies that have directly compared physical therapy outcomes using VR versus traditional VPT reveal no advantage in using virtual reality (Alahmari et al. 2014; Smaerup et al. 2017; Micarelli et al. 2017). As technology advances, VR certainly may offer unique training regimens, but for now, it appears to offer only a more engaging rehabilitation experience at best.

7.7 Atypical Medical Populations

More recently, VPT has been used in nontraditional medical populations that include stroke, attention deficit disorder, and headache. Using a controlled and randomized design, patients with hemorrhagic or ischemic strokes with onset less than 6 months were randomized into either VPT or traditional CVA rehab groups (Mitsutake et al. 2017). The VPT group included VOR gaze stability and gait stability exercises; the CVA group performed exercises focused on muscle strength, posture and gait exercises, range of motion, and walking indoors and outdoors. After 3 weeks of treatment, the VPT group showed a significantly improved fall risk score (assessed by the Dynamic Gait Index) and ability to move their head quickly and read letters (assessed by the gaze stability test), compared to the CVA group.

Children with attention deficit disorder and concurrent vestibular impairment were randomized into a VPT or control group (Lotfi et al. 2017). The VPT group did 45 min of exercise two times per week for 12 weeks that include gaze stability, postural stability, and daily living activities. The control group received no intervention. After 12 weeks, choice reaction time and spatial working memory significantly improved only for those children in the VPT group. The authors concluded that the severity of vestibular pathology should be assessed in children with ADHD and that inclusion of such children into a VPT program can improve their cognitive abilities.

Finally, a study compared the effect of VPT to reduce headache, reduce fall risk, and reduce dizziness in two types of sufferers – vestibular migraine and tension type (Sugaya et al. 2017). The VPT improved headache, fall risk, dizziness, and psychological factors as measured by the Hospital Anxiety and Depression Scale and Somatosensory Catastrophizing Scale. What was novel was the improvement of symptoms in the patients with tension-type headache. Although this study confirmed patients with vestibular migraine had the greater reduction in symptoms, literature already has confirmed benefit in this patient population (Whitney et al. 2000; Vitkovic et al. 2013).

8 Conclusion

Vestibular rehabilitation will remain useful and first line of defense in treating the functional deficits associated with reduced vestibular sensation. Dosing amounts for VPT remain unknown, though exciting new technologies such as incremental VOR adaptation holds promise that more efficient delivery is possible.

References

Agrawal, Y., Carey, J. P., Della Santina, C. C., Schubert, M. C., & Minor, L. B. (2009). Disorders of balance and vestibular function in US adults: Data from the National Health and Nutrition Examination Survey, 2001–2004. *Archives of Internal Medicine, 169*, 938–944.

Alahmari, K. A., Sparto, P. J., Marchetti, G. F., Redfern, M. S., Furman, J. M., & Whitney, S. L. (2014). Comparison of Virtual Reality Based Therapy with Customized Vestibular Physical Therapy for the Treatment of Vestibular Disorders. *IEEE Transactions on Neural Systems and Rehabilitation Engineering, 22*(2), 389–399. PMCID: PMC5527704.

Alvarez, J. C., Díaz, C., Suárez, C., Fernández, J. A., González del Rey, C., Navarro, A., & Tolivia, J. (1998). Neuron loss in human medial vestibular nucleus. *The Anatomical Record, 251*, 431–438.

Andersen, B. B., Gundersen, H. J., & Pakkenberh, B. (2003). Ageing of the human cerebellum: A stereological study. *The Journal of Comparative Neurology, 466*, 356–365.

Aw, S. T., Halmagyi, G. M., Haslwanter, T., Curthoys, I. S., Yavor, R. A., & Todd, M. J. (1996). Three-dimensional vector analysis of the human vestibuloocular reflex in response to high-acceleration head rotations. II. Responses in subjects with unilateral vestibular loss and selective semicircular canal occlusion. *Journal of Neurophysiology, 76*, 4021–4030.

Bergstrom, B. (1973). Morphology of the vestibular nerve: Part II- the number of myelinated vestibular nerve fibers in man at various ages. *Acta Oto-Laryngologica, 76*, 173–179.

Binetti, A. C., Varela, A. X., Lucarelli, D. L., & Verdecchia, D. H. (2017). Unilateral head impulses training in uncompensated vestibular hypofunction. *Case Reports in Otolaryngology, 2017*, 2145173. PMID: 28243476.

Büttner, U., & Büttner-Ennever, J. A. (2006). Present concepts of oculomotor organization. *Progress in Brain Research, 151*, 1–42.

Cohen, B., Henn, V., Raphan, T., & Dennett, D. (1981). Velocity storage, nystagmus, and visual-vestibular interactions in humans. *Annals of the New York Academy of Sciences, 374*, 421–433.

Crane, B. T., & Schubert, M. C. (2018). An adaptive vestibular rehabilitation technique. *The Laryngoscope, 128*(3), 713–718. PMID: 28543062.

Fadaee, S. B., & Migliaccio, A. A. (2016). The effect of retinal image error update rate on human vestibulo-ocular reflex gain adaptation. *Experimental Brain Research, 234*(4), 1085–1094.

Gauthier, G. M., & Robinson, D. A. (1975). Adaptation of the human vestibuloocular reflex to magnifying lenses. *Brain Research, 92*, 331–335.

Gonshor, A., & Jones, G. M. (1976). Short-term adaptive changes in the human vestibulo-ocular reflex arc. *Journal of Physiology, 256*, 361–379.

Gimmon Y, Migliaccio AA, Todd C, Figtree W, Schubert MC. (2018). Simultaneous laterally-inversive vestibulo-ocular reflex adaptation during head impulses. *American Physical Therapy Association, combined sections meeting*, New Orleans, LA.

Hall, C. D., Schubert, M. C., & Herdman, S. J. (2004). Prediction of fall risk reduction as measured by dynamic gait index in individuals with unilateral vestibular hypofunction. *Otology & Neurotology, 25*(5), 746–751.

Hall, C. D., Herdman, S. J., Whitney, S. L., Cass, S. P., Clendaniel, R. A., Fife, T. D., Furman, J. M., Getchius, T. S., Goebel, J. A., Shepard, N. T., & Woodhouse, S. N. (2016). Vestibular rehabilitation for peripheral vestibular hypofunction: An evidence-based clinical practice guideline: From the American Physical Therapy Association Neurology Section. *Journal of Neurologic Physical Therapy, 40*(2), 124–155. PMID: 26913496.

Hall, T. C., Miller, A. K. H., Corsellis, J. A. N. (1975). Variation in the human purkinje cell population according to age and sex. *Neuropathol Appl Neurobiol, 1*, 267–292.

Halmagyi, G. M., & Curthoys, I. S. (1988). A clinical sign of canal paresis. *Archives of Neurology, 45*, 737–739.

Halmagyi, G. M., Curthoys, I. S., Cremer, P. D., Todd, M. J., & Curthoys, I. S. (1990). The human horizontal vestibulo-ocular reflex in response to high-acceleration stimulation before and after unilateral vestibular neurectomy. *Experimental Brain Research, 81*, 479–490.

Hattori, K., Watanabe, S., Nakamura, T., & Kato, I. (2000). Flexibility in the adaptation of the vestibulo-ocular reflex to modified visual inputs in humans. *Nihon Jibiinkoka Gakkai Kaiho, 103*, 1186–1194.

Herdman, S. J., Schubert, M. C., Das, V. E., & Tusa, R. J. (2003). Recovery of dynamic visual acuity in unilateral vestibular hypofunction. *Archives of Otolaryngology – Head & Neck Surgery, 129*(8), 819–824.

Herdman, S. J., Hall, C. D., Schubert, M. C., Das, V. E., & Tusa, R. J. (2007). Recovery of dynamic visual acuity in bilateral vestibular hypofunction. *Archives of Otolaryngology – Head & Neck Surgery, 133*(4), 383–389.

Hillier, S. L., & McDonnell, M. (2011). Vestibular rehabilitation for unilateral peripheral vestibular dysfunction. *Cochrane Database of Systematic Reviews*, (2), CD005397. https://doi.org/10.1002/14651858.CD005397.pub3.

Hillier, S., & McDonnell, M. (2016). Is vestibular rehabilitation effective in improving dizziness and function after unilateral peripheral vestibular hypofunction? An abridged version of a Cochrane Review. *European Journal of Physical and Rehabilitation Medicine, 52*, 541–556.

Hilton, M., & Pinder, D. (2004). The Epley (canalith repositioning) manoeuvre for benign paroxysmal positional vertigo. *Cochrane Database of Systematic Reviews*, (2), CD003162. Review. Update in: Cochrane Database Syst Rev. 2014;12: CD003162.

Huang, K., Sparto, P. J., Kiesler, S., Siewiorek, D. P., & Smailagic, A. (2014). iPod-based in-home system for monitoring gaze-stabilization exercise compliance of individuals with vestibular hypofunction. *Journal of Neuroengineering and Rehabilitation, 11*, 69. https://doi.org/10.1186/1743-0003-11-69.

Hübner, P. P., Khan, S. I., & Migliaccio, A. A. (2015). The mammalian efferent vestibular system plays a crucial role in the high-frequency response and short-term adaptation of the vestibulo-ocular reflex. *Journal of Neurophysiology, 114*(6), 3154–3165.

Hübner, P. P., Khan, S. I., & Migliaccio, A. A. (2017). The mammalian efferent vestibular system plays a crucial role in vestibulo-ocular reflex compensation after unilateral labyrinthectomy. *Journal of Neurophysiology, 117*(4), 1553–1568.

Ishiyama, G. (2009). Imbalance and vertigo: The aging human vestibular periphery. *Seminars in Neurology, 29*, 491–499.

Ito, M. (1982). Cerebellar control of the vestibulo-ocular reflex—Around the flocculus hypothesis. *Annual Review of Neuroscience, 5*, 275–296.

Jacobson, G. P., McCaslin, D. L., Grantham, S. L., & Piker, E. G. (2008). Significant vestibular system impairment is common in a cohort of elderly patients referred for assessment of falls risk. *Journal of the American Academy of Audiology, 19*, 799–807.

Johnsson, L. G. (1971). Degenerative changes and anomalies of the vestibular system in man. *Laryngoscope, 81*, 1682–1694.

Khan, S. I., Hübner, P. P., Brichta, A. M., Smith, D. W., & Migliaccio, A. A. (2017). Ageing reduces the high-frequency and short-term adaptation of the vestibulo-ocular reflex in mice. *Neurobiology of Aging, 51*, 122–131.

Koganemaru, S., Goto, F., Arai, M., Toshikuni, K., Hosoya, M., Wakabayashi, T., Yamamoto, N., Minami, S., Ikeda, S., Ikoma, K., & Mima, T. (2017). Effects of vestibular rehabilitation combined with transcranial cerebellar direct current stimulation in patients with chronic dizziness: An exploratory study. *Brain Stimulation, 10*(3), 576–578.

Kording, K. P., Tenenbaum, J. B., & Shadmehr, R. (2007). The dynamics of memory as a consequence of optimal adaptation to a changing body. *Nature Neuroscience, 10*(6), 779–786. Epub 2007 May 13..

Lisberger, S. G., & Pavelko, T. A. (1986). Vestibular signals carried by pathways subserving plasticity of the vestibulo-ocular reflex in monkeys. *The Journal of Neuroscience, 6*, 346–354.

Lisberger, S. G., & Fuchs, A. F. (1978). Role of primate flocculus during rapid behavioral modification of vestibule-ocular reflex. I. Purkinje cell activity during visually guided horizontal smooth-pursuit eye movements and passive head rotation. *Journal of Neurophysiology, 41*, 733–763.

Lopez, I., Honrubia, V., & Baloh, R. W. (1997). Ageing and the human vestibular nucleus. *Journal of Vestibular Research, 7*, 77–85.

Lotfi, Y., Rezazadeh, N., Moossavi, A., Haghgoo, H. A., Rostami, R., Bakhshi, E., Badfar, F., Moghadam, S. F., Sadeghi-Firoozabadi, V., & Khodabandelou, Y. (2017). Preliminary evidence of improved cognitive performance following vestibular rehabilitation in children with combined ADHD (cADHD) and concurrent vestibular impairment. *Auris, Nasus, Larynx, 44*(6), 700–707. PMID: 28238393.

Mahfuz, M. M., Schubert, M. C., Todd, C. J., Figtree, W. V. C., Khan, S. I., & Migliaccio, A. A. (2017). The effect of visual contrast on human vestibulo-ocular reflex training. *Journal of the Association for Research in Otolaryngology, 19*(1), 113–122.

Mahfuz, M. M., Schubert, M. C., Figtree, W. V. C., Todd, C. J., Khan, S. I., & Migliaccio, A. A. (2018). Optimal human passive vestibulo-ocular reflex adaptation does not rely on passive training. *Journal of the Association for Research in Otolaryngology, 19*(3), 261–271.

Matiño-Soler, E., Esteller-More, E., Martin-Sanchez, J. C., Martinez-Sanchez, J. M., & Perez-Fernandez, N. (2015). Normative data on angular vestibule-ocular responses in the yaw axis measured using the video head impulse test. *Otology & Neurotology, 36,* 466–471.

Matsugi, A., Ueta, Y., Oku, K., Okuno, K., Tamaru, Y., Nomura, S., Tanaka, H., & Mori, N. (2017). Effect of gaze-stabilization exercises on vestibular function during postural control. *Neuroreport, 28,* 439–443.

McDonnell, M. N., & Hillier, S. L. (2015). Vestibular rehabilitation for unilateral peripheral vestibular dysfunction. *Cochrane Database of Systematic Reviews, 1,* CD005397. https://doi.org/10.1002/14651858.CD005397.pub4.

Meldrum, D., Herdman, S., Moloney, R., Murray, D., Duffy, D., Malone, K., French, H., Hone, S., Conroy, R., & McConn-Walsh, R. (2012). Effectiveness of conventional versus virtual reality based vestibular rehabilitation in the treatment of dizziness, gait and balance impairment in adults with unilateral peripheral vestibular loss: a randomised controlled trial. *BMC Ear, Nose and Throat Disorders, 12,* 3. https://doi.org/10.1186/1472-6815-12-3.

Merchant, S. N., Velazquez-Villasenor, L., Tsuji, K., Glynn, R. J., Wall, C., & Rauch, S. D. (2000). Temporal bone studies of the human peripheral vestibular system. Normative vestibular hair cell data. *The Annals of Otology, Rhinology, and Laryngology, 181,* 3–13.

Meyer, C. H., Lasker, A. G., & Robinson, D. A. (1985). The upper limit of human smooth pursuit velocity. *Vision Research, 25,* 561–563.

Micarelli, A., Viziano, A., Augimeri, I., Micarelli, D., & Alessandrini, M. (2017). Three-dimensional head-mounted gaming task procedure maximizes effects of vestibular rehabilitation in unilateral vestibular hypofunction: A randomized controlled pilot trial. *International Journal of Rehabilitation Research, 40*(4), 325–332. PMID: 28723718.

Migliaccio, A. A., & Schubert, M. C. (2013). Unilateral adaptation of the human angular vestibulo-ocular reflex. *Journal of the Association for Research in Otolaryngology, 14*(1), 29–36.

Migliaccio, A. A., & Schubert, M. C. (2014). Pilot study of a new rehabilitation tool: Improved unilateral short-term adaptation of the human angular vestibulo-ocular reflex. *Otology & Neurotology, 35*(10), e310–e316.

Mitsutake, T., Sakamoto, M., Ueta, K., Oka, S., & Horikawa, E. (2017). Effects of vestibular rehabilitation on gait performance in poststroke patients: A pilot randomized controlled trial. *International Journal of Rehabilitation Research, 40*(3), 240–245. PMID: 28542112.

Mulavara Fiedler, M. J., Kofman, I. S., Wood, S. J., Serrador, J. M., Peters, B., Cohen, H. S., Reschke, M. F., & Bloomberg, J. J. (2011). Improving balance function using vestibular stochastic resonance: Optimizing stimulus characteristics. *Experimental Brain Research, 210,* 303–312. PMID: 21442221.

Neuhauser, H. K., & Lempert, T. (2009). Vertigo: epidemiologic aspects. *Semin Neurol, 29,* 473–481. Ogata R, Ikari K, Hayashi M, Tamai K. Age related changes in the Purkinje's cells in the rat cerebellar cortex; a quantitative electron microscopic study. Folia Psychiatrica et Neurologica Japonica 1984;38.159–67.

Paige, G. D., & Sargent, E. W. (1991). Visually-induced adaptive plasticity in the human vestibulo-ocular reflex. *Experimental Brain Research, 84*(1), 25–34.

Pal, S., Rosengren, S. M., & Colebatch, J. G. (2009). Stochastic galvanic vestibular stimulation produces a small reduction in sway in Parkinson's disease. *Journal of Vestibular Research, 19,* 137–142.

Park, J. J., Tang, Y., Lopez, I., & Ishiyama, A. (2001). Age-related change in the number of neurons in the human vestibular ganglion. *The Journal of Comparative Neurology, 431,* 437–443.

Prestori, F., Bonardi, C., Mapelli, L., Lombardo, P., Goselink, R., De Stefano, M. E., Gandolfi, D., Mapelli, J., Bertrand, D., Schonewille, M., De Zeeuw, C., & D'Angelo, E. (2013). Gating of long-term potentiation by nicotinic acetylcholine receptors at the cerebellum input stage. *PLoS One, 8,* e64828.

Rauch, S. D., Velazquez-Villasenor, L., Dimitri, P. S., & Merchant, S. N. (2001). Decreasing hair cell counts in aging humans. *Annals of the New York Academy of Sciences, 942,* 220–227.

Richter, E. (1980). Quantitative study of human Scarpa's ganglion and vestibular sensory epithelia. *Acta Oto-Laryngologica, 90*, 199–208.

Rogers, J., Zornetzer, S. F., Bloom, F. E., & Mervis, R. E. (1984). Senescent microstructural changes in rat cerebellum. *Brain Research, 292*, 23–32.

Rosenhall, U. (1973). Degenerative patterns in the aging human vestibular neuro-epithelia. *Acta Oto-Laryngologica, 76*, 208–220.

Ross, M. D., Peacor, D., Johnsson, L. G., & Allard, L. F. (1976). Observation on normal and degenerating human otoconia. *The Annals of Otology, Rhinology, and Laryngology, 85*, 310–326.

Schonewille, M., Gao, Z., Boele, H. J., Veloz, M. F., Amerika, W. E., Simek, A. A., De Jeu, M. T., Steinberg, J. P., Takamiya, K., Hoebeek, F. E., Linden, D. J., Huganir, R. L., & De Zeeuw, C. I. (2011). Reevaluating the role of LTD in cerebellar motor learning. *Neuron, 70*, 43–50.

Schubert, M. C., Migliaccio, A. A., & Della Santina, C. C. (2006). Modification of compensatory saccades after VOR gain recovery. *Journal of Vestibular Research, 16*, 285–291.

Schubert, M. C., Migliaccio, A. A., Clendaniel, R. A., Allak, A., & Carey, J. P. (2008a). Mechanism of dynamic visual acuity recovery with vestibular rehabilitation. *Archives of Physical Medicine and Rehabilitation, 89*(3), 500–507. PMID: 18295629.

Schubert, M. C., Della Santina, C. C., & Shelhamer, M. (2008b). Incremental angular vestibulo-ocular reflex adaptation to active head rotation. *Experimental Brain Research, 191*(4), 435–446. PMID: 18712370.

Schubert, M. C., Migliaccio, A. A., Minor, L. B., & Clendaniel, R. A. (2008c). Retention of VOR gain following short-term VOR adaptation. *Experimental Brain Research, 187*, 117–127.

Serrador, J. M., Deegan, B. M., Geraghty, M. C., & Wood, S. J. (2018). Enhancing vestibular function in the elderly with imperceptible electrical stimulation. *Scientific Reports, 8*(1), 336. PMID: 29321511.

Shelhamer, M., Robinson, D. A., & Tan, H. S. (1992). Context-specific adaptation of the gain of the vestibulo-ocular reflex in humans. *Journal of Vestibular Research, 2*, 89–96.

Shelhamer, M., Tiliket, C., Roberts, D., Kramer, P. D., & Zee, D. S. (1994). Short-term vestibulo-ocular reflex adaptation in humans. II. Error signals. *Experimental Brain Research, 100*, 328–336.

Solomon, D., Zee, D. S., & Straumann, D. (2003). Torsional and horizontal vestibular ocular reflex adaptation: Three-dimensional eye movement analysis. *Experimental Brain Research, 152*, 150–155.

Sugaya, N., Arai, M., & Goto, F. (2017). Is the headache in patients with vestibular migraine attenuated by vestibular rehabilitation? *Frontiers in Neurology, 8*, 124.

Sumnall, J. H., Freeman, T. C., & Snowden, R. J. (2003). Optokinetic potential and the perception of head-centred speed. *Vision Research, 43*, 1709–1718.

Shayman, C. S., Earhart, G. M., & Hullar, T. E. (2017). Improvements in gait with hearing aids and cochlear implants. *Otology & Neurotology, 38*(4), 484–486. PMID:28187057.

Smaerup, M., Grönvall, E., Larsen, S. B., Laessoe, U., Henriksen, J. J., & Damsgaard, E. M. (2017). Exercise gaming – A motivational approach for older adults with vestibular dysfunction. *Disability and Rehabilitation. Assistive Technology, 12*(2), 137–144. Epub 2016 Jan 4.

Titley, H. K., & Hansel, C. (2016). Asymmetries in cerebellar plasticity and motor learning. *Cerebellum, 15*, 87–92.

Thompson, P. (1982). Perceived rate of movement depends on contrast. *Vision Research, 22*, 377–380.

Thompson, P. (1983). Discrimination of moving gratings at and above detection threshold. *Vision Research, 23*, 1533–1538.

Todd, C. J., Hübner, P. P., Hübner, P., Schubert, M. C., & Migliaccio, A. A. (2018). StableEyes – A Portable vestibular rehabilitation device. *IEEE Trans Neural Syst Rehabil Eng. 2018 Jun;26.* (6):1223–1232. https://doi.org/10.1109/TNSRE.2018.2834964. PMID: 29877847.

Velazquez-Villasenor, L., Merchant, S. N., Tsuji, K., Glynn, R. J., Wall, C., 3rd, & Rauch, S. D. (2000). Temporal bone studies of the human peripheral vestibular system. Normative Scarpa's ganglion cell data. *The Annals of Otology, Rhinology, and Laryngology, 181*, 14–19.

Vitkovic, J., Winoto, A., Rance, G., Dowell, R., & Paine, M. (2013). Vestibular rehabilitation outcomes in patients with and without vestibular migraine. *Journal of Neurology, 260*(12), 3039–3048. PMID: 24061769.

Waddington, J., & Harris, C. M. (2015). Human optokinetic nystagmus and spatial frequency. *Journal of Vision, 15*, 7. https://doi.org/10.1167/15.13.7.

Whitney, S. L., Wrisley, D. M., Brown, K. E., & Furman, J. M. (2000). Physical therapy for migraine-related vestibulopathy and vestibular dysfunction with history of migraine. *The Laryngoscope, 110*(9), 1528–1534. PMID: 10983955.

Woodruff-Pak, D. S., Foy, M. R., Akopian, G. G., Lee, K. H., Zach, J., Nguyen, K. P., Comalli, D. M., Kennard, J. A., Agelan, A., & Thompson, R. F. (2010). Differential effects and rates of normal aging in cerebellum and hippocampus. *Proceedings of the National Academy of Sciences, 107*, 1624–1629.

Wuehr, M., Nusser, E., Decker, J., Krafczyk, S., Straube, A., Brandt, T., Jahn, K., & Schniepp, R. (2016). Noisy vestibular stimulation improves dynamic walking stability in bilateral vestibulopathy. *Neurology, 86*(23), 2196–2202.

Yakushin, S. B., Raphan, T., & Cohen, B. (2003). Gravity-specific adaptation of the angular vestibuloocular reflex: Dependence on head orientation with regard to gravity. *Journal of Neurophysiology, 89*, 571–586.

Yuan, R., Tsaih, S. W., Petkova, S. B., Marin de Evsikova, C., Xing, S., Marion, M. A., Bogue, M. A., Mills, K. D., Peters, L. L., Bult, C. J., Rosen, C. J., Sundberg, J. P., Harrison, D. E., Churchill, G. A., & Paigen, B. (2009). Aging in inbred strains of mice: study design and interim report on median lifespans and circulating IGF1 levels. *Aging Cell, 8*, 277–287.

Zhang, C., Zhu, Q., & Hua, T. (2010). Aging of cerebellar Purkinje cells. *Cell and Tissue Research, 341*, 341–347.

Positional Downbeat Nystagmus

Jeong-Yoon Choi and Ji-Soo Kim

Abstract Positional downbeat nystagmus is one of the common forms of central positional nystagmus. It has been reported in various structural, metabolic, or degenerative disorders affecting the inferior cerebellum. According to the temporal characteristics, it can be divided into the paroxysmal form that is usually evident after lying down or straight head-hanging (backward pitch rotation) and the persistent form that is mostly observed in the static prone head position. The paroxysmal form of central positional downbeat nystagmus may be ascribed to a pathologically enhanced post-rotational cue. In contrast, the persistent form has been explained by hyperactive otolith-ocular reflex or by a bias in estimated gravity direction in the head coordinate frame. Positional downbeat nystagmus may be observed in benign paroxysmal positional vertigo. Characteristics of the nystagmus and associated neuro-otological findings may allow differentiating central from peripheral positional downbeat nystagmus.

Keywords Positional nystagmus · Central positional nystagmus · Downbeat nystagmus · Benign paroxysmal positional vertigo

1 Introduction

Positional nystagmus emerges when the head orientation is changed in relation to gravity direction (Buttner et al. 1999; Choi et al. 2015). The changes in head position in relation to gravity are inevitably accompanied by rotational head motion that

J.-Y. Choi · J.-S. Kim (✉)
Department of Neurology, Seoul National University College of Medicine,
Seoul, South Korea

Dizziness Center, Clinical Neuroscience Center and Department of Neurology,
Seoul National University Bundang Hospital, Seongnam, South Korea
e-mail: jisookim@snu.ac.kr

© Springer Nature Switzerland AG 2019
A. Shaikh, F. Ghasia (eds.), *Advances in Translational Neuroscience of Eye Movement Disorders*, Contemporary Clinical Neuroscience,
https://doi.org/10.1007/978-3-030-31407-1_10

stimulates the semicircular canals. In addition, there is an alteration of the estimated gravity direction in the head coordinate frame even though the gravity direction remains unchanged in the earth coordinate frame. During both tilt and translation, the otolith organs relay the same gravito-inertial acceleration (GIA) signal. This inherent ambiguity can be resolved by a "tilt-estimator circuit" in which information from the semicircular canals about head rotation is combined with otolith information about linear acceleration through the velocity-storage mechanism (see Fig. 1) (Laurens and Angelaki 2011). As a result, any erroneous signal from the semicircular canals, otolith organs, or the central vestibular system may generate positional nystagmus.

Previously, the terminology "positioning" was introduced to discriminate the vertigo and nystagmus that are induced by certain head motion (rotation and

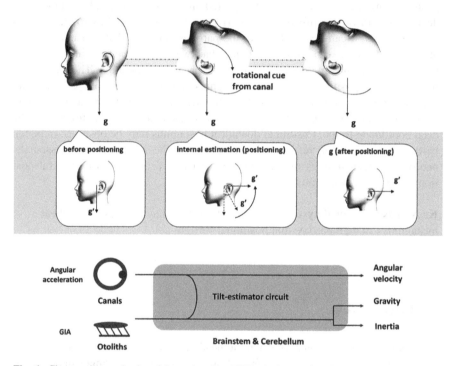

Fig. 1 Changes in gravito-inertial acceleration (GIA) during positioning. In either static head upright or supine position, there is no inertial component. Therefore, the GIA relayed from the otolith purely represents the gravity (g). Although the gravity direction remains constant irrespective of head positions in the earth coordinate frame (upper column), it is subjected to change according to head position in the head coordinate frame (g') (middle column). During both tilt and translation, the otolith organs are known to relay the same GIA signal. This inherent ambiguity is believed to be resolved by a "tilt-estimator circuit" in which information from the semicircular canals about head rotation is combined with otolith information about linear acceleration through the velocity-storage mechanism (low column). Thus, the central vestibular system updates the direction of internally estimated gravity using the rotational cue relayed from the canals during each positioning

translation) from the "positional" vertigo and nystagmus that are observed in certain head positions (Buttner et al. 1999). Even though this distinction was discarded and both terms were integrated into "positional" in a recent paper on the nomenclature of vestibular symptoms (Bisdorff et al. 2009), we may infer the actual stimuli of positional nystagmus from the duration, i.e., "paroxysmal (transient)" vs. "persistent" (Bisdorff et al. 2009). Even though positional nystagmus mostly refers to the nystagmus that is observed only in head positions other than upright, any modulation of spontaneous nystagmus, i.e., augmentation or attenuation, by changes in head position may be ascribed to positional nystagmus.

Among the several types of positional nystagmus, positional downbeat nystagmus has drawn attention since it is one of the common forms of central positional nystagmus (Macdonald et al. 2017). However, positional downbeat nystagmus may be observed in either central or peripheral vestibular disorders. Thus, differentiation of central from peripheral form of positional downbeat nystagmus is important in clinical practice (Buttner et al. 1999; Choi et al. 2015). From the neurophysiological aspect, positional downbeat nystagmus may provide information on how the central vestibular system handles the canal and otolith information during head motion. As mentioned above, positional downbeat nystagmus may be paroxysmal or persistent (Choi et al. 2015).

2   Central Positional Downbeat Nystagmus

One of the fundamental functions of the central vestibular system is to refine the information on rotational velocity relayed from the semicircular canals and to estimate the gravity direction accurately in the head coordinate frame (Laurens and Angelaki 2011). Therefore, in central vestibular dysfunction, the source of positional downbeat nystagmus would be the inaccurately estimated rotational velocity or gravity direction (Choi et al. 2015, 2018). Of interest, the paroxysmal form of central positional downbeat nystagmus was more prominent in the lying down or straight head-hanging positions, while the persistent form was mostly observed in the prone (nose-down) position in the previous studies (Marti et al. 2002; Choi et al. 2018).

3   Paroxysmal Form of Central Positional Downbeat Nystagmus

This type of nystagmus is characterized by a short-lasting, usually within 30 seconds, prominent downbeat nystagmus (see Fig. 2). It is usually provoked by changing the positions from upright to supine or straight head-hanging (Leigh and Zee 2015). The nystagmus develops immediately after the positional changes and

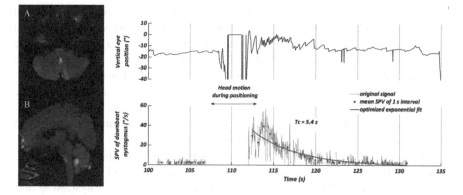

Fig. 2 Paroxysmal positional downbeat nystagmus in a nodular lesion. In a patient with posterior inferior cerebellar artery infarction (**a**, **b**) involving the nodulus, straight head-hanging invokes prominent downbeat nystagmus. The peak slow phase velocity (SPV) of the positional downbeat nystagmus is about 40°/s, and the estimated time constant (Tc) for decay is 5.4 seconds (**c**)

declines rapidly with a time constant of about 5 seconds. The rotational axis of downbeat nystagmus is closely aligned with the vector sum of the rotational axes of the canals inhibited during the positioning (Buttner et al. 1999; Choi et al. 2015). It can emerge in the absence of primary downbeat nystagmus or be added to pre-existing primary downbeat nystagmus (Choi et al. 2015).

This nystagmus has mostly been recognized and studied in patients with Chiari malformation or tumors or strokes involving the posterior fossa (Baloh and Spooner 1981; Barber 1984; Yee et al. 1984; Cho et al. 2017). It has also been reported in various degenerative or metabolic disorders such as spinocerebellar ataxia type 6 (Yabe et al. 2003; Kim et al. 2013), multisystem atrophy (Lee et al. 2009; Kim et al. 2013), episodic ataxia type 2 (Jen et al. 2004, 2007), and anti-epileptic drug intoxication (Oh et al. 2006; Choi et al. 2014b). Paroxysmal positional downbeat nystagmus may appear in paraneoplastic or parainfectious autoimmune cerebellitis (Choi et al. 2014a, c), after heat stroke (Jung et al. 2017), in X-linked adrenoleukodystrophy (Kim et al. 2016a), and even in essential tremor (Kim et al. 2016b).

Based on these reports, paroxysmal positional downbeat nystagmus has been ascribed to brainstem and cerebellar dysfunction. However, benign paroxysmal positional vertigo may also present positional downbeat nystagmus (Bertholon et al. 2002; Lopez-Escamez et al. 2006; Cambi et al. 2013). Furthermore, positional downbeat nystagmus can be observed even in normal subjects during straight head-hanging or upside-down position (Bisdorff et al. 2000; Kim et al. 2000). However, the temporal characteristics of positional downbeat nystagmus require further elucidation in normal subjects and in patients with benign paroxysmal positional vertigo.

In the paroxysmal type of central positional downbeat nystagmus while lying-down or straight head-hanging (backward head rotation), the location of the responsible lesions and the mechanism remain unclear, but dysfunction of the inferior cerebellar vermis such as the nodulus and uvula has been speculated (Choi et al. 2015). A recent study has invoked enhanced post-rotational cue to explain this type of nystagmus (Choi et al. 2015; Choi and Kim 2017). Of the two types of vestibular afferents (regular vs. irregular) (Goldberg and Fernandez 1971), the irregular afferents have a medium to large axonal caliber, more linkage with the calyx type of hair cells, and moderate to high adaptive properties (Leigh and Zee 2015). Hence, the irregular vestibular afferents may be engaged in the velocity-storage circuitry (Leigh and Zee 2015). The adaptive properties of the irregular afferents are characterized by transient up- or downregulation of the resting discharge following the acceleration or deceleration period (Goldberg and Fernandez 1971). Therefore, the irregular afferents can generate a post-rotational cue even in normal subjects (i.e., forward rotational cue after straight head-hanging) and drifts in the internally estimated gravity direction (Laurens and Angelaki 2011). This post-rotational cue, however, is dampened rapidly by the rotational feedback signals originating from the difference between the otolith inputs and the estimated gravity direction (Laurens and Angelaki 2011). The nodulus and uvula inhibit the vestibular activity through the reciprocal connections with the vestibular nucleus. Thus, the lesions involving the nodulus and uvula may enhance the responses of the neurons within the vestibular nuclei to the irregular afferents and produce prominent post-rotational cue that cannot be corrected by the rotational feedback signals in a real-time manner (see Fig. 3) (Angelaki and Perachio 1993; Clendaniel et al. 2001). Consequently, the enhanced forward rotational cue after backward head rotation may generate paroxysmal downbeat nystagmus with a time constant of about 5 seconds (Choi et al. 2015), which is similar to the time constant of the vertical vestibulo-ocular reflexes (Bertolini and Ramat 2011).

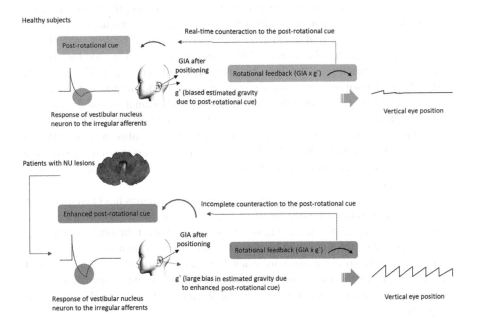

Fig. 3 Hypothetical explanation of central paroxysmal positional downbeat nystagmus. Abbreviations: GIA gravito-inertial acceleration, g' internally estimated gravity, NU nodulus and uvula

4 Persistent Positional Downbeat Nystagmus in Central Lesions

Patients with cerebellar dysfunction often show downbeat nystagmus with the head upright, which may be explained by asymmetry of the vertical smooth pursuit or vestibulo-ocular reflex (Zee et al. 1974; Baloh and Spooner 1981; Gresty et al. 1986) or by impaired gaze-holding mechanism in the pitch plane (Zee et al. 1980). There have been several reports on the intensity of downbeat nystagmus that changes according to the static head positions, augmented in the nose-down (prone) and attenuated in the nose-up (supine) positions (Gresty et al. 1986; Marti et al. 2002). Moreover, in some patients, downbeat nystagmus while head upright changes into upbeat nystagmus on assuming supine position (Helmchen et al. 2004). These patterns of modulation are opposed to those observed in paroxysmal positional downbeat nystagmus and request additional inference on the mechanisms engaged in positional modulation of primary downbeat nystagmus. In various static head positions, the gravity direction differs only in the head coordinate frame. Thus, one explanation for positional modulation of downbeat nystagmus would be that the eye velocity bias is generated by the effort to re-align the eye position along the gravity direction in the earth coordinate frame and this bias modulates the primary downbeat nystagmus in each head position (Marti et al. 2002). Hence, downbeat nystagmus may be enhanced by upward eye velocity bias generated in the nose-down position and reduced by downward eye velocity bias in the nose-up position. Indeed, the modulation of downbeat nystagmus induced by changing the head orientation relative to gravity fits on a sinusoidal waveform centered at upright position (Marti et al. 2002). This pattern of modulation was also observed in healthy subjects with a downscaled intensity (Marti et al. 2002). Downbeat nystagmus has mostly been ascribed to dysfunction of the flocculus and tonsil (Hufner et al. 2007). Therefore, the enhanced positional modulation in patients with downbeat nystagmus was explained by hyperactive otolith-ocular reflex due to floccular or nodular dysfunction (Precht et al. 1976; Snyder and King 1996). Since re-aligning the eye position along the gravity has been considered the function of the nodulus and uvula (Wearne et al. 1998; Paige and Seidman 1999), however, the performance of these structures should be remained to an extent to be able to keep the eye orientation according to this explanation.

Recently persistent form of central apogeotropic nystagmus has been ascribed to a bias in the estimated gravity direction due to lesions involving the nodulus and uvula (Choi et al. 2018). According to this preposition, the rotational feedback signals generated by the difference between the actual and estimated gravity generate persistent apogeotropic nystagmus when the head is turned to either side while supine and the bias is toward the nose in this position. Likewise, downbeat nystagmus may be simulated when the bias is toward the chin of the head while supine or prone.

5 Positional Downbeat Nystagmus in Peripheral Lesions

In peripheral vestibular lesions, positional downbeat nystagmus may occur due to a false rotational cue generated by movements of detached otoconia in the semicircular canals during and after the positioning.

In a study on 50 patients with positional downbeat nystagmus, 12 (24%) with an idiopathic cause were assumed to have lithiasis of the anterior semicircular canal (Bertholon et al. 2002). In those patients, the positional downbeat nystagmus was observed in the straight head-hanging as well as during Dix-Hallpike maneuver in either direction and was accompanied by a weak torsional component. The development of positional downbeat nystagmus during Dix-Hallpike maneuver in either direction was ascribed to the more vertically oriented anterior than posterior semicircular canal in the head upright position. Thus, even during the Dix-Hallpike maneuver in the affected side, the otolithic debris in the anterior canal may gravitate and induce the vertigo and nystagmus (Bertholon et al. 2002). In another study on 80 patients with benign paroxysmal positional vertigo, 14 (17.5%) showed positional downbeat nystagmus during straight head-hanging or Dix-Hallpike maneuvers, possibly due to lithiasis of the anterior semicircular canal (Lopez-Escamez et al. 2006). The duration of these positional downbeat nystagmus ranged from 7 to 50 seconds. During the follow-up, the nystagmus mostly disappeared within 30 days. Another study also showed a similar finding that the positional downbeat nystagmus induced in the anterior canal lithiasis lasts about 40 seconds during straight head-hanging or Dix-Hallpike maneuvers and disappears mostly within a month (Cambi et al. 2013). Its association with the posterior canal lithiasis prior to or after the diagnosis of positional downbeat nystagmus strongly supports anterior canal lithiasis (Cambi et al. 2013).

Given the orientation of the anterior canal while sitting or supine, migration of the canalith from the anterior (the segment between the cupula and isthmus) to the posterior portion (the segment between the isthmus and common crus) of the long arm (from the cupula to common crus) can hardly occur during straight head-hanging or Dix-Hallpike maneuvers. Thus, the positional downbeat nystagmus in benign paroxysmal positional vertigo has also been ascribed to anterior canal cupulolithiasis or canalolithiasis in the short arm (the segment between the utricle and cupula) (Buki 2014). Indeed, the utricle is located above the cupula of the anterior semicircular canal when the head is placed below the horizontal line. Thus, the fragile or already dislodged otoconia debris within the utricle would fall onto the cupula or into the short arm of the anterior canal during straight head-hanging. In this case, positional downbeat nystagmus can be either persistent or paroxysmal (Buki 2014).

Cupulolithiasis of the posterior canal is another hypothesis to explain positional downbeat nystagmus. During straight head-hanging, the otoconia pulled by the gravity may deflect the cupula toward the utricle when the otoconia is placed in the inferior portion of the cupula and generate inhibitory pattern of persistent downbeat nystagmus. In this instance, the absence of spontaneous nystagmus during upright position is explained by adaptation to the prolonged cupular deflection (Buki 2014). Positional downbeat nystagmus may be observed after treatment of the typical type of posterior canal lithiasis (Oh et al. 2018). After removing most of the otoconia chunks with repositioning maneuvers, a small amount of debris remained in the long arm of the posterior canal may migrate in the ampullopetal direction during subsequent positional maneuvers and cause positional downbeat nystagmus.

Taken all together, positional downbeat nystagmus may be observed in benign paroxysmal positional vertigo even though the suggested mechanisms remain largely hypothetical.

6　Differential Diagnosis of Positional Downbeat Nystagmus: Central Versus Peripheral

In patients with positional downbeat nystagmus, especially when the nystagmus is prominent during straight head-hanging or Dix-Hallpike maneuver, the presence of latency or the duration of induced nystagmus may not help differentiating central from peripheral etiologies (Lopez-Escamez et al. 2006; Cambi et al. 2013; Choi et al. 2015). However, positional downbeat nystagmus due to central lesions often accompanies apogeotropic positional nystagmus during supine roll test (Choi et al. 2015). Therefore, central mechanism(s) should be suspected in positional nystagmus that occurs in the direction of head motion during more than one positional maneuver. In this instance, careful examination may disclose other central signs, such as gaze-evoked nystagmus, saccadic dysmetria, perverted head-shaking nystagmus, or limb and truncal ataxia. In isolated positional downbeat nystagmus, especially when paroxysmal and symptomatic, canalith repositioning maneuvers for the anterior canal type should be attempted first (Leigh and Zee 2015), and central pathologies involving the nodulus or uvula should be suspected when the vertigo and nystagmus are refractory to those maneuvers (Kronenbuerger et al. 2018).

Source of Funding/Support　This research was supported by the Basic Science Research Program through the National Research Foundation of Korea (NRF) and funded by the Ministry of Education, Science and Technology (NRF-2016R1D1A1B04935568).

Conflict of Interest Statement　The authors declare they have no conflicts of interest.

Disclosure　J-Y Choi has nothing to disclose. J-S Kim serves as an Associate Editor of *Frontiers in Neuro-Otology* and on the editorial boards of the *Journal of Clinical Neurology, Frontiers in Neuro-Ophthalmology, Journal of Neuro-Ophthalmology, Journal of Vestibular Research, Journal of Neurology,* and *Medicine*.

Authors' Contributions　Jeong-Yoon Choi acquired the data, analyzed the data, and drafted the manuscript; Ji-Soo Kim conceptualized the study, analyzed the data, and revised the manuscript.

References

Angelaki, D. E., & Perachio, A. A. (1993). Contribution of irregular semicircular canal afferents to the horizontal vestibuloocular response during constant velocity rotation. *Journal of Neurophysiology, 69*(3), 996–999.

Baloh, R. W., & Spooner, J. W. (1981). Downbeat nystagmus: A type of central vestibular nystagmus. *Neurology, 31*(3), 304–310.

Barber, H. O. (1984). Positional nystagmus. *Otolaryngology and Head and Neck Surgery, 92*(6), 649–655.

Bertholon, P., Bronstein, A. M., Davies, R. A., Rudge, P., & Thilo, K. V. (2002). Positional down beating nystagmus in 50 patients: Cerebellar disorders and possible anterior semicircular canalithiasis. *Journal of Neurology, Neurosurgery, and Psychiatry, 72*(3), 366–372.

Bertolini, G., & Ramat, S. (2011). Velocity storage in the human vertical rotational vestibuloocular reflex. *Experimental Brain Research, 209*(1), 51–63.

Bisdorff, A., Sancovic, S., Debatisse, D., Bentley, C., Gresty, M., & Bronstein, A. (2000). Positional nystagmus in the dark in normal subjects. *Neuro-Ophthalmology, 24*(1), 283–290.

Bisdorff, A., Von Brevern, M., Lempert, T., & Newman-Toker, D. E. (2009). Classification of vestibular symptoms: Towards an international classification of vestibular disorders. *Journal of Vestibular Research, 19*(1–2), 1–13.

Buki, B. (2014). Benign paroxysmal positional vertigo--toward new definitions. *Otology & Neurotology, 35*(2), 323–328.

Buttner, U., Helmchen, C., & Brandt, T. (1999). Diagnostic criteria for central versus peripheral positioning nystagmus and vertigo: A review. *Acta Oto-Laryngologica, 119*(1), 1–5.

Cambi, J., Astore, S., Mandala, M., Trabalzini, F., & Nuti, D. (2013). Natural course of positional down-beating nystagmus of peripheral origin. *Journal of Neurology, 260*(6), 1489–1496.

Cho, B. H., Kim, S. H., Kim, S. S., Choi, Y. J., & Lee, S. H. (2017). Central positional nystagmus associated with cerebellar tumors: Clinical and topographical analysis. *Journal of the Neurological Sciences, 373*, 147–151.

Choi, J. Y., Glasauer, S., Kim, J. H., Zee, D. S., & Kim, J. S. (2018). Characteristics and mechanism of apogeotropic central positional nystagmus. *Brain, 141*(3), 762–775.

Choi, J. Y., Kim, J. H., Kim, H. J., Glasauer, S., & Kim, J. S. (2015). Central paroxysmal positional nystagmus: Characteristics and possible mechanisms. *Neurology, 84*(22), 2238–2246.

Choi, J. Y., & Kim, J. S. (2017). Nystagmus and central vestibular disorders. *Current Opinion in Neurology, 30*(1), 98–106.

Choi, J. Y., Kim, J. S., Jung, J. M., Kwon, D. Y., Park, M. H., Kim, C., et al. (2014a). Reversed corrective saccades during head impulse test in acute cerebellar dysfunction. *Cerebellum, 13*(2), 243–247.

Choi, J. Y., Park, Y. M., Woo, Y. S., Kim, S. U., Jung, J. M., & Kwon, D. Y. (2014b). Perverted head-shaking and positional downbeat nystagmus in pregabalin intoxication. *Journal of the Neurological Sciences, 337*(1–2), 243–244.

Choi, S. Y., Park, S. H., Kim, H. J., & Kim, J. S. (2014c). Paraneoplastic downbeat nystagmus associated with cerebellar hypermetabolism especially in the nodulus. *Journal of the Neurological Sciences, 343*(1–2), 187–191.

Clendaniel, R. A., Lasker, D. M., & Minor, L. B. (2001). Horizontal vestibuloocular reflex evoked by high-acceleration rotations in the squirrel monkey. IV. Responses after spectacle-induced adaptation. *Journal of Neurophysiology, 86*(4), 1594–1611.

Goldberg, J. M., & Fernandez, C. (1971). Physiology of peripheral neurons innervating semicircular canals of the squirrel monkey. I. Resting discharge and response to constant angular accelerations. *Journal of Neurophysiology, 34*(4), 635–660.

Gresty, M., Barratt, H., Rudge, P., & Page, N. (1986). Analysis of downbeat nystagmus. Otolithic vs semicircular canal influences. *Archives of Neurology, 43*(1), 52–55.

Helmchen, C., Sprenger, A., Rambold, H., Sander, T., Kompf, D., & Straumann, D. (2004). Effect of 3,4-diaminopyridine on the gravity dependence of ocular drift in downbeat nystagmus. *Neurology, 63*(4), 752–753.

Hufner, K., Stephan, T., Kalla, R., Deutschlander, A., Wagner, J., Holtmannspotter, M., et al. (2007). Structural and functional MRIs disclose cerebellar pathologies in idiopathic downbeat nystagmus. *Neurology, 69*(11), 1128–1135.

Jen, J., Kim, G. W., & Baloh, R. W. (2004). Clinical spectrum of episodic ataxia type 2. *Neurology, 62*(1), 17–22.

Jen, J. C., Graves, T. D., Hess, E. J., Hanna, M. G., Griggs, R. C., Baloh, R. W., et al. (2007). Primary episodic ataxias: Diagnosis, pathogenesis and treatment. *Brain, 130*(Pt 10), 2484–2493.

Jung, I., Choi, S. Y., Kim, H. J., & Kim, J. S. (2017). Delayed vestibulopathy after heat exposure. *Journal of Neurology, 264*(1), 49–53.

Kim, J. I., Somers, J. T., Stahl, J. S., Bhidayasiri, R., & Leigh, R. J. (2000). Vertical nystagmus in normal subjects: Effects of head position, nicotine and scopolamine. *Journal of Vestibular Research, 10*(6), 291–300.

Kim, J. S., Kim, J. S., Youn, J., Seo, D. W., Jeong, Y., Kang, J. H., et al. (2013). Ocular motor characteristics of different subtypes of spinocerebellar ataxia: Distinguishing features. *Movement Disorders, 28*(9), 1271–1277.

Kim, S. H., Kim, S. S., Ha, H., & Lee, S. H. (2016a). X-linked adrenoleukodystrophy presenting with positional downbeat nystagmus. *Neurology, 86*(23), 2214–2215.

Kim, Y. E., Kim, J. S., Yang, H. J., Yun, J. Y., Kim, H. J., Ehm, G., et al. (2016b). Perverted head-shaking and positional downbeat nystagmus in essential tremor. *Cerebellum, 15*(2), 152–158.

Kronenbuerger, M., Olivi, A., & Zee, D. S. (2018). Pearls & Oy-sters: Positional vertigo and vertical nystagmus in medulloblastoma: A picture is worth a thousand words. *Neurology, 90*(4), e352–e354.

Laurens, J., & Angelaki, D. E. (2011). The functional significance of velocity storage and its dependence on gravity. *Experimental Brain Research, 210*(3–4), 407–422.

Lee, J. Y., Lee, W. W., Kim, J. S., Kim, H. J., Kim, J. K., & Jeon, B. S. (2009). Perverted head-shaking and positional downbeat nystagmus in patients with multiple system atrophy. *Movement Disorders, 24*(9), 1290–1295.

Leigh, R. J., & Zee, D. S. (2015). *The neurology of eye movements.* New York: Oxford University Press.

Lopez-Escamez, J. A., Molina, M. I., & Gamiz, M. J. (2006). Anterior semicircular canal benign paroxysmal positional vertigo and positional downbeating nystagmus. *American Journal of Otolaryngology, 27*(3), 173–178.

Macdonald, N. K., Kaski, D., Saman, Y., Al-Shaikh Sulaiman, A., Anwer, A., & Bamiou, D.-E. (2017). Central positional nystagmus: A systematic literature review. *Frontiers in Neurology, 8*, 141.

Marti, S., Palla, A., & Straumann, D. (2002). Gravity dependence of ocular drift in patients with cerebellar downbeat nystagmus. *Annals of Neurology, 52*(6), 712–721.

Oh, E. H., Lee, J. H., Kim, H. J., Choi, S. Y., Choi, K. D., & Choi, J. H. (2018). Incidence and clinical significance of positional downbeat nystagmus in posterior canal benign paroxysmal positional vertigo. *Journal of Clinical Neurology, 15*(2), 143–148.

Oh, S. Y., Kim, J. S., Lee, Y. H., Lee, A. Y., Kim, J., & Kim, J. M. (2006). Downbeat, positional, and perverted head-shaking nystagmus associated with lamotrigine toxicity. *Journal of Clinical Neurology, 2*(4), 283–285.

Paige, G. D., & Seidman, S. H. (1999). Characteristics of the VOR in response to linear acceleration. *Annals of the New York Academy of Sciences, 871*, 123–135.

Precht, W., Volkind, R., Maeda, M., & Giretti, M. L. (1976). The effects of stimulating the cerebellar nodulus in the cat on the responses of vestibular neurons. *Neuroscience, 1*(4), 301–312.

Snyder, L. H., & King, W. M. (1996). Behavior and physiology of the macaque vestibulo-ocular reflex response to sudden off-axis rotation: Computing eye translation. *Brain Research Bulletin, 40*(5–6), 293–301; discussion 302.

Wearne, S., Raphan, T., & Cohen, B. (1998). Control of spatial orientation of the angular vestibuloocular reflex by the nodulus and uvula. *Journal of Neurophysiology, 79*(5), 2690–2715.

Yabe, I., Sasaki, H., Takeichi, N., Takei, A., Hamada, T., Fukushima, K., et al. (2003). Positional vertigo and macroscopic downbeat positioning nystagmus in spinocerebellar ataxia type 6 (SCA6). *Journal of Neurology, 250*(4), 440–443.

Yee, R. D., Baloh, R. W., & Honrubia, V. (1984). Episodic vertical oscillopsia and downbeat nystagmus in a Chiari malformation. *Archives of Ophthalmology, 102*(5), 723–725.

Zee, D. S., Friendlich, A. R., & Robinson, D. A. (1974). The mechanism of downbeat nystagmus. *Archives of Neurology, 30*(3), 227–237.

Zee, D. S., Leigh, R. J., & Mathieu-Millaire, F. (1980). Cerebellar control of ocular gaze stability. *Annals of Neurology, 7*(1), 37–40.

Slow Saccades

Kelsey Jensen and Aasef Shaikh

Abstract Rapid and yoked eye movements made to direct the target over fovea, saccades, are critical aspects of visual scanning behavior. Elegant physiological, anatomical, and computational studies have described the complex circuitry of saccadic eye movements. These basic science examinations have facilitated our ability to confidently localize central pathology affecting the saccade generation. Amplitude, velocity, duration, direction, latency, and accuracy are traditional measures of saccadic eye movements. The visually guided saccades are very commonly studied eye movements. In this chapter, we will outline pertinent anatomy and physiology of saccades. We will also discuss disorders affecting the velocity of visually guided saccades. Finally, we will discuss the mechanistic underpinning of slowing of saccade velocity in context of these neurological disorders.

Keywords Saccade · Cerebellum · Basal ganglia · Degenerative disorder · Eye movement

1 Introduction

Saccades are rapid movements of both eyes that position the fovea at an object of interest, thereby increasing visual acuity of the object. These movements require coordination and precision, and the complex circuitry involved in producing them can offer insight into a variety of pathologies affecting the central nervous system. There are many different types of saccades. Visually guided saccades are simultaneous movements directed toward an intended target and internally generated.

K. Jensen · A. Shaikh (✉)
Neurological Institute, University Hospitals, Cleveland, OH, USA

Department of Neurology, Case Western Reserve University, Cleveland, OH, USA

Neurology Service, Louis Stokes Cleveland VA Medical Center, Cleveland, OH, USA

© Springer Nature Switzerland AG 2019
A. Shaikh, F. Ghasia (eds.), *Advances in Translational Neuroscience of Eye Movement Disorders*, Contemporary Clinical Neuroscience,
https://doi.org/10.1007/978-3-030-31407-1_11

Reflexive saccades are aptly named re-fixations made toward an unexpected target. In contrast, memory-guided saccades are made in the direction of a remembered target. Lastly, antisaccades are of an equal but opposite amplitude from a target and suppress movement in the direction of the target. Each saccade type is generated via its distinct circuitry, although they share a final common pathway in the brainstem. They all require precision in amplitude, timing, and velocity. Disturbances of the eye or orbital muscles typically affect all three parameters, whereas central pathologies may affect only one or two of them. This means that subtle changes in saccades have utility in understanding the central pathologies affecting them.

Of the different saccade types, visually guided saccades are the best understood, making them a valuable tool for localization and diagnosis of neurological disorders affecting them. They also provide a useful metric for assessing response to treatment. This chapter will cover the neuroanatomy and physiology of visually guided saccades, as well as cerebellar and neurodegenerative disorders that affect the velocity of these saccades.

2   Generation of Visually Guided Saccades

2.1   Cerebral Cortex

Visually guided saccades begin in the cerebral hemisphere in two distinct pathways that converge on the superior colliculus (Munoz 2002; Pierrot-Deseilligny et al. 2004).

In the first pathway, the frontal eye field (FEF), supplementary eye field (SEF), and dorsolateral prefrontal cortex (dlPFC) generate the initiating signal and project to the caudate nucleus (Fox et al. 1985; Pierrot-deseilligny et al. 1991; Berman et al. 1999; Hikosaka et al. 2000). This results in direct inhibition by the caudate nucleus on the substantia nigra pars reticulata (SNpr) (Hikosaka et al. 2000; Nambu et al. 2002). Inhibition of the SNpr by the caudate nucleus results in activation of the superior colliculus, which is otherwise under tonic GABAergic inhibition by the SNpr (Fisher et al. 1986; Francois et al. 1984; Handel and Glimcher 1999; Hikosaka and Wurtz 1983). This cessation of inhibition on the superior colliculus results in saccade initiation. The caudate nucleus also sends fibers through the external segment of the globus pallidus to the subthalamic nucleus (Hikosaka et al. 2000).

In the second pathway, the parietal eye field (PEF) located in the posterior parietal cortex initiates saccades (Munoz 2002; Pierrot-Deseilligny et al. 2004). The PEF both projects onto and receives input from the FEF and also projects to the superior colliculus (Müri et al. 1996; Pouget 2015). The communication between the PEF and FEF appears to be important in visual processing, whereas the path from PEF to superior colliculus is believed to impact the expression of saccades (Pouget 2015).

Damage to these cortical structures does not directly impair saccade velocity, although chronic lesions may cause a decrease in amplitude and impaired latency in saccades (Schiller et al. 1980; Rivaud et al. 1994).

2.2 Superior Colliculus

The superior colliculus, a layered midbrain structure, receives input from both the frontal and parietal circuits involved in saccade generation and integrates both excitatory cortical input and inhibitory input from the SNpr (Pouget 2015; Hanes and Wurtz 2001; Baloh et al. 1975; Leigh and Kennard 2004). Input from the striate, extrastriate, and parietal cortex, as well as from the frontal lobes, is relayed to a motor map with information about eye movement parameters occupying the ventral layers of the superior colliculus (Illing and Graybiel 1985; Sparks and Hartwich-Young 1989; Moschovakis et al. 1996; May 2005). Primate studies using direct stimulation of cells within the superior colliculus have demonstrated neuronal populations located in deeper layers of the superior colliculus that are directly involved in saccade initiation (Schiller and Stryker 1972; Sparks 1978). These neurons project to the midbrain and pontine reticular formation, which house premotor structures engaged in the generation of saccades.

The superior colliculus is believed to play a part in movement initiation and determination of saccadic velocity, as well as in selecting a target to focus on (Krauzlis 2004; Bell 2005; Hanes et al. 2005).

2.3 Cerebellum

The cerebellum receives input from the cortical eye fields via the pontine nuclei and the superior colliculus (Büttner and Büttner-Ennever 2005; Yamada and Noda 1987; Thielert and Thier 1993; Dicke et al. 2004; Noda et al. 1990). The nucleus reticularis tegmenti pontis (NRTP) of the midbrain sends projections to the medial cortico-nuclear zone in the dorsal vermis of the cerebellum (Ohtsuka and Noda 1995). This structure sends fibers to the caudal part of the fastigial nucleus deep within the cerebellum, a region that is also known as the fastigial oculomotor (FOR). It also receives input from the FEF and superior colliculus (Noda et al. 1990). The FOR then sends projections to omnipause neurons (OPNS), excitatory burst neurons (EBNs), and inhibitory burst neurons (IBNs) as well as to the thalamus, superior colliculus, and reticular formation (May et al. 1990). FOR stimulates burst neurons during contralateral saccades and provides inhibition during ipsilateral saccades (Fuchs et al. 1993; Kleine et al. 2003).

The cerebellum plays an important role in the accuracy of saccades, calibration of amplitude, and saccadic pulse-step match (Optican and Robinson 1980; Ritchie 1976; Barash et al. 1999; Straube et al. 2001).

2.4 OPNs

OPNs are glycinergic tonic inhibitors of the horizontal and vertical saccade burst generators, located in the nucleus raphe interpositus of the midline pons (Yamada and Noda 1987; Noda et al. 1990; Ohtsuka and Noda 1995; Selhorst et al. 1976; Robinson 1974). The OPNs receive projections from the fastigial oculomotor (FOR) nucleus located in the deep cerebellum, as well as from the superior colliculus (Yamada and Noda 1987; May et al. 1990; Fuchs et al. 1993; Kleine et al. 2003). The tonic activation of these neurons results in saccade suppression, and electrical stimulation can stop a saccade in its tracks (Zee et al. 1981; Ron and Robinson 1973) – inhibition of OPNs results in the initiation of a saccade.

2.5 Premotor Burst Complex

EBNs responsible for both horizontal and vertical saccades are located in different midbrain nuclei (Horn a et al. 1994; Horn et al. 1996; Van Gisbergen et al. 1981). These neurons are only active during saccades and determine saccadic eye velocity through the strength in firing rate (Horn et al. 1996; Van Gisbergen et al. 1981; Scudder et al. 2002). EBNs located in the paramedian pontine reticular formation are responsible for conveying information to oculomotor neurons regarding horizontal saccades, while EBNs located in the medial longitudinal fasciculus control vertical saccades (Scudder et al. 2002).

3 Neurodegenerative Disorders Affecting Visually Guided Saccades

3.1 Parkinson's Disease

Parkinson's disease (PD) is a neurodegenerative disorder affecting the basal ganglia that are characterized by the triad of resting tremor, rigidity, and bradykinesia, along with postural instability (Factor and Weiner 2004). PD affects numerous forms of eye movement, including memory-guided saccades, convergence insufficiency, and gaze restriction (DeJong and Jones 1971; Herishanu and Sharpe 1981; Rascol et al. 1989; Rottach et al. 1996a; Terao et al. 2011; Otero-Millan et al. 2013). Visually guided saccades in patients with PD are marked by hypometria, especially in large amplitude (great than 20 degrees) saccades (Terao et al. 2011). The result of this hypometria is a conversion to shift gaze toward the target, resulting in a "staircase" appearance of the saccade (Kimmig et al. 2002; Blekher et al. 2009). While the actual velocity of the saccade may not be altered, the time it takes for the inefficient saccade to reach the target lengthens (Shaikh et al. 2011). In patients that have asymmetric parkinsonism, hypometria of saccades is also asymmetric and worse on the more affected side (Choi et al. 2011). Patients with PD also appear to have increased latency in saccadic initiation with verbal instruction, when compared to healthy controls (Chambers and Prescott 2010). However, saccades made to random visual targets are not altered (Lueck et al. 1990).

PD patients show increased variability in peak saccade velocity, with decreased velocity seen only in advanced cases of the disease (Rottach et al. 1996b; White et al. 1983; Vidailhet et al. 1994).

3.2 Atypical Parkinsonism Syndromes

Atypical parkinsonism syndromes include progressive supranuclear palsy (PSP), dementia with Lewy bodies, and corticobasal degeneration (CBD). Saccadic disturbances have also been demonstrated in these disorders.

PSP is a tauopathy that causes supranuclear gaze and bulbar palsies, as well as postural instability and axial rigidity (Steele et al. 1964; Williams et al. 2008). It also causes hypometria and slowing of saccades, most prominent in the vertical axis (Shaikh et al. 2017). PSP can be distinguished from PD and other forms of atypical parkinsonism by the higher horizontal velocity of saccades when compared with the vertical component (Rottach et al. 1996a). This causes abnormally pronounced curvature of oblique saccades in PSP patients. As the disease progresses, vertical saccades are lost entirely, and the vertical gaze palsy is complete (Chen et al. 2010). While horizontal saccades are faster relative to their vertical counterpart, they are often hypometric early in the disease and gradual slow as the disease progresses (Bhidayasiri et al. 2001). These changes in saccade function correlate to anatomical changes found on autopsy in typical forms of PSP. Affected areas of the brain include the substantia nigra pars compacta of the basal ganglia, mesencephalon, diencephalon, and the brainstem reticular formation (Juncos et al. 1991; Collins et al. 1995; Halliday et al. 2000).

Dementia with Lewy bodies, caused by aggregations of alpha-synuclein within neurons, is characterized by parkinsonism, cognitive impairment, visual hallucinations, autonomic dysfunction, a fluctuating mental state, and sleep disorders (Capouch et al. 2018; McKeith et al. 2005; Walker et al. 2015). It causes both slowing and hypometria of horizontal and vertical saccades (Kapoula et al. 2010). There is also increased latency of saccade initiation, decreased latency of visually guided saccades, and increased directional errors during an antisaccade task in these patients (Kapoula et al. 2010; Mosimann et al. 2005).

CBD is another tauopathy that presents with asymmetric limb rigidity, dystonia, bradykinesia, myoclonus, tremor, postural instability, and gait changes with additional cortical features (Kompoliti et al. 1998; Mahapatra et al. 2004). The slowed velocity of saccades in CBD is occasionally found, while saccade latency is consistently increased (Rottach et al. 1996a; Vidailhet et al. 1994; Mahapatra et al. 2004).

3.3 Huntington's Disease

Huntington's disease (HD) is caused by a CAG triplet repeat disorder resulting in a defect in the protein huntingtin. This autosomal dominant disorder results in degeneration of the frontal lobe and caudate nucleus, resulting in behavior changes, choreoathetosis, and cognitive decline in patients with the disease (Vonsattel et al. 1985; Andrew et al. 1993; Walker 2013). Saccade abnormalities serve as a good marker in Huntington's disease (Lasker et al. 1987). Early in the disease, saccades have increased latency and hypometria as well as slowed velocity in either the horizontal or vertical planes (Kirkwood et al. 2000; Antoniades et al. 2010; Leigh et al. 1983; Collewijn et al. 1988; Lasker et al. 1988).

4 Cerebellar Disorders

4.1 Spinocerebellar Ataxia Type 2

Spinocerebellar ataxia type 2 (SCA 2), an autosomal dominant disorder characterized by progressive cerebellar ataxia, action tremor, dysarthria, and early neuropathy, is caused by an unstable polyglutamine expansion within ataxin-2 (Pulst et al. 1996; Wadia 1998; Wadia and Swami 1971). This disorder also causes slowing of saccades, which correlates strongly with the size of polyglutamine expansion and inversely with ataxia severity (Velázquez-Pérez et al. 2004). In contrast, disease duration, patient gender, and age of onset do not correlate with the extent of saccade slowing (Velázquez-Pérez et al. 2004). Interesting, saccadic slowing can be present before other clinical manifestations of SCA 2, thereby making slowing of saccades a useful early marker for this disease (Velázquez-Pérez et al. 2009). Saccade velocity also has utility as a sensitive and specific disease activity marker and as a surrogate for disease severity (Rodríguez-Labrada et al. 2016; Seifried et al. 2005). Eventually, the progression of saccade slowing leads to a complete horizontal and vertical gaze palsy (Klostermann et al. 1997).

Quantitative brain MRI has demonstrated reduced volumes in the cerebellum, pons, midbrain, and frontal lobes of patients with SCA2 (Politi et al. 2016). Saccade slowing in SCA2 may therefore not be primarily due to cerebellar dysfunction, as brainstem abnormality affecting burst generation may be the key driver of slowing (Politi et al. 2016; Rufa and Federighi 2011). This is further supported by the significant cell and synaptic density loss found in the mesencephalic area containing EBNs in patients with SCA2, as this area leads to adequate saccade burst intensity (Geiner et al. 2008).

4.2 Spinocerebellar Ataxia Type 3

Spinocerebellar ataxia type 3 (SCA3) is an autosomal dominant disorder caused by a CAG triplet expansion on chromosome 14 that is also known as Machado-Joseph disease (Haberhausen et al. 1995; Matilla et al. 1995; Ranum et al. 1995). It presents with progressive gait, stance, limb, and truncal ataxia as well as dysarthria, dystonia and somatosensory deficits, occasional parkinsonism, and dysphagia (Twist et al. 1995; Maruyama et al. 1997; Riess et al. 2008). Eye movement abnormalities seen in SCA3 include optokinetic or gaze-evoked nystagmus, impaired smooth pursuit, saccadic dysmetria, and dysfunction of the horizontal vestibulo-ocular reflex (Dawson et al. 1982; Hotson et al. 1987; Rivaud-Pechoux et al. 1998; Gordon et al. 2003; Gordon et al. 2014; Ghasia et al. 2016). Limitation of vertical gaze, most commonly in the upward direction, may also be seen (Murofushi et al. 1995). Very commonly, abduction ophthalmoplegia with sparing of adduction is present in SCA3 (Murofushi et al. 1995). Saccades demonstrate dynamic overshoot in some SCA3 patients but have low peak velocity in patients without dynamic overshoot (Caspi et al. 2013).

MRI of SCA3 patients shows diffuse atrophic changes that specifically affect the cerebellar vermis, superior cerebellar peduncle, pontine tegmentum, and frontal lobes (Murata et al. 1998; Tokumaru et al. 2003). The reticulotegmental nucleus of the pons and omnipause neurons of nucleus also have degeneration on histopathology (Rüb et al. 2003, 2004). Slowing of saccades may be due to degeneration of mesencephalic neurons responsible for burst generation (Rüb et al. 2003, 2008).

4.3 Wernicke's Encephalopathy

Wernicke's encephalopathy, characterized by the classic triad of ophthalmoplegia, changes in mental status, and gait ataxia, is caused by thiamine deficiency (Sechi and Serra 2007). It is frequently associated with alcohol abuse but may also be seen following gastrointestinal surgery or disorders and other causes of malnutrition (Sechi and Serra 2007). Gaze-evoked and upbeat nystagmus are early ocular motor findings that may switch to downbeat nystagmus with convergence (Shin et al. 2010; Kim et al. 2012). Early on patients may experience impairment in the horizontal vestibulo-ocular response. As the disease progresses, this leads to abduction impairment, horizontal and vertical gaze palsies, and internuclear ophthalmoplegia that eventually becomes complete ophthalmoplegia (Cogan and Victor 1954; Cox et al. 1981; Delapaz et al. 1992). Wernicke's encephalopathy has rarely been associated with slowing of saccades (Hamann 1979). Without thiamine repletion, Wernicke's disease can progress to Korsakoff syndrome which is characterized by severe memory loss and psychiatric symptoms. Korsakoff syndrome may include significant abnormalities of eye movement, which include impairment of smooth pursuit, hypometria, and slowed and inaccurate saccades (Kenyon et al. 1984a, b).

An increased number of directional errors on an antisaccade task may also be seen in these papers (Van Der Stigchel et al. 2012).

Wernicke's encephalopathy affects extracerebellar brainstem regions, including those responsible for burst generation, and is not predominantly cerebellar (Kim et al. 2012; Halliday et al. 1993). Impairment of saccade burst generation is, therefore, the likely cause of the rare saccadic slowing that can be seen in Wernicke's encephalopathy. Wernicke's encephalopathy may also affect the substantia nigra, in atypical cases, which could affect saccades through lack of tectal inhibition (Kalidass et al. 2012). In these cases, parkinsonism would be expected to accompany the slowed saccades.

4.4 Syndrome of Anti-GAD Antibody

Glutamic acid decarboxylase (GAD) is responsible for catalyzing the conversion of glutamic acid to γ-aminobutyric acid (GABA) (Watanabe et al. 2002). Autoantibodies directed against this important enzyme (anti-GAD Ab) have been found in patients with insulin-dependent diabetes mellitus, epilepsy, stiff-person syndrome, and late-onset cerebellar ataxia (Solimena et al. 1988, 1990; Abele et al. 1999; Vianello et al. 2002). Patients with anti-GAD Ab may also have eye movement abnormalities that include downbeat nystagmus, loss of downward smooth pursuit, impaired ocular pursuit and cancellation of vestibulo-ocular reflex, prolonged saccade latency, saccadic dysmetria, and saccadic oscillations (Antonini et al. 2003; Economides and Horton 2005; Zivotofsky et al. 2006; Shaikh and Wilmot 2016). There have been some reports of periodic alternating nystagmus and opsoclonus myoclonus seen in this syndrome as well (Tilikete et al. 2005; Markakis et al. 2008).

The saccade abnormalities seen in this syndrome could be multifactorial. Frequent hypometria, resulting in frequent interruptions of saccades with otherwise normal velocity, is a classic cerebellar phenomenon (Goffart et al. 1998).

5 Slow Saccades in Cerebellar Disorders

The prompt cessation of the inhibition on burst neurons is responsible for the high velocities of saccades (Enderle and Engelken 1995). Sustained inhibition of omnipause neurons is responsible for the persistent inhibition of excitatory burst neurons. When a saccade is initiated, OPN inhibition is halted, and there is an abrupt increase in excitatory burst neuron firing, resulting in a rapid saccade velocity (Shaikh et al. 2007, 2010; Enderle and Wolfe 1987; Miura and Optican 2006). Multiple malfunctions along this pathway can cause slowing of saccades. OPN malfunction can lead to a slowing of saccades in the vertical and horizontal direction. Excessive excitatory burst neuron activity (as predicted in the syndrome of anti-GAD antibody) also leads to saccade slowing by decreasing the efficacy of

OPN inhibition (Shaikh and Wilmot 2016). However, decreased excitability of excitatory burst neurons also leads to slow saccade velocity via ineffective burst generation. This kind of deficit would be expected in patients with SCA2, SCA3, and Wernicke's encephalopathy.

Slowing of saccades in chronic cerebellar disorders may be due to a multitude of factors. The effect of structural changes resulting in slowing and degeneration of saccade burst neurons may in part explain inefficient velocity command generation. On the other hand, neuroeconomics is at the root of another possible explanation. Cerebellar disorders negatively affect destination gaze stability leading to increased endpoint variability. In attempting to optimize saccade accuracy, saccade velocity may be sacrificed hence measurable slowing. In reaching its destination, a goal-directed movement, such as a saccade, might have limitless trajectories. However, optimization of trajectory position and saccade speed must occur to minimize the trade-off between time and accuracy (Fitts 1954).

It has been suggested that the most important component of trajectory planning lies in minimizing the variance of eye position in the presence of constant, biological noise in the circuitry generating saccades. The amount of noise in the final common pathway leading to motor neuron activity determines deviation from the intended saccadic path. Over the course of a saccade, these deviations are amassed, leading to variability in the end position of the eye. The original control signal generating the movement does not determine the amount of noise present; however rapid movements can rapidly minimize accumulated error (Wolpert et al. 1995). There continues to be a trade-off in the system, as rapid movement requires larger control signals, which leaves more room for variability in the ultimate destination. Consequently, inaccuracy of movement causes dysmetria that requires further correction (Meyer et al. 1988).

Low control signals decrease the speed of the movement but enhance accuracy. Signal-dependent noise forces a compromise between movement accuracy, duration, and speed. It has been proposed that this is achieved through the temporal profile of neural command, which minimizes variability of the desired position with concurrent adjustment of saccade velocity. Patients with cerebellar disorders have higher endpoint variability because of a multitude of co-existent deficits. These deficits include saccadic intrusions, dysmetria, and nystagmus. In order to achieve the best possible endpoint accuracy, these patients have compromised saccade velocity. Saccade velocity is therefore frequently reduced in cerebellar disorders and a useful metric for clinicians diagnosing and treating these diseases.

Acknowledgments This work was supported by grants from Dystonia Medical Research Foundation (DMRF), American Academy of Neurology, American Parkinson's Disease Association, and Dystonia Coalition.

References

Abele, M., Weller, M., Mescheriakov, S., Burk, K., Dichgans, J., & Klockgether, T. (1999). Cerebellar ataxia with glutamic acid decarboxylase autoantibodies. *Neurology, 52*, 857–859.

Andrew, S. E., Goldberg, Y. P., Kremer, B., Telenius, H., Theilmann, J., Adam, S., et al. (1993). The relationship between trinucleotide (CAG) repeat length and clinical features of Huntington's disease. *Nature Genetics, 4*, 398.

Antoniades, C. A., Xu, Z., Mason, S. L., Carpenter, R. H. S., & Barker, R. A. (2010). Huntington's disease: Changes in saccades and hand-tapping over 3 years. *Journal of Neurology, 257*, 1890.

Antonini, G., Nemni, R., Giubilei, F., Gragnani, F., Ceschin, V., Morino, S., et al. (2003). Autoantibodies to glutamic acid decarboxylase in downbeat nystagmus. *Journal of Neurology, Neurosurgery, and Psychiatry, 74*, 998–999.

Baloh, R. W., Sills a, W., Kumley, W. E., & Honrubia, V. (1975). Quantitative measurement of saccade amplitude, duration, and velocity. *Neurology, 25*, 1065–1070.

Barash, S., Melikyan, A., Sivakov, A., Zhang, M., Glickstein, M., & Thier, P. (1999). Saccadic dysmetria and adaptation after lesions of the cerebellar cortex. *The Journal of Neuroscience, 19*, 10931.

Bell, A. H. (2005). Crossmodal integration in the primate superior colliculus underlying the preparation and initiation of saccadic eye movements. *Journal of Neurophysiology, 93*, 3659.

Berman, R. A., Colby, C. L., Genovese, C. R., Voyvodic, J. T., Luna, B., Thulborn, K. R., et al. (1999). Cortical networks subserving pursuit and saccadic eye movements in humans: An FMRI study. *Human Brain Mapping, 8*, 209–225.

Bhidayasiri, R., Riley, D. E., Somers, J. T., Lerner, A. J., Büttner-Ennever, J. A., & Leigh, R. J. (2001). Pathophysiology of slow vertical saccades in progressive supranuclear palsy. *Neurology, 57*, 2070–2077.

Blekher, T., Weaver, M., Rupp, J., Nichols, W. C., Hui, S. L., Gray, J., et al. (2009). Multiple step pattern as a biomarker in Parkinson disease. *Parkinsonism & Related Disorders, 15*, 506–510.

Büttner, U., & Büttner-Ennever, J. A. (2005). Present concepts of oculomotor organization. *Progress in Brain Research., 151*, 1–42.

Capouch, S. D., Farlow, M. R., & Brosch, J. R. (2018). A review of dementia with Lewy bodies' impact, diagnostic criteria and treatment. *Neurology and Therapy, 7*, 249.

Caspi, A., Zivotofsky, A. Z., & Gordon, C. R. (2013). Multiple saccadic abnormalities in spinocerebellar ataxia type 3 can be linked to a single deficiency in velocity feedback. *Investigative Ophthalmology and Visual Science, 54*, 731–738.

Chambers, J. M., & Prescott, T. J. (2010). Response times for visually guided saccades in persons with Parkinson's disease: A meta-analytic review. *Neuropsychologia, 48*, 887.

Chen, A. L., Riley, D. E., King, S. A., Joshi, A. C., Serra, A., Liao, K., et al. (2010). The disturbance of gaze in progressive supranuclear palsy: Implications for pathogenesis. *Frontiers in Neurology, 1*, 147.

Choi, S. M., Lee, S. H., Choi, K. H., Nam, T. S., Kim, J. T., Park, M. S., et al. (2011). Directional asymmetries of saccadic hypometria in patients with early Parkinson's disease and unilateral symptoms. *European Neurology, 66*, 170–174.

Cogan, D. G., & Victor, M. (1954). Ocular signs of wernicke's disease. *AMA Archives of Ophthalmology, 51*, 204–211.

Collewijn, H., Went, L. N., Tamminga, E. P., & Vegter-Van der Vlis, M. (1988). Oculomotor defects in patients with Huntington's disease and their offspring. *Journal of the Neurological Sciences, 86*, 307–320.

Collins, S. J., Ahlskog, J. E., Parisi, J. E., & Maraganore, D. M. (1995). Progressive supranuclear palsy: Neuropathologically based diagnostic clinical criteria. *Journal of Neurology, Neurosurgery, and Psychiatry, 58*, 167.

Cox, T. A., Corbett, J. J., Thompson, H. S., & Lennarson, L. (1981). Upbeat nystagmus changing to downbeat nystagmus with convergence. *Neurology, 31*, 891–892.

Dawson, D. M., Feudo, P., Zubick, H. H., Rosenberg, R., & Fowler, H. (1982). Electro-oculographic findings in Machado-Joseph disease 550. *Neurology, 32*, 1272–1276.

DeJong, J. D., & Jones, G. M. (1971). Akinesia, hypokinesia, and bradykinesia in the oculomotor system of patients with Parkinson's disease. *Experimental Neurology, 32*, 58.

Delapaz, M. A., Chung, S. M., & Mccrary, J. A. (1992). Bilateral internuclear ophthalmoplegia in a patient with wernickes encephalopathy. *Journal of Clinical Neuro-Ophthalmology, 12*, 116–120.

Dicke, P. W., Barash, S., Ilg, U. J., & Thier, P. (2004). Single-neuron evidence for a contribution of the dorsal pontine nuclei to both types of target-directed eye movements, saccades and smooth-pursuit. *The European Journal of Neuroscience, 19*, 609–624.

Economides, J. R., & Horton, J. C. (2005). Eye movement abnormalities in stiff person syndrome. *Neurology, 65*, 1462–1464.

Enderle, J. D., & Engelken, E. J. (1995). Simulation of oculomotor post-inhibitory rebound burst firing using a Hodgkin-Huxley model of a neuron. *Biomedical Sciences Instrumentation, 31*, 53–58.

Enderle, J. D., & Wolfe, J. W. (1987). Time-optimal control of saccadic eye movements. *IEEE Transactions on Biomedical Engineering, BME-34*, 43–55.

Factor, S. A., & Weiner, W. J. (2004). *Parkinson's disease diagnosis and clinical management*. New York: Demos.

Fisher, R. S., Buchwald, N. A., Hull, C. D., & Levine, M. S. (1986). The GABAergic striatonigral neurons of the cat: Demonstration by double peroxidase labeling. *Brain Research, 398*, 148–156.

Fitts, P. M. (1954). The information capacity of the human motor system in controlling the amplitude of movement. *Journal of Experimental Psychology, 47*, 381–391.

Fox, P. T., Fox, J. M., Raichle, M. E., & Burde, R. M. (1985). The role of cerebral cortex in the generation of voluntary saccades: A positron emission tomographic study. *Journal of Neurophysiology, 54*, 348–369. http://www.ncbi.nlm.nih.gov/pubmed/3875696.

Francois, C., Percheron, G., & Yelnik, J. (1984). Localization of nigrostriatal, nigrothalamic and nigrotectal neurons in ventricular coordinates in macaques. *Neuroscience, 13*, 61–76.

Fuchs, A. F., Robinson, F. R., & Straube, A. (1993). Role of the caudal fastigial nucleus in saccade generation. I. Neuronal discharge pattern. *Journal of Neurophysiology, 70*, 1723–1740. https://doi.org/10.1152/jn.1993.70.5.1723.

Geiner, S., Horn, A. K. E., Wadia, N. H., Sakai, H., & Büttner-Ennever, J. A. (2008). The neuroanatomical basis of slow saccades in spinocerebellar ataxia type 2 (Wadia-subtype). *Progress in Brain Research, 171*, 575–581.

Ghasia, F. F., Wilmot, G., Ahmed, A., & Shaikh, A. G. (2016). Strabismus and micro-opsoclonus in Machado-Joseph disease. *Cerebellum, 15*, 491–497.

Goffart, L., Pélisson, D., & Guillaume, A. (1998). Orienting gaze shifts during Muscimol inactivation of caudal fastigial nucleus in the cat. II. Dynamics and eye-head coupling. *Journal of Neurophysiology, 79*, 1959–1976. https://doi.org/10.1152/jn.1998.79.4.1959.

Gordon, C. R., Joffe, V., Vainstein, G., & Gadoth, N. (2003). Vestibulo-ocular arreflexia in families with spinocerebellar ataxia type 3 (Machado-Joseph disease). *Journal of Neurology, Neurosurgery, and Psychiatry, 74*, 1403–1406.

Gordon, C. R., Zivotofsky, A. Z., & Caspi, A. (2014). Impaired vestibulo-ocular reflex (VOR) in spinocerebellar ataxia type 3 (SCA3): Bedside and search coil evaluation. *Journal of Vestibular Research: Equilibrium and Orientation, 24*(5, 6), 351–355.

Haberhausen, G., Damian, M. S., Leweke, F., & Müller, U. (1995). Spinocerebellar ataxia, type 3 (SCA3) is genetically identical to Machado-Joseph disease (MJD). *Journal of the Neurological Sciences, 132*, 71–75.

Halliday, G. M., Ellis, J., Heard, R., Caine, D., & Harper, C. (1993). Brainstem serotonergic neurons in chronic alcoholics with and without the memory impairment of korsakoff's psychosis. *Journal of Neuropathology and Experimental Neurology, 52*, 567–579.

Halliday, G. M., Hardman, C. D., Cordato, N. J., Hely, M. A., & Morris, J. G. (2000). A role for the substantia nigra pars reticulata in the gaze palsy of progressive supranuclear palsy. *Brain, 123,* 724.

Hamann, K. U. (1979). Slowed saccades in various neurological disorders. *Ophthalmologica, 178,* 357–364.

Handel, A. R. I., & Glimcher, P. W. (1999). Quantitative analysis of substantia nigra pars reticulata activity during a visually guided saccade task. *Journal of Neurophysiology, 82,* 3458–3475. http://www.ncbi.nlm.nih.gov/pubmed/10601475.

Hanes, D. P., & Wurtz, R. H. (2001). Interaction of the frontal eye field and superior colliculus for saccade generation. *Journal of Neurophysiology, 85,* 804–815. https://doi.org/10.1152/jn.2001.85.2.804.

Hanes, D. P., Smith, M. K., Optican, L. M., & Wurtz, R. H. (2005). Recovery of saccadic dysmetria following localized lesions in monkey superior colliculus. *Experimental Brain Research, 160,* 312.

Herishanu, Y. O., & Sharpe, J. A. (1981). Normal square wave jerks. *Investigative Ophthalmology and Visual Science, 20*(2), 268–272.

Hikosaka, O., & Wurtz, R. H. (1983). Visual and oculomotor functions of monkey substantia nigra pars reticulata. IV. Relation of substantia nigra to superior colliculus. *Journal of Neurophysiology, 49,* 1285–1301. https://doi.org/10.1152/jn.1983.49.5.1285.

Hikosaka, O., Takikawa, Y., & Kawagoe, R. (2000). Role of the basal ganglia in the control of purposive saccadic eye movements. *Physiological Reviews, 80,* 953–978.

Horn a, K., J A, B.-E., Wahle, P., & Reichenberger, I. (1994). Neurotransmitter profile of saccadic omnipause neurons in nucleus raphe interpositus. *The Journal of Neuroscience, 14,* 2032–2046. http://www.ncbi.nlm.nih.gov/pubmed/7908956.

Horn, A. K. E., Büttner-Ennever, J. A., & Büttner, U. (1996). Saccadic premotor neurons in the brainstem: Functional neuroanatomy and clinical implications. *Neuro-Ophthalmology, 16,* 229–240.

Hotson, J. R., Langston, E. B., Louis, A. A., & Rosenberg, R. N. (1987). The search for a physiologic marker of Machado-Joseph disease 534. *Neurology, 37,* 112–116.

Illing, R. B., & Graybiel, A. M. (1985). Convergence of afferents from frontal cortex and substantia nigra onto acetylcholinesterase-rich patches of the cat's superior colliculus. *Neuroscience, 14,* 455.

Juncos, J. L., Hirsch, E. C., Malessa, S., Duyckaerts, C., Hersh, L. B., & Agid, Y. (1991). Mesencephalic cholinergic nuclei in progressive supranuclear palsy. *Neurology, 41,* 25.

Kalidass, B., Sunnathkal, R., Rangashamanna, V., & Paraswani, R. (2012). Atypical Wernicke's encephalopathy showing involvement of substantia Nigra. *Journal of Neuroimaging, 22,* 204–207.

Kapoula, Z., Yang, Q., Vernet, M., Dieudonné, B., Greffard, S., & Verny, M. (2010). Spread deficits in initiation, speed and accuracy of horizontal and vertical automatic saccades in dementia with Lewy bodies. *Frontiers in Neurology, 1,* 138. https://doi.org/10.3389/fneur.2010.00138.

Kenyon, R. V., Becker, J. T., Butters, N., & Hermann, H. (1984a). Oculomotor function in wernicke-korsakoff's syndrome: I. saccadic eye movements. *The International Journal of Neuroscience, 25,* 53–65.

Kenyon, R. V., Becker, J. T., & Butters, N. (1984b). Oculomotor function in wernicke-korsakoff's syndrome: II. Smooth pursuit eye movements. *The International Journal of Neuroscience, 25,* 67–79.

Kim, K., Shin, D. H., Lee, Y. B., Park, K. H., Park, H. M., Shin, D. J., et al. (2012). Evolution of abnormal eye movements in Wernicke's encephalopathy: Correlation with serial MRI findings. *Journal of the Neurological Sciences, 323,* 77–79.

Kimmig, H., Haußmann, K., Mergner, T., & Lücking, C. H. (2002). What is pathological with gaze shift fragmentation in Parkinson's disease? *Journal of Neurology, 249,* 683–692.

Kirkwood, S. C., Siemers, E., Hodes, M. E., Conneally, P. M., Christian, J. C., & Foroud, T. (2000). Subtle changes among presymptomatic carriers of the Huntington's disease gene. *Journal of Neurology, Neurosurgery, and Psychiatry, 69*, 773.

Kleine, J. F., Guan, Y., & Buttner, U. (2003). Saccade-related neurons in the primate fastigial nucleus: What do they encode? *Journal of Neurophysiology, 90*, 3137–3154. https://doi.org/10.1152/jn.00021.2003.

Klostermann, W., Zühlke, C., Heide, W., Kömpf, D., & Wessel, K. (1997). Slow saccades and other eye movement disorders in spinocerebellar atrophy type 1. *Journal of Neurology, 244*, 105–111.

Kompoliti, K., Goetz, C. G., Boeve, B. F., Maraganore, D. M., Ahlskog, J. E., Marsden, C. D., et al. (1998). Clinical presentation and pharmacological therapy in corticobasal degeneration. *Archives of Neurology, 55*, 957.

Krauzlis, R. J. (2004). Recasting the smooth pursuit eye movement system. *Journal of Neurophysiology, 91*(2), 591–603.

Lasker, A. G., Zee, D. S., Hain, T. C., Folstein, S. E., & Singer, H. S. (1987). Saccades in Huntington's disease: Initiation defects and distractibility. *Neurology, 37*, 364.

Lasker, A. G., Zee, D. S., Hain, T. C., Folstein, S. E., & Singer, H. S. (1988). Saccades in Huntington's disease: Slowing and dysmetria. *Neurology, 38*, 427–431. https://doi.org/10.1212/WNL.38.3.427.

Leigh, R. J., & Kennard, C. (2004). Using saccades as a research tool in the clinical neurosciences. *Brain, 127*, 460–477.

Leigh, R. J., Newman, S. A., Folstein, S. E., Lasker, A. G., & Jensen, B. A. (1983). Abnormal ocular motor control in Huntington's disease. *Neurology, 33*, 1268–1275. https://doi.org/10.1212/WNL.33.10.1268.

Lueck, C. J., Tanyeri, S., Crawford, T. J., Henderson, L., & Kennard, C. (1990). Antisaccades and remembered saccades in Parkinson's disease. *Journal of Neurology, Neurosurgery, and Psychiatry, 53*, 284.

Mahapatra, R. K., Edwards, M. J., Schott, J. M., & Bhatia, K. P. (2004). Corticobasal degeneration. *Lancet Neurology, 3*, 736.

Markakis, I., Alexiou, E., Xifaras, M., Gekas, G., & Rombos, A. (2008). Opsoclonus-myoclonus-ataxia syndrome with autoantibodies to glutamic acid decarboxylase. *Clinical Neurology and Neurosurgery, 110*, 619–621.

Maruyama, H., Kawakami, H., Kohriyama, T., Sakai, T., Doyu, M., Sobue, G., et al. (1997). CAG repeat length and disease duration in Machado-Joseph disease: A new clinical classification. *Journal of the Neurological Sciences, 152*, 166–171.

Matilla, T., McCall a, S. S. H., & Zoghbi, H. Y. (1995). Molecular and clinical correlations in spinocerebellar ataxia type 3 and Machado-Joseph disease. *Annals of Neurology, 38*, 68–72.

May, P. J. (2005). The mammalian superior colliculus: Laminar structure and connections. *Progress in Brain Research., 151*, 321–378.

May, P. J., Hartwich-Young, R., Nelson, J., Sparks, D. L., & Porter, J. D. (1990). Cerebellotectal pathways in the macaque: Implications for collicular generation of saccades. *Neuroscience, 36*, 305–324.

McKeith, I. G., Dickson, D. W., Lowe, J., Emre, M., O'Brien, J. T., Feldman, H., et al. (2005). Diagnosis and management of dementia with Lewy bodies: Third report of the DLB consortium. *Neurology, 65*, 1863.

Meyer, D. E., Abrams, R. A., Kornblum, S., Wright, C. E., & Smith, J. E. K. (1988). Optimality in human motor performance: Ideal control of rapid aimed movements. *Psychological Review, 95*, 340–370.

Miura, K., & Optican, L. M. (2006). Membrane channel properties of premotor excitatory burst neurons may underlie saccade slowing after lesions of omnipause neurons. *Journal of Computational Neuroscience, 20*, 25–41.

Moschovakis, A. K., Scudder, C. A., & Highstein, S. M. (1996). The microscopic anatomy and physiology of the mammalian saccadic system. *Progress in Neurobiology, 50*, 133–254.

Mosimann, U. P., Müri, R. M., Burn, D. J., Felblinger, J., O'Brien, J. T., & McKeith, I. G. (2005). Saccadic eye movement changes in Parkinson's disease dementia and dementia with Lewy bodies. *Brain, 128*, 1267.

Munoz, D. P. (2002). Commentary: Saccadic eye movements: Overview of neural circuitry. *Progress in Brain Research., 140*, 89–96.

Murata, Y., Yamaguchi, S., Kawakami, H., Imon, Y., Maruyama, H., Sakai, T., et al. (1998). Characteristic magnetic resonance imaging findings in Machado-Joseph disease. *Archives of Neurology, 55*, 33–37.

Müri, R. M., Iba-Zizen, M. T., Derosier, C., Cabanis, E. A., & Pierrot-Deseilligny, C. (1996). Location of the human posterior eye field with functional magnetic resonance imaging. *Journal of Neurology, Neurosurgery, and Psychiatry, 60*, 445.

Murofushi, T., Mizuno, M., Hayashida, T., Yamane, M., Osanai, R., Ito, K., et al. (1995). Neuro-otological and neuropathological findings in two cases with Machado-joseph disease. *Acta Oto-Laryngologica, 115*, 136–139.

Nambu, A., Tokuno, H., & Takada, M. (2002). Functional significance of the cortico-subthalamo-pallidal "hyperdirect" pathway. *Neuroscience Research, 43*, 111–117.

Noda, H., Sugita, S., & Ikeda, Y. (1990). Afferent and efferent connections of the oculomotor region of the fastigial nucleus in the macaque monkey. *The Journal of Comparative Neurology, 302*, 330–348.

Ohtsuka, K., & Noda, H. (1995). Discharge properties of Purkinje cells in the oculomotor vermis during visually guided saccades in the macaque monkey. *Journal of Neurophysiology, 74*, 1828–1840. http://www.ncbi.nlm.nih.gov/pubmed/8592177.

Optican, L. M., & Robinson, D. A. (1980). Cerebellar-dependent adaptive control of primate saccadic system. *Journal of Neurophysiology, 44*, 1058.

Otero-Millan, J., Schneider, R., Leigh, R. J., Macknik, S. L., & Martinez-Conde, S. (2013). Saccades during attempted fixation in parkinsonian disorders and recessive ataxia: From microsaccades to square-wave jerks. *PLoS One, 8*, e58535.

Pierrot-deseilligny, C. H., Rivaud, S., Gaymard, B., & Agid, Y. (1991). Cortical control of reflexive visually-guided saccades. *Brain, 114*, 1473–1485.

Pierrot-Deseilligny, C., Milea, D., & Müri, R. M. (2004). Eye movement control by the cerebral cortex. *Current Opinion in Neurology, 17*, 17–25.

Politi, L. S., Bianchi Marzoli, S., Godi, C., Panzeri, M., Ciasca, P., Brugnara, G., et al. (2016). MRI evidence of cerebellar and extraocular muscle atrophy differently contributing to eye movement abnormalities in SCA2 and SCA28 diseases. *Investigative Ophthalmology and Visual Science, 57*, 2714–2720.

Pouget, P. (2015). The cortex is in overall control of "voluntary" eye movement. *Eye (London, England)., 29*, 241–245.

Pulst, S.-M., Nechiporuk, A., Nechiporuk, T., Gispert, S., Chen, X.-N., Lopes-Cendes, I., et al. (1996). Moderate expansion of a normally biallelic trinucleotide repeat in spinocerebellar ataxia type 2. *Nature Genetics, 14*, 269–276. https://doi.org/10.1038/ng1196-269.

Ranum, L. P., Lundgren, J. K., Schut, L. J., Ahrens, M. J., Perlman, S., Aita, J., et al. (1995). Spinocerebellar ataxia type 1 and Machado-Joseph disease: Incidence of CAG expansions among adult-onset ataxia patients from 311 families with dominant, recessive, or sporadic ataxia. *American Journal of Human Genetics, 57*, 603–608.

Rascol, O., Clanet, M., & Montastruc, J. (1989). Abnormal ocular movements in Parkinson's disease. *Brain, 112*, 1193.

Riess, O., Rüb, U., Pastore, A., Bauer, P., & Schöls, L. (2008). SCA3: Neurological features, pathogenesis and animal models. *Cerebellum (London, England), 7*, 125–137.

Ritchie, L. (1976). Effects of cerebellar lesions on saccadic eye movements. *Journal of Neurophysiology, 39*, 1246.

Rivaud, S., Müri, R. M., Gaymard, B., Vermersch, A. I., & Pierrot-Deseilligny, C. (1994). Eye movement disorders after frontal eye field lesions in humans. *Experimental Brain Research, 102*, 110.

Rivaud-Pechoux, S., Dürr, A., Gaymard, B., Cancel, G., Ploner, C. J., Agid, Y., et al. (1998). Eye movement abnormalities correlate with genotype in autosomal dominant cerebellar ataxia type I. *Annals of Neurology, 43*, 297–302.

Robinson, D. A. (1974). The effect of cerebellectomy on the cat's vestibulo-ocular integrator. *Brain Research, 71*, 195.

Rodríguez-Labrada, R., Velázquez-Pérez, L., Auburger, G., Ziemann, U., Canales-Ochoa, N., Medrano-Montero, J., et al. (2016). Spinocerebellar ataxia type 2: Measures of saccade changes improve power for clinical trials. *Movement Disorders, 31*, 570–578.

Ron, S., & Robinson, D. A. (1973). Eye movements cerebellar evoked by in the alert monkey stimulation. *Vision Research, 36*(6), 1004–1022.

Rottach, K. G., Riley, D. E., DiScenna a, O., Zivotofsky a, Z., & Leigh, R. J. (1996a). Dynamic properties of horizontal and vertical eye movements in parkinsonian syndromes. *Annals of Neurology, 39*, 368–377. https://doi.org/10.1002/ana.410390314.

Rottach, K. G., Riley, D. E., Discenna, A. O., Zivotofsky, A. Z., & John Leigh, R. (1996b). Dynamic properties of horizontal and vertical eye movements in Parkinsonian syndromes. *Annals of Neurology, 39*, 368.

Rüb, U., Brunt, E. R., Gierga, K., Schultz, C., Paulson, H., De Vos, R. A. I., et al. (2003). The nucleus raphe interpositus in spinocerebellar ataxia type 3 (Machado-Joseph disease). *Journal of Chemical Neuroanatomy, 25*, 115–127.

Rüb, U., Bürk, K., Schöls, L., Brunt, E. R., De Vos, R. A. I., Orozco Diaz, G., et al. (2004). Damage to the reticulotegmental nucleus of the pons in spinocerebellar ataxia type 1, 2, and 3. *Neurology, 63*, 1258–1263.

Rüb, U., Brunt, E. R., & Deller, T. (2008). New insights into the pathoanatomy of spinocerebellar ataxia type 3 (Machado-Joseph disease). *Current Opinion in Neurology, 21*, 111–116.

Rufa, A., & Federighi, P. (2011). Fast versus slow: Different saccadic behavior in cerebellar ataxias. *Annals of the New York Academy of Sciences, 1233*, 148–154.

Schiller, P. H., & Stryker, M. (1972). Single-unit recording and stimulation in superior colliculus of the alert rhesus monkey. *Journal of Neurophysiology, 35*, 915–924.

Schiller, P. H., True, S. D., & Conway, J. L. (1980). Deficits in eye movements following frontal eye-field and superior colliculus ablations. *Journal of Neurophysiology, 44*, 1175.

Scudder, C. A., Kaneko, C. R., & Fuchs, A. F. (2002). The brainstem burst generator for saccadic eye movements: A modern synthesis. *Experimental Brain Research, 142*, 439.

Sechi, G., & Serra, A. (2007). Wernicke's encephalopathy: New clinical settings and recent advances in diagnosis and management. *Lancet Neurology, 6*, 442–455. https://doi.org/10.1016/S1474-4422(07)70104-7.

Seifried, C., Velázquez-Pérez, L., Santos-Falcón, N., Abele, M., Ziemann, U., Almaguer, L. E., et al. (2005). Saccade velocity as a surrogate disease marker in spinocerebellar ataxia type 2. *Annals of the New York Academy of Sciences, 1039*(1), 524–527.

Selhorst, J. B., Stark, L., Ochs, A. L., & Hoyt, W. F. (1976). Disorders in cerebellar ocular motor control: II. Macrosaccadic oscillation an oculographic, control system and clinico-anatomical analysis. *Brain, 99*, 509.

Shaikh, A. G., & Wilmot, G. (2016). Opsoclonus in a patient with increased titers of anti-GAD antibody provides proof for the conductance-based model of saccadic oscillations. *Journal of the Neurological Sciences, 362*, 169–173.

Shaikh, A. G., Miura, K., Optican, L. M., Ramat, S., Leigh, R. J., & Zee, D. S. (2007). A new familial disease of saccadic oscillations and limb tremor provides clues to mechanisms of common tremor disorders. *Brain, 130*, 3020–3031.

Shaikh, A. G., Wong, A. L., Optican, L. M., Miura, K., Solomon, D., & Zee, D. S. (2010). Sustained eye closure slows saccades. *Vision Research, 50*, 1665–1675.

Shaikh, A. G., Xu-Wilson, M., Grill, S., & Zee, D. S. (2011). "Staircase" square-wave jerks in early Parkinson's disease. *The British Journal of Ophthalmology, 95*, 705.

Shaikh, A. G., Factor, S. A., & Juncos, J. L. (2017). Saccades in progressive supranuclear palsy-maladapted, irregular, curved, and slow. *Movement Disorders Clinical Practice, 4*, 671–681. https://doi.org/10.1002/mdc3.12491.

Shin, B. S., Oh, S. Y., Kim, J. S., Lee, H., Kim, E. J., & Hwang, S. B. (2010). Upbeat nystagmus changes to downbeat nystagmus with upward gaze in a patient with Wernicke's encephalopathy. *Journal of the Neurological Sciences, 298*, 145–147.

Solimena, M., Folli, F., Denis-Donini, S., Comi, G. C., Pozza, G., De Camilli, P., et al. (1988). Autoantibodies to glutamic acid decarboxylase in a patient with stiff-man syndrome, epilepsy, and type I diabetes mellitus. *The New England Journal of Medicine, 318*, 1012–1020. https://doi.org/10.1056/NEJM198804213181602.

Solimena, M., Folli, F., Aparisi, R., Pozza, G., & De Camilli, P. (1990). Autoantibodies to GABA-ergic neurons and pancreatic beta cells in stiff-man syndrome. *The New England Journal of Medicine, 322*, 1555–1560. https://doi.org/10.1056/NEJM199005313222202.

Sparks, D. L. (1978). Functional properties of neurons in the monkey superior colliculus: Coupling of neuronal activity and saccade onset. *Brain Research, 156*, 1–16.

Sparks, D. L., & Hartwich-Young, R. (1989). The deep layers of the superior colliculus. *Reviews of Oculomotor Research, 3*, 213–255.

Steele, J., Richardson, J. C., & Olszewski, J. (1964). Progressive supranuclear palsy. *Annals of Neurology, 10*, 333–359.

Straube, A., Deubel, H., Ditterich, J., & Eggert, T. (2001). Cerebellar lesions impair rapid saccade amplitude adaptation. *Neurology, 57*, 2105.

Terao, Y., Fukuda, H., Yugeta, A., Hikosaka, O., Nomura, Y., Segawa, M., et al. (2011). Initiation and inhibitory control of saccades with the progression of Parkinson's disease – Changes in three major drives converging on the superior colliculus. *Neuropsychologia, 49*, 1794–1806.

Thielert, C.-D., & Thier, P. (1993). Patterns of projections from the pontine nuclei and the nucleus reticularis tegmenti pontis to the posterior vermis in the rhesus monkey: A study using retrograde tracers. *The Journal of Comparative Neurology, 337*, 113–126.

Tilikete, C., Veghetto, A., Trouillas, P., & Honnorat, J. (2005). Anti-GAD antibodies and periodic alternating nystagmus. *Archives of Neurology, 62*, 1300–1303.

Tokumaru, A. M., Kamakura, K., Maki, T., Murayama, S., Sakata, I., Kaji, T., et al. (2003). Magnetic resonance imaging findings of Machado-Joseph disease: Histopathologic correlation. *Journal of Computer Assisted Tomography, 27*, 241–248.

Twist, E. C., Casaubon, L. K., Ruttledge, M. H., Rao, V. S., Macleod, P. M., Radvany, J., et al. (1995). Machado Joseph disease maps to the same region of chromosome 14 as the spinocerebellar ataxia type 3 locus. *Journal of Medical Genetics, 32*, 25–31. https://doi.org/10.1136/jmg.32.1.25.

Van Der Stigchel, S., Reichenbach, R. C. L., Wester, A. J., & Nijboer, T. C. W. (2012). Antisaccade performance in Korsakoff patients reveals deficits in oculomotor inhibition. *Journal of Clinical and Experimental Neuropsychology, 34*, 876–886.

Van Gisbergen, J. A. M., Robinson, D. A., & Gielen, S. (1981). A quantitative analysis of generation of saccadic eye movements by burst neurons. *Journal of Neurophysiology, 45*, 417–442.

Velázquez-Pérez, L., Seifried, C., Santos-Falcón, N., Abele, M., Ziemann, U., Almaguer, L. E., et al. (2004). Saccade velocity is controlled by polyglutamine size in spinocerebellar ataxia 2. *Annals of Neurology, 56*, 444–447.

Velázquez-Pérez, L., Seifried, C., Abele, M., Wirjatijasa, F., Rodríguez-Labrada, R., Santos-Falcón, N., et al. (2009). Saccade velocity is reduced in presymptomatic spinocerebellar ataxia type 2. *Clinical Neurophysiology, 120*, 632–635.

Vianello, M., Tavolato, B., & Giometto, B. (2002). Glutamic acid decarboxylase autoantibodies and neurological disorders. *Neurological Sciences, 23*, 145–151. https://doi.org/10.1007/s100720200055.

Vidailhet, M., Rivaud, S., Gouider-Khouja, N., Pillon, B., Bonnet, A.-M., Gaymard, B., et al. (1994). Eye movements in parkinsonian syndromes. *Annals of Neurology, 35*, 420.

Vonsattel, J. P., Myers, R. H., Stevens, T. J., Ferrante, R. J., Bird, E. D., & Richardson, E. P. (1985). Neuropathological classification of huntington's disease. *Journal of Neuropathology and Experimental Neurology, 44*, 559.

Wadia, N. (1998). A clinicogenetic analysis of six Indian spinocerebellar ataxia (SCA2) pedigrees. The significance of slow saccades in diagnosis. *Brain, 121*, 2341–2355. https://doi.org/10.1093/brain/121.12.2341.

Wadia, N. H., & Swami, R. K. (1971). A new form of heredo-familial spinocerebellar degeneration with slow eye movements (nine families). *Brain, 94*, 359–374.

Walker, F. O. (2013). Huntington's disease: The road to progress. *The Lancet Neurology, 12*, 624.

Walker, Z., Possin, K. L., & Boeve, B. F. A. D. (2015). Lewy body dementias. *Lancet, 386*, 1683.

Watanabe, M., Maemura, K., Kanbara, K., Tamayama, T., & Hayasaki, H. G. A. B. A. (2002). GABA receptors in the central nervous system and other organs. *International Review of Cytology, 213*, 1–47.

White, O. B., Saint-cyr, J. A., Tomlinson, R. D., & Sharpe, J. A. (1983). Ocular motor deficits in parkinson's disease: II. Control of the saccadic and smooth pursuit systems. *Brain, 106*, 571.

Williams, D. R., Lees, A. J., Wherrett, J. R., & Steele, J. C. (2008). J. Clifford Richardson and 50 years of progressive supranuclear palsy. *Neurology, 70*, 566–573.

Wolpert, D. M., Ghahramani, Z., & Jordan, M. I. (1995). An internal model for sensorimotor integration. *Science, 269*, 1880–1882. https://doi.org/10.1126/science.7569931.

Yamada, J., & Noda, H. (1987). Afferent and efferent connections of the oculomotor cerebellar vermis in the macaque monkey. *The Journal of Comparative Neurology, 265*, 224–241.

Zee, D. S., Yamazaki, A., Butler, P. H., & Gucer, G. (1981). Effects of ablation of flocculus and paraflocculus of eye movements in primate. *Journal of Neurophysiology, 46*, 878.

Zivotofsky, A. Z., Siman-Tov, T., Gadoth, N., & Gordon, C. R. (2006). A rare saccade velocity profile in stiff-person syndrome with cerebellar degeneration. *Brain Research, 1093*, 135–140.

Translational Neurology of Slow Saccades

Janet C. Rucker, Todd Hudson, and John Ross Rizzo

Abstract Extensive knowledge has been gained over recent decades about anatomic and pathophysiologic mechanisms governing saccades, rapid eye movements by which gaze is shifted between visual targets. In this focused review, we highlight the physiology and anatomy of normal saccades as they pertain to pathological saccade slowing of central brainstem origin, with emphasis on excitatory and inhibitory burst neurons and omnipause neuron function. We summarize insights into saccadic dysfunction from nonhuman basic science research and further utilize these insights to review advances from translational research in common and uncommon human central neurological disease states that lead to vertical and/ or horizontal saccade slowing, such as progressive supranuclear palsy, Niemann-Pick type C, and post-cardiac surgery.

Keywords Saccades · Supranuclear gaze palsy · Burst neurons · Omnipause neurons · Brainstem · Neurological disorders · Storage disorders · Eye movements

Electronic supplementary material The online version of this chapter (https://doi.org/10.1007/978-3-030-31407-1_12) contains supplementary material, which is available to authorized users.

J. C. Rucker (✉)
Department of Neurology, New York University School of Medicine, New York, NY, USA

Department of Ophthalmology, New York University School of Medicine, New York, NY, USA
e-mail: janet.rucker@nyulangone.org

T. Hudson · J. R. Rizzo
Department of Neurology, New York University School of Medicine, New York, NY, USA

Department of Physical Medicine and Rehabilitation, New York University School of Medicine, New York, NY, USA

1 Introduction

Saccades are rapid, ballistic eye movements by which gaze is shifted between visual targets. Saccades and the other functional classes of eye movements, including smooth pursuit, vestibular-ocular reflexes (VOR), and vergence evolved to serve visual fixation by placing and maintaining the fovea, the specialized retinal region with the highest density of cone photoreceptors, on objects of visual interest. Each class of eye movement utilizes the so-called "final common pathway" of the ocular motor cranial nerve nuclei and motoneurons (e.g., abducens, trochlear, and oculomotor), the neuromuscular junction, and the extraocular muscles; however, each class of eye movements has distinct premotor, or supranuclear, neural circuits that send signals to the final common pathway.

The range of saccadic eye movements encompasses intentional goal-directed saccades, e.g., looking left and right before crossing the street (volitional); high-level frontally driven saccades to the remembered location of a previously present visual target (memory-guided); reactive saccades toward suddenly appearing stimuli, e.g., a person entering a doorway (reflexive); and the primitive reflexive fast eye movement components of optokinetic nystagmus (OKN). Saccades may be pathologically disrupted by many diseases and mechanistically in many forms, encompassing disorders of latency (i.e., time between visual stimulus onset and saccade initiation), accuracy, trajectory, and velocity; comprehensive coverage of which may be found elsewhere (Leigh and Zee 2015). Specific saccade deficits indicate dysfunction of distinct neuronal populations and assist with the first objective in clinical neurology – to localize the structures involved in a disease process (Leigh and Riley 2000). The focus here will be on centrally mediated neuroanatomic disruptions and disease states that lead to pathological slowing of saccades. Slowing of saccades may also occur in the context of peripheral ocular motor conditions affecting extraocular muscles or cranial nerves; however, these are not considered further in this review. Prior to consideration of centrally mediated pathological saccadic slowing, normal saccade anatomy and physiology with a focus on brainstem structures relevant to saccade generation and saccade slowing will be reviewed.

2 Normal Saccade Physiology and Anatomy

2.1 Saccade Characteristics and Physiology

Extensive knowledge has been gained over recent decades about anatomic and pathophysiologic mechanisms governing eye movements, including saccades – with rich interdisciplinary exchange between basic science nonhuman physiology, biological modeling, and the effects of human disease as it pertains to neurological structures involved in eye movement control (Leigh and Zee 2015; Horn and Leigh 2011). Saccades place high demands upon brain circuitry relative to other slower eye movements, as saccades must be fast (up to 600°/sec) and of brief duration (less than 100 msec) to avoid visual disruption. Saccades also require a high degree of accuracy to bring the images of target items to the small fovea. Normal saccades occur in a single smooth motion with a single-peaked velocity waveform (Fig. 1a). Despite the ideal of high saccadic accuracy and precision and the capacity of the saccadic system to achieve these under optimal laboratory conditions (Kowler and Blaser 1995), normal saccades may undershoot an intended target by a small amount and require a small secondary "corrective" saccade to land the eye on target (Fig. 1b) (Aitsebaomo and Bedell 1992; Pelisson and Prablanc 1988).

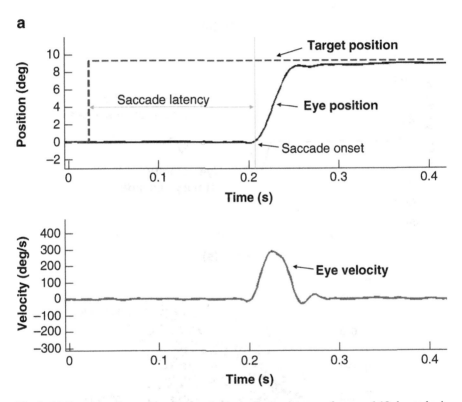

Fig. 1 (a) Representative position (top) and velocity (bottom) traces of a normal 10-degree horizontal saccade. The upward deflection of the position trace represents a rightward eye movement. The light gray line on the position trace indicates the beginning of the saccade, at which time the velocity exceeds a threshold defining saccade onset. The duration of time between target and saccade onset is the saccade latency. (b) Representative position (top) and velocity (bottom) traces of a 10-degree horizontal saccade with a large primary saccade and secondary corrective saccade to bring the eye to target in a healthy individual. Downward deflections of the position trace represent leftward eye movements. (c) Main sequence relationship of peak velocity versus amplitude for normal horizontal saccades. Data is fit with an exponential equation: peak velocity = $V\text{max} \times (1 - e^{-A/C})$, where $V\text{max}$ is the asymptotic peak velocity, A is the amplitude, and C is a constant defining the exponential rise. The 5th and 95th prediction intervals are also plotted

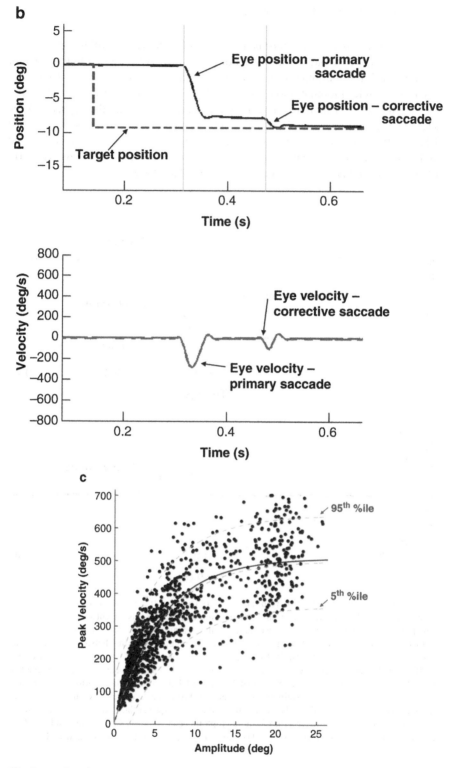

Fig. 1 (continued)

Saccades obey "main sequence" relationships between saccade amplitude and peak velocity, and between saccade amplitude and duration, by which saccade velocity and duration increase in a stereotypical manner with increasing saccade amplitude (Fig. 1c) (Bahill et al. 1975; Boghen et al. 1974; Baloh et al. 1975). Velocity increases linearly with increasing amplitude for small saccades but reaches a saturation point for large saccades. A high degree of intrasubject and intersubject variability occurs for a saccade of a given amplitude (Boghen et al. 1974), and factors such as mental state, level of alertness, and degree of target illumination may play a role in this variation (Bronstein and Kennard 1987; Leigh and Kennard 2004). Normal peak velocity ranges may also differ depending on the technique utilized for eye movement recording (i.e., infrared video versus scleral search coil versus electrooculography) (Boghen et al. 1974), by directionality and starting eye position (i.e., abducting versus adducting horizontal saccades, centripetal versus centrifugal saccades) (Pelisson and Prablanc 1988; Schneider et al. 2011; Frost and Poppel 1976), and by the frequency with which saccades are made (Lueck et al. 1991); thus, establishment of normative laboratory values for saccades in all directions is critical to proper definition and detection of pathological saccade slowing in neurological disease. Additional information may be revealed by examination of the shape, or skewness (Van Opstal and Van Gisbergen 1987), of the velocity profiles of saccades that reflects the ratio between the accelerating and decelerating portions of the movement. The acceleration phase of a saccade tends to be constant regardless of saccade size, whereas the deceleration component becomes prolonged as saccade amplitude increases. Thus, small saccades have a symmetrical velocity profile, and large saccades appeared skewed (Van Opstal and Van Gisbergen 1987).

In order to meet the demands ensuing from the requirements of saccade brevity and speed and to overcome the elastic inertia imposed by the intraorbital tissues, initiation of a saccade requires a high-frequency burst of neuronal discharge called the pulse in the agonist motoneuron that leads to vigorous agonist muscle contraction (Robinson 1970). Simultaneous with the excitatory pulse agonist innervation, the antagonist muscle relaxes due to a pause in discharge of its innervation source. Following the pulse, an exponential slide occurs, and neuronal discharge to the agonist muscle is transitioned to a lower level of tonic activity called the step to maintain the eyes in the new orbital position (Robinson 1970; Scudder et al. 2002). Each ocular motoneuron fires in this pulse-slide-step pattern (Scudder et al. 2002).

2.2 Saccade Anatomy

Anatomic structures involved in initiation, termination, and dynamic control of saccadic eye movements are diffusely distributed and include cortical, brainstem, and cerebellar regions. Excitatory inputs from cortical frontal, parietal, and supplementary eye fields and inhibitory inputs from the basal ganglia are received by the superior colliculus (SC). SC is the primary saccade initiation command source, with unilateral intermediate and deeper SC layers sending excitatory signals to

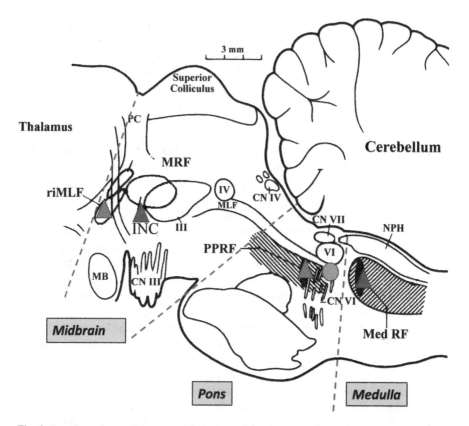

Fig. 2 Location of saccade premotor structures on a nonhuman primate mid-sagittal brainstem diagram. Premotor excitatory (green triangles) and inhibitory (red triangles) burst neurons are located in the pons and medulla for horizontal saccades and in the midbrain for vertical saccades. Omnipause neurons (yellow oval) are intermingled with the sixth cranial nerve rootlets. Abbreviations: PC posterior commissure, riMLF rostral interstitial medial longitudinal fasciculus, INC interstitial nucleus of Cajal, MB mammillary body, MRF mesencephalic reticular formation, CN III oculomotor nerve fascicle, III oculomotor nucleus, IV trochlear nucleus, CN IV trochlear nerve fascicle, CN VII facial nerve fascicle, MLF medial longitudinal fasciculus, PPRF paramedian pontine reticular formation, VI abducens nucleus, CN VI abducens nerve rootlets, NPH nucleus prepositus hypoglossi, Med RF medullary reticular formation. (Figure adapted from Büttner and Büttner-Ennever 2005)

contralateral brainstem reticular formation premotor saccade generation circuitry (Figs. 2 and 3) (Scudder et al. 2002; Raybourn and Keller 1977). A cortical spatially encoded saccade command is transformed into a temporally encoded pulse-step motor command.

The pulse of neural activity required for saccade initiation is executed by excitatory burst neurons (EBN), which fire just before a saccade occurs and are located in the brainstem reticular formation (Fig. 2) (Van Gisbergen et al. 1981; Horn 2006; Horn et al. 2003a; Strassman et al. 1986a). EBN also send signals to neural integrators,

Fig. 3 Schematic of the anatomic connections of neurons of the horizontal burst generator. Excitatory (filled circles) and inhibitory (open triangles) connections. Confidence in the connection is reflected in line thickness, from heavy solid (confirmed) to thin dashed (hypothesized). Abbreviations: cMRF central mesencephalic reticular formation, EBN excitatory burst neuron, IBN inhibitory burst neuron, LLBN long-lead burst neuron, trig trigger, MN motoneuron, nph nucleus prepositus hypoglossi, OPN omnipause neuron. (Reprinted by permission from Springer Nature: Scudder et al. 2002)

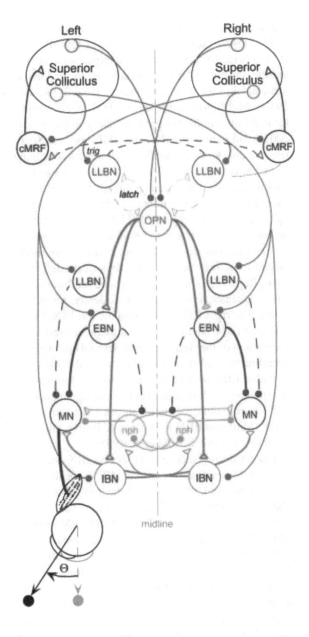

which give rise to the step innervation. Neural integrators for horizontal eye movements include the medial vestibular nucleus and nucleus prepositus hypoglossi (Langer et al. 1986), whereas vertical neural integration occurs in the interstitial nucleus of Cajal (INC). For horizontal saccades, EBN are located in the caudal paramedian pontine reticular formation (PPRF) just rostral to the abducens nucleus (Horn et al. 1995). For vertical and torsional saccades, EBN are located in the rostral interstitial medial longitudinal fasciculus (riMLF) (Horn and Büttner-Ennever 1998; Moschovakis et al. 1991a, b). A few also lie within the INC (Moschovakis et al. 1991a, b). Inhibitory glycinergic burst neurons (IBN) that relax antagonist muscles during horizontal saccades are located in the medullary reticular formation just caudal to the abducens nucleus (Fig. 2) (Scudder et al. 1988; Strassman et al. 1986b; McElligott and Spencer 2000). GABAergic IBN for vertical saccades are located in the midbrain predominantly in the area of INC in monkey (Horn et al. 2003a) and in both the INC and riMLF in cats (Spencer and Wang 1996). EBN and IBN are also termed short-lead or medium-lead burst neurons.

For horizontal saccades, EBN project to ipsilateral agonist motoneurons and IBN project to contralateral antagonist motoneurons to generate an ipsilateral saccade (Fig. 3) (Strassman et al. 1986a, b; Scudder et al. 1988). For example, for a rightward saccade, right PPRF EBN send excitatory signals to the right abducens nucleus which activates the motoneurons destined for the right lateral rectus and the interneurons destined to decussate in the medial longitudinal fasciculus and ascend to activate the motoneurons in the left medial rectus oculomotor subnucleus. Right IBN simultaneously send inhibitory signals to inhibit the left EBN and left abducens nucleus and, consequently, the left lateral rectus and right medial rectus. The right IBN also inhibit the left IBN, thereby preventing inhibition of the activated right lateral rectus. Ultimate activation in this neural pathway leads to contraction of the right lateral rectus and left medial rectus and inhibition of the left lateral rectus and right medial rectus and evokes a rightward saccade. For vertical saccades, EBN project bilaterally to elevator muscle (e.g., superior rectus and inferior oblique) motoneurons but unilaterally to depressor muscle (e.g., inferior rectus and superior oblique) motoneurons (Moschovakis et al. 1990; Bhidayasiri et al. 2000). Midbrain EBN also promote ipsidirectional rapid torsional eye movements (Villis et al. 1989; Suzuki et al. 1995).

Except during saccades, EBN and IBN are tonically and directly inhibited by glycinergic omnipause neurons (OPN) located in the nucleus raphe interpositus (RIP) in the caudal pons just off the midline on each side and intermingled with the abducens rootlets (Fig. 2) (Büttner-Ennever et al. 1988; Keller 1977; Horn et al. 1994). OPN project directly to PPRF EBN to inhibit horizontal saccades and to riMLF EBN to inhibit vertical and torsional saccades (Fig. 3) (Strassman et al. 1987; Nakao et al. 1980; King et al. 1980; Langer and Kaneko 1983; Ohgaki et al. 1989). Just prior to a saccade, OPN firing ceases and inhibition is maintained through most or all of the saccade duration, possibly due to activity of putative inhibitory latch neurons that receive signals from EBN (Fig. 3) (Optican et al. 2008; Keller and Missal 2003; Yoshida et al. 1999). OPN resume firing just before the end of a saccade, though many resume firing after saccade end; thus, the precise mechanism of saccade termination may be either OPN reactivation or more likely contralateral IBN firing via activation by the ipsilateral caudal fastigial nucleus (cFN), which chokes the ipsilateral saccade drive signal (Optican et al. 2008; Optican and Quaia 2002).

3   Basic Science of Slow Saccades – Insights from Nonhuman Studies

Based on the anatomy of the EBN, it is anticipated that lesions in the PPRF EBN for horizontal saccades would affect horizontal saccade function and lesions in the riMLF EBN for vertical and torsional saccades would affect vertical and torsional saccade function. Indeed, unilateral pharmacologic inactivation of the PPRF eradicates ipsidirectional horizontal saccades, and bilateral inactivation eradicates horizontal saccades in both directions (Henn et al. 1984). Unilateral pharmacologic inactivation of the riMLF leads to a mild deficit of downward saccades and loss of ipsidirectional torsional fast phases (Bhidayasiri et al. 2000; Villis et al. 1989; Suzuki et al. 1995), whereas bilateral inactivation eradicates vertical and torsional saccades (Suzuki et al. 1995). Bilateral pharmacologic inactivation of the caudal PPRF also leads to vertical saccade slowing, in addition to impairment of horizontal saccades in both directions (Henn et al. 1984). This impairment of vertical saccades is likely related to lesional effects on OPN.

OPN electrical stimulation during a saccade will interrupt the saccade mid-flight (Keller 1977). It might be hypothesized that inactivation of OPN would precipitate spontaneous saccades if their role is solely as an all-on or all-off gate to inhibit burst neurons. However, pharmacologic inactivation of omnipause neurons leads to saccade slowing without a change in saccade latency (Henn et al. 1984; Kaneko 1996; Soetedjo et al. 2002), which suggests that the OPN lesion reduces the EBN firing rate. Thus it is proposed that the OPN play a direct role in saccade generation, in part by creating a post-inhibitory rebound depolarization in EBN induced by sudden release from OPN inhibition that may facilitate early saccade acceleration (Miura and Optican 2006). Indeed, saccade modeling of neuronal conductance incorporating loss of EBN post-inhibitory rebound depolarization from OPN dysfunction reproduces the saccade slowing induced by OPN lesioning experiments (Miura and Optican 2006, 2003; Enderle and Engelken 1995). With modeling, post-inhibitory rebound depolarization was reduced in EBN due to reduction of low-threshold calcium and NMDA currents resulting from the reduced presence of glycine from OPN (Miura and Optican 2006). Slowing of vertical saccades from pontine lesions has also been seen in human clinical disease states (Hanson et al. 1986; Rufa et al. 2008; Johnston et al. 1993). Inhibition of OPN occurs in nonhuman primates and humans during blinks and transient eyelid closure (Mays and Morrissee 1993), and saccade slowing and depression of OKN quick-phase velocities, in turn, accompany blinks (Goossens and Van Opstal 2000; Rambold et al. 2002; Rottach et al. 1998; Shaikh et al. 2010). Reduction in saccade peak acceleration also occurs with eye closure, further supporting a direct OPN role in saccade generation secondary to reduction of EBN post-inhibitory rebound depolarization (Shaikh et al. 2010).

Saccade slowing has been reported with lesions in structures other than EBN and OPN, most notably the SC and central mesencephalic reticular formation, which plays an important role in vertical saccades (Bhidayasiri et al. 2000, 2001; Hikosaka and Wurtz 1985, 1986; Schiller et al. 1980; Waitzman et al. 2000).

Functional deafferentation of the SC with the GABA agonist muscimol (Hikosaka and Wurtz 1985) and functional removal with lidocaine (Hikosaka and Wurtz 1986) lead to reductions in saccadic velocities, possibly by effects on OPN leading to incomplete release of OPN inhibition and subsequent reductions in EBN burst frequency. Paired frontal eye field and SC lesions more substantially reduce saccadic velocities than SC lesions alone (Schiller et al. 1980).

4 Translational Neurology of Slow Saccades – Clinical Disorders

4.1 Clinical Assessment and Classic Features

A wide range of pathologic etiologies can cause slow saccades, including degenerative, inflammatory, neoplastic and paraneoplastic, ischemic, metabolic, and hereditary conditions (Table 1) – the clinical features of which have been recently reviewed (Lloyd-Smith Sequeira et al. 2017). What they hold in common is the implication that clinical saccade slowing generally indicates direct EBN involvement, either of horizontal or vertical neuronal populations or both, and/or OPN disease. As in non-human lesioning studies, human unilateral PPRF lesions involving EBN predominantly lead to loss of ipsidirectional saccades, whereas bilateral lesions lead to bilateral conjugate horizontal saccadic gaze palsy. Human lesions affecting riMLF EBN lead to vertical saccade palsy. It might be expected that unilateral riMLF lesions would predominantly impact downward saccades, given the unilateral projection of innervation to motoneurons for downward saccades and bilateral projection for upward saccades (Moschovakis et al. 1990; Bhidayasiri et al. 2000). However, reported human lesional effects do not adhere precisely to this physiology, and unilateral or more global up and down deficits have been reported with unilateral lesions (Ranalli et al. 1988; Onofrj et al. 2004; Deleu et al. 1989). This is likely because human lesions are rarely restricted to a single nucleus, often affecting nearby structures simultaneously, and may extend beyond the limits of the lesion seen on neuroimaging.

Saccades are tested at the bedside by asking the patient to make rapid eye movements between two stationary targets in the horizontal and vertical planes. A rule of thumb in the clinic is that the examiner's eye should not be able to follow the patient's eye through the trajectory of a saccade. If the patient's eye can be followed (Video 1), saccades are pathologically slow (Fig. 4). By the time saccade slowing is detectable with the examiner's unaided eye in the clinic, saccade velocity has likely fallen severely below lower normal limits of saccades on the main sequence peak velocity to amplitude relationship curve. Detection of subclinical or very mild slowing that may have diagnostic implications may be missed on bedside exam and be detectable solely with quantified ocular motor recordings.

Table 1 Etiologic causes of slow saccades and supranuclear saccadic gaze palsy

Neurodegenerative: progressive supranuclear palsy (PSP, classic cause, vertical predominant), Huntington's disease (Leigh et al. 1983; Collewijn et al. 1988), corticobasal degeneration (Shiozawa et al. 2000; Rinne et al. 1994) (if saccade slowing occurs, it is typically later in disease course and less severe than in PSP (Rivaud-Pechoux et al. 2000)), multiple system atrophy (Murphy et al. 2005), Lewy body dementia (Fearnley et al. 1991; de Bruin et al. 1992), primary pallidal degeneration (Gordon et al. 2004), amyotrophic lateral sclerosis (Averbuch-Heller et al. 1998; Ushio et al. 2009; Donaghy et al. 2010; Moss et al. 2012; Okuda et al. 1992)
Genetic/metabolic: Niemann-Pick type C (Rottach et al. 1997; Abel et al. 2009; Salsano et al. 2012; Solomon et al. 2005) (classic cause, vertical – down worse than up (Rottach et al. 1997; Cogan et al. 1981)), Gaucher disease type 3 (Benko et al. 2011; Pensiero et al. 2005; Schiffmann et al. 2008) (classic cause, horizontal[a]), genetic Parkinson's disease with leucine-rich repeat kinase 2 (LRRK2) mutations (Spanaki et al. 2006), Kufor-Rakeb syndrome (Williams et al. 2005b), Wilson's disease (Kirkham and Kamin 1974), dominant and recessive spinocerebellar ataxias (Zee et al. 1976; Burk et al. 1999; Velazquez-Perez et al. 2004; Rufa and Federighi 2011; Wadia et al. 1998; Geiner et al. 2008; Rosini et al. 2013) (SCA, classically SCA2[b] – horizontal, slowing reported with other SCAs including 1,3,6,7, and 28 (Klostermann et al. 1997; Oh et al. 2001; Ying et al. 2005; Wu et al. 2017; Moscovich et al. 2015; Zuhlke et al. 2015; Christova et al. 2008)), episodic ataxia type 2 (Kipfer et al. 2013), late-onset Tay-Sachs (Rucker et al. 2004), cerebral autosomal dominant arteriopathy with subcortical infarcts and leukoencephalopathy (Erro et al. 2014), mitochondrial disease (Gupta et al. 1995), kernicterus (Hoyt et al. 1978), medications (Thurston et al. 1984; Rothenberg and Selkoe 1981) (e.g., anticonvulsants, benzodiazepines)
Vascular: midbrain infarction (Onofrj et al. 2004; Deleu et al. 1989; Green et al. 1993; Hommel and Bogousslavsky 1991; Büttner-Ennever et al. 1982) (vertical), pontine infarction (horizontal), post-cardiac or aortic surgery[c] (Hanson et al. 1986; Eggers et al. 2008, 2015a; Bernat and Lukovits 2004; Antonio-Santos and Eggenberger 2007; Solomon et al. 2008a, b; Mokri et al. 2004; Nandipati et al. 2013; Kim et al. 2014)
Paraneoplastic (Tan et al. 2005; Baloh et al. 1993)/***autoimmune:*** Anti-Ma1 and Ma2 antibodies (Dalmau et al. 2004) (vertical), anti-glutamic acid decarboxylase (GAD) and anti-glycine receptor (GlyR) antibodies (Warren et al. 2002; Economides and Horton 2005; Tilikete et al. 2005; Pittock et al. 2006; Oskarsson et al. 2008; Peeters et al. 2012; Sarva et al. 2016), anti-IgLON5 antibodies (Dale and Ramanathan 2017; Gaig et al. 2017), demyelinating disease (Rufa et al. 2008; Quint et al. 1993; Nerrant and Tilikete 2017)
Prion disease: Creutzfeldt–Jakob disease (Wallach et al. 2017; Bertoni et al. 1983; Grant et al. 1993; Shimamura et al. 2003; Josephs et al. 2004; Prasad et al. 2007; Rowe et al. 2007; Petrovic et al. 2013) (CJD, vertical predominant)
Infection: Whipple disease (Lee 2002; Coria et al. 2000; Louis et al. 1996; Knox et al. 1995; Adams et al. 1987; Averbuch-Heller et al. 1999), infectious encephalitis

[a]Saccade speed may be utilized as a treatment outcome measure (Pensiero et al. 2005; Schiffmann et al. 2008)
[b]Saccade dysfunction correlates with number of polyglutamine repeats (Velazquez-Perez et al. 2004)
[c]Suspected to have a vascular ischemic etiology, though direct evidence of ischemia is lacking in most cases

Clinical saccade slowing typically indicates a brainstem-mediated supranuclear gaze palsy, which may be more accurately called a supranuclear saccade palsy. If slowing occurs predominantly in either the horizontal or vertical direction relative to the other, alteration of the trajectory, or "round-the-house," saccades may occur

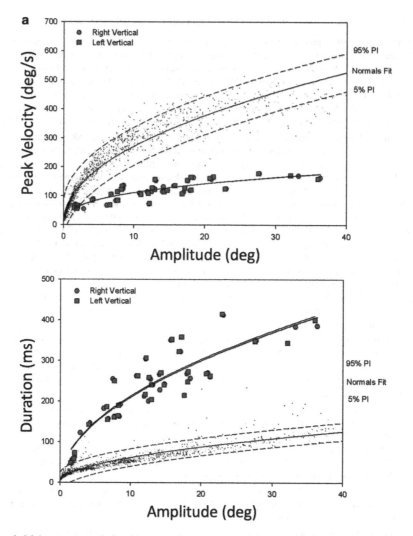

Fig. 4 Main sequence relationships of peak velocity versus amplitude (upper plots) and duration versus amplitude (lower plots) for vertical (**a**) and horizontal (**b**) saccades in a patient with ultimate autopsy diagnosis of progressive supranuclear palsy. Saccades are slow with prolonged duration, with vertical saccades more severely affected than horizontal saccades. See Video 1 for bedside saccade examination at the time of quantitative recordings shown here. Not shown: 4 years following the saccades shown here, saccades were re-recorded, and progressive slowing of saccades was demonstrated just prior to death; however, the range of eye movements remained full with no range limitation in vertical or horizontal directions

(Crespi et al. 2016; Eggink et al. 2016; Rottach et al. 1996; Garbutt et al. 2004). For example, if vertical saccades are selectively slowed, assessment of vertical saccades may result in a laterally curved vertical saccade trajectory. Examination of diagonal/oblique saccades may result in an initial horizontal movement that is then followed

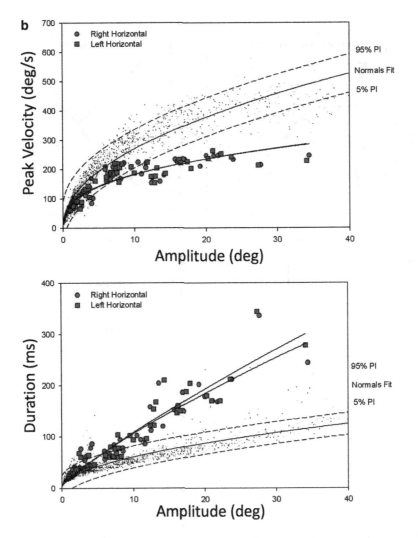

Fig. 4 (continued)

in its latter portion by a slowed vertical movement toward the target, as has been shown in patients with Niemann-Pick type C (NPC) which severely and selectively affects vertical saccades (Fig. 5a) (Rottach et al. 1997). Upon completion of the horizontal component of the saccade and prior to completion of the vertical component of the saccade, horizontal oscillations may occur (Fig. 5b) (Rottach et al. 1997). Saccade slowing may be accompanied by impaired range of vertical or horizontal eye movements (Video 2) that may be more severe during saccade testing than during smooth pursuit testing and can be improved by the use of larger visual targets (Seemungal et al. 2003). However, saccade slowing with a full range of eye movements

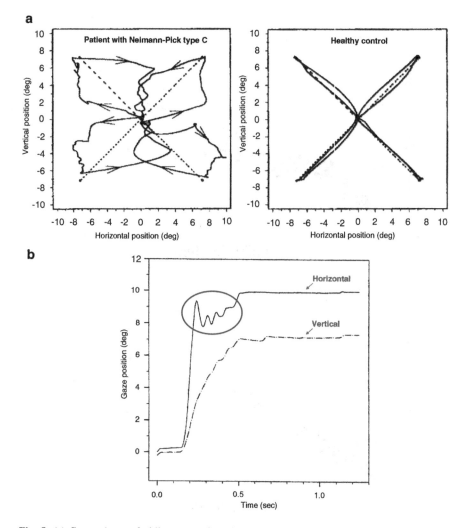

Fig. 5 (a) Comparisons of oblique saccade trajectories made to and from four target positions starting at primary position in a patient with Niemann-Pick type C (NPC) and a healthy control. The direction of the eye movement is indicated by the arrowheads. Saccade trajectory is shown as a dotted line. The patient's saccade trajectories are strongly curved, indicating an initial faster horizontal component and a later slowed vertical component. (b) Horizontal oscillations (blue oval) occur after completion of the horizontal component of an oblique saccade, while the vertical saccade component is ongoing in a patient with NPC. (Reprinted by permission from Elsevier Science Ltd.: Rottach et al. 1997)

is often seen and is equally indicative of a brainstem-mediated supranuclear saccade palsy (Fig. 4 and Video 1) (Hardwick et al. 2009). Bedside exam should include assessment of smooth pursuit and VOR, as a brainstem-mediated supranuclear saccade palsy will typically affect saccades to a greater extent than smooth pursuit, and impaired ocular motor range should be overcome with VOR (Video 2) until very late in the disease course, since vestibular pathways have different supranuclear inputs to the ocular motor nuclei than saccade premotor pathways. Assessment of OKN is particularly useful in early detection of a supranuclear saccade palsy, given that saccades and OKN quick phases share the same brainstem neural circuitry for generation (Garbutt et al. 2003). Loss of saccadic OKN fast phases may indicate early EBN disease and might be more easily seen at the bedside than very subtle early saccade slowing. With a midbrain supranuclear saccade palsy involving the riMLF, torsional quick phases may also be lost since riMLF EBN control both vertical and torsional fast eye movements.

Of critical importance to accurate neurological localization and diagnosis is differentiation between a brainstem-mediated supranuclear saccade palsy and cortically mediated ocular motor apraxia (Table 2). With the former, anatomic localization of the eye movement disorder is generally confined to brainstem EBN and OPN. The hallmarks include saccade slowing and loss of OKN fast phases with or without ocular motor range limitation, smooth pursuit ocular motor range superior to that obtained with saccades if range limitation is present, and ocular motor range limitations overcome by VOR. Ocular motor apraxia (OMA), in contrast, localizes to cortical eye fields and basal ganglionic circuits and is, therefore, less precisely localizing than clinically detectable saccade slowing (Leigh and Riley 2000). OMA is a disorder of impaired initiation of voluntary eye movements with retained reflexive eye movements. Hallmark features include prolonged latency of voluntary saccades that have normal speed once generated, difficulty with initiation and maintenance of smooth pursuit, and normal primitive reflexive VOR and OKN (Video 3) (Rottach et al. 1996). Patients with OMA tend to utilize head thrusting and blinking behaviors to facilitate voluntary eye movements; however, these behaviors are not pathognomonic for OMA and occur also in some brainstem-mediated supranuclear saccade palsies with saccade slowing (Eggers et al. 2008, 2015a). The majority of human disease states that lead to a supranuclear gaze palsy cause either a brainstem saccadic gaze palsy or OMA. However, Huntington's disease (HD) classically causes both simultaneously in the same patient, along with distractibility manifested as inability to suppress saccades to novel-appearing stimuli (Leigh et al. 1983; Lasker et al. 1987). Increased saccadic latencies particularly for voluntary saccades to command are a manifestation of OMA, whereas saccadic slowing represents dysfunction of EBN or their inputs (Leigh et al. 1985). Further, saccadic slowing may be more common in mildly affected HD patients who have disease onset at age less than

Table 2 Clinical features of brainstem saccadic gaze palsy versus cortical ocular motor apraxia

	Brainstem saccadic gaze palsy	Cortical ocular motor apraxia
Saccades	Slow with normal or limited range of motion	Impaired initiation, typically with normal speed, range of motion may appear impaired, but full range can be elicited often with stronger visual stimulus (i.e., $20 bill)
Smooth pursuit	Normal or, if range limitation present, range less severely impaired than with saccades	Impaired initiation and fixation maintenance during pursuit
Optokinetic nystagmus	Absent quick phases	Normal
Head thrust/blinks to initiate saccades	Uncommon[a]	Common
Overview	Affects saccades most predominantly	Affects voluntarily initiated eye movements, spares reflexive eye movements
Localization	Brainstem saccadic burst neurons	Cortical eye fields
Etiologic example	Progressive supranuclear palsy	Balint's syndrome with bilateral parietal brain lesions

[a]Head thrusting and blinking to facilitate saccades occasionally occur with brainstem saccadic gaze palsy and are especially reported with post-cardiac or aortic surgery saccadic palsies (Hanson et al. 1986; Eggers et al. 2015a)

30 years, and increased saccadic latencies may be more common in patients with disease onset after age 30 years (Lasker et al. 1988), as well as in patients over 30 who have genetically confirmed preclinical HD (Golding et al. 2006). Distractibility is common in all patients with HD (Lasker et al. 1988).

Performance of quantified ocular motor recordings in clinical patients greatly enhances accurate identification of saccade slowing and has the capacity to identify early subclinical saccade slowing (Figs. 4a and 6). Eye movement recordings with large field OKN stimuli may also precipitate saccadic quick phases that can be examined for slowing in the context of advanced supranuclear gaze palsy when larger saccades can no longer be initiated (Garbutt et al. 2004; Garbutt and Harris 2000). Similar to saccades, OKN quick phases obey "main sequence" relationships between saccade amplitude and peak velocity, and between saccade amplitude and duration, by which saccade velocity and duration increase in a stereotypical manner with increasing saccade amplitude. However, velocity profiles differ, with slightly higher peak velocities for saccades than OKN quick phases in both the horizontal (Garbutt et al. 2001) and vertical (Garbutt et al. 2003) planes.

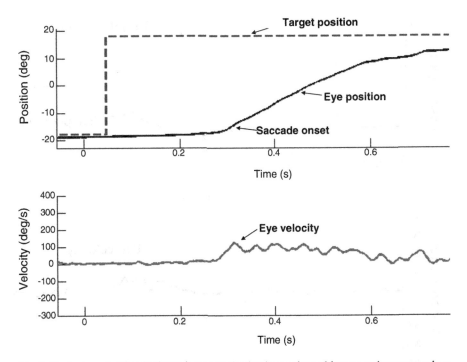

Fig. 6 Vertical greater than horizontal saccade slowing in a patient with progressive supranuclear palsy. Bedside examination revealed a 50% limitation in the range of elevation of each eye, full depression of each eye, vertical greater than horizontal saccade slowing, and absent vertical opto-kinetic quick phases. (**a**) Representative position (top) and velocity (bottom) traces of a severely slow horizontal saccade directed toward a target jump of nearly 40 degrees. The upward deflection of the position trace represents a hypometric rightward eye movement. (**b**) Representative position (top) and velocity (bottom) traces of a severely slow and hypometric vertical saccade directed toward a target jump of 20 degrees. For comparison with a normal saccade position and velocity profile, see Fig. 1a

4.2   Progressive Supranuclear Palsy: Insights into Saccade Slowing and Translation to the Clinic – EBN Versus OPN

The prototypical disorder of gradually progressive vertical saccade slowing in humans is progressive supranuclear palsy (PSP), which has an incidence of up to 18 cases per 100,000 people (Coyle-Gilchrist et al. 2016; Takigawa et al. 2016). PSP is a neurodegenerative tauopathy that, in its classic form of Richardson's syndrome (PSP-RS) (Williams and Lees 2009; Steele et al. 1964), shares some clinical features with idiopathic Parkinson's disease (PD) but is more rapidly progressive and not responsive to levodopa treatment (Goetz et al. 2003; Litvan et al. 1996; Diroma et al. 2003). The hallmark feature is slowing of vertical saccades that typically occurs early in the disease course. It has been a common clinical misconception that involvement of downward saccades is more common than upward saccades in PSP (Chen et al. 2010). Slowing of both downward and upward saccades is common.

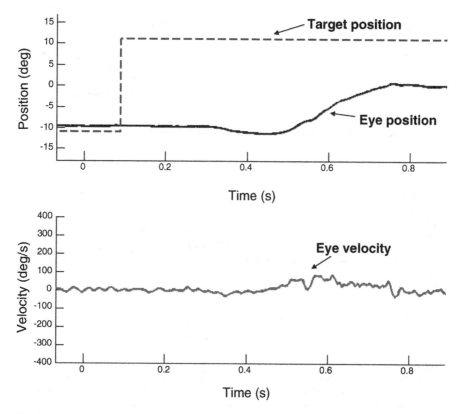

Fig. 6 (continued)

Limitation of upward ocular motor range is more common than simultaneous range limitation in both vertical directions, and both of the prior are more common than selective downward range limitation (Chen et al. 2010). Early in the disease course, horizontal saccades are also slowed but much less so than vertical saccades. Late in the disease course, saccades and smooth pursuit may be lost both vertically and horizontally, though VOR still tends to overcome the range limitations. In end-stage disease, even the VOR may be lost.

Saccade slowing in PSP could potentially be attributed to either EBN or OPN dysfunction, though most evidence points to early direct riMLF EBN involvement (Bhidayasiri et al. 2001). Midbrain regions containing riMLF neurons have been shown to be affected neuropathologically (Hardwick et al. 2009; Steele et al. 1964; Juncos et al. 1991; Daniel et al. 1995), as has the RIP containing OPN, presumably late in the disease course (Revesz et al. 1996). Marked vertical more than horizontal saccade (Figs. 4 and 6) slowing implicates direct riMLF EBN dysfunction, especially given its presence in a patient cohort with a median disease duration of 3 years (Bhidayasiri et al. 2001; Chen et al. 2010). Further supporting riMLF EBN dysfunction are experiments utilizing the Mueller paradigm, which combines a saccadic eye movement with a vergence movement (Bhidayasiri et al. 2001). Given that OPN govern the timing of, and are turned off during, both saccades and vergence movements, rapid completion of a saccade during a combined saccade-vergence movement leads to an interval of time during which OPN are off while the saccade is completed and the slower vergence movement continues to occur. Horizontal ocular oscillations of saccadic origin typically occur during this interval (Ramat et al. 1999). PSP patients generate these oscillations similar to healthy control subjects, indicating that OPN are off; however, vertical saccades are still pathologically slow, suggesting that OPN involvement is not the primary mechanism of vertical saccade slowing in early PSP (Bhidayasiri et al. 2001). Similar horizontal oscillations may occur during oblique saccade testing when vertical saccade palsy leads to earlier completion of the horizontal than the vertical component of an oblique saccade (Fig. 5b) (Rottach et al. 1997). Combination of saccades with blinks or sustained eye closure, which also turn off OPN (Mays and Morrissee 1993), can cause transient horizontal ocular oscillations in healthy individuals as well (Rottach et al. 1998; Shaikh et al. 2007, 2010). When this occurs in PSP patients, there is no change in vertical saccade velocity, again suggesting that the primary pathology involves the riMLF rather than the OPN (Bhidayasiri et al. 2001). Though large amplitude saccades fall below the limits of normal in PSP, small saccades less than 5 degrees in patients with early PSP (mean disease duration of less than 2 years) tend to have normal velocity (Averbuch-Heller et al. 2002). This leads to questions regarding whether a frequency-dependent saturation in the riMLF (i.e., EBN discharge normally for small saccades but discharge intensity saturates for large saccades) or abnormal feedback control of EBN due to SC involvement early in PSP may play a role in this size differential effect (Averbuch-Heller et al. 2002).

A second form of PSP, PSP with parkinsonism (PSP-P), was recognized as a distinct phenotype in 2005 (Williams et al. 2005a). PSP-P typically mimics PD in that it presents with asymmetric parkinsonism with tremor and partial levodopa responsivity, has a more rapid course than PD but a more indolent course than PSP-RS, and does not develop clinical ocular motor involvement until much later in the disease course (Respondek et al. 2014). In a series of pathologically proven PSP-P, none had supranuclear gaze palsy in the first 2 years of disease, but over 70% developed it as a later clinical feature (Williams et al. 2005a). Interestingly, some of these patients are considered for years to have idiopathic Parkinson's disease until they reach a point when progression occurs more rapidly than anticipated and saccade slowing begins and may gradually progress to a more complete supranuclear gaze palsy (Williams and Lees 2010). Despite the absence of clinically detectable saccade slowing early in the course of PSP-P, video-oculography has demonstrated pathological subclinical vertical greater than horizontal saccadic slowing in patients with PSP-P that does not statistically differ from that seen in patients with PSP-RS. Saccade slowing in both PSP-P and PSP-RS, in contrast, is significantly different than the normal saccadic velocities seen in PD, which are the same as in healthy controls (Fig. 7) (Pinkhardt et al. 2008). In this study, 20-degree vertical saccade peak velocities reached less than 30% of normal in PSP-RS patients and less than 35% of normal in PSP-P patients with no difference in either group

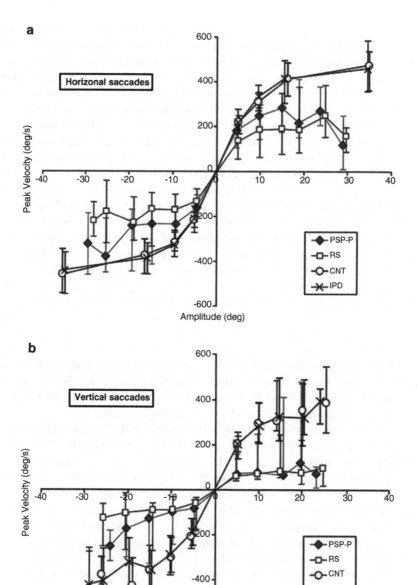

Fig. 7 Population-based main sequence relationships of median peak velocity versus amplitude for horizontal (**a**) and vertical (**b**) saccades in patients with progressive supranuclear palsy with parkinsonism (PSP-P), progressive supranuclear palsy Richardson's syndrome (RS), idiopathic Parkinson's disease (IPD), and healthy control participants (CNT). Note that saccade peak velocity values are slower for both PSP-P and RS groups but normal and overlapping for IPD and CNT groups. (Reprinted by permission from Springer Nature: Pinkhardt et al. 2008)

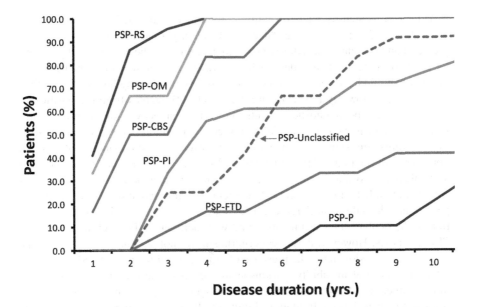

Fig. 8 Frequency of supranuclear gaze palsy as a function of time after disease onset in the different PSP-predominance types in 100 definitive PSP cases. Abbreviations: PSP progressive supranuclear palsy, RS Richardson's syndrome, OMD ocular motor dysfunction, CBS corticobasal syndrome, PI postural instability, FTD frontotemporal dementia, P parkinsonism. (Reprinted by permission from John Wiley and Sons: Respondek et al. 2014)

between upward and downward saccades; 20-degree horizontal saccade peak velocities reached 60% of normal in PSP-RS and 70% of normal in PSP-P. Thus, eye movement recordings of saccade peak velocities cannot differentiate PSP-RS from PSP-P but can differentiate PSP-P and PSP-RS from PD (Pinkhardt et al. 2008). Other than PSP-RS and PSP-P, additional variant PSP phenotypes are described in which eye movement involvement is less frequent, though the prevalence of supranuclear gaze palsy has been shown to increase in all PSP types with increasing disease duration (Fig. 8) (Respondek et al. 2014).

The importance of early establishment of the diagnosis of PSP, for which early detection of subclinical saccade slowing is very high yield, has increased substantially in recent years, given the advent of clinical trials of new anti-tau-directed therapies (Boxer et al. 2017; West et al. 2017). PSP diagnostic criteria have recently been revised and include ocular motor dysfunction as one of four functional domains of core clinical features (the others being postural instability, akinesia, and cognitive dysfunction) (Hoglinger et al. 2017). Certainty levels with regard to ocular motor dysfunction are graded: (1) vertical supranuclear gaze palsy, (2) slow velocity of vertical saccades, and (3) frequent macro-square wave jerks or "eyelid opening apraxia." Vertical supranuclear gaze palsy is defined as "a clear limitation in the range of voluntary gaze in the vertical more than in the horizontal plane, affecting both up- and downgaze, more than expected for age, which is overcome...with the vestibulo-ocular reflex; at later stages, the vestibulo-ocular reflex may be lost, or the maneuver prevented by nuchal rigidity" (Hoglinger et al. 2017). Slow velocity of vertical saccades is defined as "decreased velocity (and amplitude) of vertical greater than horizontal saccadic eye movements; this may be established by quantitative measurements of saccades, such as infrared oculography, or by bedside testing; gaze should be assessed by command ("look at the flicking finger") rather than by pursuit ("follow my finger"), with the target >20 degrees from the position of primary gaze; to be diagnostic, saccadic movements are slow enough for the examiner to see their movement…; a delay in saccade initiation is not considered slowing; findings are supported by slowed or absent fast components of vertical optokinetic nystagmus…" (Hoglinger et al. 2017). Prior diagnostic criteria included vertical gaze palsy and slowing of saccades; however, such precise definition details were not included. This new detailed instruction set for recognition of saccadic gaze palsies and slowed saccades represents definite advancement in diagnostic application of the ocular motor examination and will facilitate earlier and more sensitive PSP diagnosis. However, these new criteria still do not take into account that true vertical saccade slowing in the presence of other clinical features of PSP (i.e., early falls with postural instability, parkinsonism), even in the presence of full ocular motor range, has identical localization and prognostic significance as supranuclear palsy with vertical range limitation (Leigh and Riley 2000). Indeed, some patients with pathologically proven PSP have progressive saccade slowing over time but never develop a vertical range limitation prior to death (Hardwick et al. 2009).

4.3   Observations from Other Clinical Disorders

Selective Involvement of Up Versus Down Burst Neurons

Predominant impairment of either horizontal or vertical saccades implies involvement of different populations of EBN and informs neurological differential diagnosis. For example, the ocular motor hallmark of spinocerebellar atrophy type 2 (SCA2) is selective slowing of horizontal saccades (Zee et al. 1976; Burk et al. 1999; Velazquez-Perez et al. 2004; Rufa and Federighi 2011; Wadia et al. 1998; Geiner et al. 2008), whereas PSP and NPC (Rottach et al. 1997; Abel et al. 2009; Salsano et al. 2012; Solomon et al. 2005; Cogan et al. 1981) cause vertical greater than horizontal saccade slowing. This is due to selective vulnerability of saccade burst neurons (or their connections) to the disease process and, indeed, horizontal EBN in the PPRF in SCA2 (Geiner et al. 2008) and vertical EBN in the riMLF in NPC (Solomon et al. 2005) have been shown via immunohistochemical techniques to be selectively affected. Though the precise mechanisms of selective vulnerability of horizontal versus vertical EBN populations in a specific disease may not be fully understood, at least these are two distinct neuronal populations that are located at different brainstem levels.

More difficult to explain is the mechanism of selective vulnerability of either downward or upward saccades in conditions such as NPC, which has a more profound effect on downward saccades (Rottach et al. 1997; Cogan et al. 1981), and motor neuron disease in which upward worse than downward saccade involvement has been shown (Averbuch-Heller et al. 1998). Hypotheses to explain this downward saccade selective vulnerability in NPC include possible additional damage to the axons of riMLF neurons responsible for downward saccades or a less robust projection of the unilateral riMLF downward saccade neurons to the oculomotor nucleus than the bilateral projections serving upward gaze (Solomon et al. 2005). riMLF neuronal loss has been demonstrated in motor neuron disease, interestingly with "...loss of 60% of the riMLF neurons result(ing) in about 80–90% decrease in velocity for upward saccades but only about 40–60% decrease of velocity for downward saccades..." (Averbuch-Heller et al. 1998). Hypotheses to explain relative upward saccade involvement in motor neuron disease include the following: (1) upward and downward EBN may have unique neurochemical properties, (2) less neuronal activation may be required to elicit downward than upward saccades, (3) downward EBN may be more numerous and thus better preserved in this disease than upward EBN, or (4) concurrent damage to medium-lead burst neurons for upward saccades in the posterior commissure may exist (Averbuch-Heller et al. 1998).

Expanding Anatomic Mechanisms: Inhibitory Burst Neurons and Perineuronal Nets

Most clinical disorders involving saccade slowing directly implicate EBN with or without OPN involvement via selective directionality of the saccade disorder or via direct pathological evidence. Careful attention to the details of eye position and velocity traces generated by eye movement recordings may offer additional insights. For example, saccade slowing of some large saccades occurs in late-onset Tay-Sachs (LOTS) (Rucker et al. 2004), and saccade slowing may rarely occur with a selective saccade palsy after cardiac or aortic surgery, most typically after aortic valve surgery requiring cardiopulmonary bypass and hypothermic circulatory arrest and usually with normal MRI of the brain (Hanson et al. 1986; Eggers et al. 2008, 2015a; Bernat and Lukovits 2004; Antonio-Santos and Eggenberger 2007; Solomon et al. 2008a, b). In addition to slow saccades, many patients with these disorders generate a "staircase" (Rucker et al. 2004; Solomon et al. 2008b) series of small saccades to reach the target (Fig. 9), suggesting that EBN firing is prematurely and intermittently inhibited during the course of an intended large saccade. This leads to transient decelerations mid-saccade (Fig. 9), during which the eyes may (Solomon et al. 2008b) or may not (Rucker et al. 2004) come to a complete stop. Generation of premature saccade "choke" signals by IBN is considered to be the probable mechanism of transient decelerations in these "staircase" saccades (Optican et al. 2008; Solomon et al. 2008b; Rucker et al. 2011).

Early in a normal rightward saccade, left ocular motor vermis Purkinje cells cease firing, thereby allowing the left fastigial nucleus to excite the right IBN, which in turn facilitates the saccade by inhibiting the antagonist muscles (Optican

Horizontal saccades

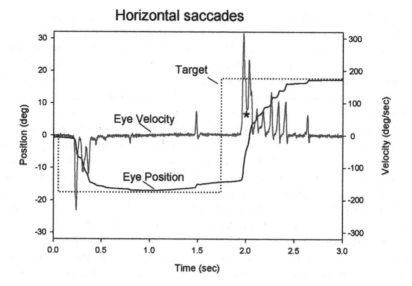

Fig. 9 Representative record of horizontal saccades made by a patient with late-onset Tay-Sachs disease. The eye position trace shows hypometric, multistep saccades. The velocity record shows transient decelerations during which eye velocity transiently declines, but not to zero, and then increases again (indicated by *). (Reprinted by permission from Wolters Kluwers: Rucker et al. 2004)

and Quaia 2002). Toward the end of a rightward saccade, right ocular motor vermis Purkinje cells cease firing, thereby allowing the right fastigial nucleus to excite the left IBN, which chokes the saccade drive. With "staircase" saccades, loss of cerebellar ocular motor vermis Purkinje cell inhibition on ipsilateral fastigial nucleus neurons may lead to premature contralateral IBN "choke" signals, while an intact EBN/OPN network continues to drive completion of the eye movement to the target. Indeed, in both LOTS and iatrogenic surgical ischemic brain injury, pathology has confirmed severe loss of Purkinje cells in the ocular motor vermis (Welsh et al. 2002; Rucker et al. 2008). Further supporting this concept of the effects of Purkinje cell loss is autopsy confirmation of unaffected EBN and OPN in a patient who died 8 years after the development of post-aortic surgery selective saccade palsy (Eggers et al. 2008, 2015a). Despite this, the story may not be so straightforward, as brainstem regions involved in saccade generation may also be pathologically abnormal in these conditions, with neurons in the OPN region showing massive inclusions in LOTS (Rucker et al. 2008) and a different case of post-cardiac surgery selective saccade palsy showing PPRF neuronal necrosis and astrocytosis (Hanson et al. 1986).

Given discrepant and non-uniform pathologic findings in the EBN and OPN in post-cardiac or aortic surgery saccadic palsy, an alternative hypothesis to direct EBN injury as the cause of saccadic slowing is injury to perineuronal nets (Eggers et al. 2015a). These nets are networks of chondroitin sulfate proteoglycans that ensheath the highly metabolically active and fast-firing EBN and OPN neurons

(Horn et al. 2003b) and are thought to provide a supportive specialized microenvironment in the extracellular matrix that may serve several roles, including synaptic contact support, homeostasis of local ions, and neuronal protection (Eggers et al. 2015a; Hartig et al. 1999). Perineuronal nets have been shown to be highly vulnerable to ischemic injury (Hobohm et al. 2005) and, indeed, perineuronal nets have also been shown to demonstrate severe loss and fragmentation in a patient with post-aortic surgery saccadic gaze palsy in whom EBN and OPN, themselves, were histologically normal with synapse preservation (Eggers et al. 2015b).

5 Conclusions

Saccadic slowing is most often due directly to metabolic and or structural dysfunction of brainstem EBN and/or OPN, though various other mechanisms may also be involved, such as disruption of inputs to these structures or injury to the supportive surrounding extracellular perineuronal nets. A wide range of human disease states may cause saccadic slowing, and careful attention to the details of the ocular motor disorder can assist with diagnosis and prognosis and provide ample opportunities for advancement in the neuroscientific understanding of neuroanatomic and neurophysiologic ocular motor control.

References

Abel, L. A., Walterfang, M., Fietz, M., Bowman, E. A., & Velakoulis, D. (2009). Saccades in adult Niemann-Pick disease type C reflect frontal, brainstem, and biochemical deficits. *Neurology, 72*, 1083–1086.

Adams, M., Rhyner, P. A., Day, J., DeArmond, S., & Smuckler, E. A. (1987). Whipple's disease confined to the central nervous system. *Annals of Neurology, 21*, 104–108.

Aitsebaomo, A. P., & Bedell, H. E. (1992). Psychophysical and saccadic information about direction for briefly presented visual targets. *Vision Research, 32*, 1729–1737.

Antonio-Santos, A., & Eggenberger, E. R. (2007). Asaccadia and ataxia after repair of ascending aortic aneurysm. *Seminars in Ophthalmology, 22*, 33–34.

Averbuch-Heller, L., Helmchen, C., Horn, A. K., Leigh, R. J., & Büttner-Ennerver, J. A. (1998). Slow vertical saccades in motor neuron disease: Correlation of structure and function. *Annals of Neurology, 44*, 641–648.

Averbuch-Heller, L., Paulson, G. W., Daroff, R. B., & Leigh, R. J. (1999). Whipple's disease mimicking progressive supranuclear palsy: The diagnostic value of eye movement recording. *Journal of Neurology, Neurosurgery, and Psychiatry, 66*, 532–535.

Averbuch-Heller, L., Gordon, C., Zivotofsky, A., et al. (2002). Small vertical saccades have normal speeds in progressive supranuclear palsy (PSP). *Annals of the New York Academy of Sciences, 956*, 434–437.

Bahill, A. T., Clark, M. R., & Stark, L. (1975). The main sequence, a tool for studying human eye movements. *Mathematical Biosciences, 24*, 191–204.

Baloh, R. W., Sills, A. W., Kumley, W. E., & Honrubia, V. (1975). Quantitative measurement of saccade amplitude, duration, and velocity. *Neurology, 25*, 1065–1070.

Baloh, R. W., DeRossett, S. E., Cloughesy, T. F., et al. (1993). Novel brainstem syndrome associated with prostate carcinoma. *Neurology, 43*, 2591–2596.

Benko, W., Ries, M., Wiggs, E. A., Brady, R. O., Schiffmann, R., & Fitzgibbon, E. J. (2011). The saccadic and neurological deficits in type 3 Gaucher disease. *PLoS One, 6*, e22410.

Bernat, J. L., & Lukovits, T. G. (2004). Syndrome resembling PSP after surgical repair of ascending aorta dissection or aneurysm. *Neurology, 63*, 1141–1142; author reply -2.

Bertoni, J. M., Label, L. S., Sackelleres, J. C., & Hicks, S. P. (1983). Supranuclear gaze palsy in familial Creutzfeldt-Jakob disease. *Archives of Neurology, 40*, 618–622.

Bhidayasiri, R., Plant, G. T., & Leigh, R. J. (2000). A hypothetical scheme for the brainstem control of vertical gaze. *Neurology, 54*, 1985–1993.

Bhidayasiri, R., Riley, D. E., Somers, J. T., Lerner, A. J., Büttner-Ennever, J. A., & Leigh, R. J. (2001). Pathophysiology of slow vertical saccades in progressive supranuclear palsy. *Neurology, 57*, 2070–2077.

Boghen, D., Troost, B. T., Daroff, R. B., Dell'Osso, L. F., & Birkett, J. E. (1974). Velocity characteristics of normal human saccades. *Investigative Ophthalmology, 13*, 619–623.

Boxer, A. L., Yu, J. T., Golbe, L. I., Litvan, I., Lang, A. E., & Hoglinger, G. U. (2017). Advances in progressive supranuclear palsy: New diagnostic criteria, biomarkers, and therapeutic approaches. *Lancet Neurology, 16*, 552–563.

Bronstein, A. M., & Kennard, C. (1987). Predictive eye saccades are different from visually triggered saccades. *Vision Research, 27*, 517–520.

Burk, K., Fetter, M., Abele, M., et al. (1999). Autosomal dominant cerebellar ataxia type I: Oculomotor abnormalities in families with SCA1, SCA2, and SCA3. *Journal of Neurology, 246*, 789–797.

Büttner, U., & Büttner-Ennever, J. A. (2005). Present concepts of oculomotor organization. *Progress in Brain Research, 151*, 1–42.

Büttner-Ennever, J. A., Büttner, U., Cohen, B., & Baumgartner, G. (1982). Vertical glaze paralysis and the rostral interstitial nucleus of the medial longitudinal fasciculus. *Brain, 105*, 125–149.

Büttner-Ennever, J. A., Cohen, B., Pause, M., & Fries, W. (1988). Raphe nucleus of the pons containing omnipause neurons of the oculomotor system in the monkey, and its homologue in man. *The Journal of Comparative Neurology, 267*, 307–321.

Chen, A. L., Riley, D. E., King, S. A., et al. (2010). The disturbance of gaze in progressive supranuclear palsy: Implications for pathogenesis. *Frontiers in Neurology, 1*, 147.

Christova, P., Anderson, J. H., & Gomez, C. M. (2008). Impaired eye movements in presymptomatic spinocerebellar ataxia type 6. *Archives of Neurology, 65*, 530–536.

Cogan, D. G., Chu, F. C., Reingold, D., & Barranger, J. (1981). Ocular motor signs in some metabolic diseases. *Archives of Ophthalmology, 99*, 1802–1808.

Collewijn, H., Went, L. N., Tamminga, E. P., & Vegter-Van der Vlis, M. (1988). Oculomotor defects in patients with Huntington's disease and their offspring. *Journal of the Neurological Sciences, 86*, 307–320.

Coria, F., Cuadrado, N., Velasco, C., et al. (2000). Whipple's disease with isolated central nervous system symptomatology diagnosed by molecular identification of Tropheryma whippelii in peripheral blood. *Neurología, 15*, 173–176.

Coyle-Gilchrist, I. T., Dick, K. M., Patterson, K., et al. (2016). Prevalence, characteristics, and survival of frontotemporal lobar degeneration syndromes. *Neurology, 86*, 1736–1743.

Crespi, J., Brathen, G., Quist-Paulsen, P., Pagonabarraga, J., & Roig-Arnall, C. (2016). Facial dystonia with facial grimacing and vertical gaze palsy with "round the houses" sign in a 29-year-old woman. *Neuroophthalmology, 40*, 31–34.

Dale, R. C., & Ramanathan, S. (2017). Cell surface antibody-associated neurodegeneration: The case of anti-IgLON5 antibodies. *Neurology, 88*, 1688–1690.

Dalmau, J., Graus, F., Villarejo, A., et al. (2004). Clinical analysis of anti-Ma2-associated encephalitis. *Brain, 127*, 1831–1844.

Daniel, S. E., de Bruin, V. M., & Lees, A. J. (1995). The clinical and pathological spectrum of Steele-Richardson-Olszewski syndrome (progressive supranuclear palsy): A reappraisal. *Brain, 118*(Pt 3), 759–770.

de Bruin, V. M., Lees, A. J., & Daniel, S. E. (1992). Diffuse Lewy body disease presenting with supranuclear gaze palsy, parkinsonism, and dementia: A case report. *Movement Disorders, 7,* 355–358.

Deleu, D., Buisseret, T., & Ebinger, G. (1989). Vertical one-and-a-half syndrome. Supranuclear downgaze paralysis with monocular elevation palsy. *Archives of Neurology, 46,* 1361–1363.

Diroma, C., Dell'Aquila, C., Fraddosio, A., et al. (2003). Natural history and clinical features of progressive supranuclear palsy: A clinical study. *Neurological Sciences, 24,* 176–177.

Donaghy, C., Pinnock, R., Abrahams, S., et al. (2010). Slow saccades in bulbar-onset motor neurone disease. *Journal of Neurology, 257,* 1134–1140.

Economides, J. R., & Horton, J. C. (2005). Eye movement abnormalities in stiff person syndrome. *Neurology, 65,* 1462–1464.

Eggers, S. D., Moster, M. L., & Cranmer, K. (2008). Selective saccadic palsy after cardiac surgery. *Neurology, 70,* 318–320.

Eggers, S. D., Horn, A. K., Roeber, S., et al. (2015a). Saccadic palsy following cardiac surgery: A review and new hypothesis. *Annals of the New York Academy of Sciences, 1343,* 113–119.

Eggers, S. D., Horn, A. K., Roeber, S., et al. (2015b). Saccadic palsy following cardiac surgery: Possible role of Perineuronal nets. *PLoS One, 10,* e0132075.

Eggink, H., Brandsma, R., van der Hoeven, J. H., Lange, F., de Koning, T. J., & Tijssen, M. A. (2016). Teaching Video neuroimages: The "round the houses" sign as a clinical clue for Niemann-Pick disease type C. *Neurology, 86,* e202.

Enderle, J. D., & Engelken, E. J. (1995). Simulation of oculomotor post-inhibitory rebound burst firing using a Hodgkin-Huxley model of a neuron. *Biomedical Sciences Instrumentation, 31,* 53–58.

Erro, R., Lees, A. J., Moccia, M., et al. (2014). Progressive parkinsonism, balance difficulties, and supranuclear gaze palsy. *JAMA Neurology, 71,* 104–107.

Fearnley, J. M., Revesz, T., Brooks, D. J., Frackowiak, R. S., & Lees, A. J. (1991). Diffuse Lewy body disease presenting with a supranuclear gaze palsy. *Journal of Neurology, Neurosurgery, and Psychiatry, 54,* 159–161.

Frost, D., & Poppel, E. (1976). Different programming modes of human saccadic eye movements as a function of stimulus eccentricity: Indications of a functional subdivision of the visual field. *Biological Cybernetics, 23,* 39–48.

Gaig, C., Graus, F., Compta, Y., et al. (2017). Clinical manifestations of the anti-IgLON5 disease. *Neurology, 88,* 1736–1743.

Garbutt, S., & Harris, C. M. (2000). Abnormal vertical optokinetic nystagmus in infants and children. *The British Journal of Ophthalmology, 84,* 451–455.

Garbutt, S., Harwood, M. R., & Harris, C. M. (2001). Comparison of the main sequence of reflexive saccades and the quick phases of optokinetic nystagmus. *The British Journal of Ophthalmology, 85,* 1477–1483.

Garbutt, S., Han, Y., Kumar, A. N., Harwood, M., Harris, C. M., & Leigh, R. J. (2003). Vertical optokinetic nystagmus and saccades in normal human subjects. *Investigative Ophthalmology & Visual Science, 44,* 3833–3841.

Garbutt, S., Riley, D. E., Kumar, A. N., Han, Y., Harwood, M. R., & Leigh, R. J. (2004). Abnormalities of optokinetic nystagmus in progressive supranuclear palsy. *Journal of Neurology, Neurosurgery, and Psychiatry, 75,* 1386–1394.

Geiner, S., Horn, A. K., Wadia, N. H., Sakai, H., & Büttner-Ennever, J. A. (2008). The neuroanatomical basis of slow saccades in spinocerebellar ataxia type 2 (Wadia-subtype). *Progress in Brain Research, 171,* 575–581.

Goetz, C. G., Leurgans, S., Lang, A. E., & Litvan, I. (2003). Progression of gait, speech and swallowing deficits in progressive supranuclear palsy. *Neurology, 60,* 917–922.

Golding, C. V., Danchaivijitr, C., Hodgson, T. L., Tabrizi, S. J., & Kennard, C. (2006). Identification of an oculomotor biomarker of preclinical Huntington disease. *Neurology, 67,* 485–487.

Goossens, H. H., & Van Opstal, A. J. (2000). Blink-perturbed saccades in monkey. I. Behavioral analysis. *Journal of Neurophysiology, 83,* 3411–3429.

Gordon, P. H., Fahn, S., Chin, S., Golbe, L. I., Lynch, T., & Eidelberg, D. (2004). Woman with a 26-year history of parkinsonism, supranuclear ophthalmoplegia, and loss of postural reflexes. *Movement Disorders, 19*, 950–961.

Grant, M. P., Cohen, M., Petersen, R. B., et al. (1993). Abnormal eye movements in Creutzfeldt-Jakob disease. *Annals of Neurology, 34*, 192–197.

Green, J. P., Newman, N. J., & Winterkorn, J. S. (1993). Paralysis of downgaze in two patients with clinical-radiologic correlation. *Archives of Ophthalmology, 111*, 219–222.

Gupta, S. R., Brigell, M., Gujrati, M., & Lee, J. M. (1995). Supranuclear eye movement dysfunction in mitochondrial myopathy with tRNA(LEU) mutation. *Journal of Neuro-Ophthalmology, 15*, 20-5.

Hanson, M. R., Hamid, M. A., Tomsak, R. L., Chou, S. S., & Leigh, R. J. (1986). Selective saccadic palsy caused by pontine lesions: Clinical, physiological, and pathological correlations. *Annals of Neurology, 20*, 209–217.

Hardwick, A., Rucker, J. C., Cohen, M. L., et al. (2009). Evolution of oculomotor and clinical findings in autopsy-proven Richardson syndrome. *Neurology, 73*, 2122–2124.

Hartig, W., Derouiche, A., Welt, K., et al. (1999). Cortical neurons immunoreactive for the potassium channel Kv3.1b subunit are predominantly surrounded by perineuronal nets presumed as a buffering system for cations. *Brain Research, 842*, 15–29.

Henn, V., Lang, W., Hepp, K., & Reisine, H. (1984). Experimental gaze palsies in monkeys and their relation to human pathology. *Brain, 107*(Pt 2), 619–636.

Hikosaka, O., & Wurtz, R. H. (1985). Modification of saccadic eye movements by GABA-related substances. I. Effect of muscimol and bicuculline in monkey superior colliculus. *Journal of Neurophysiology, 53*, 266–291.

Hikosaka, O., & Wurtz, R. H. (1986). Saccadic eye movements following injection of lidocaine into the superior colliculus. *Experimental Brain Research, 61*, 531–539.

Hobohm, C., Gunther, A., Grosche, J., Rossner, S., Schneider, D., & Bruckner, G. (2005). Decomposition and long-lasting downregulation of extracellular matrix in perineuronal nets induced by focal cerebral ischemia in rats. *Journal of Neuroscience Research, 80*, 539–548.

Hoglinger, G. U., Respondek, G., Stamelou, M., et al. (2017). Clinical diagnosis of progressive supranuclear palsy: The movement disorder society criteria. *Movement Disorders, 32*, 853–864.

Hommel, M., & Bogousslavsky, J. (1991). The spectrum of vertical gaze palsy following unilateral brainstem stroke. *Neurology, 41*, 1229–1234.

Horn, A. K. (2006). The reticular formation. *Progress in Brain Research, 151*, 127–155.

Horn, A. K., & Büttner-Ennever, J. A. (1998). Premotor neurons for vertical eye movements in the rostral mesencephalon of monkey and human: Histologic identification by parvalbumin immunostaining. *The Journal of Comparative Neurology, 392*, 413–427.

Horn, A. K., & Leigh, R. J. (2011). The anatomy and physiology of the ocular motor system. *Handbook of Clinical Neurology, 102*, 21–69.

Horn, A. K., Büttner-Ennever, J. A., Wahle, P., & Reichenberger, I. (1994). Neurotransmitter profile of saccadic omnipause neurons in nucleus raphe interpositus. *The Journal of Neuroscience, 14*, 2032–2046.

Horn, A. K., Büttner-Ennever, J. A., Suzuki, Y., & Henn, V. (1995). Histological identification of premotor neurons for horizontal saccades in monkey and man by parvalbumin immunostaining. *The Journal of Comparative Neurology, 359*, 350–363.

Horn, A. K., Helmchen, C., & Wahle, P. (2003a). GABAergic neurons in the rostral mesencephalon of the macaque monkey that control vertical eye movements. *Annals of the New York Academy of Sciences, 1004*, 19–28.

Horn, A. K., Bruckner, G., Hartig, W., & Messoudi, A. (2003b). Saccadic omnipause and burst neurons in monkey and human are ensheathed by perineuronal nets but differ in their expression of calcium-binding proteins. *The Journal of Comparative Neurology, 455*, 341–352.

Hoyt, C. S., Billson, F. A., & Alpins, N. (1978). The supranuclear disturbances of gaze in kernicterus. *Annals of Ophthalmology, 10*, 1487–1492.

Johnston, J. L., Sharpe, J. A., Ranalli, P. J., & Morrow, M. J. (1993). Oblique misdirection and slowing of vertical saccades after unilateral lesions of the pontine tegmentum. *Neurology, 43,* 2238–2244.

Josephs, K. A., Tsuboi, Y., & Dickson, D. W. (2004). Creutzfeldt-Jakob disease presenting as progressive supranuclear palsy. *European Journal of Neurology, 11,* 343–346.

Juncos, J. L., Hirsch, E. C., Malessa, S., Duyckaerts, C., Hersh, L. B., & Agid, Y. (1991). Mesencephalic cholinergic nuclei in progressive supranuclear palsy. *Neurology, 41,* 25–30.

Kaneko, C. R. (1996). Effect of ibotenic acid lesions of the omnipause neurons on saccadic eye movements in rhesus macaques. *Journal of Neurophysiology, 75,* 2229–2242.

Keller, E. L. (1977). Control of saccadic eye movements by midline brain stem neurons. In R. Baker & A. Berthoz (Eds.), *Control of gaze by brain stem neurons* (pp. 319–326). Amsterdam: Elsevier.

Keller, E. L., & Missal, M. (2003). Shared brainstem pathways for saccades and smooth-pursuit eye movements. *Annals of the New York Academy of Sciences, 1004,* 29–39.

Kim, E. J., Choi, K. D., Kim, J. E., et al. (2014). Saccadic palsy after cardiac surgery: Serial neuroimaging findings during a 6-year follow-up. *Journal of Clinical Neurology, 10,* 367–370.

King, W. M., Precht, W., & Dieringer, N. (1980). Afferent and efferent connections of cat omnipause neurons. *Experimental Brain Research, 38,* 395–403.

Kipfer, S., Jung, S., Lemke, J. R., et al. (2013). Novel CACNA1A mutation(s) associated with slow saccade velocities. *Journal of Neurology, 260,* 3010–3014.

Kirkham, T. H., & Kamin, D. F. (1974). Slow saccadic eye movements in Wilson's disease. *Journal of Neurology, Neurosurgery, and Psychiatry, 37,* 191–194.

Klostermann, W., Zuhlke, C., Heide, W., Kompf, D., & Wessel, K. (1997). Slow saccades and other eye movement disorders in spinocerebellar atrophy type 1. *Journal of Neurology, 244,* 105–111.

Knox, D. L., Green, W. R., Troncoso, J. C., Yardley, J. H., Hsu, J., & Zee, D. S. (1995). Cerebral ocular Whipple's disease: A 62-year odyssey from death to diagnosis. *Neurology, 45,* 617–625.

Kowler, E., & Blaser, E. (1995). The accuracy and precision of saccades to small and large targets. *Vision Research, 35,* 1741–1754.

Langer, T. P., & Kaneko, C. R. (1983). Efferent projections of the cat oculomotor reticular omnipause neuron region: An autoradiographic study. *The Journal of Comparative Neurology, 217,* 288–306.

Langer, T., Kaneko, C. R., Scudder, C. A., & Fuchs, A. F. (1986). Afferents to the abducens nucleus in the monkey and cat. *The Journal of Comparative Neurology, 245,* 379–400.

Lasker, A. G., Zee, D. S., Hain, T. C., Folstein, S. E., & Singer, H. S. (1987). Saccades in Huntington's disease: Initiation defects and distractibility. *Neurology, 37,* 364–370.

Lasker, A. G., Zee, D. S., Hain, T. C., Folstein, S. E., & Singer, H. S. (1988). Saccades in Huntington's disease: Slowing and dysmetria. *Neurology, 38,* 427–431.

Lee, A. G. (2002). Whipple disease with supranuclear ophthalmoplegia diagnosed by polymerase chain reaction of cerebrospinal fluid. *Journal of Neuro-Ophthalmology, 22,* 18–21.

Leigh, R. J., & Kennard, C. (2004). Using saccades as a research tool in the clinical neurosciences. *Brain, 127,* 460–477.

Leigh, R. J., & Riley, D. E. (2000). Eye movements in parkinsonism: It's saccadic speed that counts. *Neurology, 54,* 1018–1019.

Leigh, R. J., & Zee, D. S. (2015). *The neurology of eye movements* (5th ed.). New York: Oxford University Press.

Leigh, R. J., Newman, S. A., Folstein, S. E., Lasker, A. G., & Jensen, B. A. (1983). Abnormal ocular motor control in Huntington's disease. *Neurology, 33,* 1268–1275.

Leigh, R. J., Parhad, I. M., Clark, A. W., Buettner-Ennever, J. A., & Folstein, S. E. (1985). Brainstem findings in Huntington's disease. Possible mechanisms for slow vertical saccades. *Journal of the Neurological Sciences, 71,* 247–256.

Litvan, I., Mangone, C. A., McKee, A., et al. (1996). Natural history of progressive supranuclear palsy (Steele-Richardson-Olszewski syndrome) and clinical predictors of survival: A clinicopathological study. *Journal of Neurology, Neurosurgery, and Psychiatry, 60,* 615–620.

Lloyd-Smith Sequeira, A., Rizzo, J. R., & Rucker, J. C. (2017). Clinical approach to supranuclear brainstem saccadic gaze palsies. *Frontiers in Neurology, 8*, 429.

Louis, E. D., Lynch, T., Kaufmann, P., Fahn, S., & Odel, J. (1996). Diagnostic guidelines in central nervous system Whipple's disease. *Annals of Neurology, 40*, 561–568.

Lueck, C. J., Crawford, T. J., Hansen, H. C., & Kennard, C. (1991). Increase in saccadic peak velocity with increased frequency of saccades in man. *Vision Research, 31*, 1439–1443.

Mays, L. E., & Morrissee, D. W. (1993). Activity of omnipause neurons during blinks. *Society for Neuroscience – Abstracts, 19*, 1404.

McElligott, J. G., & Spencer, R. F. (2000). Neuropharmacological aspects of the vestibulo-ocular reflex. In J. H. Anderson & A. J. Beitz (Eds.), *Neurochemistry of the vestibular system* (pp. 199–222). Boca Raton: CRC Press.

Miura, K., & Optican, L. M. (2003). Membrane properties of medium-lead burst neurons may contribute to dynamical properties of saccades. In *Proceedings of 1st international IEEE EMBS conference on neural engineering* (pp. 20–23). https://doi.org/10.1109/CNE.2003.1196745

Miura, K., & Optican, L. M. (2006). Membrane channel properties of premotor excitatory burst neurons may underlie saccade slowing after lesions of omnipause neurons. *Journal of Computational Neuroscience, 20*, 25–41.

Mokri, B., Ahlskog, J. E., Fulgham, J. R., & Matsumoto, J. Y. (2004). Syndrome resembling PSP after surgical repair of ascending aorta dissection or aneurysm. *Neurology, 62*, 971–973.

Moschovakis, A. K., Scudder, C. A., & Highstein, S. M. (1990). A structural basis for Hering's law: Projections to extraocular motoneurons. *Science, 248*, 1118–1119.

Moschovakis, A. K., Scudder, C. A., & Highstein, S. M. (1991a). Structure of the primate oculomotor burst generator. I. Medium-lead burst neurons with upward on-directions. *Journal of Neurophysiology, 65*, 203–217.

Moschovakis, A. K., Scudder, C. A., Highstein, S. M., & Warren, J. D. (1991b). Structure of the primate oculomotor burst generator. II. Medium-lead burst neurons with downward on-directions. *Journal of Neurophysiology, 65*, 218–229.

Moscovich, M., Okun, M. S., Favilla, C., et al. (2015). Clinical evaluation of eye movements in spinocerebellar ataxias: A prospective multicenter study. *Journal of Neuro-Ophthalmology, 35*, 16–21.

Moss, H. E., McCluskey, L., Elman, L., et al. (2012). Cross-sectional evaluation of clinical neuro-ophthalmic abnormalities in an amyotrophic lateral sclerosis population. *Journal of the Neurological Sciences, 314*, 97–101.

Murphy, M. A., Friedman, J. H., Tetrud, J. W., & Factor, S. A. (2005). Neurodegenerative disorders mimicking progressive supranuclear palsy: A report of three cases. *Journal of Clinical Neuroscience, 12*, 941–945.

Nakao, S., Curthoys, I. S., & Markham, C. H. (1980). Direct inhibitory projection of pause neurons to nystagmus-related pontomedullary reticular burst neurons in the cat. *Experimental Brain Research, 40*, 283–293.

Nandipati, S., Rucker, J. C., & Frucht, S. J. (2013). Progressive supranuclear palsy-like syndrome after aortic aneurysm repair: A case series. *Tremor and other Hyperkinetic Movements, 3*. https://doi.org/10.7916/D8N29VNW

Nerrant, E., & Tilikete, C. (2017). Ocular motor manifestations of multiple sclerosis. *Journal of Neuro-Ophthalmology, 37*, 332–340.

Oh, A. K., Jacobson, K. M., Jen, J. C., & Baloh, R. W. (2001). Slowing of voluntary and involuntary saccades: An early sign in spinocerebellar ataxia type 7. *Annals of Neurology, 49*, 801–804.

Ohgaki, T., Markham, C. H., Schneider, J. S., & Curthoys, I. S. (1989). Anatomical evidence of the projection of pontine omnipause neurons to midbrain regions controlling vertical eye movements. *The Journal of Comparative Neurology, 289*, 610–625.

Okuda, B., Yamamoto, T., Yamasaki, M., Maya, K., & Imai, T. (1992). Motor neuron disease with slow eye movements and vertical gaze palsy. *Acta Neurologica Scandinavica, 85*, 71–76.

Onofrj, M., Iacono, D., Luciano, A. L., Armellino, K., & Thomas, A. (2004). Clinically evidenced unilateral dissociation of saccades and pursuit eye movements. *Journal of Neurology, Neurosurgery, and Psychiatry, 75*, 1048–1050.

Optican, L. M., & Quaia, C. (2002). Distributed model of collicular and cerebellar function during saccades. *Annals of the New York Academy of Sciences, 956*, 164–177.

Optican, L. M., Rucker, J. C., Keller, E. L., & Leigh, R. J. (2008). Mechanism of interrupted saccades in patients with late-onset Tay-Sachs disease. *Progress in Brain Research, 171*, 567–570.

Oskarsson, B., Pelak, V., Quan, D., Hall, D., Foster, C., & Galetta, S. (2008). Stiff eyes in stiff-person syndrome. *Neurology, 71*, 378–380.

Peeters, E., Vanacker, P., Woodhall, M., Vincent, A., Schrooten, M., & Vandenberghe, W. (2012). Supranuclear gaze palsy in glycine receptor antibody-positive progressive encephalomyelitis with rigidity and myoclonus. *Movement Disorders, 27*, 1830–1832.

Pelisson, D., & Prablanc, C. (1988). Kinematics of centrifugal and centripetal saccadic eye movements in man. *Vision Research, 28*, 87–94.

Pensiero, S., Accardo, A., Pittis, M. G., Ciana, G., Bembi, B., & Perissutti, P. (2005). Saccade testing in the diagnosis and treatment of type 3 Gaucher disease. *Neurology, 65*, 1837.

Petrovic, I. N., Martin-Bastida, A., Massey, L., et al. (2013). MM2 subtype of sporadic Creutzfeldt-Jakob disease may underlie the clinical presentation of progressive supranuclear palsy. *Journal of Neurology, 260*, 1031–1036.

Pinkhardt, E. H., Jurgens, R., Becker, W., Valdarno, F., Ludolph, A. C., & Kassubek, J. (2008). Differential diagnostic value of eye movement recording in PSP-parkinsonism, Richardson's syndrome, and idiopathic Parkinson's disease. *Journal of Neurology, 255*, 1916–1925.

Pittock, S. J., Yoshikawa, H., Ahlskog, J. E., et al. (2006). Glutamic acid decarboxylase autoimmunity with brainstem, extrapyramidal, and spinal cord dysfunction. *Mayo Clinic Proceedings, 81*, 1207–1214.

Prasad, S., Ko, M. W., Lee, E. B., Gonatas, N. K., Stern, M. B., & Galetta, S. (2007). Supranuclear vertical gaze abnormalities in sporadic Creutzfeldt-Jakob disease. *Journal of the Neurological Sciences, 253*, 69–72.

Quint, D. J., Cornblath, W. T., & Trobe, J. D. (1993). Multiple sclerosis presenting as Parinaud syndrome. *AJNR American Journal of Neuroradiology, 14*, 1200–1202.

Ramat, S., Somers, J. T., Das, V. E., & Leigh, R. J. (1999). Conjugate ocular oscillations during shifts of the direction and depth of visual fixation. *Investigative Ophthalmology & Visual Science, 40*, 1681–1686.

Rambold, H., Sprenger, A., & Helmchen, C. (2002). Effects of voluntary blinks on saccades, vergence eye movements, and saccade-vergence interactions in humans. *Journal of Neurophysiology, 88*, 1220–1233.

Ranalli, P. J., Sharpe, J. A., & Fletcher, W. A. (1988). Palsy of upward and downward saccadic, pursuit, and vestibular movements with a unilateral midbrain lesion: Pathophysiologic correlations. *Neurology, 38*, 114–122.

Raybourn, M. S., & Keller, E. L. (1977). Colliculoreticular organization in primate oculomotor system. *Journal of Neurophysiology, 40*, 861–878.

Respondek, G., Stamelou, M., Kurz, C., et al. (2014). The phenotypic spectrum of progressive supranuclear palsy: A retrospective multicenter study of 100 definite cases. *Movement Disorders, 29*, 1758–1766.

Revesz, T., Sangha, H., & Daniel, S. E. (1996). The nucleus raphe interpositus in the Steele-Richardson-Olszewski syndrome (progressive supranuclear palsy). *Brain, 119*(Pt 4), 1137–1143.

Rinne, J. O., Lee, M. S., Thompson, P. D., & Marsden, C. D. (1994). Corticobasal degeneration. A clinical study of 36 cases. *Brain, 117*(Pt 5), 1183–1196.

Rivaud-Pechoux, S., Vidailhet, M., Gallouedec, G., Litvan, I., Gaymard, B., & Pierrot-Deseilligny, C. (2000). Longitudinal ocular motor study in corticobasal degeneration and progressive supranuclear palsy. *Neurology, 54*, 1029–1032.

Robinson, D. A. (1970). Oculomotor unit behavior in the monkey. *Journal of Neurophysiology, 33*, 393–403.

Rosini, F., Federighi, P., Pretegiani, E., et al. (2013). Ocular-motor profile and effects of memantine in a familial form of adult cerebellar ataxia with slow saccades and square wave saccadic intrusions. *PLoS One, 8*, e69522.

Rothenberg, S. J., & Selkoe, D. (1981). Specific oculomotor deficit after diazepam. I. Saccadic eye movements. *Psychopharmacology (Berl), 74*, 232–236.

Rottach, K. G., Riley, D. E., DiScenna, A. O., Zivotofsky, A. Z., & Leigh, R. J. (1996). Dynamic properties of horizontal and vertical eye movements in parkinsonian syndromes. *Annals of Neurology, 39*, 368–377.

Rottach, K. G., von Maydell, R. D., Das, V. E., et al. (1997). Evidence for independent feedback control of horizontal and vertical saccades from Niemann-Pick type C disease. *Vision Research, 37*, 3627–3638.

Rottach, K. G., Das, V. E., Wohlgemuth, W., Zivotofsky, A. Z., & Leigh, R. J. (1998). Properties of horizontal saccades accompanied by blinks. *Journal of Neurophysiology, 79*, 2895–2902.

Rowe, D. B., Lewis, V., Needham, M., et al. (2007). Novel prion protein gene mutation presenting with subacute PSP-like syndrome. *Neurology, 68*, 868–870.

Rucker, J. C., Shapiro, B. E., Han, Y. H., et al. (2004). Neuro-ophthalmology of late-onset Tay-Sachs disease (LOTS). *Neurology, 63*, 1918–1926.

Rucker, J. C., Leigh, R. J., Optican, L. M., Keller, E. L., & Büttner-Ennever, J. A. (2008). Ocular motor anatomy in a case of interrupted saccades. *Progress in Brain Research, 171*, 563–566.

Rucker, J. C., Ying, S. H., Moore, W., et al. (2011). Do brainstem omnipause neurons terminate saccades? *Annals of the New York Academy of Sciences, 1233*, 48–57.

Rufa, A., & Federighi, P. (2011). Fast versus slow: Different saccadic behavior in cerebellar ataxias. *Annals of the New York Academy of Sciences, 1233*, 148–154.

Rufa, A., Cerase, A., De Santi, L., et al. (2008). Impairment of vertical saccades from an acute pontine lesion in multiple sclerosis. *Journal of Neuro-Ophthalmology, 28*, 305–307.

Salsano, E., Umeh, C., Rufa, A., Pareyson, D., & Zee, D. S. (2012). Vertical supranuclear gaze palsy in Niemann-Pick type C disease. *Neurological Sciences, 33*, 1225–1232.

Sarva, H., Deik, A., Ullah, A., & Severt, W. L. (2016). Clinical spectrum of stiff person syndrome: A review of recent reports. *Tremor and other Hyperkinetic Movements, 6*, 340.

Schiffmann, R., Fitzgibbon, E. J., Harris, C., et al. (2008). Randomized, controlled trial of miglustat in Gaucher's disease type 3. *Annals of Neurology, 64*, 514–522.

Schiller, P. H., True, S. D., & Conway, J. L. (1980). Deficits in eye movements following frontal eye-field and superior colliculus ablations. *Journal of Neurophysiology, 44*, 1175–1189.

Schneider, R., Chen, A. L., King, S. A., et al. (2011). Influence of orbital eye position on vertical saccades in progressive supranuclear palsy. *Annals of the New York Academy of Sciences, 1233*, 64–70.

Scudder, C. A., Fuchs, A. F., & Langer, T. P. (1988). Characteristics and functional identification of saccadic inhibitory burst neurons in the alert monkey. *Journal of Neurophysiology, 59*, 1430–1454.

Scudder, C. A., Kaneko, C. S., & Fuchs, A. F. (2002). The brainstem burst generator for saccadic eye movements: A modern synthesis. *Experimental Brain Research, 142*, 439–462.

Seemungal, B. M., Faldon, M., Revesz, T., Lees, A. J., Zee, D. S., & Bronstein, A. M. (2003). Influence of target size on vertical gaze palsy in a pathologically proven case of progressive supranuclear palsy. *Movement Disorders, 18*, 818–822.

Shaikh, A. G., Miura, K., Optican, L. M., Ramat, S., Leigh, R. J., & Zee, D. S. (2007). A new familial disease of saccadic oscillations and limb tremor provides clues to mechanisms of common tremor disorders. *Brain, 130*, 3020–3031.

Shaikh, A. G., Wong, A. L., Optican, L. M., Miura, K., Solomon, D., & Zee, D. S. (2010). Sustained eye closure slows saccades. *Vision Research, 50*, 1665–1675.

Shimamura, M., Uyama, E., Hirano, T., et al. (2003). A unique case of sporadic Creutzfeldt-Jacob disease presenting as progressive supranuclear palsy. *Internal Medicine, 42*, 195–198.

Shiozawa, M., Fukutani, Y., Sasaki, K., et al. (2000). Corticobasal degeneration: An autopsy case clinically diagnosed as progressive supranuclear palsy. *Clinical Neuropathology, 19*, 192–199.

Soetedjo, R., Kaneko, C. R., & Fuchs, A. F. (2002). Evidence that the superior colliculus participates in the feedback control of saccadic eye movements. *Journal of Neurophysiology, 87*, 679–695.

Solomon, D., Winkelman, A. C., Zee, D. S., Gray, L., & Büttner-Ennever, J. (2005). Niemann-Pick type C disease in two affected sisters: Ocular motor recordings and brain-stem neuropathology. *Annals of the New York Academy of Sciences, 1039*, 436–445.

Solomon, D., Ramat, S., Tomsak, R. L., et al. (2008a). Saccadic palsy after cardiac surgery: Characteristics and pathogenesis. *Annals of Neurology, 63*, 355–365.

Solomon, D., Ramat, S., Leigh, R. J., & Zee, D. (2008b). A quick look at slow saccades after cardiac surgery: Where is the lesion? *Progress in Brain Research, 171*, 587–590.

Spanaki, C., Latsoudis, H., & Plaitakis, A. (2006). LRRK2 mutations on Crete: R1441H associated with PD evolving to PSP. *Neurology, 67*, 1518–1519.

Spencer, R. F., & Wang, S. F. (1996). Immunohistochemical localization of neurotransmitters utilized by neurons in the rostral interstitial nucleus of the medial longitudinal fasciculus (riMLF) that project to the oculomotor and trochlear nuclei in the cat. *The Journal of Comparative Neurology, 366*, 134–148.

Steele, J. C., Richardson, J. C., & Olszewski, J. (1964). Progressive supranuclear palsy. A heterogeneous degeneration involving the brain stem, basal ganglia and cerebellum with vertical gaze and pseudobulbar palsy, nuchal dystonia and dementia. *Archives of Neurology, 10*, 333–359.

Strassman, A., Highstein, S. M., & McCrea, R. A. (1986a). Anatomy and physiology of saccadic burst neurons in the alert squirrel monkey. I. Excitatory burst neurons. *The Journal of Comparative Neurology, 249*, 337–357.

Strassman, A., Highstein, S. M., & McCrea, R. A. (1986b). Anatomy and physiology of saccadic burst neurons in the alert squirrel monkey. II. Inhibitory burst neurons. *The Journal of Comparative Neurology, 249*, 358–380.

Strassman, A., Evinger, C., McCrea, R. A., Baker, R. G., & Highstein, S. M. (1987). Anatomy and physiology of intracellularly labelled omnipause neurons in the cat and squirrel monkey. *Experimental Brain Research, 67*, 436–440.

Suzuki, Y., Büttner-Ennever, J. A., Straumann, D., Hepp, K., Hess, B. J., & Henn, V. (1995). Deficits in torsional and vertical rapid eye movements and shift of Listing's plane after uni- and bilateral lesions of the rostral interstitial nucleus of the medial longitudinal fasciculus. *Experimental Brain Research, 106*, 215–232.

Takigawa, H., Kitayama, M., Wada-Isoe, K., Kowa, H., & Nakashima, K. (2016). Prevalence of progressive supranuclear palsy in Yonago: Change throughout a decade. *Brain and Behavior: A Cognitive Neuroscience Perspective, 6*, e00557.

Tan, J. H., Goh, B. C., Tambyah, P. A., & Wilder-Smith, E. (2005). Paraneoplastic progressive supranuclear palsy syndrome in a patient with B-cell lymphoma. *Parkinsonism & Related Disorders, 11*, 187–191.

Thurston, S. E., Leigh, R. J., Abel, L. A., & Dell'Osso, L. F. (1984). Slow saccades and hypometria in anticonvulsant toxicity. *Neurology, 34*, 1593–1596.

Tilikete, C., Vighetto, A., Trouillas, P., & Honnorat, J. (2005). Potential role of anti-GAD antibodies in abnormal eye movements. *Annals of the New York Academy of Sciences, 1039*, 446–454.

Ushio, M., Iwasaki, S., Sugasawa, K., & Murofushi, T. (2009). Atypical motor neuron disease with supranuclear vertical gaze palsy and slow saccades. *Auris Nasus Larynx, 36*, 85–87.

Van Gisbergen, J. A., Robinson, D. A., & Gielen, S. (1981). A quantitative analysis of generation of saccadic eye movements by burst neurons. *Journal of Neurophysiology, 45*, 417–442.

Van Opstal, A. J., & Van Gisbergen, J. A. (1987). Skewness of saccadic velocity profiles: A unifying parameter for normal and slow saccades. *Vision Research, 27*, 731–745.

Velazquez-Perez, L., Seifried, C., Santos-Falcon, N., et al. (2004). Saccade velocity is controlled by polyglutamine size in spinocerebellar ataxia 2. *Annals of Neurology, 56*, 444–447.

Villis, T., Hepp, K., Schwarz, U., & Henn, V. (1989). On the generation of vertical and torsional rapid eye movements in the monkey. *Experimental Brain Research, 77*, 1–11.

Wadia, N., Pang, J., Desai, J., Mankodi, A., Desai, M., & Chamberlain, S. (1998). A clinicogenetic analysis of six Indian spinocerebellar ataxia (SCA2) pedigrees. The significance of slow saccades in diagnosis. *Brain, 121*(Pt 12), 2341–2355.

Waitzman, D. M., Silakov, V. L., DePalma-Bowles, S., & Ayers, A. S. (2000). Effects of reversible inactivation of the primate mesencephalic reticular formation. II. Hypometric vertical saccades. *Journal of Neurophysiology, 83*, 2285–2299.

Wallach, A. I., Park, H., Rucker, J. C., & Kaufmann, H. (2017). Supranuclear gaze palsy and horizontal ocular oscillations in Creutzfeldt-Jakob disease. *Neurology, 89*, 749.

Warren, J. D., Scott, G., Blumbergs, P. C., & Thompson, P. D. (2002). Pathological evidence of encephalomyelitis in the stiff man syndrome with anti-GAD antibodies. *Journal of Clinical Neuroscience, 9*, 328–329.

Welsh, J. P., Yuen, G., Placantonakis, D. G., et al. (2002). Why do Purkinje cells die so easily after global brain ischemia? Aldolase C, EAAT4, and the cerebellar contribution to posthypoxic myoclonus. *Advances in Neurology, 89*, 331–359.

West, T., Hu, Y., Verghese, P. B., et al. (2017). Preclinical and clinical development of ABBV-8E12, a humanized anti-tau antibody, for treatment of Alzheimer's disease and other tauopathies. *The Journal of Prevention of Alzheimer's Disease, 4*, 236–241.

Williams, D. R., & Lees, A. J. (2009). Progressive supranuclear palsy: Clinicopathological concepts and diagnostic challenges. *Lancet Neurology, 8*, 270–279.

Williams, D. R., & Lees, A. J. (2010). What features improve the accuracy of the clinical diagnosis of progressive supranuclear palsy-parkinsonism (PSP-P)? *Movement Disorders, 25*, 357–362.

Williams, D. R., de Silva, R., Paviour, D. C., et al. (2005a). Characteristics of two distinct clinical phenotypes in pathologically proven progressive supranuclear palsy: Richardson's syndrome and PSP-parkinsonism. *Brain, 128*, 1247–1258.

Williams, D. R., Hadeed, A., al-Din, A. S., Wreikat, A. L., & Lees, A. J. (2005b). Kufor Rakeb disease: Autosomal recessive, levodopa-responsive parkinsonism with pyramidal degeneration, supranuclear gaze palsy, and dementia. *Movement Disorders, 20*, 1264–1271.

Wu, C., Chen, D. B., Feng, L., et al. (2017). Oculomotor deficits in spinocerebellar ataxia type 3: Potential biomarkers of preclinical detection and disease progression. *CNS Neuroscience & Therapeutics, 23*, 321–328.

Ying, S. H., Choi, S. I., Lee, M., et al. (2005). Relative atrophy of the flocculus and ocular motor dysfunction in SCA2 and SCA6. *Annals of the New York Academy of Sciences, 1039*, 430–435.

Yoshida, K., Iwamoto, Y., Chimoto, S., & Shimazu, H. (1999). Saccade-related inhibitory input to pontine omnipause neurons: An intracellular study in alert cats. *Journal of Neurophysiology, 82*, 1198–1208.

Zee, D. S., Optican, L. M., Cook, J. D., Robinson, D. A., & Engel, W. K. (1976). Slow saccades in spinocerebellar degeneration. *Archives of Neurology, 33*, 243–251.

Zuhlke, C., Mikat, B., Timmann, D., Wieczorek, D., Gillessen-Kaesbach, G., & Burk, K. (2015). Spinocerebellar ataxia 28: A novel AFG3L2 mutation in a German family with young onset, slow progression and saccadic slowing. *Cerebellum Ataxias, 2*, 19.

Fusion Maldevelopment (Latent) Nystagmus: How Insights from Nonhuman Primate Experiments Have Benefitted Clinical Practice

Lawrence Tychsen

Abstract Binocular fusion is blending of corresponding images from each eye to form a single percept. Maldevelopment of fusion (binocular non-correspondence) in infancy causes a specific type of lifelong ocular instability: fusion maldevelopment nystagmus (FMN). Because fusion maldevelopment – in the form of strabismus and amblyopia – is common, FMN is the most prevalent pathologic nystagmus encountered in clinical practice. Experiments on nonhuman primates (NHP) with strabismus and amblyopia have revealed that loss of binocular connections within area V1 (striate cortex) in the first months of life is the necessary and sufficient cause of FMN. The severity of FMN increases with greater losses of V1 connections. No manipulation of brainstem motor pathways is required. The binocular maldevelopment originating in area V1 is passed on to downstream, extrastriate regions of cerebral cortex that drive conjugate gaze (notably MSTd). Conjugate gaze is stable when MSTd neurons of the right vs. left cerebral hemisphere have balanced, binocular activity. Fusion maldevelopment causes unbalanced, monocular activity. If input from one eye dominates and the other is suppressed, MSTd in one hemisphere becomes more active. Downstream projections to the ipsilateral nucleus of the optic tract (NOT) drive the eyes conjugately to that side. The unbalanced MSTd drive is evident as nasalward slow-phase nystagmus when viewing with either eye.

Experiments on NHPs have provided the functional-structural correlations needed to explain the pathophysiology of LN. The translational value of NHP studies cannot be overstated. The NHP studies provide pivotal facts necessary to explain one of the most common ocular motor disorders encountered by eye care providers worldwide. The NHP studies have motivated pediatric ophthalmologists to repair fusion earlier in infancy (Tychsen, J AAPOS 9:510–21, 2005), thereby preventing FMN or reducing its severity.

L. Tychsen (✉)
St. Louis Children's Hospital at Washington University Medical Center,
Department of Ophthalmology and Visual Sciences,
St. Louis, MO, USA
e-mail: tychsen@wustl.edu

© Springer Nature Switzerland AG 2019
A. Shaikh, F. Ghasia (eds.), *Advances in Translational Neuroscience of Eye Movement Disorders*, Contemporary Clinical Neuroscience,
https://doi.org/10.1007/978-3-030-31407-1_13

Keywords Amblyopia · Strabismus · Binocular control · Double vision · Visual cortex

1 FMN Signs, Terminology, and Adverse Effects

FMN is characterized by a conjugate, horizontal slow-phase drift of eye position that is directed nasalward with respect to the viewing eye (Tychsen 2007; Dell'Osso et al. 1979; Dell'Osso 1985). When viewing switches from eye to eye, the directions of the slow phases reverse instantaneously: leftward when the right eye is fixating and rightward when the left eye is fixating (Fig. 1). The severity of the nystagmus

RE viewing

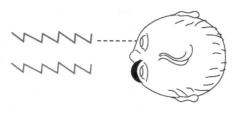

LE viewing

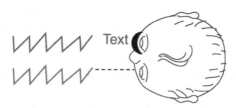

BE viewing

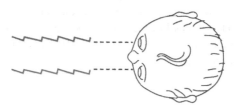

Fig. 1 Fusion maldevelopment nystagmus in child, as in NHP. *RE viewing*: viewing with the right eye (left eye covered), both eyes have a nasalward slow-phase drift with respect to the viewing RE, followed by temporalward fast phases. The movement of the two eyes is similar (conjugate). *LE viewing*: The direction of the nystagmus reverses instantaneously when viewing changes to the LE (RE covered), so that the slow phase is nasalward with respect to the viewing LE. *BE viewing*: Velocity of the slow phase and the amplitude of the fast phase reduce when both eyes (BE) are uncovered. In the child shown, the RE is dominant and the LE – though open – is suppressed partially as a consequence of strabismus and/or amblyopia

(and its conspicuity during clinical examination) increases when one eye is covered, hence the entrenched clinical term "latent nystagmus (LN)." When the nystagmus is evident with both eyes open, it was called "manifest LN (MLN)."

Infantile esotropia (convergent strabismus) is the leading clinical association with FMN. But any disorder that perturbs development of binocular fusion in infancy causes FMN, such as monocular or severe binocular deprivation of spatial vision (amblyopia) (Tychsen and Vision 1992; Tusa et al. 2002). In 2001, the NIH Committee on Eye Movement and Strabismus (CEMAS) classification (N.E.I. / N.I.H. Committee 2001) recommended that the terms LN/MLN be replaced by the etiologic descriptor FMN.

FMN is distinguished easily during clinical examination from congenital ("idiopathic infantile") nystagmus (CN, or the infantile nystagmus syndrome (INS)) by the feature of instantaneous reversal of direction with alternating fixation. By eye movement recording, it is distinguished also in waveform. The waveform of FMN is of decreasing velocity and linear trajectory, whereas that of CN/INS is of increasing velocity and pendular trajectory (Dell'Osso et al. 1979). Eye movement recordings or high magnification inspection of patients with a slit-lamp biomicroscope or ophthalmoscope reveals a superimposed small torsional movement. A minority of patients (Dell'Osso 1985) and NHP (Wong and Tychsen 2002) with FMN may show a small pendular component.

Contributions to our understanding of the clinical features of LN have been made by Dell'Osso, Daroff, Abel, and their colleagues (Dell'Osso et al. 1979, 1983), who have also clarified the historical origins of FMN/LN's various terms. In 1872 Faucon (Faucon 1872) first described what we now appreciate as MLN. In 1912, Fromaget and Fromaget (Fromaget and Fromaget 1912) introduced the term "nystagmus latent." These early reports of LN were reviewed by Sorsby (Sorsby 1931) in 1931. The oxymoron "manifested latent" nystagmus was introduced by Kestenbaum (Kestenbaum 1961) in 1947, who emphasized that LN is often observed in patients with strabismus when they view with both eyes open.

FMN degrades visual acuity due to "retinal slip" (instability, oscillation) of gaze (Tychsen 2012; Pirdankar and Das 2016). The nystagmus-imposed degradation of visual function is superimposed on the disruptions caused by strabismus and/or amblyopia. Patients and NHP with FMN must use abnormal head postures (turns and tilts, ocular torticollis) to help damp FMN and stabilize vision. FMN also causes other subtypes of vertical and horizontal strabismus behaviors, adding to the baseline strabismus; in clinical parlance: "dissociated vertical and horizontal deviations (DVD, DHD)." (Tychsen 2012; Pirdankar and Das 2016; Das 2016) The constellation of visual degradation, wandering eyes, subnormal binocularity, and odd head postures impacts quality of life (McLean et al. 2016). School performance, sports participation, vocational choices, and psychosocial interactions are impaired.

2 Innate Nasalward Visual Cortex Biases Disappear if Fusion Develops Normally

Behavioral studies have shown that the postnatal development of binocular sensory and motor functions in normal infant NHP parallels that of normal infant humans, but on a compressed time scale; 1 week of NHP development approximates 1 month of human (Atkinson 1979; Boothe et al. 1985; O'Dell and Boothe 1997; Brown et al. 1998). Binocular disparity sensitivity and binocular fusion are absent in human and NHP neonates. Stereopsis emerges abruptly in humans during the first 3–5 months of postnatal life (Fox et al. 1980; Birch et al. 1982, 1983, 1985; Gwiazda et al. 1989) and in NHP during the first 3–5 weeks (O'Dell and Boothe 1997), achieving adult-like levels of sensitivity.

V1 horizontal axonal connections are key components of fusion development and maldevelopment (Fig. 2). Binocularity in NHP begins with horizontal connections between V1 ocular dominance columns (ODCs) of opposite ocularity (Hubel and Wiesel 1977; Tychsen and Burkhalter 1995; Tychsen et al. 2004). These connections are immature in the first weeks of life, conveying crude, weak binocular responses (Chino et al. 1996, 1997; Hatta et al. 1998). Maturation of binocular

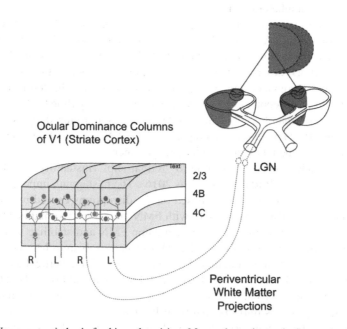

Fig. 2 Neuroanatomic basis for binocular vision. Monocular retinogeniculate projections from the left eye (temporal retina-nasal visual hemifield) and right eye (nasal retina-temporal hemifield) remain segregated up to and within the input layer of ocular dominance columns (ODCs) in V1, layer 4C (striate visual cortex). Binocular vision is made possible by horizontal connections between ODCs of opposite ocularity in upper layers 4B and 2/3 (as well as lower layers 5/6, not shown). RE inputs, red; LE inputs, blue

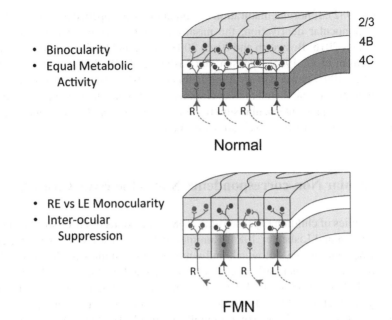

- Binocularity
- Equal Metabolic
  Activity

Normal

- RE vs LE Monocularity
- Inter-ocular
  Suppression

FMN

Fig. 3 Horizontal connections for binocular vision and ODC metabolic activity in V1 of normal NHP vs. NHP with FMN. V1 of normal primates is characterized by (**a**) a rich plexus of binocular connections between R- and L-eye ODCs in layers 2–4B and (**b**) uniform metabolic activity in R- and L-eye ODCs layer 4C. In NHP with FMN, the connections are predominantly monocular due to a paucity of binocular connections. Interocular suppression caused by binocular non-correspondence results in higher metabolic activity in ODCs of dominant eye, here shown as L eye. RE inputs and connections, red; LE, blue; binocular, violet

connections requires corresponding (synchronous) activity between right-eye and left-eye geniculostriate inputs (Lowel and Singer 1992; Lowel and Engelmann 2002). Non-correspondence of inputs causes loss of horizontal connections over a period of days in V1 of kittens (Lowel and Singer 1992; Trachtenberg and Stryker 2001). The inference from NHP experiments and clinical studies is that similar losses occur over a period of weeks in V1 of NHP and over a period of months in V1 of children. Binocular non-correspondence also promotes interocular suppression as a further hindrance to fusion (Fig. 3) (Tychsen 2007).

In the first months of life in humans and weeks of life in NHP, monocular motion VEPs reveal a nasotemporal asymmetry (Norcia et al. 1991; Norcia 1996; Brown and Norcia 1997; Birch et al. 2000). Monocular preferential-looking testing reveals greater perceptual sensitivity to nasalward motion (Bosworth and Birch 2003). Monocular pursuit and optokinetic tracking reveal biases favoring nasalward target motion (Atkinson 1979; Naegele and Held 1982; Wattam-Bell et al. 1987; Jacobs et al. 1994; Tychsen 2001). These nasalward motion biases are pronounced before onset of sensorial fusion and stereopsis, but systematically diminish thereafter. They are retained in subtle form in normal adult humans and can be unmasked using contrived, monocular stimuli (van Dalen 1981; van Die and

Collewijn 1982). If normal maturation of binocularity is impeded by eye misalignment or monocular deprivation, the nasalward biases persist and become pronounced (Das 2016; Bosworth and Birch 2003; Schor and Levi 1980; Maurer et al. 1983; Tychsen et al. 1985; Tychsen and Lisberger 1986; Tychsen et al. 1996a; Westall et al. 1998; Fawcett et al. 2000; Joshi et al. 2017). The nasalward gaze bias is the key feature of the fusion maldevelopment syndrome. Other common features are loss of stereopsis, interocular suppression, strabismus, amblyopia, and smaller amplitude torsional/vertical oscillations of the eyes.

3 Binocular Non-correspondence Starts the FMN Cascade

Clinical studies of children (Tychsen et al. 1985) and adults (Dell'Osso et al. 1979; 1983; Tychsen and Lisberger 1986; Kommerell 1982) with FMN have inspired a series of behavioral, physiological, and neuroanatomic studies in NHPs who had FMN associated with naturally occurring (Tychsen and Burkhalter 1995; Tychsen et al. 2004; Matsumoto et al. 1991; Tychsen et al. 1996b; Tychsen and Boothe 1996; Tychsen et al. 1996c; Tychsen and Burkhalter 1997; Tychsen et al. 2000; Tychsen and Scott 2003) or experimentally induced (Tychsen 2007; Tusa et al. 2002; Das 2016; Joshi et al. 2017; Tusa et al. 1991; Tychsen et al. 1991, 1996d; Kiorpes et al. 1996; Tychsen et al. 1999; Horton et al. 1999; Tusa et al. 2001; Angelucci et al. 2002; Richards et al. 2008) infantile strabismus. The common finding of these experiments is that the prevalence and severity of FMN (Fig. 4) correlate systematically with the age of onset and duration of binocular non-correspondence in infancy.

The most common clinical cause of binocular non-correspondence is strabismus, which in human infants is overwhelmingly esotropic (convergent) (Tychsen 1999). Early-onset esotropia exceeds exotropia by a ratio of 9:1. Esotropia is also the most common form of naturally occurring strabismus in NHPs (Kiorpes and Boothe 1981; Kiorpes et al. 1985). It may therefore be considered the paradigmatic form of strabismus in human and NHP. However, any prolonged deprivation of normal binocular experience in early infancy can cause binocular non-correspondence, including monocular or binocular deprivation (unilateral or bilateral amblyopia) (Tusa et al. 1991, 2001, 2002). An important fact to note is that loss of spatial vision is not required; the majority of human and NHP infants with strabismus alternate fixation initially and have no amblyopia (Panel AAOQOCCPO 1992). The necessary and sufficient factor is binocular non-correspondence, not lack of sharp visual acuity.

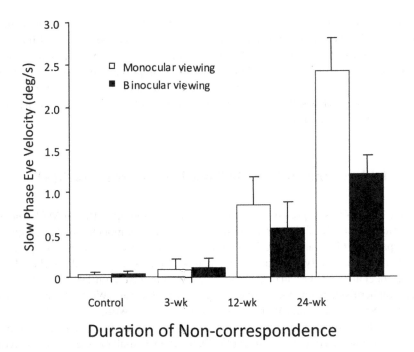

Duration of Non-correspondence

Fig. 4 Slow-phase eye velocity (SPEV) of FMN during monocular vs. binocular viewing as a function of duration of binocular non-correspondence in four groups of NHPs. NHPs were reared wearing prism goggles to induce optical strabismus for durations of 3, 12, or 24 weeks after birth. Control NHPs were reared with plano (no prism) goggles. All NHPs in the 12- and 24-week groups manifested permanent, convergent (esotropic) strabismus after cessation of goggle wear. The severity of SPEV FMN was more pronounced under conditions of monocular vs. binocular (both eyes open) viewing. Means ± SD; pooled responses of the right and left eye. (Adapted from (Kiorpes and Boothe 1981))

4 Translational Value of FMN Experiments in NHP

Using an NHP model of infantile strabismus, we found that non-correspondence durations that exceed the equivalent of 3 months in human infant result in an FMN prevalence of 100% (Tychsen 2007; Richards et al. 2008; Wong et al. 2003). Perturbing these inputs from week 1 of life causes FMN, but delaying the perturbation to the time of onset of normal fusion and stereopsis (the equivalent of age 2–4 months in human) is equally effective (Foeller et al. 2008). The severity of the ocular motor defect – measured as either slow-phase eye velocity or nystagmus intensity – increased systematically with duration of non-correspondence (Richards et al. 2008). The primate model allowed us to (a) impose strict periods of binocular non-correspondence and (b) record precisely large blocks of trials to compare the behavioral deficits. This information would be difficult or impossible to obtain from human infants in clinical trials. Extrapolated to humans, these results imply that timely surgery for infantile strabismus prevents or lessens nystagmus, while delayed surgery promotes nystagmus.

The severity of FMN corresponds to the severity of loss of binocular connections between ODCs of opposite ocularity in visual area V1 and the severity of interocular suppression (Tychsen 2007; Tychsen et al. 2008). Area V1 feeds forward to extrastriate areas (areas MT/MST) known to be important for gaze-holding and gaze-tracking, such as smooth pursuit, OKN, and the short-latency ocular following response (OFR) (Dursteler and Wurtz 1988; Komatsu and Wurtz 1988a; Kawano et al. 1994; Inoue et al. 1998).

5   Maldevelopment in V1 Is Passed on to Areas MT/MST

Visual areas V1, V2 (pre-striate cortex), MT (medial temporal), and MST (medial superior temporal) of the cerebral cortex are major components of the conjugate gaze pathway (Tusa and Ungerleider 1988). Each of these areas in normal primates contains directionally selective, binocular neurons (Poggio and Fischer 1977; Albright et al. 1984; Orban et al. 1986; DeAngelis and Newson 1999). MST in each cerebral hemisphere encodes ipsiversive gaze (Dursteler and Wurtz 1988; Newsome et al. 1988; Komatsu and Wurtz 1988b; Yamasaki and Wurtz 1991). MST in turn projects downstream to the brainstem visuomotor nuclei that generate eye movements: the nucleus of the optic tract (NOT), medial vestibular nucleus, and interconnected abducens and ocular motor nuclei (Tusa and Ungerleider 1988; Mustari et al. 1994). In primates, subcortical inputs to NOT may play a minor role (Tychsen and Vision 1992; Tusa et al. 2002; Hoffmann and Distler 1992). But the dominant pathway is cerebral – from MST to brainstem. The dominant role of the cortical pathway, and the minimal role of a subcortical pathway, is reinforced by studies of children. Neuroimaging of visual cortex, combined with eye movement recordings, has shown absence of visually driven pursuit or OKN in cerebrally blind infants (Tychsen 1996; Werth 2007).

One mechanism for the gaze-holding asymmetry would be over-representation of nasalward neurons within visual areas V1 through MT in the immature/strabismic cortex. However, directional and binocular responses of neurons in V1, V2, and MT have been investigated in infant monkeys, as well as in monkeys with early-onset strabismus, and no over-representation of neurons selective for nasalward motion has been found (Hatta et al. 1998; Tychsen et al. 1999; Tychsen 1996; Werth 2007). In the strabismic animals, binocular (excitatory) responses are reduced, and interocular suppression is increased (Watanabe et al. 2005; Mustari et al. 2008; Endo et al. 2000). These physiological abnormalities have neuroanatomic correlates. In V1 of strabismic monkeys, binocular connections are deficient (Tychsen and Burkhalter 1995; Tychsen et al. 2004), and interocular metabolic activity is suppressed (Tychsen and Burkhalter 1997; Horton et al. 1999; Wong et al. 2005).

6 An Innate Nasalward Monocular Bias Is Engrained by Binocular Non-correspondence: Hypothetical Signal Flow for FMN

A key implication emerging from the NHP studies is that the visual cortex in each cerebral hemisphere is wired innately for nasalward motion. The innate wiring is monocular. To generate temporalward gaze-holding, signals must traverse binocular connections, unimpeded by interocular suppression. If normal binocularity fails to develop, the system remains predominantly monocular and asymmetric, incapable of driving temporalward gaze-holding or robust, temporalward pursuit/OKN (Tusa et al. 2002; Das 2016; Tychsen et al. 1985; Tychsen and Lisberger 1986; Joshi et al. 2017; Kiorpes et al. 1996; Kiorpes and Boothe 1981; Mustari et al. 2008). FMN may be interpreted as an abnormal monocular bias added on to a normal, ipsiversive hemispheric gaze bias.

Figure 5 illustrates a mechanism for FMN, showing the circuit mediating gaze-holding in primates and the role of binocular connections. Shaded structures indicate less active visual and motor neurons caused by occlusion of one eye and/or interocular suppression. The circuit on the left depicts the pathways and visuomotor component structures in a primate with FMN.

The flow is from top to bottom, starting from the monocular visual field of the fixating (or viewing) right eye. The nasal and temporal visual fields (VF) in primates are unequal in area, with a bias favoring the larger, temporal hemifield. Retinal ganglion cell fibers (RGC) from the nasal and temporal hemiretinas decussate at the optic chiasm (chi), synapse at the lateral geniculate nucleus (LGN), and project to alternating, monocular right-eye (RE) and left-eye (LE) ODCs in V1. During development, RGCs from the nasal retina outnumber and establish connections earlier than those from the temporal retina (not shown). The LGN lamina corresponding to the nasal retina – lamina 1, 3, and 5 (also not shown) – contain more neurons and develop earlier than those from the temporal hemiretina, 2, 4, and 6. Within the LGN, the neurons remain monocular, with little or no binocular interlaminar interaction.

The monocular bias, favoring nasal hemiretina inputs, is passed on to the ODCs of area V1. In each V1, ODCs representing the nasal hemiretina (temporal visual hemifield) occupy slightly more cortical territory (in the diagram are larger) than those representing the temporal hemiretinas (nasal hemifield), but each ODC contains neurons sensitive to nasalward (in this case rightward) versus temporalward

---

Fig. 5 (continued) motion are wired innately – through monocular connections – to the pursuit area. *Normal:* Binocular connections are present, linking neurons with similar orientation/directional preferences within ODCs of opposite ocularity (diagonal lines between columns). Viewing with the right eye, visual neurons preferring nasally directed motion project to the left hemisphere pursuit area; visual neurons preferring temporally directed motion project to the right hemisphere pursuit area. Temporally directed visual area neurons gain access to pursuit area neurons only through binocular connections. Call = corpus callosum, through which visual area neurons in each hemisphere project to opposite pursuit area. Bold lines = active neurons and neuronal projections. (Adapted from (Kiorpes and Boothe 1981))

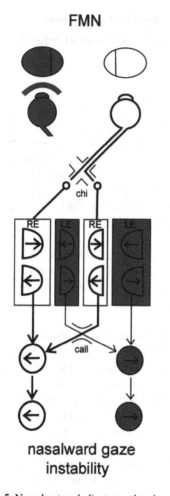 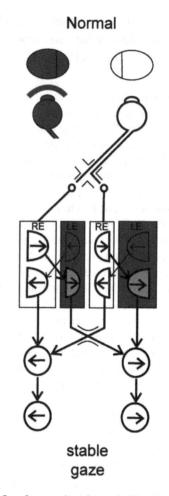

Fig. 5 Neural network diagrams showing visual signal flow for pursuit and gaze-holding in strabismic vs. normal primates. Paucity of mature binocular connections explains behavioral asymmetries evident as asymmetric pursuit/OKN and latent fixation nystagmus. Note that in all primates, pursuit area neurons in each hemisphere encode ipsilaterally directed pursuit. Signal flow is initiated by a moving stimulus in the monocular visual field, which evokes a response in visual area neurons (i.e., V1/MT). Each eye at birth has access – through innate, monocular connections – to the pursuit area neurons (e.g., MSTd) of the contralateral hemisphere. Access to pursuit neurons of the ipsilateral hemisphere requires mature, binocular connections. *FMN:* Moving from top to bottom, starting with target motion in monocular visual field of right eye. Retinal ganglion cell fibers from the nasal and temporal hemiretinas (eye) decussate at the optic chiasm (chi), synapse at the lateral geniculate nucleus (LGN), and project to alternating rows of ODCs in V1 (visual area rectangles). In each V1, ODCs representing the nasal hemiretinas (temporal visual hemifield) occupy slightly more cortical territory than those representing the temporal hemiretinas (nasal hemifield), but each ODC contains neurons sensitive to nasally directed vs. temporally directed motion (half circles shaped like the matching hemifield; arrows indicate directional preference). Visual area neurons (including those beyond V1 in area MT) are sensitive to both nasally directed and temporally directed motion, but only those encoding nasally directed

(leftward) visual motion. Receptive field neurons in V1 and MT are simplified here as half circles to match their corresponding hemifields. The arrows indicate the directional preference of the neurons. Note that visual area neurons (including those beyond V1 in area MT) are sensitive to both nasalward and temporalward motion (Hatta et al. 1998; Kiorpes et al. 1996; Watanabe et al. 2005), but only those encoding nasalward motion are wired innately, through monocular connections, to gaze (eye motion) neurons in area MST (congregated in the dorsal-medial portion or MSTd). MSTd in each cerebral hemisphere encodes ipsiversive gaze (Dursteler and Wurtz 1988; Newsome et al. 1988; Komatsu and Wurtz 1988b; Yamasaki and Wurtz 1991), which is nasalward gaze in relation to the contralateral eye (leftward for MST in the left cerebral hemisphere and rightward for MST in the right hemisphere) (Mustari et al. 2008).

The only difference between the FMN primate's visual cortex and the normal primate's visual cortex is a paucity of binocular horizontal connections (Tychsen et al. 2004, 2008) (compounded by interocular suppression (Tychsen et al. 1996c; Horton et al. 1999; Wong et al. 2005)). The paucity is depicted as a lack of diagonal right-eye ODC to left-eye ODC connections, absent in the FMN cortex (left side of figure) and present in the normal cortex (right side of figure). In the cortex of normal primates, access to MSTd for temporalward gaze requires binocular connections to homoversive neurons within neighboring ODCs that have opposite ocularity (LE ODC neurons when viewing with the RE). The pathway from V1/MT to MSTd requires efferent projections through the splenium of the corpus callosum (labeled as "call") (Pasik and Pasik 1964; Hoffmann et al. 1992).

MSTd efferents project to the ipsilateral brainstem nucleus of the optic tract (NOT) (Mustari et al. 1994; Mustari and Fuchs 1990) and to ipsiversive-related brainstem structures (medial vestibular nucleus, dorsolateral pontine nucleus, and ocular motor nuclei of cranial nerves 3 and 6).

References

Albright, T. D., Desimone, R., & Gross, C. G. (1984). Columnar organization of directionally selective cells in visual area MT of the macaque. *Journal of Neurophysiology, 51*, 16–31.

Angelucci, A., Levitt, J. B., Walton, E. J., et al. (2002). Circuits for local and global signal integration in primary visual cortex. *The Journal of Neuroscience, 22*, 8633–8646.

Atkinson, J. (1979). Development of optokinetic nystagmus in the human infant and monkey infant: an analogue to development in kittens. In R. D. Freeman (Ed.), *Developmental neurobiology of vision* (pp. 277–287). New York: Plenum.

Birch, E. E., Gwiazda, J., & Held, R. (1982). Stereoacuity development for crossed and uncrossed disparities in human infants. *Vision Research, 22*, 507–513.

Birch, E. E., Gwiazda, J., & Held, R. (1983). The development of vergence does not account for the onset of stereopsis. *Perception, 12*, 331–336.

Birch, E. E., Shimojo, S., & Held, R. (1985). Preferential-looking assessment of fusion and stereopsis in infants aged 1 to 6 months. *Investigative Ophthalmology & Visual Science, 26*, 366–370.

Birch, E. E., Fawcett, S., & Stager, D. (2000). Co-development of VEP motion response and binocular vision in normal infants and infantile esotropes. *Investigative Ophthalmology & Visual Science, 41*, 1719–1723.

Boothe, R. G., Dobson, V., & Teller, D. Y. (1985). Postnatal development of vision in human and nonhuman primates. *Annual Review of Neuroscience (Palo Alto, CA), 8*, 495–546.

Bosworth, R. G., & Birch, E. E. (2003). Nasal-temporal asymmetries in motion sensitivity and motion VEPs in normal infants and patients with infantile esotropia. *AAPOS Abstr, 38*, 61.

Brown, R. J., & Norcia, A. M. (1997). A method for investigating binocular rivalry in real-time with the steady-state VEP. *Vision Research, 37*, 2401–2408.

Brown, R. J., Wilson, J. R., Norcia, A. M., et al. (1998). Development of directional motion symmetry in the monocular visually evoked potential of infant monkeys. *Vision Research, 38*, 1253–1263.

Chino, Y., Smith, E. L., Hatta, S., et al. (1996). Suppressive binocular interactions in the primary visual cortex (V1) of infant rhesus monkeys. *Society for Neuroscience – Abstracts, 255*, 13.

Chino, Y. M., Smith, E. L., III, Hatta, S., et al. (1997). Postnatal development of binocular disparity sensitivity in neurons of the primate visual cortex. *The Journal of Neuroscience, 17*, 296–307.

Das, V. E. (2016). Strabismus and the oculomotor system: Insights from macaque models. *Annual Review of Vision Science, 2*, 37–59.

DeAngelis, G. C., & Newsome, W. T. (1999). Organization of disparity selectivity in macaque area MT. *Journal of Neuroscience, 19*, 1398–1415.

Dell'Osso, L. F. (1985). Congenital, latent and manifest latent nystagmus - similarities, differences and relation to strabismus. *Japanese Journal of Ophthalmology, 29*, 351–368.

Dell'Osso, L. F., Schmidt, D., & Daroff, R. B. (1979). Latent, manifest latent, and congenital nystagmus. *Archives of Ophthalmology, 97*, 1877–1885.

Dell'Osso, L. F., Traccis, S., & Abel, L. A. (1983). Strabismus: A necessary condition for latent and manifest latent nystagmus. *Neuro-Ophthalmology, 3*, 247–257.

Dursteler, M. R., & Wurtz, R. H. (1988). Pursuit and optokinetic deficits following chemical lesions of cortical areas MT and MST. *Journal of Neurophysiology, 60*, 940–965.

Endo, M., Kaas, J. H., Jain, N., et al. (2000). Binocular cross-orientation suppression in the primary visual cortex (V1) of infant rhesus monkeys. *Investigative Ophthalmology & Visual Science, 41*, 4022–4031.

Faucon, A. (1872). Nystagmus par insuffisance des droits externes. *J Ophtal (Paris), 1*, 233.

Fawcett, S. L., Birch, E. E., & Motion, V. E. P. (2000). Stereopsis, and bifoveal fusion in children with strabismus. *Investigative Ophthalmology & Visual Science, 41*, 411–416.

Foeller, P. E., Bradley, D., & Tychsen, L. (2008). Shorter vs longer durations of later-onset infantile strabismus in macaque monkeys. *Investigative Ophthalmology & Visual Science, 49*, 1121.

Fox, R., Aslin, R. N., Shea, S. L., et al. (1980). Stereopsis in human infants. *Science, 207*, 323–324.

Fromaget, C., & Fromaget, H. (1912). Nystagmus latent. *Annales d'Oculistique, 147*, 344.

Gwiazda, J., Bauer, J. A. J., & Held, R. (1989). Binocular function in human infants: Correlation of stereoptic and fusion-rivalry discriminations. *Journal of Pediatric Ophthalmology and Strabismus, 26*, 128–132.

Hatta, S., Kumagami, T., Qian, J., et al. (1998). Nasotemporal directional bias of V1 neurons in young infant monkeys. *Investigative Ophthalmology & Visual Science, 39*, 2259–2267.

Hoffmann, K. P., & Distler, C. (1992). The optokinetic reflex in macaque monkeys with congenital strabismus. IV symposium of the Bielschowsky Society for the Study of. *Strabismus, 1*, 46.

Hoffmann, K. P., Distler, C., & Ilg, U. (1992). Callosal and superior temporal sulcus contributions to receptive field properties in the macaque monkey's nucleus of the optic tract and dorsal terminal nucleus of the accessory optic tract. *The Journal of Comparative Neurology, 321*, 150–162.

Horton, J. C., Hocking, D. R., & Adams, D. L. (1999). Metabolic mapping of suppression scotomas in striate cortex of macaques with experimental strabismus. *The Journal of Neuroscience, 19*, 7111–7129.

Hubel, D. H., & Wiesel, T. N. (1977). Ferrier lecture. Functional architecture of macaque monkey visual cortex. *Proceedings of the Royal Society of London Series B. Biological Sciences, 198*, 1–59.

Inoue, Y., Takemura, A., Kawano, K., et al. (1998). Dependence of short-latency ocular following and associated activity in the medial superior temporal area (MST) on ocular vergence. *Experimental Brain Research, 121*, 135–144.

Jacobs, M., Harris, C., & Taylor, D. (1994). The development of eye movements in infancy. In G. Lennerstrand (Ed.), *Update on strabismus and pediatric ophthalmology. Proceedings of the Joint ISA and AAPO&S Meeting. Vancouver, Canada. June 19 to 23, 1994* (pp. 140–143). Boca Raton: CRC Press.

Joshi, A. C., Agaoglu, M. N., & Das, V. E. (2017). Comparison of Naso-temporal asymmetry during monocular smooth pursuit, optokinetic nystagmus, and ocular following response in strabismic monkeys. *Strabismus, 25*, 47–55.

Kawano, J., Shirdara, M., Watanabe, Y., et al. (1994). Neural activity in cortical area MST of alert monkey during ocular following responses. *Journal of Neurophysiology, 71*, 2305–2324.

Kestenbaum, A. (1961). *Clinical methods of neuro-ophthalmologic examination* (2nd ed.). New York: Grune and Stratton.

Kiorpes, L., & Boothe, R. G. (1981). Naturally occurring strabismus in monkeys (Macaca nemestrina). *Investigative Ophthalmology & Visual Science, 20*, 257–263.

Kiorpes, L., Boothe, R. G., Carlson, M. R., et al. (1985). Frequency of naturally occurring strabismus in monkeys. *Journal of Pediatric Ophthalmology and Strabismus, 22*, 60–64.

Kiorpes, L., Walton, P. J., O'Keefe, L. P., et al. (1996). Effects of artificial early-onset strabismus on pursuit eye movements and on neuronal responses in area MT of macaque monkeys. *The Journal of Neuroscience, 16*, 6537–6553.

Komatsu, H., & Wurtz, R. H. (1988a). Relation of cortical areas MT and MST to pursuit eye movements.I. localization and visual properties of neurons. *Journal of Neurophysiology, 60*, 580–603.

Komatsu, H., & Wurtz, R. H. (1988b). Relation of cortical areas MT and MST to pursuit eye movements. III. Interaction with full-field visual stimulation. *Journal of Neurophysiology, 60*, 621–644.

Kommerell, G. (1982). Mehdorn E. is an optokinetic defect the cause of congenital and latent nystagmus? In G. Lennerstrand, D. S. Zee, & E. L. Keller (Eds.), *Functional basis of ocular motility disorders* (pp. 159–168). New York: Pergamon Press.

Lowel, S., & Engelmann, R. (2002). Neuroanatomical and neurophysiological consequences of strabismus: Changes in the structural and functional organization of the primary visual cortex in cats with alternating fixation and strabismic amblyopia. *Strabismus, 10*, 95–105.

Lowel, S., & Singer, W. (1992). Selection of intrinsic horizontal connections in the visual cortex by correlated neuronal activity. *Science, 255*, 209–212.

Matsumoto, B., MacDonald, R., & Tychsen, L. (1991). Constellation of ocular motor findings in naturally-strabismic macaque: Animal model for human infantile strabismus. *Investigative Ophthalmology & Visual Science Supplement, 32*, 820.

Maurer, D., Lewis, T. L., & Brent, H. P. (1983). Peripheral vision and optokinetic nystagmus in children with unilateral congenital cataract. *Behavioural Brain Research, 10*, 151–161.

McLean, R. J., Maconachie, G. D., Gottlob, I., et al. (2016). The development of a nystagmus-specific quality-of-life questionnaire. *Ophthalmology, 123*, 2023–2027.

Mustari, M. J., & Fuchs, A. F. (1990). Discharge patterns of neurons in the pretectal nucleus of the optic tract (NOT) in the behaving primate. *Journal of Neurophysiology, 64*, 77–90.

Mustari, M. J., Fuchs, A. F., Kaneko, C. R. S., et al. (1994). Anatomical connections of the primate pretectal nucleus of the optic tract. *The Journal of Comparative Neurology, 349*, 111–128.

Mustari, M. J., Ono, S., & Vitorello, K. C. (2008). How disturbed visual processing early in life leads to disorders of gaze-holding and smooth pursuit. *Progress in Brain Research, 171*, 487–495.

N.E.I. / N.I.H. Committee. (2001). A classification of eye movement abnormalities and strabismus (CEMAS). Report of a National Eye Institute sponsored workshop.

Naegele, J. R., & Held, R. (1982). The postnatal development of monocular optokinetic nystagmus in infants. *Vision Research, 22*, 341–346.

Newsome, W. T., Wurtz, R. H., & Komatsu, H. (1988). Relation of cortical areas MT and MST to pursuit eye movements. II. Differentiation of retinal from extraretinal inputs. *Journal of Neurophysiology, 60*, 604–620.

Norcia, A. M. (1996). Abnormal motion processing and binocularity: Infantile esotropia as a model system for effects of early interruptions of binocularity. *Eye, 10*, 259–265.

Norcia, A. M., Garcia, H., Humphry, R., et al. (1991). Anomalous motion VEPs in infants and in infantile esotropia. *Investigative Ophthalmology & Visual Science, 32*, 436–439.

O'Dell, C., & Boothe, R. G. (1997). The development of stereoacuity in infant rhesus monkeys. *Vision Research, 37*, 2675–2684.

Orban, G. A., Kennedy, H., & Buillier, J. (1986). Velocity sensitivity and direction selectivity of neurons in area V1 and V2 of the monkey: Influence of eccentricity. *Journal of Neurophysiology, 56*, 462–480.

Panel AAOQOCCPO. (1992). Amblyopia. Preferred practice patterns. 1–24.

Pasik, T., & Pasik, P. (1964). Optokinetic nystagmus: An unlearned response altered by section of chiasma and corpus callosum in monkeys. *Nature, 203*, 609–611.

Pirdankar, O. H., & Das, V. E. (2016). Influence of target parameters on fixation stability in normal and strabismic monkeys. *Investigative Ophthalmology & Visual Science, 57*, 1087–1095.

Poggio, G. F., & Fischer, B. (1977). Binocular interaction and depth sensitivity in striate and prestriate cortex of behaving rhesus monkey. *Journal of Neurophysiology, 40*, 1392–1405.

Richards, M., Wong, A., Foeller, P., et al. (2008). Duration of binocular decorrelation predicts the severity of latent (fusion maldevelopment) nystagmus in strabismic macaque monkeys. *Investigative Ophthalmology & Visual Science, 49*, 1872–1878.

Schor, C. M., & Levi, D. M. (1980). Disturbances of small-field horizontal and vertical optokinetic nystagmus in amblyopia. *Investigative Ophthalmology & Visual Science, 19*, 668–683.

Sorsby, A. (1931). Latent nystagmus. *The British Journal of Ophthalmology, 15*, 1–8.

Trachtenberg, J. T., & Stryker, M. P. (2001). Rapid anatomical plasticity of horizontal connections in the developing visual cortex. *The Journal of Neuroscience, 21*, 3476–3482.

Tusa, R. J., & Ungerleider, L. G. (1988). Fiber pathway of cortical areas mediating smooth pursuit eye movements in monkeys. *Annals of Neurology, 23*, 174.

Tusa, R. J., Repka, M. X., Smith, C. B., et al. (1991). Early visual deprivation results in persistent strabismus and nystagmus in monkeys. *Investigative Ophthalmology & Visual Science, 32*, 134–141.

Tusa, R. J., Mustari, M. J., Burrows, A. F., et al. (2001). Gaze-stabilizing deficits and latent nystagmus in monkeys with brief, early-onset visual deprivation: Eye movement recordings. *Journal of Neurophysiology, 86*, 651–661.

Tusa, R. J., Mustari, M. J., Das, V. E., et al. (2002). Animal models for visual deprivation-induced strabismus and nystagmus. *Annals of the New York Academy of Sciences, 956*, 346–360.

Tychsen, L. (1996). Absence of subcortical pathway optokinetic eye movements in an infant with cortical blindness. *Strabismus, 4*, 11–14.

Tychsen, L. (1999). Infantile esotropia: Current neurophysiologic concepts. In A. L. Rosenbaum & A. P. Santiago (Eds.), *Clinical strabismus management* (pp. 117–138). Philadelphia: WB Saunders.

Tychsen, L. (2001). Critical periods for development of visual acuity, depth perception and eye tracking. In D. B. Bailey Jr., J. T. Bruer, F. J. Symons, et al. (Eds.), *Critical thinking about critical periods* (pp. 67–80). Baltimore: Paul H. Brookes Publishing Co.

Tychsen, L. (2005). Can ophthalmologists repair the brain in infantile esotropia? Early surgery, stereopsis, monofixation syndrome, and the legacy of Marshall Parks. *Journal of AAPOS, 9*, 510–521.

Tychsen, L. (2007). Causing and curing infantile esotropia in primates: The role of decorrelated binocular input (an American Ophthalmological Society thesis). *Transactions of the American Ophthalmological Society, 105*, 564–593.

Tychsen, L. (2012). Latent nystagmus and dissociated vertical-horizontal deviation. In C. S. Hoyt & D. Taylor (Eds.), *Pediatric ophthalmology and strabismus* (pp. 901–908). London: Elsevier Science Ltd.

Tychsen, L., & Boothe, R. G. (1996). Latent fixation nystagmus and nasotemporal asymmetries of motion visually evoked potentials in naturally strabismic primate. *Journal of Pediatric Ophthalmology and Strabismus, 33*, 148–152.

Tychsen, L., & Burkhalter, A. (1995). Neuroanatomic abnormalities of primary visual cortex in macaque monkeys with infantile esotropia: Preliminary results. *Journal of Pediatric Ophthalmology and Strabismus, 32*, 323–328.

Tychsen, L., & Burkhalter, A. (1997). Nasotemporal asymmetries in V1: Ocular dominance columns of infant, adult, and strabismic macaque monkeys. *The Journal of Comparative Neurology, 388*, 32–46.

Tychsen, L., & Lisberger, S. G. (1986). Maldevelopment of visual motion processing in humans who had strabismus with onset in infancy. *The Journal of Neuroscience, 6*, 2495–2508.

Tychsen, L., & Scott, C. (2003). Maldevelopment of convergence eye movements in macaque monkeys with small- and large-angle infantile esotropia. *Investigative Ophthalmology & Visual Science, 44*, 3358–3368.

Tychsen, L., & Vision, B. (1992). In W. M. Hart (Ed.), *Adler's Physiology of the Eye: Clinical Applications* (pp. 773–853). St. Louis: CV Mosby.

Tychsen, L., Hurtig, R. R., & Scott, W. E. (1985). Pursuit is impaired but the vestibulo-ocular reflex is normal in infantile strabismus. *Archives of Ophthalmology, 103*, 536–539.

Tychsen, L., Quick, M., & Boothe, R. G. (1991). Alternating monocular input from birth causes stereoblindness, motion processing asymmetries, and strabismus in infant macaque. *Investigative Ophthalmology & Visual Science Supplement, 32*, 1044.

Tychsen, L., Rastelli, A., Steinman, S., et al. (1996a). Biases of motion perception revealed by reversing gratings in humans who had infantile-onset strabismus. *Developmental Medicine and Child Neurology, 38*, 408–422.

Tychsen, L., Leibole, M., & Drake, D. (1996b). Comparison of latent nystagmus and nasotemporal asymmetries of optokinetic nystagmus in adult humans and macaque monkeys who have infantile strabismus. *Strabismus, 4*, 171–177.

Tychsen, L., Burkhalter, A., & Boothe, R. (1996c). Functional and structural abnormalities of visual cortex in infantile strabismus. *Klinische Monatsblätter für Augenheilkunde, 208*, 18–22.

Tychsen, L., Burkhalter, A., & Boothe, R. G. (1996d). Neural mechanisms in infantile esotropia: What goes wrong? *The American Orthoptic Journal, 46*, 18–28.

Tychsen, L., Yildirim, C., & Foeller, P. (1999). Effect of infantile strabismus on visuomotor development in squirrel monkey (Saimiri Sciureus): Optokinetic nystagmus, motion VEP and spatial sweep VEP. *Investigative Ophthalmology & Visual Science, 40*, S405.

Tychsen, L., Yildirim, C., Anteby, I., et al. (2000). Macaque monkey as an ocular motor and neuroanatomic model of human infantile strabismus. In G. Lennerstrand & J. Ygge (Eds.), *Advances in strabismus research: basic and clinical aspects* (pp. 103–119). London: Wenner-Gren International Series, Portland Press Ltd..

Tychsen, L., Wong, A. M., & Burkhalter, A. (2004). Paucity of horizontal connections for binocular vision in V1 of naturally strabismic macaques: Cytochrome oxidase compartment specificity. *The Journal of Comparative Neurology, 474*, 261–275.

Tychsen, L., Richards, M., Wong, A., et al. (2008). Spectrum of infantile esotropia in primates: Behavior, brains, and orbits. *Journal of AAPOS, 12*, 375–380.

van Dalen, J. T. W. (1981). Flash-induced nystagmus: Its relation to latent nystagmus. *Japanese Journal of Ophthalmology, 25*, 370–376.

van Die, G., & Collewijn, H. (1982). Optokinetic nystagmus in man. *Human Neurobiology, 1*, 111–119.

Watanabe, I., Bi, H., Zhang, B., et al. (2005). Directional bias of neurons in V1 and V2 of strabismic monkeys: Temporal-to-nasal asymmetry? *Investigative Ophthalmology & Visual Science, 46*, 3899–3905.

Wattam-Bell, J., Braddick, O., Atkinson, J., et al. (1987). Measures of infant binocularity in a group at risk for strabismus. *Clinical Vision Science, 4*, 327–336.

Werth, R. (2007). Residual visual functions after loss of both cerebral hemispheres in infancy. *Investigative Ophthalmology & Visual Science, 48*, 3098–3106.

Westall, C. A., Eizenman, M., Kraft, S. P., et al. (1998). Cortical binocularity and monocular optokinetic asymmetry in early-onset esotropia. *Investigative Ophthalmology & Visual Science, 39*, 1352–1360.

Wong, A. M., & Tychsen, L. (2002). Effects of extraocular muscle tenotomy on congenital nystagmus in macaque monkeys. *Journal of AAPOS, 6*, 100–107.

Wong, A. M. F., Foeller, P., Bradley, D., et al. (2003). Early versus delayed repair of infantile strabismus in macaque monkeys: I. Ocular motor effects. *Journal of AAPOS, 7*, 200–209.

Wong, A. M., Burkhalter, A., & Tychsen, L. (2005). Suppression of metabolic activity caused by infantile strabismus and strabismic amblyopia in striate visual cortex of macaque monkeys. *Journal of AAPOS, 9*, 37–47.

Yamasaki, D. S., & Wurtz, R. H. (1991). Recovery of function after lesions in the superior temporal sulcus in the monkey. *Journal of Neurophysiology, 66*, 651–673.

Pattern Strabismus: Where Does the Role of the Brain End and the Role of Muscles Begin?

Nataliya Pyatka and Fatema Ghasia

Abstract Pattern strabismus is a vertically incomitant horizontal strabismus comprising 50% of its infantile forms. High-resolution orbit imaging and contemporary physiology literature suggested the role of oblique muscle dysfunction as its pathophysiology. The role of the central nervous system was also outlined. Here, we will discuss contemporary theories with specific focus on pathophysiological concepts including oblique muscle dysfunction, loss of fusion with altered recti muscle pull, displacements and instability in connective tissue pulleys of the recti muscles, vestibular hypofunction, and abnormal neural connections.

Keywords Strabismus · Extraocular muscle · Orbit · Brainstem · Cerebellum

1 Background

Vertically incomitant horizontal strabismus, which makes up about 12–50% of strabismus cases, has been described to have either "A" or "V" pattern based on the horizontal deviation from midline in downgaze or upgaze (Harley and Manley 1969; Knapp 1959; Urist 1958; Costenbader 1964). In "A" pattern, horizontal deviation demonstrates more divergence in downgaze and more convergence in upgaze. In "V" pattern, divergence is more prominent in upgaze and convergence is more in downgaze. "A" and "V" patterns are clinically significant when the measurement difference between downgaze and upgaze is more than 10 and 15 prism diopters (Δ), respectively. Although the focus of this chapter will be on "A" and "V" pattern strabismus, it should be mentioned that other rarer patterns exist, including "X" where both downgaze and upgaze lead to divergence from primary

N. Pyatka (✉)
Department of Neurology, Neurological Institute, University Hospitals Cleveland Medical Center, Cleveland, OH, USA
e-mail: Natalia.Pyatka@Uhhospitals.org

F. Ghasia
Cole Eye Institute, Cleveland Clinic, Cleveland, OH, USA

© Springer Nature Switzerland AG 2019
A. Shaikh, F. Ghasia (eds.), *Advances in Translational Neuroscience of Eye Movement Disorders*, Contemporary Clinical Neuroscience,
https://doi.org/10.1007/978-3-030-31407-1_14

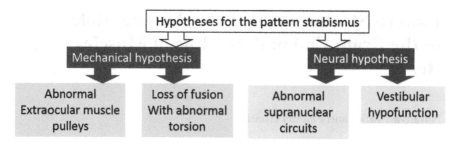

Fig. 1 Proposed theories for the pattern strabismus

position, lambda ("λ") where downgaze leads to exodeviation, and "Y" where upgaze leads to exodeviation.

Early theories proposed that dysfunction of vertical and horizontal recti muscles causes pattern strabismus (Urist 1951; Brown 1953). More recent theories argue that oblique muscle dysfunction leads to pattern strabismus since over-depression in adduction is seen in "A" pattern strabismus (possibly due to superior oblique over-action or inferior oblique underaction). Similarly, over-elevation in adduction in "V" pattern may be due to inferior oblique overaction or superior oblique underaction (Knapp 1959; Hertle 2002). However, in some cases of "A" and "V" pattern strabismus, there is no obvious oblique muscle dysfunction.

Several theories have been proposed to describe the cause of isolated extraocular muscle (EOM) dysfunction (Fig. 1). These theories focus on two general mechanisms: peripheral and central. The peripheral mechanism hypothesis proposes that mechanical orbital factors lead to ocular misalignment based on the direction of gaze. On the other hand, the central mechanism hypothesis suggests that abnormal neurophysiology is the cause of pattern strabismus. Ultimately, pattern strabismus may be multifactorial, and both central and peripheral theories may be valid.

2 Mechanical (Peripheral) Etiology of Pattern Strabismus

Abnormal orientation of superior and inferior oblique muscles can cause pattern strabismus due to misdirected muscle force. Hence, "A" and "V" pattern strabismus is common in plagiocephaly, hydrocephalus, and craniofacial dysmorphisms (such as upslanting or downslanting palpebral fissures) (Fink 1955; Biglan 1990; France 1975; Urrets-Zavalia et al. 1961). However, the mechanism may be more complex due dysfunction of other EOMs, such as the medial and lateral recti. Orbital magnetic resonance imaging (MRI) has revealed that the path of recti muscles is actually not straight, but instead inflected in the orbit on eccentric gaze due to EOM pulleys (Miller 1989).

The pulleys assure a smooth posterior muscle path during globe rotation and act as the "origin" of recti muscles (Demer et al. 1995, 1997). Each rectus muscle is composed of two layers: orbital and global. The pulleys are attached to the orbital layers and therefore are moved posteriorly when the recti muscles contract. The global layer assists with rotation of the eye.

In vivo imaging and neurophysiological research support the role of pulleys in stabilizing posterior EOM paths (van den Bedem et al. 2005; Schutte et al. 2006; Demer 2006). Studies using MR imaging demonstrated that sharp rectus EOM path inflections and resistance to sideslips are supported by orbitally coupled connective tissues (Demer et al. 1995, 2000; Clark et al. 2000; Demer 2004; Kono et al. 2002). Furthermore, EOM path inflections persist even after globe enucleation or surgical transpositions (Detorakis et al. 2003; Miller et al. 1993). Moreover, experiments with nonhuman primates showed that pulleys can change EOM pulling direction upon eccentric gaze (Ghasia and Angelaki 2005; Klier et al. 2006).

Incomitant strabismus can be caused by pulley location displacement perpendicular to the plane of action of the muscle, such as in horizontal displacement of vertical rectus muscle pulley (Demer et al. 1996; Clark et al. 1998). Clinical oblique dysfunction can be caused by heterotopic pulley displacement, especially in patients with connective tissue disorders (Oh et al. 2002; Miller and Folk 1975; Robb and Boger 3rd 1983). However, pattern strabismus is not present in all cases of craniofacial anomalies, bringing up the concern that pattern strabismus may have multiple etiologies such as dysfunction in central neurophysiology (Dickmann et al. 2012; Tychsen et al. 2008).

3 Neural (Central) Mechanisms of Pattern Strabismus

Abnormal neural connections may cause pattern strabismus by altering the direction of the EOM pull.

4 Loss of Fusion with Abnormal Torsion

Some have argued that loss of fusion can lead to pattern strabismus (Miller and Guyton 1994). One study compared patients with intermittent exotropia overcorrected with horizontal muscle surgery versus controls who maintained fusion postoperatively. After 28 months, 43% of subjects with consecutive esotropia developed "A" or "V" pattern strabismus versus only 5% of controls (Miller and Guyton 1994). Therefore, it was proposed that loss of fusion results in torsional drift, similar to exotropic drift, as seen in cases of monocular sensory deprivation. Torsional drift alters recti muscles' pulling direction, resembling oblique muscle overaction in the non-viewing eye. For instance, in "V" pattern strabismus, the roles of inferior rectus as partial adductor and superior rectus as partial abductor are altered due to a tonic torsional eye shift with clockwise torsion in the left eye and counterclockwise torsion in the right (Fig. 2a). Hence, force vectors of the vertical recti are altered,

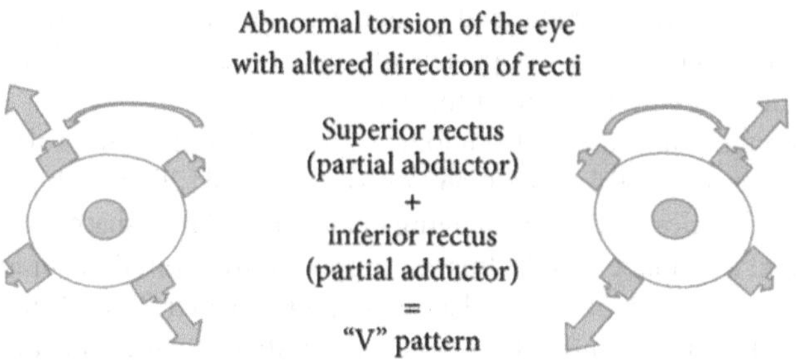

Abnormal torsion of the eye
with altered direction of recti

Superior rectus
(partial abductor)
+
inferior rectus
(partial adductor)
=
"V" pattern

Fig. 2 Abnormal muscle forces leading to abnormal pull causing "V" pattern strabismus

resulting in esoshift in downgaze and exoshift in upgaze, leading to "V" pattern strabismus. Similarly, lateral recti can act as partial depressors and medial recti as partial elevators, leading to vertical deviation on eccentric gaze imitating inferior oblique overaction (elevation in adduction) (Fig. 2b). However, prospective research has revealed abnormal torsion on fundoscopy years before inferior oblique overaction (adduction in elevation) is seen in infants with esotropia (Eustis and Nussdorf 1996). Therefore, the contribution of static torsion to the etiology of pattern strabismus remains debatable.

5 Abnormalities in Supranuclear Circuits

Nonhuman primate research has revealed a role of abnormal supranuclear circuits in pattern strabismus (Das et al. 2005; Das and Mustari 2007; Joshi and Das 2011). Macaques develop "A" or "V" pattern strabismus if raised with alternate monocular occlusion (AMO) during the periods of visual development (Das et al. 2005). These animals exhibit improper cross-axis movement of the non-viewing eye in the plane orthogonal to the appropriate eye movement trajectory. For instance, the non-viewing eye has a simultaneous vertical component (in addition to horizontal movement) during horizontal action task. As a result, the covered eye moves in an oblique trajectory (Das et al. 2005). This cross-coupled response likely leads to ocular misalignment in eccentric gaze in pattern strabismus.

One possible etiology of cross-coupled movements may be altered recti pull due to peripheral eye pathology or abnormal static torsion. However, in animal models of strabismus, there is no evidence of neural correlate of vertical cross-coupled response in the activity of the horizontal motoneurons and vice versa. For example, in response to vertical cross-axis and vertical eye movements, research shows that only vertical (not horizontal) motoneurons fire (Das and Mustari 2007; Joshi and Das 2011). Therefore, this research disproves the altered recti pull hypothesis.

Consequently, other researchers hypothesized that cross-coupled responses may be due to crosstalk between the vertical and horizontal eye movement structures. The source of cross-coupled eye movements may be the projections from vertical and torsional eye movement structures to the medial rectus motoneurons (Ugolini et al. 2006). Of note, this theory is based on models of strabismic animals reared with AMO. However, unlike infants with strabismus, these animals do not exhibit the characteristic oculomotor deficits such as nasotemporal asymmetry of optokinetic nystagmus (OKN) or latent nystagmus (LN). The animals reared with AMO exhibit impaired binocularity despite having good visual acuity from each eye. On the other hand, when reared with prism goggles, animals develop image non-correspondence which interferes with normal binocular vision. Furthermore, if animals are reared with binocular lid suture without tarsal plate, they develop thin translucent eyelids for perceiving diffuse luminance but not spatial vision.

Research in monkeys reared with binocular deprivation has shown that OKN asymmetry and LN are the results of dysfunction in the nucleus of optic tract (NOT). LN in these monkeys can be terminated by inactivating the NOT with muscimol injection (Mustari et al. 2001). At birth contralateral retina provides direct input to NOT (Hoffmann and Stone 1985). Normal visual experience allows for development of ipsilateral eye inputs from striate and extra-striate visual cortical areas (Hoffmann et al. 1992; Mustari et al. 1994). In monkeys with early-onset strabismus, binocular cells are absent in the NOT (Cynader and Hoffmann 1981; Distler and Hoffmann 1996).

Anatomically, striate cortex of animals reared with AMO has no binocular connections between ocular dominance columns of opposite ocularity. However, other areas used for motion processing (such as sub-cortical area-NOT and extra-striate visual cortex-medial temporal area) have preserved binocular cells (Mustari et al. 2001; Maunsell and Van Essen 1983; Tusa et al. 2002). Therefore, LN is not present in these monkeys. In contrast, animals reared with binocular lid suture without tarsal plate exhibit OKN asymmetry and LN due to the NOT being driven by the contralateral eye with a sparsity of cells that respond to both eyes.

Prism goggles reared animals have strabismus, LN, and OKN asymmetry (Hasany et al. 2008; Richards et al. 2008). These animals do not develop binocular connections between ocular dominance columns of opposite ocularity in visual area V1 (Tychsen et al. 2010; Tychsen 2007). Future research should be done to address whether the cross-coupled responses seen in pattern strabismic animals reared with AMO are also evident in monkeys with prism goggles (infantile onset strabismus model) or those with binocular lid suture (media opacity model).

Research by Ghasia validated primate models of pattern strabismus in human subjects by quantitatively assessing eye movements in pattern strabismus to delineate the role of neural circuits versus orbital etiologies (Ghasia et al. 2015). Cross-coupling of saccades was seen in all pattern strabismus patients with inferior oblique overaction with abnormal excyclotorsion. No correlation was found between the severity of cross-coupling and the amount of fundus torsion or the grade of oblique overaction. Therefore, cross-coupling could not be attributed strictly to the orbital etiology and is likely related to abnormalities in the abnormal saccade generators (Ghasia et al. 2015).

A recent study by Walton and Mustari tested three different models to simulate the known abnormalities of saccades in strabismus in order to determine where the dysfunctional horizontal-vertical crosstalk occurs. They were able to predict the dynamics of saccades only with the model that was based on crosstalk occurring at multiple nodes (Walton and Mustari 2017).

Another study proposed that in strabismus brainstem, circuits specific to saccadic eye movements are abnormal. Researchers found that microstimulation of pontine paramedian reticular formation (PPRF) in strabismic monkeys evoked disconjugate movements. Furthermore, increased connections between PPRF and the vertical burst generator were present in pattern strabismus (Walton and Mustari 2015).

6 Vestibular Hypofunction Theory

Another study proposed that "A" pattern strabismus is a variant form of "skew deviation" as evident in cerebellar or bilateral brainstem disease (Donahue and Itharat 2010). In such case, brainstem lesions diminish the otolith-equivalent of the anterior semicircular canal signal. As a result, posterior semicircular canal's otolith-equivalent would dominate. Thus, superior oblique and inferior rectus would be activated, and superior rectus and inferior oblique would be inhibited, leading to intorsion. These premotor inputs would cause relative overaction of superior oblique and resulting ocular tilt. This would precipitate increased head tilt and produce otolith stimulation. Such responses are not present in "A" pattern strabismus, leading to the conclusion that "A" pattern must be a special form of skew deviation. The patients in this study had other neurological deficits (such as spina bifida, hydrocephalus, or neural tube defects) which may not be present in all cases of "A" pattern strabismus (Donahue and Itharat 2010; Lee and Rosenbaum 2003).

7 Surgical Approach and Outcomes in Pattern Strabismus

Knowledge of the mechanism of pattern strabismus is important for successful surgical intervention. For instance, surgical weakening can correct overacting oblique muscles. On the other hand, if no oblique muscle dysfunction is seen clinically, then pattern strabismus can be fixed with horizontal recti transposition. A study by Kushner found that rectus muscle transposition can lead to increased ocular torsion. Transposition surgery to address torsion may lead to pattern strabismus (Kushner 2013). Therefore, abnormal torsion is less likely to be the driver of pattern strabismus.

Furthermore, strabismus surgery outcomes differ in pattern versus non-pattern exotropia. Postoperatively, the drift is less severe in patients with pattern intermittent exotropia and under-corrected exotropia. Therefore, the underlying etiology of pattern versus non-pattern exotropia is likely different (Pineles et al. 2008, 2009).

8 Conclusion

Multiple underlying mechanisms of pattern strabismus exist. In infants, the etiology is often abnormal neural connections. In patients that are older or have craniofacial anomalies, abnormal static torsion or orbital pulley instability often causes pattern strabismus.

References

Biglan, A. W. (1990). Ophthalmologic complications of meningomyelocele: A longitudinal study. *Transactions of the American Ophthalmological Society, 88*, 389–462.

Brown, H. W. (1953). Symposium; strabismus; vertical deviations. *Transactions-American Academy of Ophthalmology and Otolaryngology. American Academy of Ophthalmology and Otolaryngology, 57*(2), 157–162.

Clark, R. A., Miller, J. M., Rosenbaum, A. L., & Demer, J. L. (1998). Heterotopic muscle pulleys or oblique muscle dysfunction? *Journal of American Association for Pediatric Ophthalmology and Strabismus, 2*(1), 17–25.

Clark, R. A., Miller, J. M., & Demer, J. L. (2000). Three-dimensional location of human rectus pulleys by path inflections in secondary gaze positions. *Investigative Ophthalmology & Visual Science, 41*(12), 3787–3797.

Costenbader, F. D. (1964). Symposium: The A and V patterns in strabismus. The physiopathology of the A and V patterns. *Transactions-American Academy of Ophthalmology and Otolaryngology. American Academy of Ophthalmology and Otolaryngology, 68*, 354–355.

Cynader, M., & Hoffmann, K. P. (1981). Strabismus disrupts binocular convergence in cat nucleus of the optic tract. *Brain Research, 227*(1), 132–136.

Das, V. E., & Mustari, M. J. (2007). Correlation of cross-axis eye movements and motoneuron activity in non-human primates with "A" pattern strabismus. *Investigative Ophthalmology & Visual Science, 48*(2), 665–674. https://doi.org/10.1167/iovs.06-0249. [published Online First: Epub Date].

Das, V. E., Fu, L. N., Mustari, M. J., & Tusa, R. J. (2005). Incomitance in monkeys with strabismus. *Strabismus, 13*(1), 33–41. https://doi.org/10.1080/09273970590910298. [published Online First: Epub Date].

Demer, J. L. (2004). Pivotal role of orbital connective tissues in binocular alignment and strabismus: The Friedenwald lecture. *Investigative Ophthalmology & Visual Science, 45*(3), 729–738. 28.

Demer, J. L. (2006). Regarding van den Bedem, Schutte, van der Helm, and Simonsz: Mechanical properties and functional importance of pulley bands or 'Faisseaux Tendineux'. *Vision Research, 46*(18), 3036–3038; author reply 39–40. https://doi.org/10.1016/j.visres.2005.10.010. [published Online First: Epub Date].

Demer, J. L., Miller, J. M., Poukens, V., Vinters, H. V., & Glasgow, B. J. (1995). Evidence for fibromuscular pulleys of the recti extraocular muscles. *Investigative Ophthalmology & Visual Science, 36*(6), 1125–1136.

Demer, J. L., Miller, J. M., & Poukens, V. (1996). Surgical implications of the rectus extraocular muscle pulleys. *Journal of Pediatric Ophthalmology and Strabismus, 33*(4), 208–218.

Demer, J. L., Poukens, V., Miller, J. M., & Micevych, P. (1997). Innervation of extraocular pulley smooth muscle in monkeys and humans. *Investigative Ophthalmology & Visual Science, 38*(9), 1774–1785.

Demer, J. L., Oh, S. Y., & Poukens, V. (2000). Evidence for active control of rectus extraocular muscle pulleys. *Investigative Ophthalmology & Visual Science, 41*(6), 1280–1290.

Detorakis, E. T., Engstrom, R. E., Straatsma, B. R., & Demer, J. L. (2003). Functional anatomy of the anophthalmic socket: Insights from magnetic resonance imaging. *Investigative Ophthalmology & Visual Science, 44*(10), 4307–4313.

Dickmann, A., Parrilla, R., Aliberti, S., et al. (2012). Prevalence of neurological involvement and malformative/systemic syndromes in A- and V-pattern strabismus. *Ophthalmic Epidemiology, 19*(5), 302–305. https://doi.org/10.3109/09286586.2012.694553. [published Online First: Epub Date].

Distler, C., & Hoffmann, K. P. (1996). Retinal slip neurons in the nucleus of the optic tract and dorsal terminal nucleus in cats with congenital strabismus. *Journal of Neurophysiology, 75*(4), 1483–1494.

Donahue, S. P., & Itharat, P. (2010). A-pattern strabismus with overdepression in adduction: A special type of bilateral skew deviation? *Journal of American Association for Pediatric Ophthalmology and Strabismus, 14*(1), 42–46. https://doi.org/10.1016/j.jaapos.2009.11.009. [published Online First: Epub Date].

Eustis, H. S., & Nussdorf, J. D. (1996). Inferior oblique overaction in infantile esotropia: Fundus extorsion as a predictive sign. *Journal of Pediatric Ophthalmology and Strabismus, 33*(2), 85–88.

Fink, W. H. (1955). The role of developmental anomalies in vertical muscle defects. *American Journal of Ophthalmology, 40*(4), 529–553.

France, T. D. (1975). Strabismus in hydrocephalus. *The American Orthoptic Journal, 25,* 101–105.

Ghasia, F. F., & Angelaki, D. E. (2005). Do motoneurons encode the noncommutativity of ocular rotations? *Neuron, 47*(2), 281–293. https://doi.org/10.1016/j.neuron.2005.05.031. [published Online First: Epub Date].

Ghasia, F. F., Shaikh, A. G., Jacobs, J., & Walker, M. F. (2015). Cross-coupled eye movement supports neural origin of pattern strabismus. *Investigative Ophthalmology & Visual Science, 56*(5), 2855–2866.

Harley, R. D., & Manley, D. R. (1969). Bilateral superior oblique tenectomy in A-pattern exotropia. *Transactions of the American Ophthalmological Society, 67,* 324–338.

Hasany, A., Wong, A., Foeller, P., Bradley, D., & Tychsen, L. (2008). Duration of binocular decorrelation in infancy predicts the severity of nasotemporal pursuit asymmetries in strabismic macaque monkeys. *Neuroscience, 156*(2), 403–411. https://doi.org/10.1016/j.neuroscience.2008.06.070. [published Online First: Epub Date].

Hertle, R. W. (2002). A next step in naming and classification of eye movement disorders and strabismus. *Journal of American Association for Pediatric Ophthalmology and Strabismus (JAAPOS), 6*(4), 201–202. https://doi.org/10.1067/mpa.2002.126491. [published Online First: Epub Date].

Hoffmann, K. P., & Stone, J. (1985). Retinal input to the nucleus of the optic tract of the cat assessed by antidromic activation of ganglion cells. *Experimental Brain Research, 59*(2), 395–403.

Hoffmann, K. P., Distler, C., & Ilg, U. (1992). Callosal and superior temporal sulcus contributions to receptive field properties in the macaque monkey's nucleus of the optic tract and dorsal terminal nucleus of the accessory optic tract. *The Journal of Comparative Neurology, 321*(1), 150–162. https://doi.org/10.1002/cne.903210113. [published Online First: Epub Date].

Joshi, A. C., & Das, V. E. (2011). Responses of medial rectus motoneurons in monkeys with strabismus. *Investigative Ophthalmology & Visual Science, 52*(9), 6697–6705. https://doi.org/10.1167/iovs.11-7402. [published Online First: Epub Date].

Klier, E. M., Meng, H., & Angelaki, D. E. (2006). Three-dimensional kinematics at the level of the oculomotor plant. *The Journal of Neuroscience, 26*(10), 2732–2737. https://doi.org/10.1523/JNEUROSCI.3610-05.2006. [published Online First: Epub Date].

Knapp, P. (1959). Vertically incomitant horizontal strabismus: The so-called "A" and "V" syndromes. *Transactions of the American Ophthalmological Society, 57,* 666–699.

Kono, R., Clark, R. A., & Demer, J. L. (2002). Active pulleys: Magnetic resonance imaging of rectus muscle paths in tertiary gazes. *Investigative Ophthalmology & Visual Science, 43*(7), 2179–2188.

Kushner, B. J. (2013). Torsion and pattern strabismus: Potential conflicts in treatment. *JAMA Ophthalmology, 131*(2), 190–193. https://doi.org/10.1001/2013.jamaophthalmol.199. [published Online First: Epub Date].

Lee, S. Y., & Rosenbaum, A. L. (2003). Surgical results of patients with A-pattern horizontal strabismus. *Journal of American Association for Pediatric Ophthalmology and Strabismus, 7*(4), 251–255. https://doi.org/10.1016/mpa.2003.S1091853103001150. [published Online First: Epub Date].

Maunsell, J. H., & Van Essen, D. C. (1983). Functional properties of neurons in middle temporal visual area of the macaque monkey. I. Selectivity for stimulus direction, speed, and orientation. *Journal of Neurophysiology, 49*(5), 1127–1147.

Miller, J. M. (1989). Functional anatomy of normal human rectus muscles. *Vision Research, 29*(2), 223–240.

Miller, M., & Folk, E. (1975). Strabismus associated with craniofacial anomalies. *The American Orthoptic Journal, 25*, 27–37.

Miller, M. M., & Guyton, D. L. (1994). Loss of fusion and the development of A or V patterns. *Journal of Pediatric Ophthalmology and Strabismus, 31*(4), 220–224.

Miller, J. M., Demer, J. L., & Rosenbaum, A. L. (1993). Effect of transposition surgery on rectus muscle paths by magnetic resonance imaging. *Ophthalmology, 100*(4), 475–487.

Mustari, M. J., Fuchs, A. F., Kaneko, C. R., & Robinson, F. R. (1994). Anatomical connections of the primate pretectal nucleus of the optic tract. *The Journal of Comparative Neurology, 349*(1), 111–128. https://doi.org/10.1002/cne.903490108. [published Online First: Epub Date].

Mustari, M. J., Tusa, R. J., Burrows, A. F., Fuchs, A. F., & Livingston, C. A. (2001). Gaze-stabilizing deficits and latent nystagmus in monkeys with early-onset visual deprivation: Role of the pretectal not. *Journal of Neurophysiology, 86*(2), 662–675.

Oh, S. Y., Clark, R. A., Velez, F., Rosenbaum, A. L., & Demer, J. L. (2002). Incomitant strabismus associated with instability of rectus pulleys. *Investigative Ophthalmology & Visual Science, 43*(7), 2169–2178.

Pineles, S. L., Rosenbaum, A. L., & Demer, J. L. (2008). Changes in binocular alignment after surgery for concomitant and pattern intermittent exotropia. *Strabismus, 16*(2), 57–63. https://doi.org/10.1080/09273970802020292. [published Online First: Epub Date].

Pineles, S. L., Rosenbaum, A. L., & Demer, J. L. (2009). Decreased postoperative drift in intermittent exotropia associated with A and V patterns. *Journal of American Association for Pediatric Ophthalmology and Strabismus, 13*(2), 127–131. https://doi.org/10.1016/j.jaapos.2008.10.013. [published Online First: Epub Date].

Richards, M., Wong, A., Foeller, P., Bradley, D., & Tychsen, L. (2008). Duration of binocular decorrelation predicts the severity of latent (fusion maldevelopment) nystagmus in strabismic macaque monkeys. *Investigative Ophthalmology & Visual Science, 49*(5), 1872–1878. https://doi.org/10.1167/iovs.07-1375. [published Online First: Epub Date].

Robb, R. M., & Boger, W. P., 3rd. (1983). Vertical strabismus associated with plagiocephaly. *Journal of Pediatric Ophthalmology and Strabismus, 20*(2), 58–62.

Schutte, S., van den Bedem, S. P., van Keulen, F., van der Helm, F. C., & Simonsz, H. J. (2006). A finite-element analysis model of orbital biomechanics. *Vision Research, 46*(11), 1724–1731. https://doi.org/10.1016/j.visres.2005.11.022. [published Online First: Epub Date].

Tusa, R. J., Mustari, M. J., Das, V. E., & Boothe, R. G. (2002). Animal models for visual deprivation-induced strabismus and nystagmus. *Annals of the New York Academy of Sciences, 956*, 346–360.

Tychsen, L. (2007). Causing and curing infantile esotropia in primates: The role of decorrelated binocular input (an American Ophthalmological Society thesis). *Transactions of the American Ophthalmological Society, 105*, 564–593.

Tychsen, L., Richards, M., Wong, A., et al. (2008). Spectrum of infantile esotropia in primates: Behavior, brains, and orbits. *Journal of American Association for Pediatric Ophthalmology and Strabismus, 12*(4), 375–380. https://doi.org/10.1016/j.jaapos.2007.11.010. [published Online First: Epub Date].

Tychsen, L., Richards, M., Wong, A., Foeller, P., Bradley, D., & Burkhalter, A. (2010). The neural mechanism for Latent (fusion maldevelopment) nystagmus. *Journal of Neuro-Ophthalmology, 30*(3), 276–283. https://doi.org/10.1097/WNO.0b013e3181dfa9ca. [published Online First: Epub Date].

Ugolini, G., Klam, F., Doldan Dans, M., et al. (2006). Horizontal eye movement networks in primates as revealed by retrograde transneuronal transfer of rabies virus: Differences in monosynaptic input to "slow" and "fast" abducens motoneurons. *The Journal of Comparative Neurology, 498*(6), 762–785. https://doi.org/10.1002/cne.21092. [published Online First: Epub Date].

Urist, M. J. (1951). Horizontal squint with secondary vertical deviations. *AMA Archives of Ophthalmology, 46*(3), 245–267.

Urist, M. J. (1958). The etiology of the so-called A and V syndromes. *American Journal of Ophthalmology, 46*(6), 835–844.

Urrets-Zavalia, A., Solares-Zamora, J., & Olmos, H. R. (1961). Anthropological studies on the nature of cyclovertical squint. *The British Journal of Ophthalmology, 45*(9), 578–596.

van den Bedem, S. P., Schutte, S., van der Helm, F. C., & Simonsz, H. J. (2005). Mechanical properties and functional importance of pulley bands or 'faisseaux tendineux'. *Vision Research, 45*(20), 2710–2714. https://doi.org/10.1016/j.visres.2005.04.016. [published Online First: Epub Date].

Walton, M. M., & Mustari, M. J. (2015). Abnormal tuning of saccade-related cells in pontine reticular formation of strabismic monkeys. *Journal of Neurophysiology, 114*(2), 857–868.

Walton, M. M. G., & Mustari, M. J. (2017). Comparison of three models of saccade disconjugacy in strabismus. *Journal of Neurophysiology, 118*(6), 3175–3193.

Part III
Clinical Science

Video-Oculography in the Emergency Department: An "ECG" for the Eyes in the Acute Vestibular Syndrome

Georgios Mantokoudis, Daniel R. Gold, and David E. Newman-Toker

Abstract Dizziness and vertigo are among the most common symptoms evaluated in the emergency department (ED). Patients with the acute vestibular syndrome often have vestibular neuritis, but some have dangerous posterior circulation strokes. Rapid eye movement–based diagnosis by experts is possible using the "HINTS" exam (Head Impulse test of vestibular reflexes, assessment of Nystagmus direction in different fields of gaze, and measuring Skew deviation by alternate cover testing). Unfortunately, knowledge of bedside techniques among ED providers is limited, and misdiagnosis is frequent. Commercially available, portable video-oculography (VOG) systems can now be used to instantaneously assess eye and head movements in ED patients with acute dizziness and vertigo. This noninvasive approach can bring expertise to the bedside by quantitatively measuring the HINTS parameters. When coupled with remote access to specialists via telemedicine or computer-based decision support, such VOG testing has the potential to transform diagnosis of acute dizziness and vertigo in the way electrocardiography transformed diagnosis of chest pain—making accurate, timely, and efficient ED diagnosis routinely possible.

Keywords Dizziness · Vertigo · Acute vestibular syndrome · Diagnosis · Vestibular neuritis · Stroke · Video-oculography · Head impulse test · Vestibulo-ocular reflex · Emergency department

G. Mantokoudis
University Department of Otorhinolaryngology, Head and Neck Surgery, Inselspital, Bern University Hospital, University of Bern, Bern, Switzerland

D. R. Gold · D. E. Newman-Toker (✉)
Department of Neurology, Johns Hopkins University School of Medicine, Baltimore, MD, USA
e-mail: toker@jhu.edu

© Springer Nature Switzerland AG 2019
A. Shaikh, F. Ghasia (eds.), *Advances in Translational Neuroscience of Eye Movement Disorders*, Contemporary Clinical Neuroscience,
https://doi.org/10.1007/978-3-030-31407-1_15

1   Background

Optimizing clinical assessment of patients with acute dizziness in frontline care settings presents an important public health challenge (Newman-Toker 2016). Dizziness is common with a 1-year prevalence of 0.3% of the adult population (Neuhauser et al. 2008), resulting in over 4 million emergency department (ED) visits annually in the USA alone (Newman-Toker et al. 2008a). One in five patients leaves the ED given only a "diagnosis" of the original symptom of which they complained (i.e., "dizziness" or "vertigo") (Newman-Toker et al. 2008a). An astonishing 80% of patients with benign vestibular disorders may be misdiagnosed (Kerber et al. 2011). More than one third of strokes presenting dizziness or vertigo are initially missed (Kerber et al. 2006; Tarnutzer et al. 2017), sometimes resulting in devastating consequences for patients (Savitz et al. 2007). Assessment of current ED practice reveals a waste of critical diagnostic resources, including gross overuse of unnecessary brain CT scans resulting in rising costs over time (Saber Tehrani et al. 2013), without a corresponding improvement in neurologic diagnosis (Kim et al. 2012).

To be sure, diagnosing acute dizziness is not easy. There is a wide spectrum of diseases that cause dizziness, including general medical diseases, benign inner ear disorders, and dangerous central nervous system pathologies such as stroke (Newman-Toker et al. 2008a). In the ED, the goals are simple—rule out the worst-case scenario (in this case, stroke), treat what you can treat easily, and move on quickly to the next patient. Time is limited, and testing resources are precious—so rapid bedside assessment is at a premium. Unfortunately, stroke is uncommon (3–5% of ED dizziness) (Newman-Toker and Edlow 2015), and classical approaches to differentiating stroke from benign inner ear disorders such as vestibular neuritis are largely ineffective, resulting in frequent misdiagnosis (Kerber and Newman-Toker 2015). The symptom "type" (vertigo vs. dizziness vs. presyncope vs. dysequilibrium), often relied upon by ED physicians for diagnosis (Stanton et al. 2007), is typically vague and unreliable (Newman-Toker et al. 2007). Vascular risk stratification helps identify elderly and medically infirm patients at higher risk for stroke, but lacks specificity (Newman-Toker et al. 2013a) and, from the outset, sacrifices young patients with treatable diseases such as vertebral artery dissection (Gottesman et al. 2012). Neuroimaging by CT remains the community standard (Saber Tehrani et al. 2013), despite the fact that sensitivity for acute ischemic stroke is <15% (Newman-Toker et al. 2016) and having undergone a negative head CT actually *increases* the risk of a missed stroke in the ED more than twofold (Grewal et al. 2015), presumably because of false reassurance by the negative result. MRI is expensive, time-consuming, frequently unavailable, and wrong surprisingly often— false negatives in the first 24–48 hours occur ~15–20% of the time overall in patients with dizziness (Newman-Toker et al. 2016) and ~50% of the time when strokes causing acute symptoms are small (Saber Tehrani et al. 2014).

All of this might suggest abandoning the task entirely. There is, however, a clinical subgroup of patients at known high risk for stroke—those with the so-called acute vestibular syndrome (AVS) (Kim and Lee 2012; Tarnutzer et al. 2011). AVS patients suffer from severe, continuous dizziness or vertigo, nausea or vomiting, gait disturbance, and head-motion intolerance. They can be identified by careful history taking focused on "timing and triggers" rather than "type" (Newman-Toker and Edlow 2015). Nearly one in four of these patients has a stroke, but these strokes often closely mimic vestibular neuritis or other benign inner ear disorders (Kim and Lee 2012; Tarnutzer et al. 2011). Fewer than 20% of these patients have neurological signs likely to be obvious to an ED physician (Kattah et al. 2009). Fortunately, however, careful inspection of eye movements in these patients often reveals a "hint" or two about the nature of the underlying cause. Subspecialty experts in neuro-otology can detect stroke in AVS patients more accurately than even MRI (Newman-Toker et al. 2015).

If there were enough neuro-otologists to evaluate tens of millions of patients with acute dizziness seen each year around the world, this might suffice, but such experts are in short supply and unevenly distributed geographically. Recent advantages in portable, quantitative video-oculography (VOG), however, have opened new vistas for rapid diagnosis and treatment of these patients that were heretofore unimaginable. In this chapter we first review eye movement–based diagnosis in AVS patients, before addressing critical technical aspects of VOG-based recordings. We then elaborate on the transformative concept of ocular motor electrophysiology-based acute stroke diagnosis in acute dizziness using VOG. We do this using the analogy of cardiac electrophysiology-based diagnosis of acute myocardial infarction in acute chest pain using an electrocardiogram (ECG)—what we call the "eye ECG" approach.

2 Clinical "HINTS"

Once an AVS clinical picture is established, the "HINTS" exam, first introduced in 2009 (Kattah et al. 2009), forms the basis of eye movement–based diagnosis at the bedside. The "HINTS" acronym describes a battery of three ocular motor/vestibular tests: Head Impulse test, Nystagmus, and Test of Skew. Figure 1 illustrates the HINTS battery, and a video can be viewed at https://collections.lib.utah.edu/details?id=177180.

When experts perform this three-step exam in AVS patients with nystagmus and no hearing loss, it has a sensitivity of 99% and specificity of 97% for detection of central lesions, most of which are strokes (Newman-Toker et al. 2013a). This is substantially better than MRI with diffusion-weighted imaging (MRI-DWI), which misses up to ~30% of posterior fossa infarctions in the first 24 hours (Marx et al. 2004), particularly when strokes are small and located in the brainstem (Saber Tehrani et al. 2014). Superiority of the HINTS battery over MRI-DWI for stroke detection has been confirmed in multiple studies and systematic reviews (Newman-Toker et al. 2013a; Tarnutzer et al. 2011; Cohn 2014; Carmona et al. 2016). The approach performs less well in the context of AVS with hearing loss, where the risks

Fig. 1 HINTS bedside exam. A previous study has shown that eye exams differentiate stroke from benign causes of dizziness better than MRI. Three specific eye movement tests ("HINTS": Head Impulse, Nystagmus, Test of Skew) are the most potent. (With permission from Gianni Pauciello, Department of Otorhinolaryngology, Head and Neck Surgery, Inselspital, Bern University Hospital, University of Bern, CH-3010 Bern, Switzerland)

of labyrinthine stroke rise precipitously, which has ocular motor physiology effectively indistinguishable from that of labyrinthitis (Huh et al. 2013). Fortunately, such combined audiovestibular presentations of AVS are relatively uncommon among ED patients with acute dizziness or vertigo (Huh et al. 2013).

2.1 Head Impulse Test

The horizontal head impulse test (HIT or hHIT; Fig. 1a), first described by Halmagyi and Curthoys in 1988, evaluates the vestibulo-ocular reflex (VOR) in the horizontal plane (horizontal semicircular canals) using a rapid head rotation to assess whether visual fixation can be maintained. Some authors use the term cHIT for the non-quantitative clinical HIT to distinguish it from the quantitative video HIT (vHIT; see below), although better terms might be "nqHIT" and "qHIT" since both are clinical bedside tests. The HIT is the most potent of the three HINTS tests to detect stroke in AVS and alone likely outperforms MRI-DWI (Newman-Toker et al. 2015, 2008b).

To perform the test, the physician asks the patient to fixate the examiner's nose while (s)he is turning the head rapidly from 10 to 20° lateral to center (for video example of a normal HIT, see https://collections.lib.utah.edu/details?id=187678). The head movements have to be brisk (impulse-like) at high accelerations and unpredictable. High head velocities are important, since both labyrinths contribute to neural signals during head rotations, but the ipsilateral vestibular signal is dominant at higher head velocities. If the peak head velocity is too low during the test in a patient with unilateral vestibular loss, the healthy (unaffected) ear can compensate. During high peak head velocity HITs, asymmetry between the affected and unaffected sides is maximized with the contralesional ear contributing only 10–20% of the vestibular signal through neural inhibition and commissural pathways. With a significant unilateral or bilateral vestibulopathy, the examiner will observe an overt (visible) corrective (refixation) saccade immediately after the head comes to rest, suggesting an insufficient ipsilateral VOR (for video example of an abnormal HIT, see https://collections.lib.utah.edu/ark:/87278/s6x398q2). These abnormal (pathological) HITs are seen in patients with deficient vestibular function provided that the degree of canal paresis is substantial (Perez and Rama-Lopez 2003). The test is best performed using unpredictable, passive (examiner-led) head movements. This minimizes the patient's unconscious use of predictive (anticipatory) compensatory saccades that occur *during* head movements and result in *covert* (invisible) saccades and a normal-appearing HIT to visual inspection (see example in recovering vestibular neuritis at https://collections.lib.utah.edu/details?id=1278691). In contrast, overt saccades occur *after* head movements, so are evident to inspection, usually even without VOG recording equipment. Fortunately, effective compensatory covert saccades in the context of new vestibular loss usually take days or weeks to develop, so most patients presenting with new AVS have mostly overt refixations.

The VOR is usually intact (normal head impulses bilaterally) in patients with strokes causing AVS, which is why the hHIT remains the most potent test with a sensitivity of 90% to detect stroke in the posterior fossa (Newman-Toker et al. 2013a). Counterintuitively, in AVS a *normal* HIT bilaterally is concerning and usually indicates a central etiology, while a unilaterally *abnormal* HIT should reassure the clinician that the etiology is probably benign (except as determined by the other HINTS tests). Since healthy subjects and some patients with benign spontaneous or triggered causes of the episodic vestibular syndrome such as vestibular migraine, Menière's disease, or horizontal canal BPPV usually have a normal VOR and a normal HIT, the application of this test to differentiate between central and peripheral localization or etiology is only valid in patients fulfilling criteria for AVS.

A small fraction of vestibular strokes, generally those in the anterior inferior cerebellar artery (AICA) distribution, cause legitimately "peripheral" vestibular and cochlear deficits based on compromise of the internal auditory artery (fed by AICA) and resultant labyrinthine ischemia. This results in an abnormal HIT, usually with accompanying ipsilateral sudden hearing loss. There may be concomitant infarction of the pontine vestibular nucleus or vestibular nerve root entry zone (fascicle of the eighth cranial nerve, often in association with a lower motor neuron seventh cranial nerve palsy or long-tract signs) (Kim and Lee 2012), each of which can also cause an abnormal HIT on a central basis. Since an abnormal HIT is generally a finding suggestive of a benign peripheral process (vestibular neuritis), such strokes might lead to a false diagnostic impression of a non-serious etiology. This is why HINTS must be performed and interpreted as a three-step test. The additional two steps of the HINTS battery increase the overall sensitivity in the 10–15% of stroke cases where the head impulse test is abnormal. Extra care should be taken in patients with combined audiovestibular loss (Chang et al. 2018), as suggested by the use of a fourth step to assess for unilateral hearing loss (so-called HINTS plus) (Newman-Toker et al. 2013a).

2.2   Nystagmus and Gaze Holding

When spontaneous nystagmus is present, the second test of the "HINTS" battery evaluates the direction of the fast phase during primary and eccentric gaze positions (Fig. 1b). If the spontaneous nystagmus changes direction in lateral gaze away from the fast phase (gaze-evoked nystagmus), this suggests an impaired gaze-holding mechanism. The combination is referred to as mixed vestibular and gaze-holding nystagmus, historically sometimes known as "Bruns' nystagmus" (for example, see video at https://collections.lib.utah.edu/details?id=177176). Gaze-holding failure is a central finding and indicates neural integrator dysfunction in the brainstem or cerebellum (Cnyrim et al. 2008). Note that such nystagmus is present in only ~38% of stroke patients presenting AVS (Tarnutzer et al. 2011), so its *absence* is unhelpful in diagnosis (although its presence is a fairly clear indication of central pathology in this context).

Although the slow phase of jerk nystagmus is the pathologic phase and is directed toward the affected ear, the nystagmus is named for the fast phase which beats toward the healthy ear when the lesion is destructive. For example, acute right-sided vestibular neuritis results in an abnormal HIT when the head is moved rightward (e.g., from left-of-center to center) and spontaneous left-beating nystagmus (for example, see video at https://collections.lib.utah.edu/details?id=1277126). Frenzel goggles improve the visualization of vestibular nystagmus (whether peripheral or central (Lee and Kim 2013)) by accentuating it through removal of fixation and magnification with +20–30 diopter lenses, whereas gaze-evoked nystagmus is less influenced by the presence or absence of visual fixation. Peripheral vestibular nystagmus follows Alexander's law—a unidirectional nystagmus increases in the direction of the fast phase and decreases (or disappears) in the direction of the slow phase, without reversing direction with gaze in the slow-phase direction. However, small strokes frequently mimic this type of nystagmus (Newman-Toker et al. 2008b). A "direction-changing" nystagmus can also be seen in horizontal canal BPPV, but the nystagmus direction changes with head position, not gaze position, and symptoms are episodic.

2.3   Test of Skew

The alternate cover "test of skew" seeks vertical misalignment of the eyes of vestibular origin (so-called skew deviation, which is typically a central ocular motor sign (Brodsky et al. 2006)). During alternate cover testing, the examiner looks for vertical refixation eye movements that can be seen after an occluder (or examiner's hand) is moved from one eye to the other while the patient is fixating on a target (Fig. 1c; see video examples of normal [skew absent] at https://collections.lib.utah.edu/details?id=187677 and abnormal [skew present] at https://collections.lib.utah.edu/details?id=187730). When no vertical refixation is seen, the test of skew is negative or normal (i.e., there is no vertical ocular misalignment). It is important to note that healthy individuals almost never have a visible vertical misalignment by alternate cover testing, so the presence of a vertical misalignment is usually pathologic and, if not long-standing, likely to be diagnostically relevant to the presenting AVS clinical syndrome.

By contrast, if a horizontal refixation (termed either eso- [cross-eyed] or exodeviation [wall-eyed]) is seen, this could be a normal finding, since mild horizontal misalignments are common among healthy individuals. Furthermore, beats of horizontal nystagmus could easily be mistaken for horizontal refixations that result from misalignment. Thus, although eso- and exodeviations could theoretically be due to failures of central vestibular control over horizontal gaze pathways (Brodsky 2003), any horizontal refixations observed during alternate cover testing should be ignored for the purposes of HINTS-based diagnostic assessment, unless they reflect an obvious sixth cranial nerve palsy or internuclear ophthalmoplegia [INO].

When a vertical refixation is seen in the acutely vertiginous patient, a skew deviation is the usual cause, absent a long-standing history of prior known strabismus or eye muscle surgery. This is because other causes of new vertical strabismus are usually either obvious (e.g., third cranial nerve palsy) or generally unassociated with dizziness/vertigo (e.g., fourth nerve palsy).

The misalignment of skew occurs because of a deficit in otolithic function, specifically disruption of the otolithic (graviceptive) pathways from utricle to vertical extraocular muscles in the midbrain. As a result, skew deviation is often seen with one or more features of the pathologic ocular tilt reaction (OTR), which is made up of the clinical triad of skew deviation, head tilt (toward the side opposite the hyperdeviated eye), and ocular counterroll (with the 12-o'clock pole of both eyes rotated toward the ear opposite the hyperdeviated eye) (Brodsky et al. 2006).

The most obvious and lasting pathologic OTRs and skews tend to occur with lesions of the vestibular nuclei in the lateral medulla and the interstitial nucleus of Cajal (INC) in the medial rostro-dorsal midbrain (Brandt and Dieterich 1994). Destructive medullary lesions tend to produce ipsiversive OTR (head tilt and ocular counterroll toward the side of the lesion, with hyperdeviation in the contralesional eye), while destructive midbrain lesions tend to produce contraversive OTR (tilt and counterroll away from the side of the lesion, with hyperdeviation in the ipsilesional eye [barring the comorbid presence of a third nerve palsy] (Halmagyi et al. 1990)). Smaller, more transient otolithic signs are common in patients with lesions of the medial longitudinal fasciculus (MLF) but are usually overshadowed by an obvious INO, so are often ignored by clinical examiners. Because the graviceptive pathway from the utricle to INC crosses in the mid-pons, MLF lesions tend to produce ipsiversive OTR when caudally located (prior to decussation) and contraversive OTR when rostrally located (after decussation). Exceptions to these directional rules, however, have been reported, including a recent report of contraversive OTR with lateral medullary stroke attributed to disruption of inhibitory cerebellovestibular pathways in the inferior cerebellar peduncle (Cho et al. 2015).

Both skew and the full ocular tilt triad can occur with peripheral lesions (Halmagyi et al. 1979). However, when skews are caused by vestibular neuritis, they tend to be of small amplitude; as a result, they are usually not visible to the examiner acutely, when spontaneous nystagmus may be brisk. Skew can be quantified "manually" at the bedside by the application of prisms to neutralize the skew. It is difficult for an examiner to see a deviation smaller than 2–4 prism diopters (PD) (1.15–2.29 degrees, where $PD = 100 \times tan\ (\delta)$ and δ is the deviation in degrees) (Newman-Toker et al. 2015). Skew deviations of that size and smaller are probably fairly common among patients with vestibular neuritis, and larger ones can occasionally be seen in patients undergoing vestibular neurectomy (up to ~8 PD, 4.6 degrees) (Mantokoudis et al. 2014). By comparison, midbrain lesions cause skews with mean deviations of 7.5 degrees (~13 PD) (Brandt and Dieterich 1994), and some cases are much larger (e.g., 20 degrees, ~36 PD) (Halmagyi et al. 1990), such that they might easily be seen from across the room without any testing. Regardless of size, however, when present in acute dizziness or vertigo (without a recent neurectomy), visible skew by alternate cover test should be assumed to represent a central sign until proven otherwise, since the sign is specific for central pathology an estimated 98% of the time in AVS (Tarnutzer et al. 2011). Note that skew is present in only ~30% of stroke patients presenting AVS (Tarnutzer et al. 2011), so its *absence* is unhelpful in diagnosis (although its presence is a reasonably clear indication of central pathology in this context).

Skew deviation in AVS is typically a *comitant* form of strabismus (Brodsky et al. 2006), meaning that the degree of vertical misalignment does not usually change appreciably in different gaze positions. It is sometimes *incomitant* with greater vertical deviation to one side or the other, in which some have attributed to asymmetrical involvement of anterior and posterior canal projections to the ocular motor systems (Brodsky et al. 2006). Especially when incomitant, skews can mimic fourth nerve (superior oblique) palsies by standard bedside testing, but the hyperdeviated eye in skew is generally intorted while the hyperdeviated (affected) eye in fourth nerve palsy should be extorted (Brodsky et al. 2006). Otolithic eye movements causing skew and torsion may also be more position dependent than those of superior oblique palsies, with skew deviations more likely to diminish in the supine position relative to upright posture (Parulekar et al. 2008). Chronic skew deviations are more frequently incomitant, and this is most often manifest as "lateral alternating skew," in which there is hypertropia of the abducting (or, less often, adducting) eye on lateral gaze to either side; this is usually chronic and typically caused by bilateral cerebellar conditions (Brodsky et al. 2006) that do not usually present with AVS.

2.4   Interpretation of HINTS

The acronym INFARCT (Impulse Normal, Fast-phase Alternating, Refixation on Cover Test) summarizes the interpretation of central "HINTS," with any one of these three dangerous signs pointing toward stroke in AVS (Kattah et al. 2009). Care should be taken not to use terms such as "positive" and "negative" HINTS, as this is only likely to breed confusion (Thomas and Newman-Toker 2016). Instead, "central" vs. "peripheral" HINTS should be noted and the specific components described. Some authors use the abbreviations cHINTS versus pHINTS to distinguish central from peripheral "HINTS;" however, this nomenclature might be confusing with the term "clinical HINTS" versus "VOG HINTS." It is important to emphasize that any single central sign of the HINTS battery is sufficient for suspecting a vestibular stroke.

3 Video-Oculography: From Frenzel Goggles to Modern Eye-Tracking Systems

Techniques for recording eye movements in vestibular laboratories have been used for decades, even before the digital era. Historically, the use of Frenzel goggles allowed the investigators to observe eye movements with and without visual fixation, but no recordings or quantitative measurements were possible. Scleral search coils placed inside magnetic fields were introduced by Robinson (Robinson 1963) in 1963, and these recordings continue to offer the highest temporal and spatial resolution. Two- and three-dimensional (the latter including torsional eye movements) eye movement recordings are possible with this technique. However, it is a semi-invasive method using local anesthetics in order to place these search coils on the sclera of the eye, and few functioning systems still exist around the world. This method is mainly used in selected laboratories and research facilities, yet remains the gold standard for accurate eye position and eye movement recording.

With the development of electronystagmography (ENG) and electrooculography (EOG), physicians could objectively record eye movements by placing electrodes around the eyes, especially for recording caloric testing (Furman and Jacob 1993). The ENG measures electrical fields around the eyes in relation to the ground electrode. Recordings were visualized on graph paper with a 1 mm scale, and the analysis was performed manually.

Video Frenzels (also known as videonystagmoscopy, VNS) were introduced to observe eye movements on television screens and later on personal computer screens, but the resolution and the frame rates were limited for many decades. The digital revolution in data transfer and storage along with progressive miniaturization capabilities paved the way for modern eye-tracking systems.

Modern eye-tracking algorithms in conjunction with high-performance infrared cameras allow for an accurate recording of eye movements. Such video recording systems are also called "videonystagmography" (VNG) or, more generally, "video-oculography" (i.e., VOG) systems. Initially, frame rates of 50–100 frames per second (fps) were sufficient for high-resolution video Frenzels and especially for recording of nystagmus, but they were insufficient for recording head impulses at high velocities or for tracking saccades. With the creation of firewire technology in 1995, high-speed data transmission up to 800 Mbit/s was possible while the connected camera device was supplied with power at the same time. This improved connectivity in conjunction with powerful central processing units and led to the development of camera system recordings at frame rates of 250 fps or even higher.

The size, weight, and fit of VOG goggles determine whether they can be used for static or dynamic vestibular tests. High-performance binocular VOG goggles allow static measures of vestibular function including nystagmus (spontaneous nystagmus or induced by head positioning or thermal stimulation in calorics). During these tests, the head remains stationary, and the risk of goggle slippage or other artifacts is small compared to dynamic tests of vestibular function. Algorithms recognize the fast and slow-phase component of nystagmus and calculate the angular velocity. These measurements are quite accurate, provided that there are clean recordings with clear and direct view of the pupil (region of interest) without intrusive eye movements or blinks. Additionally, VOG allows accurate recording of visually induced reflexes such as optokinetic nystagmus and visual smooth pursuit. Some high image resolution systems are even capable of detecting torsion based on the individual's iris morphology, which is useful to detect cyclorotation due to ocular counterroll during the normal dynamic or static OTR. These semicircular canal and utricle-mediated reflexes are intended to stabilize images on the retina during a variety of head movements. Another class of eye movements are saccades, which are fast eye movements intended to quickly shift gaze from one target to another (for example, see video at https://collections.lib.utah.edu/details?id=187674). These eye movements require a tracking system capable of high frame rates (>100–250 fps) to achieve accurate quantification.

However, dynamic tests of vestibular function, such as pendular chairs, rotational tests, head shaking, head positioning, and head impulses, represent a greater challenge for VOG systems. Head movements need to be recorded simultaneously while avoiding both goggle slippage and unwanted head movements. While low-frequency movements and low accelerations allow for stability of the goggles on test subjects, high accelerations during head impulses disturb the eye and head signals and may lead to inaccurate recordings. This requires a much smaller, more lightweight apparatus to maintain measurement precision during head motion. Miniaturization of cameras and gyroscopes capable of recording both eye and head movements was driven by the mobile phone industry in the first decade of the twenty-first century. Lightweight, compact, and affordable goggles connected via USB3 ports paved the way for portable systems that could record at high head accelerations.

Since 2009, monocular VOG devices have been used to record and quantify the horizontal semicircular canal VOR using the HIT (Weber et al. 2009; MacDougall et al. 2009), and since 2013, it has been possible to record and quantify the vertical (anterior and posterior) semicircular canal VOR (Halmagyi et al. 2017). Figure 2 shows two modern devices approved by the FDA and certified for use in the European Economic Area (CE-marked) for head impulse recordings. They consist of one lightweight infrared camera connected via USB3 port to a laptop computer. The eyes are recorded by reflecting mirrors, since the camera is positioned above or on the side of the frame. Built-in infrared light sources allow for recordings in darkness. The goggles contain gyroscopes which detect head accelerations in all

Fig. 2 VOG devices (ICS impulse, Otometrics and EyeSeeCam, EyeSeeTec GmbH). (With permission from Gianni Pauciello, Department of Otorhinolaryngology, Head and Neck Surgery, Inselspital, Bern University Hospital, University of Bern, CH-3010 Bern, Switzerland)

three dimensions. Eye-tracking algorithms collect data at 250 fps and calculate VOR gain based on eye and head velocities during the hHIT. Calibration is performed either by projecting laser dots on a wall or by presenting a matrix on a connected monitor.

Such VOG devices have revolutionized eye movement recordings and have paved the way for dissemination of this technology to frontline care settings.

4 Quantitative VOG "HINTS"

4.1 Summary of Studies

Studies evaluating the accuracy and efficacy of the "HINTS" battery have mainly involved subspecialty eye movement experts such as neuro-ophthalmologists or neuro-otologists (Cohn 2014). Despite studies suggesting that frontline providers can be trained to perform HINTS and related examinations (Vanni et al. 2015), these ocular motor and vestibular examination skills remain challenging for most ED physicians (Kene et al. 2015), which is why any solution that could help support frontline providers is of great importance. This problem is recognized by emergency physicians around the world as one that needs solving—an international survey of priorities for new clinical decision rules found that deciding when to image patients with vertigo was the top-ranked need in adult emergency medicine (Eagles et al. 2008). This is not surprising, since patients presenting with dizziness/vertigo may have potentially life-threatening posterior fossa strokes, and tens of thousands of such strokes are likely missed each year in US EDs alone (Newman-Toker 2016).

The challenges of clinical HINTS testing for ED physicians are substantial. First, the correct patients (i.e., AVS) must be identified for testing—but this falls outside of traditional teaching and routine practice (Stanton et al. 2007; Newman-Toker 2007). Second, in performing the clinical HIT, non-specialists may have difficulties achieving sufficient head acceleration and velocity. Third, there may be difficulty interpreting results—in patients presenting with spontaneous nystagmus, distinguishing between a corrective saccade (which suggests unilateral vestibular loss) and a fast phase of nystagmus during the HIT can be challenging, even for experts. ED providers at major dizziness research centers have not yet achieved a high level of precision either (Newman-Toker 2015). Thus, although clinical, non-quantitative HINTS works for experts to diagnose stroke in AVS (even better than MRI acutely), it cannot be directly applied in most EDs because of the challenges associated with non-subspecialists performing and interpreting the three-step exam. As a result, the concept of a VOG-based HINTS exam was introduced (Newman-Toker et al. 2013b).

A first report and feasibility study of device-based (VOG) "HINTS" showed that diagnosis of vertebrobasilar stroke in patients with AVS was possible (Newman-Toker et al. 2013b). In a prospective cross-sectional study of AVS patients, VOR gain was found to be a useful and accurate quantitative parameter for the detection of vestibular stroke with a sensitivity of 88% and specificity of 92% (Mantokoudis et al. 2015a), similar to the clinical HIT. However, only head impulses were recorded, while the other HINTS exams were performed clinically. AICA strokes, which often result in pathological, low VOR gains (due to labyrinthine ischemia or ischemia of the fascicle/root entry zone of the eighth cranial nerve), would have been missed if the clinician were to rely on VOR gain alone, as has been seen previously with non-quantitative HINTS (Tarnutzer et al. 2011).

Newer-generation VOG devices now have the capacity to record the complete HINTS battery including nystagmus in different gaze positions and skew deviation. These devices are now being studied in clinical trial applications (Newman-Toker 2014; Mantokoudis 2017) to improve diagnosis of vestibular disorders and stroke. However, technical challenges are important for the advanced user to understand before applying such techniques.

4.2 Pitfalls and Barriers

Technical Aspects

A head impulse consists of two components—first, the slow phase, which is the eye movement opposite the head rotation during the HIT and, if pathological, the fast phase, which consists of a fast corrective eye movement (so-called corrective, compensatory, or refixation saccade). Figure 3 shows an example of an abnormal vHIT trace with commonly used nomenclature with regard to the various waveforms.

The slow phase of the eye movement represents a bell-shaped time-velocity profile opposite to the head movement. Velocity profiles are often mirrored in order to show eye and head profiles as an overlaid plot, thus making the profiles easier to compare and interpret. If the velocity profiles of eye and head movements are identical (i.e., mirror images of one another), then the calculated ratio (VOR gain) is 1.0 and considered normal (for the horizontal canal VOR). Corrective saccades may occur during the head movement, usually making them invisible to the examiner (covert corrective saccade), or they may occur after the head movement (overt

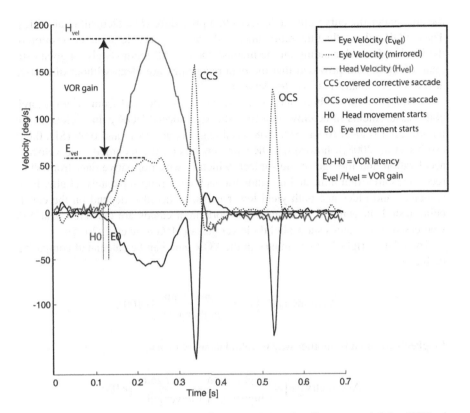

Fig. 3 Typical abnormal video HIT trace from a single impulse. Shown are a deficient VOR gain (representing the gap between slow eye and head movements) followed by two refixation saccades (fast eye movements)

corrective saccade). The direction of the corrective saccades is identical to the direction of the fast phase of spontaneous nystagmus in a unilateral destructive vestibulopathy. These differ from fast eye movements in the opposite direction occurring near the end of a HIT, also called covert anti-compensatory quick eye movements (CAQEM) (Heuberger et al. 2014).

There are different methods for calculating gain, and each method has its advantages and disadvantages. It is therefore not possible to achieve a complete understanding of all aspects of unilateral VOR dysfunction with a single gain number. There is also currently no international consensus on how to report VOR gain. Data are sometimes read and calculated at given time points (e.g., ratio of peak eye velocity to peak head velocity) or over a period of time (e.g., ratio of area under the eye velocity curve to area under the head velocity curve). This time period can be limited either to the ascending time-velocity profile just before peak velocity or span the whole HIT period until the head comes to rest. Gain can be calculated at single time points (Glasauer et al. 2004) (instantaneous gain), estimated by linear regression of eye and head velocity (Collewijn and Smeets 2000) or by computing the ratio of the area under the curve after a de-saccading procedure (MacDougall et al. 2013). These VOR gains reflect the ratio either of the velocity, acceleration, or position of the head and eye depending on the method chosen. All methods give a good estimate of VOR gain provided that the data recordings are clean without disruptive artifacts (see "Data Analysis" for details).

Normative data for horizontal semicircular canals derived from scleral search coil measurements in healthy subjects indicate a normal VOR gain is close to the theoretical 1.0 ideal (0.94 (SD, 0.08) (Halmagyi et al. 2001) and 0.98 (SD, 0.06) (Weber et al. 2008)); however, VOR gain depends on the recorded head velocity/ acceleration and the age of the subject, which is why normative data from VOG recordings are often indicated in bands for each age group separately. Higher head velocities and older age both yield lower gains. Gains also depend on the canals being tested. In general, VOR gains for horizontal canals are clustered near 1.0, whereas vertical canals show slightly lower gains (McGarvie et al. 2015).

The relative right-left asymmetry in the VOR gain can be calculated comparing both ears:

$$\text{Asymmetry}\left[\%\right] = \left(1 - \frac{\text{lower gain}}{\text{higher gain}}\right) \times 100$$

Jongkee's formula is another way to calculate asymmetry:

$$\text{Assymetry}\left[\%\right] = \left(\frac{\text{higher gain} - \text{lower gain}}{\text{higher gain} + \text{lower gain}}\right) \times 100$$

Jongkee's formula uses a normalization procedure, as generally done in calculating asymmetry for comparing right and left VOR function during caloric testing. However, there is no need to normalize data because the head impulse input (peak head velocity) and output (peak eye velocity) is well defined (i.e., for a gain of 1.0, the expected output = input). By contrast, this is not the case in caloric testing, where the effective applied thermal energy varies across subjects and ears, so results must be normalized to have meaning.

VOG devices record nystagmus and detect the slow-phase velocity (°/s) of nystagmus at different gaze positions. Gaze holding can also be assessed by calculating the time constant.

Eye position changes can be recorded during the alternate cover test. VOG goggles indicate skew deviation as degree of eye position changes. As an example, in a patient with a left lateral medullary stroke who has a skew deviation and a right hypertropia, when the right eye is uncovered and the occluder is moved to the left eye, both eyes would move downward so that the right eye (which is too high, relative to the fixation target) can take up fixation of the target. Likewise, when the left eye is then uncovered and the occluder is moved back to the right eye, the eyes would then move upward so that the left eye (which is now too low, relative to the fixation target) can take up fixation of the target. These vertical movements are quantified in degrees by the VOG device.

Data Acquisition

The acquisition of VOG data is a critical task because data quality has a direct impact on data analysis and therefore also on the triage and diagnostic workup process in dizzy patients. VOG is more susceptible to measurement errors than other systems such as ENG or scleral search coils. The patient needs to cooperate well by keeping his eyes open and to fixate on the predefined target during calibration. Video recordings and eye-tracking systems need a clean, well-centered image of the pupil, and the region of interest needs to be adjusted either digitally through the provided software or mechanically through adequate placement of the camera on the frame of the goggles (EyeSeeCam; Fig. 2). The goggles should be worn firmly with the strap tight to avoid skin movement or goggle slippage. Eyelashes should remain outside of the region of interest, and eyelash makeup should ideally be removed because of reflections of the infrared light source that may interfere with pupil detection. The examiner has to make sure that another nearby light source does not create additional reflections on the goggle mirrors.

Unlike non-quantitative HITs, which should generally be performed passively, VOG-based HITs can be performed either actively or passively. In active HITs, the patient moves his own head, while in passive HITs the examiner moves the patient's head. VOR gains are higher in active HITs compared to passive HITs and covert corrective saccades are more likely (Black et al. 2005). In general, VOG device manufacturers recommend that recordings be done with the patient sitting in front of a target that is 1–2 m away while the examiner stands behind the patient and performs passive horizontal impulses (unlike the non-quantitative HIT, where the examiner must stand in front of the patient in order to view the eyes). This setup avoids the examiner obstructing the fixation target and is often associated with fewer artifacts since impulses are applied with the examiner's hands on top (rather than side) of the head. Artifacts occur more often as a result of touching the strap when the examiner holds the patient's jaw (Mantokoudis et al. 2015b). However, many clinicians still prefer the classical HIT technique of being in front of the patient, because they can also directly observe corrective (overt) saccades that suggest unilateral vestibular loss.

The head direction, angle of displacement, velocity, and initial head position are each important because they influence predictability of the passive HIT and secondarily therefore affect VOR gain, latency, and amplitude of refixation saccades (Mantokoudis et al. 2016a). The initial reports of HIT described head displacements from an eccentric position back to the center (Halmagyi and Curthoys 1988). Small head displacements at different head velocities (150–300°/s) and high accelerations (~4000°/s^2) are sufficient for visualization of small-amplitude refixation saccades, provided that the impulses are applied unpredictably (Tjernstrom et al. 2012). When the head movements are of very small amplitude, the need for "lateral to center" movements as opposed to "center to lateral" movements matters less, because the risk of injury with an overzealous rotation from center to lateral is less of a concern. Nevertheless, when instructing novice examiners, we uniformly recommend a "center to lateral" rotation; this is because novices often struggle to accurately stop a rapid "lateral to center" impulse at 10–20 degrees of lateral rotation, but do not struggle to identify where the "center" (mid-sagittal plane) is located.

Spontaneous nystagmus may interfere with the recorded slow-phase velocities of the eye resulting in lower or higher gain; however, modern eye-tracking algorithms take nystagmus into account and correct for slow-phase eye movements (Mantokoudis et al. 2016b).

After an acute onset of dizziness with associated loss of vestibular function, central compensation and adaptative processes occur (Mantokoudis et al. 2014, 2016c). These central processes have an impact on the recorded VOR gain and saccade latency, which is why the time between examination and symptom onset matters. Saccade latency becomes shorter over time and the proportion of covert saccades increases (Mantokoudis et al. 2014). VOR gain might change over time, including on the healthy side, since adaptation can cause a transient downregulation of vestibular function even on the contralateral side (Mantokoudis et al. 2016c). Direction-changing, gaze-holding nystagmus may even begin to emerge as a result of central compensation (Hess and Reisine 1984).

The number of impulses necessary to obtain an accurate mean gain estimation depends on the quality of data. In theory, one single perfectly recorded head impulse would be sufficient for a correct interpretation. However, absolutely clean recordings and optimal lab conditions are not often realistic, especially among ED patients with AVS. Since artifacts may occur (see "Data Analysis"), we recommend performing 10–20 HITs per side to get a reliable average gain estimation, provided that the physician uses the direct device output (unfiltered data).

Data Analysis

Modern VOG devices already have an automated VOR gain calculation algorithm and a data filtering process that works in two steps: (1) a first instantaneous filtering process (less accurate) that rejects improperly performed impulses and gives immediate, direct feedback to the examiner and (2) a second, more extended post-processing analysis after the recording. Data that do not meet the predefined quality criteria of a HIT are discarded automatically. We recommend that the clinician not rely solely on the automated mean VOR gain output alone (which reflects one specific part of the HIT). Analysis should ideally consist of review of overall data quality by evaluating the time-velocity profile, clusters of gain values at various head velocities, qualitative appearance of corrective saccades, and the standard deviation of obtained gain values. A gain-velocity plot gives an overview of the collected data and possible outliers.

Algorithms for automated HIT analysis are being constantly improved by manufacturers, and most of the traces containing known artifacts are recognized and discarded. Figure 4 shows, however, that HIT recordings may contain different kinds of artifacts such as pseudosaccades (fast phases of nystagmus, blinks, and miniblinks), oscillations from pupil tracking loss, or artifacts resulting from the examiner inadvertently touching the goggles. Artifacts have to be distinguished

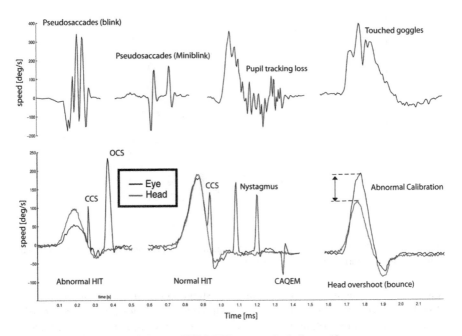

Fig. 4 Types of normal and abnormal VOG HIT patterns, including artifacts

from physiological fast eye movements such as nystagmus, covert corrective saccades, overt corrective saccades, or covert anti-compensatory quick eye movements (CAQEM; (Heuberger et al. 2014)) (Fig. 4). Other artifacts induced by the examiner or the subject during the head impulses include goggle slippage (eye and head traces are out of phase) or patient inattentiveness. Suboptimal calibration can result from spontaneous nystagmus, poor vision, or patient inattentiveness (Mantokoudis et al. 2015b).

Depending on the degree of artifactual data, the exam may have to be repeated.

Although single traces can be deleted manually if they do not meet quality criteria, manual data filtering/cleaning is not a realistic procedure for non-expert use of VOG devices. Fortunately, a recent study investigating the impact of artifacts on VOR gain and its resulting diagnostic implications showed no significant influence of unfiltered data on central (i.e., stroke) vs. peripheral vestibular classification in AVS patients, provided that the number of performed HITs exceeded ten HITs per tested ear (Mantokoudis et al. 2015a, 2016d).

A recent study showed that a VOR gain cutoff point of 0.7 accurately discriminates PICA strokes from vestibular neuritis (Mantokoudis et al. 2015a). Figure 5 shows a modeled population density plot derived from VOR gains in AVS patients recorded within 72 hours after symptom onset. Patients with PICA stroke have a VOR gain clustering close to 0.9; patients with a vestibular neuritis have a lower gain clustering around gain of 0.5 depending on the degree of unilateral vestibular paresis.

One additional VOR gain parameter discriminating vestibular strokes from vestibular neuritis is gain asymmetry. Analogous to bithermal caloric testing, an asym-

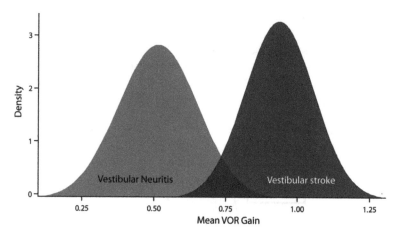

Fig. 5 Probability density plots for mean VOR gain among patients presenting AVS who have either vestibular neuritis or PICA-territory stroke

metry >20% (relative right-left asymmetry) suggests a unilateral vestibulopathy. If Jongkee's formula is used instead, the asymmetry cutoff lies between 8 and 10%. Note that for both absolute gain values and right-left asymmetries, AICA-territory strokes (particularly those involving the labyrinth) cause patterns of VOR loss that can be indistinguishable from that seen in vestibular neuritis or labyrinthitis. In such cases, additional information from the history (e.g., hearing loss, neurological symptoms), other HINTS eye movements, or neurological exam may be needed.

Another approach to use VOG data for stroke discrimination in AVS patients is an automated saccade analysis. There is evidence that both VOR gain and cumulative corrective saccade amplitudes are strong predictors of stroke, and one coil-based study found cumulative saccade amplitudes a slightly better predictor than gain (Chen et al. 2011). Some commercial VOG products now offer cumulative saccade calculations, but this saccade-based approach has not yet been validated using VOG devices. Thus, caution should be exercised, and it should not be relied upon pending further study.

5 Potential Utility and Impact of "Eye ECG" Concept

5.1 The Eye ECG Concept

The concept of VOG as an "eye ECG" to diagnose stroke in dizziness is by analogy to the use of ECG to diagnose myocardial infarction in acute chest pain. Both rely on acute perturbations in electrophysiology to identify the pathology, and, in theory, both could be used as routine and standard care in the ED. ECG is performed by ED technicians, who could also readily be trained to conduct VOG tests. Analogous to the standard ECG approach, the final goal of the eye ECG approach is the widespread use of this technology by non-experts in the ED.

If VOG-based devices were used as the standard of care for patients presenting with the AVS, there would be a significant potential for early and accurate stroke detection. Currently, 35% of vestibular strokes are missed at the initial assessment with perhaps one third of these patients suffering serious harms or death as a result (Tarnutzer et al. 2017). Even more importantly, if extended to include diagnosis of patients with triggered, episodic vestibular syndrome (e.g., by adding the Dix-Hallpike maneuver), the approach could potentially speed accurate, on-site diagnosis of the two most common peripheral vestibular disorders seen in the ED—BPPV and vestibular neuritis. This is important because an estimated 80% of patients with peripheral vestibular disorders are misclassified in current practice and so do not receive appropriate treatment (Kerber et al. 2011). The eye ECG approach could significantly reduce misdiagnosis leading to more timely, correct diagnosis, which could, in turn, allow for early administration of targeted and effective treatments, including tPA for stroke, canalith repositioning for BPPV, and early vestibular therapy with or without steroids for vestibular neuritis. Although rarely used in current ED practice (Kerber et al. 2013), diagnostic and treatment maneuvers for BPPV could be facilitated by VOG devices, which now offer visually guided placement of the head in space for users as a routine part of diagnosis and treatment. Finally, those with non-vestibular causes for dizziness could benefit by "ruling out" most central and peripheral vestibular causes through the absence of findings of spontaneous, gaze-evoked, or positional nystagmus during VOG testing and bilaterally normal HIT VOR results.

In addition to rapid and accurate diagnosis of patients presenting with the AVS, there are other benefits of using VOG in this way in the ED. Most ED clinicians are unfamiliar with the eye movement examination and in the evaluation of dizziness and vertigo patients since training in this area is very limited (Newman-Toker and Perry 2015). VOG devices can serve as an educational tool by giving immediate feedback (e.g., rejecting an improperly performed HIT). They also allow for standardized and reliable recordings of eye and head movements. In turn, such recordings can facilitate providing expert feedback on physician interpretation of eye movements, improving diagnostic calibration among ED providers, which currently is poor (Omron et al. 2018).

Some general preconditions should be met before considering VOG use in routine ED care: VOG systems should (1) be robust and easy to use by nurses and non-experts, (2) offer standardized test batteries and data analyses with high accuracy and reliability, and (3) be affordable for general use. However, actual implementation of VOG in routine practice will still require access to diagnostic expertise. In the short term, VOG systems might be used as a platform for telemedicine-based access to subspecialists. In the long term, automated diagnostic decision support is a realistic possibility. There are observational diagnostic accuracy studies and clinical trials currently underway to assess accuracy of this "eye ECG" approach in the ED (Newman-Toker 2014; Mantokoudis 2017).

5.2 Telemedicine Application

In the short term, VOG technology can be used as a telemedicine application to deliver subspecialty diagnostic expertise to the bedside in the ED. VOG technology supports the teleconsultation approach in two major ways: (1) creating a tangible record of the eye exam and (2) facilitating remote, digital review.

VOG ensures a standardized way of recording even by non-experts, and these recordings can then be reviewed as many times as necessary by experts and non-experts alike. VOG devices store high-resolution video data of the eyes and of the head/body position via a room camera, in conjunction with tracked data from head and eye movements (including time-velocity profiles or eye position data). This enables quality assurance by making sure that procedures are being done appropriately such as the HIT or Dix-Hallpike maneuver.

The VOG recordings also enable subspecialists to significantly broaden their geographic reach. Subspecialists with ocular motor and vestibular expertise are relatively few in number, and they are generally concentrated in academic medical centers. VOG enables remote, off-site consultation and assessment of quantitative head and eye movement exam once the VOG system is connected to an online network.

This approach could be used in EDs or even ambulances to improve tele-stroke diagnosis (Hubert et al. 2014) for patients with posterior fossa strokes presenting with dizziness who are at high risk for misdiagnosis (Newman-Toker et al. 2014). Specialists such as stroke neurologists, neuro-ophthalmologists, or neuro-otologists would be able to triage and differentiate central from peripheral causes of dizziness based on this remote VOG data. This kind of tele-diagnosis may someday be conducted in real time (with appropriate network connectivity bandwidth), but it is already possible in near real time using a store-forward methodology. Frontline clinical staff gather data on local computers and then digitally transfer images to servers for remote, expert review within minutes. This approach has already been successfully piloted at the Johns Hopkins Hospital ED as part of a "Tele-Dizzy™" consultation program (Gold et al. 2018). Preliminary results from the first 180 cases are extremely promising, cutting undiagnosed cases in half (~40% base rate; ~20% for those with consults) and reducing unnecessary imaging more than tenfold relative to the base rate (~40% base rate CT; ~3% for those with consults, without any increase in the rate of recommended MRIs for those with consults). But this approach still remains fundamentally constrained by the small number of specialists relative to the massive number of patients, which suggests that effective scaling will require something more—automated decision support.

5.3 Automated Decision Support

As yet, there are no commercially available diagnostic decision support tools to assist in VOG interpretation for the purpose of differentiating peripheral from central vestibular disorders, but clinical trials are underway to develop early prototypes

(Newman-Toker 2014). Such point-of-care tools could be added to existing VOG software to assist physicians in diagnosing vestibular neuritis versus stroke by recognizing a peripheral or central HINTS physiologic pattern, respectively. Likewise, the nystagmus of BPPV could also be detected automatically.

A similar approach is already in place in the ED for diagnosing heart attacks in acute chest pain by using an automated ECG interpretation. In the past, cardiologists needed to be immediately available, but now, as soon as an ECG is performed by an ED technician, automated computerized ECG interpretation algorithms (Daudelin and Selker 2005) assist frontline providers in searching for acute changes in cardiac electrophysiology. Today, ECGs are routinely interpreted in real time by ED physicians who identify ST elevation in acute myocardial infarction—but this is only commonplace now after decades of engagement by cardiologists and a change in expectations among ED physicians. Although ECGs may still sometimes be reviewed later by cardiologists and overread, this still permitted widespread scaling of the approach.

Full automation based on machine learning and artificial intelligence will require massive training data sets (at least 10,000 to begin the process) of acute ED-based eye movements with confirmed, correct diagnoses. This will not be a trivial task, but a clinical teleconsultation service as described above, if deployed in a large ED network, could be a springboard to accumulate thousands or even tens of thousands of such cases each year. A typical mid-sized ED (~60,000 visits per year) sees roughly 1000–1500 dizzy patients per year, so a hospital network of just 5 EDs could probably get 10,000 recordings in just 2 years, if it were routinely performed. We expect such decision support systems to appear in the relatively near future.

6 Conclusions

Current ED care for acute dizziness and vertigo is both low quality and high cost. Accurate, efficient, eye movement–based diagnosis that is routinely accomplished by experts might be disseminated at scale using VOG. The physiological basis of this "eye ECG" approach is sound and well-supported by scientific evidence. Recent technological advances have made immediate ED teleconsultation with a remote specialist feasible, and this approach has been implemented in a pilot program with promising initial results. While full automation with minimal need for specialist backup is still years away, there is reason to believe that real-time, point-of-care diagnostic decision support for stroke and vestibular disorder diagnosis in patients with acute dizziness and vertigo is possible, feasible, and likely realistic in the coming decade.

References

Black, R. A., et al. (2005). The active head-impulse test in unilateral peripheral vestibulopathy. *Archives of Neurology, 62*(2), 290–293.

Brandt, T., & Dieterich, M. (1994). Vestibular syndromes in the roll plane: Topographic diagnosis from brainstem to cortex. *Annals of Neurology, 36*(3), 337–347.

Brodsky, M. C. (2003). Three dimensions of skew deviation. *The British Journal of Ophthalmology, 87*(12), 1440–1441.

Brodsky, M. C., et al. (2006). Skew deviation revisited. *Survey of Ophthalmology, 51*(2), 105–128.

Carmona, S., et al. (2016). The diagnostic accuracy of truncal Ataxia and HINTS as cardinal signs for acute vestibular syndrome. *Frontiers in Neurology, 7*, 125.

Chang, T. P., et al. (2018). Sudden hearing loss with Vertigo portends greater stroke risk than sudden hearing loss or Vertigo alone. *Journal of Stroke and Cerebrovascular Diseases, 27*(2), 472–478.

Chen, L., et al. (2011). Diagnostic accuracy of acute vestibular syndrome at the bedside in a stroke unit. *Journal of Neurology, 258*(5), 855–861.

Cho, K. H., et al. (2015). Contraversive ocular tilt reaction after the lateral medullary infarction. *The Neurologist, 19*(3), 79–81.

Cnyrim, C. D., et al. (2008). Bedside differentiation of vestibular neuritis from central "vestibular pseudoneuritis". *Journal of Neurology, Neurosurgery, and Psychiatry, 79*(4), 458–460.

Cohn, B. (2014). Can bedside oculomotor (HINTS) testing differentiate central from peripheral causes of vertigo? *Annals of Emergency Medicine, 64*(3), 265–268.

Collewijn, H., & Smeets, J. B. (2000). Early components of the human vestibulo-ocular response to head rotation: Latency and gain. *Journal of Neurophysiology, 84*(1), 376–389.

Daudelin, D. H., & Selker, H. P. (2005). Medical error prevention in ED triage for ACS: Use of cardiac care decision support and quality improvement feedback. *Cardiology Clinics, 23*(4), 601–614. ix.

Eagles, D., et al. (2008). International survey of emergency physicians' priorities for clinical decision rules. *Academic Emergency Medicine, 15*(2), 177–182.

Furman, J. M., & Jacob, R. G. (1993). Jongkees' formula re-evaluated: Order effects in the response to alternate binaural bithermal caloric stimulation using closed-loop irrigation. *Acta Oto-Laryngologica, 113*(1), 3–10.

Glasauer, S., et al. (2004). Vertical vestibular responses to head impulses are symmetric in downbeat nystagmus. *Neurology, 63*(4), 621–625.

Gold, D., et al. 2018. A novel tele-dizzy consultation program in the emergency department using portable video-oculography to improve peripheral vestibular and stroke diagnosis. In *Diagnostic error in medicine, 11th annual conference*. New Orleans, LA.

Gottesman, R. F., et al. (2012). Clinical characteristics of symptomatic vertebral artery dissection: A systematic review. *The Neurologist, 18*(5), 245–254.

Grewal, K., et al. (2015). Missed strokes using computed tomography imaging in patients with vertigo: Population-based cohort study. *Stroke, 46*(1), 108–113.

Halmagyi, G. M., & Curthoys, I. S. (1988). A clinical sign of canal paresis. *Archives of Neurology, 45*(7), 737–739.

Halmagyi, G. M., Gresty, M. A., & Gibson, W. P. (1979). Ocular tilt reaction with peripheral vestibular lesion. *Annals of Neurology, 6*(1), 80–83.

Halmagyi, G. M., et al. (1990). Tonic contraversive ocular tilt reaction due to unilateral mesodiencephalic lesion. *Neurology, 40*(10), 1503–1509.

Halmagyi, G. M., et al. (2001). Impulsive testing of individual semicircular canal function. *Annals of the New York Academy of Sciences, 942*, 192–200.

Halmagyi, G. M., et al. (2017). The video head impulse test. *Frontiers in Neurology, 8*, 258.

Hess K., & Reisine H. (1984). Counterdrifting of the Eyes: Additional findings and hypothesis. *ORL 46*, 1–6.

Heuberger, M., et al. (2014). Covert anti-compensatory quick eye movements during head impulses. *PLoS One, 9*(4), e93086.

Hubert, G. J., Muller-Barna, P., & Audebert, H. J. (2014). Recent advances in TeleStroke: A systematic review on applications in prehospital management and stroke unit treatment or TeleStroke networking in developing countries. *International Journal of Stroke, 9*(8), 968–973.

Huh, Y. E., et al. (2013). Head-shaking aids in the diagnosis of acute audiovestibular loss due to anterior inferior cerebellar artery infarction. *Audiology & Neuro-Otology, 18*(2), 114–124.

Kattah, J. C., et al. (2009). HINTS to diagnose stroke in the acute vestibular syndrome: Three-step bedside oculomotor examination more sensitive than early MRI diffusion-weighted imaging. *Stroke, 40*(11), 3504–3510.

Kene, M. V., et al. (2015). Emergency physician attitudes, preferences, and risk tolerance for stroke as a potential cause of Dizziness symptoms. *The Western Journal of Emergency Medicine, 16*(5), 768–776.

Kerber, K. A., & Newman-Toker, D. E. (2015). Misdiagnosing dizzy patients: Common pitfalls in clinical practice. *Neurologic Clinics, 33*(3), 565–575. viii.

Kerber, K. A., et al. (2006). Stroke among patients with dizziness, vertigo, and imbalance in the emergency department: A population-based study. *Stroke, 37*(10), 2484–2487.

Kerber, K. A., et al. (2011). Nystagmus assessments documented by emergency physicians in acute dizziness presentations: A target for decision support? *Academic Emergency Medicine, 18*(6), 619–626.

Kerber, K. A., et al. (2013). Use of BPPV processes in emergency department dizziness presentations: A population-based study. *Otolaryngology and Head and Neck Surgery, 148*(3), 425–430.

Kim, H. A., & Lee, H. (2012). Recent advances in central acute vestibular syndrome of a vascular cause. *Journal of the Neurological Sciences, 321*(1–2), 17–22.

Kim, A. S., et al. (2012). Practice variation in neuroimaging to evaluate dizziness in the ED. *The American Journal of Emergency Medicine, 30*(5), 665–672.

Lee, H., & Kim, H. A. (2013). Nystagmus in SCA territory cerebellar infarction: Pattern and a possible mechanism. *Journal of Neurology, Neurosurgery, and Psychiatry, 84*(4), 446–451.

MacDougall, H. G., et al. (2009). The video head impulse test: Diagnostic accuracy in peripheral vestibulopathy. *Neurology, 73*(14), 1134–1141.

MacDougall, H. G., et al. (2013). Application of the video head impulse test to detect vertical semicircular canal dysfunction. *Otology & Neurotology, 34*(6), 974–979.

Mantokoudis, G. *DETECT – Dizziness evaluation tool for emergent clinical triage*. 2017–2021, Swiss National Science Foundation (Grant #320030_173081): Department of Otorhinolaryngology, Head and Neck Surgery, Inselspital, Bern University Hospital, University of Bern, Switzerland.

Mantokoudis, G., et al. (2014). Early adaptation and compensation of clinical vestibular responses after unilateral vestibular deafferentation surgery. *Otology & Neurotology, 35*(1), 148–154.

Mantokoudis, G., et al. (2015a). VOR gain by head impulse video-oculography differentiates acute vestibular neuritis from stroke. *Otology & Neurotology, 36*(3), 457–465.

Mantokoudis, G., et al. (2015b). Quantifying the vestibulo-ocular reflex with video-oculography: Nature and frequency of artifacts. *Audiology & Neuro-Otology, 20*(1), 39–50.

Mantokoudis, G., et al. (2016a). Compensatory saccades benefit from prediction during head impulse testing in early recovery from vestibular deafferentation. *European Archives of Oto-Rhino-Laryngology, 273*(6), 1379–1385.

Mantokoudis, G., et al. (2016b). The video head impulse test during post-rotatory nystagmus: Physiology and clinical implications. *Experimental Brain Research, 234*(1), 277–286.

Mantokoudis, G., et al. (2016c). Adaptation and compensation of vestibular responses following superior canal dehiscence surgery. *Otology & Neurotology, 37*(9), 1399–1405.

Mantokoudis, G., et al. (2016d). Impact of artifacts on VOR gain measures by video-oculography in the acute vestibular syndrome. *Journal of Vestibular Research, 26*(4), 375–385.

Marx, J. J., et al. (2004). Diffusion-weighted MRT in vertebrobasilar ischemia. Application, sensitivity, and prognostic value. *Nervenarzt, 75*(4), 341–346.

McGarvie, L. A., et al. (2015). The video head impulse test (vHIT) of semicircular canal function – Age-dependent normative values of VOR gain in healthy subjects. *Frontiers in Neurology, 6,* 154.

Neuhauser, H. K., et al. (2008). Burden of dizziness and vertigo in the community. *Archives of Internal Medicine, 168*(19), 2118–2124.

Newman-Toker, D. E. (2007). Charted records of dizzy patients suggest emergency physicians emphasize symptom quality in diagnostic assessment. *Annals of Emergency Medicine, 50*(2), 204–205.

Newman-Toker, D. *AVERT Phase II Trial: Acute video-oculography for vertigo in emergency rooms for rapid triage*. 2014–2019, National Institute on Deafness and Other Communication Disorders (NIDCD 1U01DC013778). ClinicalTrials.gov ID NCT02483429. Johns Hopkins University.

Newman-Toker, D. E. (2015). Comment: Diagnosing stroke in acute dizziness-do the "eyes" still have it? *Neurology, 85*(21), 1877.

Newman-Toker, D. E. (2016). Missed stroke in acute vertigo and dizziness: It is time for action, not debate. *Annals of Neurology, 79*(1), 27–31.

Newman-Toker, D. E., & Edlow, J. A. (2015). TiTrATE: A novel, evidence-based approach to diagnosing acute dizziness and vertigo. *Neurologic Clinics, 33*(3), 577–599. viii.

Newman-Toker, D. E., & Perry, J. J. (2015). Acute diagnostic neurology: Challenges and opportunities. *Academic Emergency Medicine, 22*(3), 357–361.

Newman-Toker, D. E., et al. (2007). Imprecision in patient reports of dizziness symptom quality: A cross-sectional study conducted in an acute care setting. *Mayo Clinic Proceedings, 82*(11), 1329–1340.

Newman-Toker, D. E., et al. (2008a). Spectrum of dizziness visits to US emergency departments: Cross-sectional analysis from a nationally representative sample. *Mayo Clinic Proceedings, 83*(7), 765–775.

Newman-Toker, D. E., et al. (2008b). Normal head impulse test differentiates acute cerebellar strokes from vestibular neuritis. *Neurology, 70*(24 Pt 2), 2378–2385.

Newman-Toker, D. E., et al. (2013a). HINTS outperforms ABCD2 to screen for stroke in acute continuous vertigo and dizziness. *Academic Emergency Medicine, 20*(10), 986–996.

Newman-Toker, D. E., et al. (2013b). Quantitative video-oculography to help diagnose stroke in acute vertigo and dizziness: Toward an ECG for the eyes. *Stroke, 44*(4), 1158–1161.

Newman-Toker, D. E., et al. (2014). Missed diagnosis of stroke in the emergency department: A cross-sectional analysis of a large population-based sample. *Diagnosis (Berl), 1*(2), 155–166.

Newman-Toker, D. E., Curthoys, I. S., & Halmagyi, G. M. (2015). Diagnosing stroke in acute Vertigo: The HINTS family of eye movement tests and the future of the "eye ECG". *Seminars in Neurology, 35*(5), 506–521.

Newman-Toker, D. E., Della Santina, C. C., & Blitz, A. M. (2016). Vertigo and hearing loss. *Handbook of Clinical Neurology, 136*, 905–921.

Omron, R., et al. (2018). The diagnostic performance feedback "calibration gap": Why clinical experience alone is not enough to prevent serious diagnostic errors. *AEM Education and Training, 2*(4), 339–342.

Parulekar, M. V., et al. (2008). Head position-dependent changes in ocular torsion and vertical misalignment in skew deviation. *Archives of Ophthalmology, 126*(7), 899–905.

Perez, N., & Rama-Lopez, J. (2003). Head-impulse and caloric tests in patients with dizziness. *Otology & Neurotology, 24*(6), 913–917.

Robinson, D. A. (1963). A method of measuring eye movement using a scleral search coil in a magnetic field. *IEEE Transactions on Bio-medical Electronics, 10*(4), 137–145.

Saber Tehrani, A. S., et al. (2013). Rising annual costs of dizziness presentations to U.S. emergency departments. *Academic Emergency Medicine, 20*(7), 689–696.

Saber Tehrani, A. S., et al. (2014). Small strokes causing severe vertigo: Frequency of false-negative MRIs and nonlacunar mechanisms. *Neurology, 83*(2), 169–173.

Savitz, S. I., Caplan, L. R., & Edlow, J. A. (2007). Pitfalls in the diagnosis of cerebellar infarction. *Academic Emergency Medicine, 14*(1), 63–68.

Stanton, V. A., et al. (2007). Overreliance on symptom quality in diagnosing dizziness: Results of a multicenter survey of emergency physicians. *Mayo Clinic Proceedings, 82*(11), 1319–1328.

Tarnutzer, A. A., et al. (2011). Does my dizzy patient have a stroke? A systematic review of bedside diagnosis in acute vestibular syndrome. *CMAJ, 183*(9), E571–E592.

Tarnutzer, A. A., et al. (2017). ED misdiagnosis of cerebrovascular events in the era of modern neuroimaging: A meta-analysis. *Neurology, 88*(15), 1468–1477.

Thomas, D., & Newman-Toker, D. (2016). Avoiding "HINTS positive/negative" to minimize diagnostic confusion in acute vertigo and dizziness. *Journal of Acute Care Physical Therapy, 7*(4), 129–131.

Tjernstrom, F., Nystrom, A., & Magnusson, M. (2012). How to uncover the covert saccade during the head impulse test. *Otology & Neurotology, 33*(9), 1583–1585.

Vanni, S., et al. (2015). Can emergency physicians accurately and reliably assess acute vertigo in the emergency department? *Emergency Medicine Australasia, 27*(2), 126–131.

Weber, K. P., et al. (2008). Head impulse test in unilateral vestibular loss: Vestibulo-ocular reflex and catch-up saccades. *Neurology, 70*(6), 454–463.

Weber, K. P., et al. (2009). Impulsive testing of semicircular-canal function using video-oculography. *Annals of the New York Academy of Sciences, 1164*, 486–491.

Neuroimaging of the Acute Vestibular Syndrome and Vascular Vertigo

Jorge C. Kattah

Abstract The focus of this chapter is to provide a review of the main diagnostic imaging modalities utilized in the examination of peripheral and central acute vestibular syndrome (AVS). Computed tomography (CT) and magnetic resonance imaging (MRI) scan will be the two main modalities discussed. The chapter begins with a discussion of imaging of peripheral vestibulopathy associated with an AVS, followed by imaging of central lesions. The most common cause of a central lesion causing an AVS is by far cerebrovascular; thus, our secondary aim is to review the role of imaging in the evaluation of cerebrovascular lesions (vascular vertigo) associated with an acute AVS.

Keywords Vertigo · Cerebellum · Neuroimaging · Stroke

1 Introduction

The acute onset of vertigo, nausea, vomiting, and head movement intolerance defines the "acute vestibular syndrome" (AVS). In the past 30 years, there have been significant advances in the understanding of the pathophysiology and clinical findings that identify the "lesion responsible" for an AVS. Clinical localization, based on a detailed history and physical examination, provides information as to vestibular pathway location and lateralization; the ultimate goal is to differentiate peripheral vestibular lesions from lesions that involve the central vestibular pathways or the cerebellum. Imaging provides an opportunity to verify the location and size of the lesion, to categorize the etiology, and to determine the effect of the lesion on surrounding structures.

Pure isolated vertigo may develop in patients with acute, unilateral vestibular disorders, principally vestibular neuritis (VN), and leads to prolonged, continuous

J. C. Kattah (✉)
University of Illinois College of Medicine at Peoria, Peoria, IL, USA
e-mail: kattahj@uic.edu

© Springer Nature Switzerland AG 2019
A. Shaikh, F. Ghasia (eds.), *Advances in Translational Neuroscience of Eye Movement Disorders*, Contemporary Clinical Neuroscience,
https://doi.org/10.1007/978-3-030-31407-1_16

vertigo, lasting longer than 24 h (Leigh and Zee 2015; Baloh and Halmagyi 1996; Bisdorff et al. 2009). Likewise, ischemic lesions involving the brainstem and cerebellum may present with isolated continuous vertigo (pseudo-neuritis). Clinical examination often identifies characteristic findings generally confirmed by magnetic resonance imaging (MRI) (Leigh and Zee 2015; Baloh and Halmagyi 1996; Hotson and Baloh 1998; Duncan et al. 1975; Newman-Toker et al. 2008; Kattah et al. 2009; Lee et al. 2009; Choi et al. 2014a; Kim et al. 2016a; Newman-Toker and Edlow 2015). The most common forms of vertigo are episodic, classified by virtue of their duration and potential triggers, frequently of peripheral origin and often with a benign clinical course. The major clinical challenge we currently face is the diagnosis of transient, non-triggered episodic, isolated vertigo that lasts several minutes or hours in patients with normal interictal examination. In this group, transient ischemic attacks (TIAs) are short-term predictors of a stroke (Newman-Toker and Edlow 2015). Vertigo, sometimes described by patients as dizziness, occurs in 10–20% of patients with vertebrobasilar ischemia/insufficiency (VBI) and often precedes a posterior circulation stroke (Newman-Toker and Edlow 2015; Caplan 2015; Newman-Toker 2015; Kerber et al. 2014). Dizziness is the most common symptom of posterior circulation TIA (Newman-Toker and Edlow 2015; Caplan 2015).

The focus of this chapter is to provide a review of the main diagnostic imaging modalities utilized in the examination of peripheral and central AVS. CT scan and MRI will be the two main modalities discussed. I will begin with a discussion of imaging of peripheral vestibulopathy associated with an AVS, followed by imaging of central lesions. The most common cause of a central lesion causing an AVS is by far cerebrovascular; thus, our secondary aim is to review the role of imaging in the evaluation of cerebrovascular lesions (vascular vertigo) associated with an acute AVS.

2 Peripheral Vestibulopathy

2.1 Imaging of Peripheral Vestibulopathy

1. *Benign Paroxysmal Positional Vertigo (BPPV)*: BPPV is an episodic syndrome that only rarely is associated with an AVS; when it does, it is infrequent and potentially overlooked if a Dix-Hallpike or horizontal head roll maneuver is not a routine test. This acute variant of BPPV may be associated with intense vertigo, nausea, and vomiting, and, as a result, the patient is afraid to move and is initially not aware of positional triggers. Evidence-based knowledge of the pathogenesis and management of the most common types of BPPV led to the institution of consensus recommendations; the standard of care is clinical diagnosis, followed by expeditious treatment; and imaging of the brain or the ear is not necessary (Furman and Cass 1999; Kim and Zee 2014; von Brevern et al. 2015). Central positional vertigo (CPV), on the other hand, is usually associated

with additional neurologic abnormalities and requires neuroimaging. The usual nystagmus pattern of concern during positional testing involves provoked paroxysmal or sustained downbeat positional nystagmus, in addition to different types of intractable horizontal cupulolithiasis; in these cases, the fast phase of the horizontal nystagmus beats away from the direction of the head roll (apogeotropic horizontal nystagmus). Not infrequently, both patterns of downbeat nystagmus (DBN) and apogeotropic horizontal nystagmus may be found in one patient; imaging usually detects a stroke and less commonly other etiologies (Kim et al. 2012) to account for acute CPV (Choi et al. 2015a). Migraine may be associated with an AVS and positional DBN (El-Badry et al. 2017). Imaging in migraine cases might be necessary during the first vertigo episode, particularly if a previous history is negative; in migraine, imaging results are usually normal.

2. *Vestibular Neuritis (VN)*: Acute vestibular neuritis (VN) is the second most common cause of vertigo, after BPPV, which affects about 5.6% of acutely dizzy patients routinely evaluated in the ED, every year, and roughly 150,000 annual patients in the USA. The average age affected is the sixth decade. Imaging in VN is generally normal; in a previous series involving acute VN patients that utilized an MRI slice thickness of 2 mm and no gap, 60 patients had normal findings (Strupp et al. 1998). There are isolated reports of acute VN that demonstrate focal superior vestibular nerve enhancement (Karlberg et al. 2004; Park et al. 2014), but these reports are infrequent. The lack of alterations of blood nerve barrier in VN, after contrast enhancement, argues against inflammation (Strupp et al. 1998) and does not support the proposed viral/post-viral (herpes simplex virus: HSV) etiology. Similarly, the lack of imaging changes in idiopathic VN stands in sharp distinction with the overt inflammatory findings observed in the Ramsay Hunt syndrome due to varicella zoster virus (VZV) infection of the vestibular nerve (Fig. 1), which sometimes may be associated with a brainstem encephalitis (Shen et al. 2015). The lack of contrast enhancement is also different from the facial nerve enhancement seen in Bell's palsy (Tien et al. 1990), a viral/post-viral syndrome, presumably related to HSV that frequently affects the intralabyrinthine facial nerve.

Fig. 1 Coronal T1 MRI of the brain after the administration of contrast in a patient with AVS due to VZV infection. She also had a left peripheral facial paresis and a typical vesicular rash in the left auricle. Hearing was normal. The arrow points to contrast enhancement of the seventh/eighth nerve complex

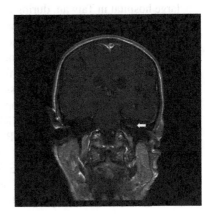

3. In our personal experience with 100 VN cases, we did not find inflammatory eighth nerve complex changes in MRI. The only abnormality found in our VN cohort was a transient radiographic conjugate horizontal eye deviation, in the direction of the slow phase of the nystagmus, noted in either MRI or CT scan as a consistent VN lateralized finding (Kattah et al. 2011).

   An acute, ischemic vestibular neuropathy has, only rarely, been associated with focal restricted diffusion in the vestibular nerve (Choi et al. 2015b). This is also a rather unusual MRI finding; the lack of consistent signal changes in the diffusion-weighted image (DWI) obtained from VN patients also questions the proposed ischemic mechanism hypothesis as the cause of VN. An autoimmune etiology, however, is a strong possibility (Greco et al. 2014) and may be even axonal, rather than demyelinating, which could explain the paucity of imaging findings.

4. *Acute cochleovestibular (ACV) loss*: Unlike VN, the combined acute continuous cochleovestibular loss (with sudden deafness or severe hearing loss) is likely to show ischemic parenchymal brainstem findings, rather than isolated VN, even in cases with a positive HIT and nystagmus characteristics that suggest a peripheral localization. This is predictable, in consideration of the fact that the entire arterial circulation of the inner ear is provided by branches of the anterior-inferior cerebellar artery (AICA) (Hotson and Baloh 1998). Acute DWI signal changes involving the lateral pons, anterior cerebellum, and flocculus may be present along with cochleo-labyrinthine ischemia (Lee et al. 2009; Kim et al. 1999; Kim and Lee 2010, 2017), and less commonly, posterior inferior cerebellar artery (PICA) territory or both AICA and PICA may be found (Kim et al. 2016a). In addition, AICA TIAs are typically combined intermittent cochleovestibular symptoms with high risk for subsequent infarction (Lee and Cho 2003). We added hearing loss to the HINTS triad (Kattah et al. 2009; Newman-Toker et al. 2013a), because acute cochleo-vestibular loss in our cohort pointed to acute labyrinthine ischemia, often in AICA territory and less commonly in PICA territory ischemia; this relationship is found in previous reports, as well (Lee et al. 2009; Kim et al. 1999, 2016a; Kim and Lee 2017; Amarenco and Hauw 1990; Huang et al. 1993). Severe cochleovestibular loss was found in 7 ($n = 503$) patients, at a large hospital in Taiwan, during a five-year period of time (Huang et al. 1993).

   Recent studies identified poor hearing prognosis among patients with acute cochlear loss concurrent with acute posterior canal BPPV; the cause of acute deafness/posterior canal BPPV may relate to utricular ischemia with dislodging otoconia and posterior canalolithiasis (Kim et al. 2014).

   Acute cochleo-vestibular loss in patients receiving chronic anticoagulant therapy or patients with coagulopathies, an intralabyrinthine hemorrhage (ILH) is a possible cause of an AVS. MRI is critical to diagnose ILH, and subtle hyperintense T1-weighted MRI and GRE signal in the affected labyrinth confirm diagnosis. The post-hemorrhage by-product of hemoglobin metabolism (met-hemoglobin) is responsible for these changes and yields increased signal within the cochlea and adjacent horizontal semicircular canal, about 48 h after the clinical onset of symptoms (Fig. 2); in addition, there is lack of post-contrast

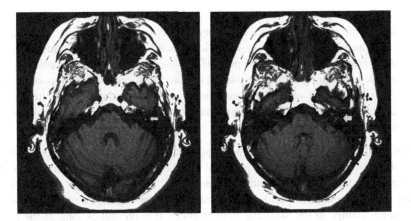

Fig. 2 Intralabyrinthine hemorrhage. Axial T1 MRI of the left cerebellopontine angle in a patient with an acute left cochleovestibular loss. The arrows in both panels point to hyperintense signal within the left cochlea and left horizontal canal in a patient on rivaroxaban (Xarelto). In contrast, the right cochlea and horizontal semicircular canal show normal hypointensity. Mild signal changes are present in the right mastoid

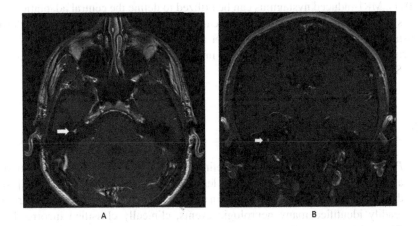

Fig. 3 Panel A. Right axial, post-contrast T1 MRI obtained in a patient with acute right cochleovestibular loss. The arrow points to an enhancing lesion within the cochlea. Panel B. Left, post-contrast coronal MRI. The arrow points to the same enhancing lesion. Two subsequent follow-up MRIs show similar findings. This is most consistent with a cochlear schwannoma

MRI enhancement (Whitehead et al. 1998; Sugiura et al. 2006). A repeat, limited met-hemoglobin-specific MRI scan may detect evolving met-hemoglobin if an earlier MRI was nondiagnostic. Cochlear schwannoma (Fig. 3) may also associate with acute cochleo-vestibular loss; the MRI scan typically shows chronic, persistent cochlear post-gadolinium enhancement (Tieleman et al. 2008). Purulent otitis may involve the labyrinth and be associated with an abnormal MRI (Kim et al. 2016b). Finally, CT scan of the temporal bone is usually not required in AVS patients; the majority of pathologic processes occurring in the adult temporal bone are inflammatory or neoplastic in nature and not likely to cause an AVS. Inflammatory conditions such as cholesteatoma and malignant otitis externa could be an exception if they are rapidly evolving. Acute intraneoplastic bleeding in tumors of the internal auditory canal/cerebellopontine angle cistern, including vestibular schwannoma and meningioma, glomus tumors, and CP angle vascular malformation, could bleed and cause an AVS and could be identified with MRI.

MRI-Induced Nystagmus and Vertigo A discussion of neuroimaging and vestibulopathies must factor in the potential effect of vestibular stimulation by the MR magnet (magnetic vestibular stimulation: MVN). This stimulation varies in intensity with the magnetic field strength and does not require head movement nor dynamic change in the magnetic field strength; it is present as long as the patient stays within the magnet but may change with the degree of head pitch (Roberts et al. 2011; Mian et al. 2015). It is absent in patients with vestibular areflexia (Roberts et al. 2011) and may be associated with vertigo and robust nystagmus (Ward and Zee 2016). The MVS results from a Lorentz magnetic force and the endolymphatic ionic current within the cupula of the horizontal semicircular canal (Roberts et al. 2011). MRI-induced nystagmus can be utilized to define the central adaptation process to a sustained vestibular bias, revealing multiple adaptation set points during sustained stimulation (Jareonsettasin et al. 2016). The effect of MVS in AVS of peripheral and central origin is under investigation. MRI in peripheral or central AVS is under investigation.

3 Vascular Vertigo

1. *Posterior Fossa TIAs*: The classic definition of TIA is an arbitrary designation of a neurologic deficit that resolves in less than 24 h. However, DWI and adjusted diffusion coefficient (ADC) studies in acute, TIA-like, CNS ischemic syndromes readily identified many neurologic events, clinically classified incorrectly as TIAs with associated MRI signal changes (Oppenheim et al. 2000). A "tissue definition of TIA" therefore seems more appropriate, when the MRI-positive clinical deficit resolves before the previous 24-h duration criterion (Albers et al. 2002). Given pre-existing literature included in this review, it is best to classify TIAs into two types: DWI/ADC-negative and DWI/ADC-positive TIAs; the latter are more frequently associated with subsequent stroke than their DWI/TIA-negative counterpart (Engelter et al. 2008).

   The principal mechanisms of posterior circulation TIAs (with vertigo and dizziness as the main symptom) include large artery atherosclerosis with artery-to-artery emboli and hemodynamic failure triggering ischemic symptoms, from hypoperfusion to arterial dissection. Any of these mechanisms may trigger reversible, asymmetric vestibular hypofunction, due to ischemia of the vestibular labyrinth, nerve, fascicle, or nucleus or due to cerebellar parenchymal ischemia with transient loss of normal inhibitory modulation of the vestibular nucleus (nodulus and uvula). The medial vestibular nucleus is particularly susceptible to ischemia in experimental animals (Lee et al. 2014a). Small vessel disease in the brainstem and cerebellum usually causes stroke without TIAs; however, with early imaging, there is a risk of a false-negative study, and the clinician is at risk of missing a structural lesion with an incorrect TIA label. Less common arteriopathies may also cause vertebrobasilar TIAs. On occasion, TIAs in PICA territory have unique characteristics such as tilting of the environment (tortopsia) or a self-perception of tilt and falls (drop attacks).

Transient isolated vascular vertigo is abrupt and lasts several minutes (Lee and Cho 2003; Grad and Baloh 1989), while spontaneous, recurrent episodic vertigo may precede a posterior circulation stroke, 1–10 days prior to stroke onset (Caplan 2015). In addition, auditory symptoms may be present in AICA vascular territory and thus mimic Meniere's disease (Lee and Cho 2003). In the NEMC-PCR, two broad groups of patients had vertebral artery (VA)-related TIAs. In the extracranial vertebral artery (ECVA) group, 80 ($n = 407$) patients 50% presented with a stroke after a TIA (26%) and 24% had isolated TIAs. In the group with intracranial vertebral artery (ICVA) disease, 132 ($n = 407$) patients had more than 50% stenosis of both intracranial vertebral arteries most related to atherosclerotic lesions; this is the largest ICVA series reported to date (Caplan 2015). In addition, 20 ($n = 407$) patients also had proximal basilar artery (BA) stenosis; among these patients, 25% had TIAs and 25% had TIAs followed by stroke, while several of the remaining patients had a stroke followed by TIAs. Both TIA and stroke affected the proximal and middle territories. About one-half of the patients with very severe (greater than 75%) stenosis had multiple TIAs, often triggered by standing, hypovolemia, and hypotension; syncope, diplopia, vertigo, and ataxia were frequent symptoms (Caplan 2015). Symptoms associated with advanced ICVA and BA disease include recurrent syncope, drop attacks, and vertigo, particularly in individuals with stroke risk factors. Curiously, TIAs in this vascular setting recurred for months and even years (Caplan 2015). Eventually, however, the strokes that developed in these high-stenosis patients were large and involved both the proximal and middle territories, without any instance of isolated proximal territory stroke. Hypoperfusion was the most common cause of TIA and stroke mechanism, followed by artery-to-artery emboli in this NEMC-PCR registry. Small perforator vessel strokes were not preceded by TIAs (Caplan 2015). Vestibular symptoms triggered by critical ischemia can potentially be very elusive as exemplified in a recent report (Halmagyi 2017). Perhaps, the best approach here is vascular imaging even if the MRI is unremarkable.

2. *False-Negative DWI/ADC MRI in Ischemic Stroke*: Following introduction of the DWI/ADC map, it became clear that DWI/ADC lesions detected approximately 95% of all strokes (Provenzale and Sorensen 1999). However, an early report, shortly thereafter, featured eight stroke patients with false-negative MRI, all identified in the posterior circulation. Early imaging performed within the first 6 h after the onset of symptoms was the theoretical explanation for the initial

lack of sensitivity (Oppenheim et al. 2000). Although DWI/ADC MRI may show false-negative signal in anterior circulation stroke, it is evident that this possibility is more frequent in the posterior circulation. Medullary strokes are perhaps the most likely to show delayed ischemic signal changes (Engelter et al. 2008; Lee et al. 2015). In a subsequent series of eight AVS patients with small- or medium-size cerebellar strokes, false-negative MRI in stroke patients with central nystagmus types and neurologic examination defined the precise diagnosis, and a repeat MRI showed either a cerebellar parenchymal or a cerebellar peduncle small- or mid-size stroke (Morita et al. 2011). In our series, HINTS and HINTS-plus predicted stroke in eight patients with initially false-negative MRI scans, and some performed as late as 48 h after symptom onset (Kattah et al. 2009; Newman-Toker et al. 2013a). Subsequent studies show similar results (Batuecas-Caletrío et al. 2014; Chen et al. 2011); however, encouraging results from a recent series showed only one, initially false-negative MRI (Kerber et al. 2015). Although it is likely that improved imaging techniques will detect a greater number of strokes in the posterior circulation, it is appropriate to avoid overreliance on diagnostic imaging.

A ready explanation for the lack of MRI tissue infarction detection in these cases is unknown; factors associated with the MRI technology itself or an inadequate stroke protocol are possibilities. Additionally, the small size of the lesion, coupled with an insufficient signal-to-noise ratio or with magnetic susceptibility artifacts, may interfere with the image analysis. The ideal cerebellar/brainstem stroke MRI should be multiplanar and without slice gap, which, to my knowledge, is not the usual protocol; besides technical factors, it is important to consider the transition tissue signal threshold from reversible ischemia to irreversible DWI/ADC map; the ischemic penumbra may be large and eludes detection during initial imaging scans (Engelter et al. 2008; Lefkowitz et al. 1999). This is an interesting possibility for potential therapeutic intervention, particularly when a symptomatic patient has a normal parenchymal MRI, coupled with evidence of "arterial occlusion" manifested by loss of the normal arterial signal void in a routine MRI (Rathakrishnan et al. 2008). Occasionally, in patients with large posterior circulation arterial occlusion, serial imaging is necessary to monitor progression (Fig. 4).

3. *Rotational Vertebral Artery Occlusion (RVAS)*: This rare syndrome is characterized by recurrent attacks of vertigo, nystagmus, or syncope, all triggered by horizontal head rotation, but also by head tilt or extension. Most RVAS patients have unilateral congenital VA hypoplasia, or acquired stenosis, and the normal "patent" dominant VA is compressed at the C1–C2 level, leading to symptomatic ischemia of the posterior fossa structures (Choi et al. 2013). Other less common compressive syndromes may also occur. The dominant nystagmus is composed of horizontal, torsional beating in the direction of the head turn with a mixed downbeat component. Typically, the symptoms subside when the head returns to the neutral position. In 23 patients from a citywide series treated at several institutions in Seoul, Korea, conservative approach led to a favorable outcome, in

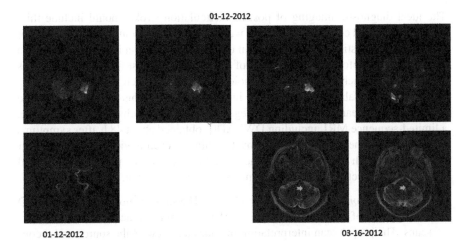

Fig. 4 Axial DWI MRI (upper four panels) obtained on 01-1-2-2012 in a patient with an AVS, ataxia and mild left hemiparesis, and normal hearing; initially his manual HIT was normal. In the follow-up, it became abnormal and an obvious explanation for this change was unclear. The embolic ischemic strokes involved both cerebellar hemispheres and the left striate cortex, which did not explain the cause for a positive HIT. The MRA showed marked flow attenuation of the left vertebral and occlusion of the proximal basilar artery (left lower panel). The manual HIT remained abnormal, and on 03/06/2012, a repeat MRI (axial T2, lower right panels) showed increased signal at the left root entry and the fascicle of the left vestibular nerve, thus providing a delayed explanation for the positive manual HIT

contrast to previous recommendations urging immediate decompression (Choi et al. 2013). Obviously, each case should have individual consideration.

3.1 Diagnostic Modalities Utilized in the Diagnosis of Stroke and TIA

At this point, we will review different diagnostic modalities utilized in patients with isolated peripheral/central or purely central vestibular pathway abnormalities or in association with other neurologic abnormalities. Visualization of the affected brainstem structures and the status of the entire posterior fossa arterial circulation, by utilizing noninvasive methods, is the clinician's goal.

Stroke MRI Brain Imaging Prior to the introduction of MRI technologies, stroke management was a significant challenge, and treatment delay, due to diagnostic uncertainty, was prevalent. DWI/ADC emerged as the earliest, most reliable test of acute ischemic cerebral infarction. For practical reasons, MRI stroke protocols are frequently designed with anterior circulation in mind; a possible explanation is the lower incidence of posterior circulation stroke, which represents only about 20% of all ischemic strokes (Newman-Toker 2015; Kerber et al. 2014; Searls et al. 2012).

The ideal diagnostic imaging of posterior circulation stroke should include thin, axial, sagittal, and coronal DWI scans, without gap, throughout the entire posterior fossa. The diagnostic yield of a CT scan of the posterior fossa is very low (Chalela et al. 2007). Therefore, MRI is the test of choice. The greatest value of a CT scan is detection of parenchymal or subarachnoid hemorrhage, which is less conspicuous, though still detectable by MRI. Future imaging improvements will likely decrease the possibility of early, initially false-negative MRI scan results. In our experience, a limited-sequence MRI, including DWI/ADC, obtained about 48 h after symptoms onset, detected the stroke predicted by the clinical examination (Newman-Toker et al. 2008; Kattah et al. 2009). In a more recent series of AVS patients, MRI failed to detect the expected lesion in only one instance (Kerber et al. 2015).

1. *Magnetic Resonance Angiography (MRA)*: The classic modalities include 2D and 3D time-of-flight and multiple overlapping thin slab acquisition (MOTSA) scans. The proper scan interpretation includes a review of the source and reconstruction images. MRA in the posterior fossa, when compared with DSA and ultrasound (US), showed high degree of sensitivity/specificity (Caplan 2015). In general, MRA overestimates the degree of arterial obstruction, and germane to this chapter, its sensitivity is low in the AICA arterial territory. The use of combined MRA and transcranial Doppler (TCD) is complementary (Caplan 2015). We evaluated patients with pacemakers and other ferromagnetic devices who cannot undergo MR testing with ultrasound (preferably TCD) and CT angiography.

2. *CT and CT Angiography (CTA)*: The sensitivity of CT scan to detect an ischemic lesion in the posterior circulation is low (Chalela et al. 2007). CT detection of a cerebellar or brainstem hemorrhage associated with isolated vertigo is also low; therefore, CT is not an adequate screening test for a vascular cause of vertigo (Kerber et al. 2012). CTA is, in general, comparable to MRA, and it may have greater sensitivity in the detection of intracranial VA stenosis/occlusion. In imaging of smaller vessels such as branches of PICA and SCA and more importantly in AVS patients, AICA is more difficult. CTA may also miss intracranial VA dissection (Fig. 5). CTA, for example, is far more sensitive than MRA in detecting vertebral or basilar artery hypoplasia or stenosis, which is greater than 50% of the lumen, as a cause of transient vertigo without stroke (Pasaoglu 2017). It is faster and cheaper; it is also more accessible and provides greater

---

Fig. 5 (continued) showed multiple PICA territory cerebellar infarcts, but no evidence of AICA infarction; we considered the possibility of a common AICA/PICA artery occlusion. (**b**) CTA showed patent vertebral arteries and there was no opacification of the left PICA. This finding was consistent with occlusion of a common artery for both the PICA and the AICA vascular territories; it further pointed to the presence of PICA/AICA artery. The patient started antiplatelet agents. Clinically, he had permanent left cochleovestibular loss, left upper extremity ataxia, and positional paroxysmal downbeat nystagmus. (**c**) Three weeks later, he developed acute left peripheral facial weakness, and we performed a DSA that confirmed a diagnosis of an ICVA dissection. Occlusion of the ostium of AICA caused a new infarct in the left middle cerebellar peduncle (not shown) that was responsible for the new stroke. He started Coumadin and did well, except for permanent deafness; his vestibular adaptation has been excellent

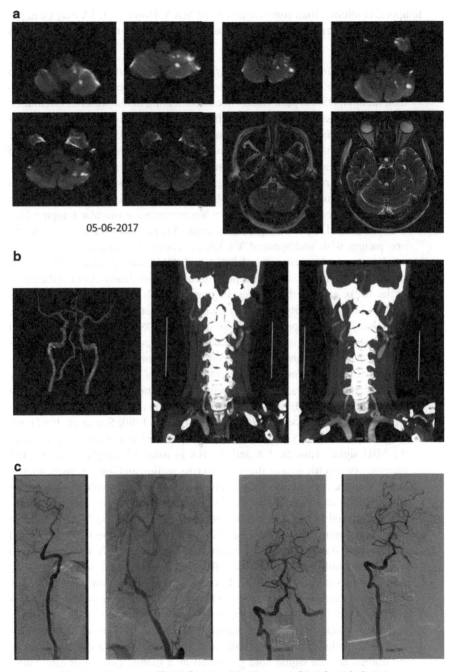

Fig. 5 Intracranial V4-VA dissection. (**a**) Axial DWI (first six panels) and axial T2 MRI (lower panels) obtained in a patient who had an acute left cochleovestibular loss and ataxia. The MRI

temporal resolution than contrast-enhanced MRA. However, CTA may be neph-rotoxic, involves radiation exposure, and may miss heavily classified stenosis (Pasaoglu 2017).

- *Vascular Imaging by Catheter Digital Subtraction Angiography (DSA) as Compared with CTA, MRA, and Ultrasound*: Considered once the gold standard method, DSA is performed via catheterization of the posterior circulation arteries with instillation of contrast. Its value in identifying the exact source of a stroke has been extensively demonstrated (Caplan 2015) and is, by far, the more precise method of arterial imaging. The introduction of noninvasive methods such as ultrasound (US), magnetic resonance angiography (MRA), and computerized tomographic angiography (CTA) subsequently enabled high-quality screening options. Dealing with this subject, L. Caplan (Caplan 2015) recommends the use of DSA when noninvasive vascular imaging fails to explain the clinical findings; for example, I applied this recommendation to one patient with undiagnosed V4-VA dissection. He had recurrent strokes involving vascular territories of PICA and AICA over a three-week period, and the MRA failed to detect the precise cause. DSA finally clarified the cause of the stroke and ruled out other diagnostic possibilities (Fig. 5a–c). DSA is useful when the severity of the vaso-occlusion is not ascertained or not precise enough to make therapeutic decisions. Certainly, DSA is necessary for arterial thrombolysis and angioplasty, as well as other angiographic intervention procedures. MRA/CTA and US can expeditiously identify the cause of the neurologic dysfunction, in most cases. Occluded vessels have high signal intensity in spin-echo sequences, with the absence of flow void or no flow enhancement in gradient-echo (GRE) images. Although initially described primarily in vertebral artery (VA) dissection (Quint and Spickler 1990; Sue et al. 1992), we frequently found this sign in atherosclerotic occlusion and routinely assessed the MRI signal from the VA and the BA in axial T2-weighted and FLAIR images; asymmetric loss of the normal cross section and low-intensity arterial signal void are early signs of arterial occlusion in patients with an AVS (target sign) (Kattah and Pula 2015). Basilar artery (BA) hyperintensity was previously identified as a sign of BA occlusion (Biller et al. 1988). We conducted a preliminary retrospective analysis of cases from a prospectively compiled database and found a "hyperintense arterial sign" among AVS patients due to stroke with initially false-negative MRIs; this sign predicted subsequent DWI/ADC changes in follow-up MRI. In contrast, we did not find this sign in VN patients (*unpublished observation*). Our main limitations rest on the fact that vertebral artery (VA) size asymmetry is common in normal subjects; in addition, the retrospective nature of our study requires further prospective investigation. Notwithstanding these limitations, in the proper clinical setting, loss of the normal arterial signal void generated by a mobile column of circulating blood may represent an additional diagnostic clue for a stroke in evolution.

3. *Evaluation of Perfusion/Diffusion Mismatch*: Perfusion/diffusion mismatch in any ischemic tissue represents a potentially great tool to determine final infarct

size; theoretically, it is an important parameter to investigate those patients with continuous vertigo and those with episodic, non-triggered vertigo/dizziness episodes (possible TIAs). Vestibular symptoms may be either isolated or one of the principal symptoms of posterior fossa arterial disease. Perfusion MRI may be performed following gadolinium administration. However, the contrast dose is high and potentially toxic; the recently introduced arterial spin labeling is a promising future tool (continuous arterial spin labeling perfusion MRI: CASL pMRI) (Alsop and Detre 1996). Perfusion CTA (p-CTA) may be of limited use in the posterior fossa because stroke detection in the posterior fossa is low and the assessment of the size of the stroke core may be potentially underestimated.

4. *Point-of-Care Duplex Sonography*: In a recent report studying VA duplex sonography in AVS patients with presumed central etiologies, the authors report a sensitivity of 53.6% and specificity of 94.9% (Nazerian et al. 2018). Point-of-care ultrasound is a novel diagnostic tool, now utilized by frontline providers; however, there is potential from this limited experience (Nazerian et al. 2018) to be fostered, as long as it does not replace the value of a careful clinical evaluation. In general, US in combination with MRA is powerful to detect circulation abnormalities in the posterior circulation (Caplan 2015).

3.2 Imaging of Central Vestibulopathy due to Stroke

To begin this discussion, the vascular territory classification used in the New England Medical Center Posterior Circulation Registry (NEMC-PCR) provides a practical scaffold for imaging-clinical correlation (Caplan et al. 2005). This study thoroughly evaluated 407 patients, between 1988 and 1996, and established three vascular territories: (1) *proximal territory* that includes the intracranial vertebral arteries, the vertebrobasilar junction, and their branches – posterior-inferior cerebellar artery (PICA) and the anterior spinal artery (ASA); (2) *middle territory* that includes the mid-basilar artery and its branches and the anterior inferior cerebellar artery (AICA); and (3) *distal territory* that includes the tip of the basilar artery and the origin of the posterior cerebral arteries (PCA). In addition, we will also discuss findings related to perforator artery ischemic lesions and lacunar strokes.

Causes for posterior circulation stroke in the NEMC-PCR included embolism in 40% of cases (14% artery to artery), large artery occlusion (32%), and in situ small vessel occlusion (28%). The possibility of cardiac emboli to the vertebrobasilar circulation is relatively low but is an integral workup in patients with central vestibular pathway lesions and a normal vascular tree. In the NEMC-PCR, 24% of the strokes originated from a cardiac source (Searls et al. 2012; Caplan et al. 2005).

Even though ischemic AVS may occasionally be isolated in small cerebellar and brainstem strokes (Choi et al. 2014b; Saber Tehrani et al. 2014), more commonly, it is an early, though dominant, presentation, in combination with additional neurologic abnormalities. Occlusion of the VA, PICA, or AICA may be found in isolated AVS cases (Choi et al. 2014a; Saber Tehrani et al. 2014); however, perforator vessel with in situ thrombosis may occur in any location in the pons, medulla, and cerebellar peduncles with prominent central vestibular signs (Choi et al. 2014a; b; Saber Tehrani et al. 2014). Lacunar brainstem strokes may also cause a central AVS. Cerebellar and brainstem strokes are frequent. Pure superior cerebellar artery (SCA) strokes present frequently with dysarthria and limb ataxia; however, they may develop concurrent with PICA/AICA territory infarcts, and thus, an AVS with severe nausea and vomiting may be present.

Table 1 summarizes common neurotologic and eye movement abnormalities associated with lacunar strokes causing focal lesions of the brainstem, inferior and middle cerebellar peduncles, and vestibular cerebellum. Timely imaging, coupled with thorough clinical examination, offers an interesting combination of findings, valuable for lesion localization, and facilitates lesion classification based on a "set" of vestibular and oculomotor findings. In reality, the majority of strokes that we confront are larger, and the neurologic findings provide helpful localization; however, the small lesions are more likely to be overlooked, and the early MRI tests are more likely to be initially false negative. In addition, we found on serial examination that localizing neurologic findings might initially be either absent or subtle and develop in the first few hours.

4   Proximal Territory

4.1   Vertebral and Posterior Inferior Cerebellar Artery (PICA)

Lateral Medullary Syndrome (LMS)

Gaspard Vieusseux, in 1808, provided the first clinical description of a lateral medullary syndrome (LMS), and the first pathologic examination and clinicopathologic correlation was described by Wallenberg (Wallenberg 1901). LMS accounts for about 2% of all strokes (Norrving and Cronqvist 1991). LMS is very heterogeneous in its clinical manifestations (Kim 2003) and prognosis (Kim 2003; Caplan et al. 1986). An AVS, on presentation, is common, and it may resolve prior to other neurologic abnormalities (Kattah et al. 2017). LMS is usually the result of an ipsilesional VA occlusion (Sacco et al. 1993); however, in Wallenberg's case, the occluded vessel was PICA (Wallenberg 1895). On occasion, a lateral medullary infarct coincides with an ipsilesional cerebellar stroke. The main stroke mechanisms leading to LMS include large vessel occlusion (VA or PICA), VA dissection, embolism, and infrequently small perforator branch artery lacunar stroke. Charles Foix described the normal irrigation of the medulla by paramedian and short circumferential and long circumferential branches (Caplan 1990). In a large series that examined 123 angiographic procedures, 95% of patients had abnormalities (83 had VA disease and 12 isolated PICA disease) (Kim 2003). The neurotologic findings vary in relation to the structures affected (Table 1). In a 2003 series, the symptom combination of vertigo, diplopia, and nausea/vomiting correlated with clinical evidence of skew, primary position nystagmus, and gait ataxia related to vestibular nuclear ischemia/stroke

Table 1 Central vestibular findings in brainstem and cerebellar lesions

Stroke location	Number of patients	Nystagmus and main neurotologic findings	Head impulse test	Skew head tilt OTR	Other common findings
Prepositus Hypoglossi Nucleus (Kim et al. 2016c) NPH	8	Horizontal Ipsilesional Fast phase 1 reversed direction in dark Direction change GEN[a] 1 perverted HSN[c]	Abnormal HIT Contralesional Normal calorics	Controversive 4 Ipsiversive 3	Ipsilesional abnormal pursuit Ipsilesional body lateropulsion
Medial Vestibular Nucleus (Kim and Lee 2010) MVN	1	1 horizontal Contralesional Asymmetric GEN[a]	Abnormal Ipsilesional HIT Ipsilesional Canal paresis vHIT shows bilaterally decreased gain	Excyclotorsion of the contralateral Eye Skew	Truncal balance mildly affected Ipsilesional
Middle Cerebellar Peduncle (Kim and Kim 2019) MCP	23	18 horizontal 14 torsional Component 15 GEN[a] 10 vertical components	14 strokes Ipsilesional Abnormal HIT 14 caloric CP 3 hemorrhages Normal hit and 2 abnormal	7 head tilt 14 skew 16 ipsiversive OTR	9 had auditory symptoms 5 had hearing loss > than 10 dB
Inferior cerebellar peduncle (ICP) (Choi et al. 2015c)	8 6 stroke 2 others	7 horizontal Ipsilesional 1 apogeotropic positional	Normal HIT Normal calorics	Contraversive OTR	SVV/OTR dissociation Ipsilesional direction of fall
Cerebellar nodulus (Moon et al. 2009)	8 6 unilateral 2 bilateral	Horizontal Ipsilesional Fast phase With DBN and torsional component 1 PAN[b] 3 perverted HSN[c]: DBN 1 positional apogeotropic	Normal calorics Normal HIT	No OTR 2 skew 1 ocular torsion	5 fell in a contralesional direction Forward head tilt did not suppress the post-rotary nystagmus in $n = 2/4$ patients
Cerebellar tonsil (Lee et al. 2014b)	1	Minimal Ipsilesional Fast phase Asymmetric GEN Rebound nystagmus	Normal	No OTR	Bilateral abnormal pursuit, > ipsilesional SVV contraversive

(continued)

Table 1 (continued)

Stroke location	Number of patients	Nystagmus and main neurotologic findings	Head impulse test	Skew head tilt OTR	Other common findings
Cerebellar flocculus (Park et al. 2013; Yacovino et al. 2018)	2	Asymmetric GEN[a] Minimal rebound	Abnormal HIT contralesional in acute phase but caloric test Ipsilesional Canal paresis 42%	Skew	Ipsilesional pursuit abnormal No DBN in unilateral lesions Truncal pulsion Contralesional SVV contralesional
Lateral medullary syndrome (Kim 2003)	130	Vertigo 74 Nystagmus on fixation (horizontal, torsional, or mixed) 73 Skew: 53 Hiccups: 33 Ocular Lateropulsion 8	Not described in this series	Skew No mention of full OTR	Ipsipulsion 79
Medial medullary syndrome Hemi-medullary syndrome (Kim et al. 2005; Kim and Han 2009)	24 First paper 86 Second paper	First paper 12 had nystagmus 5 h-GEN[a] 4 horizontal Ipsilesional 4 upbeat 1 hemi-seesaw Second paper 9 UBN	Normal HIT	Contraversive skew or head tilt or subjective visual vertical	Contrapulsion
Dorsal medullary syndrome (Lee et al. 2015)	18	*Peripheral VN type 5* Horizontal (h) nystagmus Contralesional h-GEN[a] *NPH type* 2 h nystagmus Ipsilesional 3 GEN[a] *ICP type* 3 h ipsilesional 9 mixed Combined findings	Abnormal HIT Ipsilesional in 5 *VN* type vHIT bilaterally abnormal *NPH type* Abnormal HIT Contralesional in *ICP type* HIT: normal	VN type Ipsiversive OTR NPH type No OTR or skew	Not described

[a]*GEN* gaze-evoked nystagmus
[b]*PAN* periodic alternating nystagmus
[c]*HSN* head-shaking nystagmus

(Kim 2003) (Table 1). Medial vestibular nucleus infarction[82] is one structure frequently responsible for the AVS in LMS (Lee et al. 2015; Dieterich et al. 2005). It is important to emphasize that dorsal medullary strokes may present with an isolated AVS; in addition, early ischemia in this location is frequently undetected by MRI (Lee et al. 2015). Thus, a thorough evaluation of the HINTS triad and careful follow-up monitoring are critical for proper diagnosis and management of these patients (Newman-Toker et al. 2013a; Lee et al. 2015; Saber Tehrani et al. 2014; Kheradmand and Zee 2012), MVN transient ischemia in LMS or dorsal medullary strokes may improve, while other LMS signs still remain (Kattah et al. 2017). Infarction of the ICP or the prepositus hypoglossi nucleus or the cerebellum, supplied by branches of the PICA arterial territory, may contribute to the neurotologic abnormalities (Table 1) (Engelter et al. 2008). Ocular and saccade ipsipulsion (lateropulsion) (Kommerell and Hoyt 1973), skew deviation, conjugate ocular torsion, and ocular tilt reaction (OTR) may be present (Dieterich and Brandt 1992; Brandt and Dieterich 1993). Different types of nystagmus include principally horizontal, torsional, or mixed types (Morrow and Sharpe 1988; Kim 2003; Kim et al. 1994), but not upbeat. In addition, one may find other neurologic signs that may help with the localization diagnosis such as an ipsilesional Horner's syndrome, unilateral lower cranial nerve palsies, unilateral limb ataxia, ipsilateral severe truncal lateropulsion, and crossed pinprick sensation loss in one side of the face and the contralateral trunk and limbs. The incidence of initially false-negative stroke MRI in the LMS depends on the timing of the MRI (Kattah et al. 2009; Oppenheim et al. 2000; Engelter et al. 2008; Saber Tehrani et al. 2014); however, when the clinical examination is strongly suggestive of stroke and a head MRI does not show an abnormality, the search for additional imaging signs of ongoing medullary dysfunction in CTA or MRI may be particularly helpful (Saber Tehrani et al. 2017). Small lacunar LMS strokes (Miller Fisher size criterion: 0.5–15 mm) (Fisher 1965), due to occlusion of small penetrating VA branches, may cause discrete deficits, for example, isolated truncal lateropulsion may be an infrequent finding in LMS (Hommel et al. 1985; Bertholon et al. 1996). In the NEMC-PCR, 3 ($n = 407$) patients had a lacunar stroke of the medulla (Caplan 2015). A recent review of main oculomotor abnormalities summarizes the main findings seen in selective lesions of the NPH, MVN, and ICP (Lee et al. 2018).

Medial Medullary Syndrome (MMS)

MMS is less common than LMS. The etiology is frequently VA occlusion or dissection. The structures affected by dorsal medial medullary infarcts involve the prepositus hypoglossi nuclear complex (see Table 1), which includes the nucleus of Roller and the nucleus intercalatus (Staderini nucleus) (Kim et al. 2005); lesions in this location are associated with UBN and rarely hemi-seesaw. The patients also have horizontal nystagmus with ipsilesional fast phases, normal HIT, contraversive skew, and ocular and truncal contrapulsion.

Dorsal Medullary Syndrome (DMS)

DMS is also an uncommon stroke; however, it is of particular neurotologic interest because it targets the MVN and the nucleus prepositus hypoglossi (NPH). As a result, a lesion involving the MVN will be associated with vertigo and will have a primary gaze contralesional nystagmus and a positive ipsilesional HIT; these patients often have direction-changing horizontal gaze-evoked nystagmus (GEN) and often have an ipsiversive skew or OTR (Table 1). Lesions located primarily in the NPH have an ipsilesional h-nystagmus, and the HIT is abnormal to the contralesional side (Lee et al. 2015). In this particular group of patients, the MRI may be initially false negative; thus, the HINTS triad is quite helpful (Kattah et al. 2009). The use of video-HIT to quantitate the VOR in these patients is extremely important, particularly when both MVN and NPH are simultaneously affected (Chen et al. 2014; Mantokoudis et al. 2015; Newman-Toker et al. 2013b).

Medial PICA and Lateral PICA Strokes

The clinical picture observed in infarcts involving the territory of medial PICA (mPICA) depends on the structures affected by the infarction and thus is heterogeneous. In addition, combined medial and lateral PICA (lPICA) strokes are large and complicated by ischemic cerebellar edema. The most common symptoms include an AVS with vertigo, dizziness, vomiting, and abnormal gait (Caplan 2015; Kerber et al. 2015; Tarnutzer et al. 2011). In the NEMC-PCR, 33 ($n = 84$) cerebellar stroke patients had prodromal TIAs (39%), particularly those with middle territory infarcts, 13 ($n = 19$) (68%) patients had middle territory plus strokes, and 3 ($n = 4$) patients had AICA strokes (Caplan 2015). Isolated medial or lateral PICA strokes have good prognosis, in general. In a 28-PICA stroke autopsy cohort, clinicopathologic correlation revealed pure PICA strokes in 15 patients, and PICA combined with additional vascular territories, referred to as PICA-plus, affected the remaining cases. Only two had combined medial and lateral PICA lesions (Amarenco et al. 1989). Five patients had one PICA stroke and all were incidental autopsy findings, as the cause of death was non-neurologic. None of the reported mPICA stroke cases had pathologic evidence of brainstem compression; in fact, the coexistent dorsal medullary syndrome (generally supplied by a branch of the PICA) dominated their clinical picture; all had vertigo as a prominent symptom (Amarenco et al. 1989). In contrast, the 13 PICA-plus cerebellar infarcts were associated with ischemic edema, brainstem compression, and hydrocephalus, which were eventually the presumed cause of death. This clinicopathologic correlation is particularly relevant to this chapter. Pseudotumoral cerebellar strokes may present with vestibular pseudoneuritis syndrome and require close monitoring and potentially surgical intervention to avoid brainstem compression, hydrocephalus, and tonsillar herniation (Duncan et al. 1975; Caplan 2015; Amarenco et al. 1989). At a later date, larger clinical series of mPICA stroke patients showed only 6 ($n = 36$) combined lateral or dorsal medulla/cerebellar strokes (Kase et al. 1993).

In the NEMC-PCR, 80 ($n = 407$) patients had proximal (extracranial) VA disease, 37 had occlusion, 34 had severe stenosis, and 12 had bilateral extracranial vertebral artery (ECVA) disease; 50% presented with a stroke after a TIA, and 24% had isolated TIAs. In general, patients with isolated V1-VA segment disease did well; the majority of those who did not do well had severe intracranial multifocal atherosclerosis. The most common stroke distribution was the proximal arterial territory in 44% of cases, followed by the combined proximal/middle territory in 56% of cases. V1-VA stenosis lesions are typically atherosclerotic, and when they are responsible for a stroke, embolism is the most likely mechanism; multifocal intracranial atherosclerosis is rather frequent (Caplan 2015). In contrast to atherosclerosis as the primary cause of the ECVA V1 segment, VA dissections were the most frequent pathology involving the V2 and V3 segments; neck pain actually precedes the neurologic symptoms by days and rarely weeks; on average, the time interval between headache and ischemia was 3.7 days, ranging from 1 h to 14 days (Silbert et al. 1995). The most common infarcts are in the distribution of PICA, either cerebellar or LMS. In general, MRA has lower sensitivity to detect VA dissections. A hyperintense sign in the wall of the VA with an eccentric compressed lumen is a typical finding (Levy et al. 1994). Dissection of the ICVA is uncommon (Fig. 5).

The most consistent bedside indicator of a central AVS is a normal HIT (Newman-Toker et al. 2008). In PICA strokes affecting the MVN, there is a potential for an abnormal HIT, and in such cases, the HINTS triad would be sensitive to proper clinical localization (Kattah et al. 2009). I still recall the diagnostic uncertainty in the first HINTS-central cases with initially false-negative MRIs, requiring debate about implementing a revised stroke protocol. In the end, we obtained the repeat MRI with the same magnet and without change in the stroke protocol. I believe that the change to a positive DWI signal represents a transition within the affected tissue, from ischemia to infarction. Direction-changing GEN was an excellent discriminator, when present (Hotson and Baloh 1998; Newman-Toker et al. 2013a; Chen et al. 2011; Lee 2014). Finally, it is important to keep in mind that unidirectional GEN is associated with lesions in several cerebellar locations including the vermal pyramid, the uvula, the tonsil (Table 1) also, the biventer, and the inferior semilunar lobe (Baier and Dieterich 2011). Skew deviation may occur with cerebellar infarcts, particularly in the nodulus (Moon et al. 2009). In this clinical setting, perverted ipsilesional horizontal nystagmus or DBN after horizontal head shaking points to central localization (Huh and Kim 2011).

In conclusion, VA and PICA strokes may mimic peripheral lesions; therefore, this group needs a careful neurotologic and neurologic evaluation.

5 Middle Territory

5.1 Pontine Ischemia and the Middle Posterior Circulation Intracranial Territory

Anterior-Inferior Cerebellar Artery (AICA) Strokes AICA, in most individuals, is a bilateral symmetric artery, but there are frequent variations in origin and size. It supplies the inner ear, facial nerve, middle cerebellar peduncle and lateral pons, and the anteromedial cerebellum and flocculus. Adams provided the first detailed clinicopathologic description of an AICA infarct (Adams 1943), which was further studied about 50 years later (Amarenco et al. 1993). I discussed the different patterns of vestibular and cochleovestibular loss earlier in the chapter. In AICA occlusions, the HIT is positive if the infarct involves the peripheral labyrinthine, the vestibular fascicle, or the MVN (Newman-Toker et al. 2008; Kattah et al. 2009). In addition, it is also positive with infarcts of the cerebellar flocculus (Park et al. 2013; Yacovino et al. 2018). In our first series, we reported an edematous PICA cerebellar stroke with compression of CP angle structures and brainstem; nevertheless, direction-changing horizontal nystagmus localized the lesion in the central vestibular pathways (Newman-Toker et al. 2008). Whereas in VN, the nystagmus direction does not change with positional testing, in a central lesion, the primary position nystagmus may change direction with positional testing (Shaikh et al. 2014). It is important to remark that not all AICA strokes are associated with an abnormal HIT (Kattah et al. 2013).

Penetrator Pontine Branch Artery Strokes The clinical findings in pontine strokes due to basilar artery stenosis are quite eloquent and are often associated with altered mental status, tetra paresis, conjugate, horizontal gaze palsy, small miotic pupils that react to a bright penlight, abducens and facial paresis, and contralateral limb ataxia. An internuclear ophthalmoplegia (INO) and all its variants are frequently present (Bronstein et al. 1990). An AVS may present initially, followed by rapid neurologic deterioration. Horizontal gaze paretic nystagmus is often asymmetric (more prominent on the side of the affected pontine tegmentum). A combined INO and ipsilesional gaze palsy causes the classic one and a half syndrome (Fisher 1967). Upbeat nystagmus in primary straight gaze is common (Fisher 1967). The vestibular nuclei are often involved in pontine lesions, particularly if the BA occlusion is progressive and nystagmus patterns, similar to those described with medullary lesions, might be present. Skew deviation is frequent and usually of large amplitude (> than three-prism diopters). In HINTS, the average skew was nine-prism diopters (Kattah et al. 2009). Vascular ocular tilt reactions (OTRs) are most common with strokes in the medulla and pons and less common with midbrain and cerebellar lesions. In INO with skew, the lesion is usually on the side of the hypertropic (uppermost) eye (Keane 1975).

In the NEMC-PCR, the arterial pathology involved the middle third of the BA, most commonly severe atherosclerotic BA disease in 55 ($n = 87$) cases and artery-to-artery or cardiac embolism in 32 ($n = 87$) patients. In patients with BA stenosis, 58 ($n = 87$) patients first had TIAs, followed by strokes. In contrast, cases with BA embolism had acute strokes, not preceded by TIAs. The TIAs were frequent and the average period prior to a stroke was 2.1 months (Caplan 2015). In this group, the stroke most frequently affected the pons and the cerebellum supplied by AICA. An important fact is the duration of recurrent TIAs without infarction (Caplan 2015). Among 87 patients with severe basilar artery disease, 2 died, 35 had no acute neurologic deficits, 27 had minor neurologic abnormalities, and 23 had moderately severe neurologic findings (Caplan 2015). In general, BA embolism was associated with poor prognosis.

Pontine lacunes are frequent, particularly among diabetic patients (Ichikawa et al. 2012). The distribution and size of the pontine infarcts vary with the anatomy of the perforator arteries. The topographic classification includes four basic territories: anterolateral, anteromedian, lateral tegmental, and posterior (Bassetti et al. 1996). Anteromedial infarcts (the most common and largest pontine infarcts), usually the stroke, is limited to the basis pontis and spares the tegmentum; in some instances, however, it may affect both. Anterolateral infarcts may cause conjugate gaze palsies and ipsilesional peripheral facial palsy. There is a large variety of possible lacunar stroke distributions in the pons (Kattah et al. 2009; Caplan 2015; Saber Tehrani et al. 2014; Bassetti et al. 1996), many of which may be responsible for an AVS and may have prominent ocular motor abnormalities and nystagmus (Leigh and Zee 2015; Lee et al. 2018).

In the NEMC-PCR, penetrator artery strokes in the pons accounted for 26 ($n = 58$) patients ($n = 401$, represents total registry number of patients). The VA and BA arterial flow voids of these patients were normal (Caplan 2015). Isolated INO was frequent, and in some cases, symptoms of an AVS were present. We have seen an AVS in vascular INO patients in a number of occasions, probably due to MVN ischemia or infarction (Newman-Toker et al. 2008); alternatively, involvement of vestibular fibers that travel within the MLF may be responsible. In penetrator strokes, the atheromatous plaque may block the orifice of the branch artery; to localize the abnormal penetrator artery, L. Caplan recommends one noninvasive method, which consists of obtaining and juxtaposing a sagittal T2-weighted MRI with a sagittal MRA (Caplan 2015).

Superior cerebellar artery strokes generally do not have an associated AVS, unless multiple territory infarcts are affected (Lee 2009). Lesions in the insula (Lee 2014) have been sporadically reported in patients with vertigo; this point remains still debatable or at least very infrequent (Baier et al. 2013). Cortical lesions are associated with vestibular perceptual abnormalities affecting posture and navigation but not an AVS in its strict definition (Brandt et al. 2014).

To conclude, I focused this chapter on vascular vertigo and ischemic strokes as the most common cause of central vertigo; however, other etiologies are less frequent (Kattah et al. 2009) and often identified with neuroimaging. Multiple sclerosis plaques and inflammatory, infectious, and paraneoplastic autoimmune disorders probably followed. I emphasized throughout this chapter the importance of a thorough clinical examination, as well as recent technological advances that allow precise recording of the eye movements and the head impulse test, ultimately leading to the precise clinical classification of the lesion and a target imaging study.

Acknowledgments Although this is a single-authored chapter, I shared many of the comments and personal experiences detailed within with the OSF HealthCare Illinois Neurological Institute Stroke Network, with Jeffrey R. DeSanto, MD, Department of Neuroradiology and with my collaborator, David E. Newman-Toker, MD, PhD.

References

Adams, R. D. (1943). Occlusion of the anterior inferior cerebellar artery. *Archives of Neurology, 43*, 765–770.

Albers, G. W., Caplan, L. R., Easton, J. D., et al. (2002). Transient ischemic attack – Proposal for a new definition. *The New England Journal of Medicine, 347*, 1713–1716.

Alsop, D. C., & Detre, J. A. (1996). Reduced transit-time sensitivity in noninvasive magnetic resonance imaging of human cerebral blood flow. *Journal of Cerebral Blood Flow and Metabolism: Official Journal of the International Society of Cerebral Blood Flow and Metabolism, 16*, 1236–1249.

Amarenco, P., & Hauw, J. J. (1990). Cerebellar infarction in the territory of the anterior and inferior cerebellar artery. A clinicopathological study of 20 cases. *Brain: A Journal of Neurology, 113*(Pt 1), 139–155.

Amarenco, P., Hauw, J. J., Henin, D., et al. (1989). Cerebellar infarction in the area of the posterior cerebellar artery. [Clinicopathology of 28 cases]. *Revue Neurologique, 145*, 277–286.

Amarenco, P., Rosengart, A., DeWitt, L. D., Pessin, M. S., & Caplan, L. R. (1993). Anterior inferior cerebellar artery territory infarcts. Mechanisms and clinical features. *Archives of Neurology, 50*, 154–161.

Baier, B., & Dieterich, M. (2011). Incidence and anatomy of gaze-evoked nystagmus in patients with cerebellar lesions. *Neurology, 76*, 361–365.

Baier, B., Conrad, J., Zu Eulenburg, P., et al. (2013). Insular strokes cause no vestibular deficits. *Stroke: A Journal of Cerebral Circulation, 44*, 2604–2606.

Baloh, R. W., & Halmagyi, G. M. (1996). *Disorders of the vestibular system.* Oxford University Press.

Bassetti, C., Bogousslavsky, J., Barth, A., & Regli, F. (1996). Isolated infarcts of the pons. *Neurology, 46*, 165–175.

Batuecas-Caletrío, A., Yáñez-González, R., Sanchez, B. C., et al. (2014). Application of the HINTS protocol. *Revista de Neurologia, 59*, 349–353.

Bertholon, P., Michel, D., Convers, P., Antoine, J. C., & Barral, F. G. (1996). Isolated body lateropulsion caused by a lesion of the cerebellar peduncles. *Journal of Neurology, Neurosurgery, and Psychiatry, 60*, 356–357.

Biller, J., Yuh, W. T., Mitchell, G. W., Bruno, A., & Adams, H. P., Jr. (1988). Early diagnosis of basilar artery occlusion using magnetic resonance imaging. *Stroke: A Journal of Cerebral Circulation, 19*, 297–306.

Bisdorff, A., Von Brevern, M., Lempert, T., & Newman-Toker, D. E. (2009). Classification of vestibular symptoms: Towards an international classification of vestibular disorders. *Journal of Vestibular Research: Equilibrium & Orientation, 19*, 1–13.

Brandt, T., & Dieterich, M. (1993). Skew deviation with ocular torsion: A vestibular brainstem sign of topographic diagnostic value. *Annals of Neurology, 33*, 528–534.

Brandt, T., Strupp, M., & Dieterich, M. (2014). Towards a concept of disorders of "higher vestibular function". *Frontiers in Integrative Neuroscience, 8*, 47.

Bronstein, A. M., Rudge, P., Gresty, M. A., Du Boulay, G., & Morris, J. (1990). Abnormalities of horizontal gaze. Clinical, oculographic and magnetic resonance imaging findings. II. Gaze palsy and internuclear ophthalmoplegia. *Journal of Neurology, Neurosurgery, and Psychiatry, 53*, 200–207.

Caplan, L. R. (1990). Charles Foix--the first modern stroke neurologist. *Stroke: A Journal of Cerebral Circulation, 21*, 348–356.

Caplan, L. R. (2015). *Vertebrobasilar ischemia and hemorrhage: Clinical findings. Diagnosis and management of posterior circulation disease.* Cambridge: Cambridge University Press.

Caplan, L. R., Pessin, M. S., Scott, R. M., & Yarnell, P. (1986). Poor outcome after lateral medullary infarcts. *Neurology, 36*, 1510–1513.

Caplan, L., Chung, C. S., Wityk, R., et al. (2005). New England medical center posterior circulation stroke registry: I. methods, data base, distribution of brain lesions, stroke mechanisms, and outcomes. *Journal of Clinical Neurology, 1*, 14–30.

Chalela, J. A., Kidwell, C. S., Nentwich, L. M., et al. (2007). Magnetic resonance imaging and computed tomography in emergency assessment of patients with suspected acute stroke: A prospective comparison. *Lancet, 369*, 293–298.

Chen, L., Lee, W., Chambers, B. R., & Dewey, H. M. (2011). Diagnostic accuracy of acute vestibular syndrome at the bedside in a stroke unit. *Journal of Neurology, 258*, 855–861.

Chen, L., Todd, M., Halmagyi, G. M., & Aw, S. (2014). Head impulse gain and saccade analysis in pontine-cerebellar stroke and vestibular neuritis. *Neurology, 83*, 1513–1522.

Choi, K. D., Choi, J. H., Kim, J. S., et al. (2013). Rotational vertebral artery occlusion: Mechanisms and long-term outcome. *Stroke: A Journal of Cerebral Circulation, 44*, 1817–1824.

Choi, J.-H., Kim, H.-W., Choi, K.-D., et al. (2014a). Isolated vestibular syndrome in posterior circulation stroke: Frequency and involved structures. *Neurology: Clinical Practice, 4*, 410–418.

Choi, J. H. K. H., Dong, K., Kim, M. J., Choi, Y. R., Cho, H. J., Sung, M. S., Kim, H. J., Kim, J. S., & Jung, D. S. (2014b). Isolated vestibular syndrome in posterior circulation stroke. *Neurology Clinical Practice, 4*, 410–418.

Choi, J. Y., Kim, J. H., Kim, H. J., Glasauer, S., & Kim, J. S. (2015a). Central paroxysmal positional nystagmus: Characteristics and possible mechanisms. *Neurology, 84*, 2238–2246.

Choi, S. Y., Park, J. H., Kim, H. J., & Kim, J. S. (2015b). Vestibulocochlear nerve infarction documented with diffusion-weighted MRI. *Journal of Neurology, 262*, 1363–1365.

Choi, J. H., Seo, J. D., Choi, Y. R., et al. (2015c). Inferior cerebellar peduncular lesion causes a distinct vestibular syndrome. *European Journal of Neurology, 22*, 1062–1067.

Dieterich, M., & Brandt, T. (1992). Wallenberg's syndrome: Lateropulsion, cyclorotation, and subjective visual vertical in thirty-six patients. *Annals of Neurology, 31*, 399–408.

Dieterich, M., Bense, S., Stephan, T., Brandt, T., Schwaiger, M., & Bartenstein, P. (2005). Medial vestibular nucleus lesions in Wallenberg's syndrome cause decreased activity of the contralateral vestibular cortex. *Annals of the New York Academy of Sciences, 1039*, 368–383.

Duncan, G. W., Parker, S. W., & Fisher, C. M. (1975). Acute cerebellar infarction in the PICA territory. *Archives of Neurology, 32*, 364–368.

El-Badry, M. M., Samy, H., Kabel, A. M., Rafat, F. M., & Sanyelbhaa, H. (2017). Clinical criteria of positional vertical nystagmus in vestibular migraine. *Acta Oto-Laryngologica, 137*, 720–722.

Engelter, S. T., Wetzel, S. G., Bonati, L. H., Fluri, F., & Lyrer, P. A. (2008). The clinical significance of diffusion-weighted MR imaging in stroke and TIA patients. *Swiss Medical Weekly, 138*, 729–740.

Fisher, C. M. (1965). Lacunes: Small, deep cerebral infarcts. *Neurology, 15*, 774–784.

Fisher, C. M. (1967). Some neuro-ophthalmological observations. *Journal of Neurology, Neurosurgery, and Psychiatry, 30*, 383–392.

Furman, J. M., & Cass, S. P. (1999). Benign paroxysmal positional vertigo. *The New England Journal of Medicine, 341*, 1590–1596.

Grad, A., & Baloh, R. W. (1989). Vertigo of vascular origin. Clinical and electronystagmographic features in 84 cases. *Archives of Neurology, 46*, 281–284.

Greco, A., Macri, G. F., Gallo, A., et al. (2014). Is vestibular neuritis an immune related vestibular neuropathy inducing vertigo? *Journal of Immunology Research, 2014*, 459048.

Halmagyi, G. M. (2017). Brainstem stroke preceded by transient isolated vertigo attacks. *Journal of Neurology, 264*, 2170–2172.

Hommel, M., Borgel, F., Gaio, J. M., Lavernhe, G., & Perret, J. (1985). Isolated ipsilateral lateropulsion caused by bulbar hematoma. *Revue Neurologique, 141*, 53–54.

Hotson, J. R., & Baloh, R. W. (1998). Acute vestibular syndrome. *The New England Journal of Medicine, 339*, 680–685.

Huang, M. H., Huang, C. C., Ryu, S. J., & Chu, N. S. (1993). Sudden bilateral hearing impairment in vertebrobasilar occlusive disease. *Stroke: A Journal of Cerebral Circulation, 24*, 132–137.

Huh, Y. E., & Kim, J. S. (2011). Patterns of spontaneous and head-shaking nystagmus in cerebellar infarction: Imaging correlations. *Brain: A Journal of Neurology, 134*, 3662–3671.

Ichikawa, H., Kuriki, A., Kinno, R., Katoh, H., Mukai, M., & Kawamura, M. (2012). Occurrence and clinicotopographical correlates of brainstem infarction in patients with diabetes mellitus. *Journal of Stroke and Cerebrovascular Diseases: The Official Journal of National Stroke Association, 21*, 890–897.

Jareonsettasin, P., Otero-Millan, J., Ward, B. K., Roberts, D. C., Schubert, M. C., & Zee, D. S. (2016). Multiple time courses of vestibular set-point adaptation revealed by sustained magnetic field stimulation of the labyrinth. *Current Biology, 26*, 1359–1366.

Karlberg, M., Annertz, M., & Magnusson, M. (2004). Acute vestibular neuritis visualized by 3-T magnetic resonance imaging with high-dose gadolinium. *Archives of Otolaryngology – Head & Neck Surgery, 130*, 229–232.

Kase, C. S., Norrving, B., Levine, S. R., et al. (1993). Cerebellar infarction. Clinical and anatomic observations in 66 cases. *Stroke: A Journal of Cerebral Circulation, 24*, 76–83.

Kattah, J. C., & Pula, J. (2015). New windows into the brain: Technological advances in forntline neurologic diagnosis via the visual and oculomotor systems. *Future Neurology, 10*, 301–303.

Kattah, J. C., Talkad, A. V., Wang, D. Z., Hsieh, Y. H., & Newman-Toker, D. E. (2009). HINTS to diagnose stroke in the acute vestibular syndrome: Three-step bedside oculomotor examination more sensitive than early MRI diffusion-weighted imaging. *Stroke: A Journal of Cerebral Circulation, 40*, 3504–3510.

Kattah, J. C., Pula, J., & Newman-Toker, D. E. (2011). Ocular lateropulsion as a central oculomotor sign in acute vestibular syndrome is not posturally dependent. *Annals of the New York Academy of Sciences, 1233*, 249–255.

Kattah, J. C., Nair, D., Talkad, A., Wang, D. Z., & Fraser, K. (2013). A case of vestibular and oculomotor pathology from bilateral AICA watershed infarcts treated with basilar artery stenting. *Clinical Neurology and Neurosurgery, 115*, 1098–1101.

Kattah, J. C., Saber Tehrani, A. S., Roeber, S., et al. (2017). Transient Vestibulopathy in Wallenberg's syndrome: Pathologic analysis. *Frontiers in Neurology, 8*, 191.

Keane, J. R. (1975). Ocular skew deviation. Analysis of 100 cases. *Archives of Neurology, 32*, 185–190.

Kerber, K. A., Burke, J. F., Brown, D. L., et al. (2012). Does intracerebral haemorrhage mimic benign dizziness presentations? A population based study. *Emergency Medicine Journal: EMJ, 29*, 43–46.

Kerber, K. A., Zahuranec, D. B., Brown, D. L., et al. (2014). Stroke risk after nonstroke emergency department dizziness presentations: A population-based cohort study. *Annals of Neurology, 75*, 899–907.

Kerber, K. A., Meurer, W. J., Brown, D. L., et al. (2015). Stroke risk stratification in acute dizziness presentations: A prospective imaging-based study. *Neurology, 85*, 1869–1878.

Kheradmand, A., & Zee, D. S. (2012). The bedside examination of the vestibulo-ocular reflex (VOR): An update. *Revue Neurologique, 168*, 710–719.

Kim, J. S. (2003). Pure lateral medullary infarction: Clinical-radiological correlation of 130 acute, consecutive patients. *Brain: A Journal of Neurology, 126*, 1864–1872.

Kim, J. S., & Han, Y. S. (2009). Medial medullary infarction: Clinical, imaging, and outcome study in 86 consecutive patients. *Stroke: A Journal of Cerebral Circulation, 40*, 3221–3225.

Kim, S. H., & Kim, J. S. (2019). Eye movement abnormalities in middle cerebellar peduncle strokes. *Acta Neurologica Belgica, 119*(1), 37–45.

Kim, H. A., & Lee, H. (2010). Isolated vestibular nucleus infarction mimicking acute peripheral vestibulopathy. *Stroke: A Journal of Cerebral Circulation, 41*, 1558–1560.

Kim, H. A., & Lee, H. (2017). Recent advances in understanding audiovestibular loss of a vascular cause. *Journal of Stroke, 19*, 61–66.

Kim, J. S., & Zee, D. S. (2014). Clinical practice. Benign paroxysmal positional vertigo. *The New England Journal of Medicine, 370*, 1138–1147.

Kim, J. S., Lee, J. H., Suh, D. C., & Lee, M. C. (1994). Spectrum of lateral medullary syndrome. Correlation between clinical findings and magnetic resonance imaging in 33 subjects. *Stroke: A Journal of Cerebral Circulation, 25*, 1405–1410.

Kim, J. S., Lopez, I., DiPatre, P. L., Liu, F., Ishiyama, A., & Baloh, R. W. (1999). Internal auditory artery infarction: Clinicopathologic correlation. *Neurology, 52*, 40–44.

Kim, J. S., Choi, K. D., Oh, S. Y., et al. (2005). Medial medullary infarction: Abnormal ocular motor findings. *Neurology, 65*, 1294–1298.

Kim, H. A., Yi, H. A., & Lee, H. (2012). Apogeotropic central positional nystagmus as a sole sign of nodular infarction. *Neurological Sciences: Official Journal of the Italian Neurological Society and of the Italian Society of Clinical Neurophysiology, 33*, 1189–1191.

Kim, C. H., Choi, J. M., Jung, H. V., Park, H. J., & Shin, J. E. (2014). Sudden sensorineural hearing loss with simultaneous positional vertigo showing persistent geotropic direction-changing positional nystagmus. *Otology & Neurotology: Official Publication of the American Otological Society, American Neurotology Society [and] European Academy of Otology and Neurotology, 35*, 1626–1632.

Kim, H. A., Yi, H. A., & Lee, H. (2016a). Recent advances in cerebellar ischemic stroke syndromes causing vertigo and hearing loss. *Cerebellum, 15*, 781–788.

Kim, C. H., Yang, Y. S., Im, D., & Shin, J. E. (2016b). Nystagmus in patients with unilateral acute otitis media complicated by serous labyrinthitis. *Acta Oto-Laryngologica, 136*, 559–563.

Kim, S. H., Zee, D. S., du Lac, S., Kim, H. J., & Kim, J. S. (2016c). Nucleus prepositus hypoglossi lesions produce a unique ocular motor syndrome. *Neurology, 87*, 2026–2033.

Kommerell, G., & Hoyt, W. F. (1973). Lateropulsion of saccadic eye movements. Electro-oculographic studies in a patient with Wallenberg's syndrome. *Archives of Neurology, 28*, 313–318.

Lee, H. (2009). Neuro-otological aspects of cerebellar stroke syndrome. *Journal of Clinical Neurology, 5*, 65–73.

Lee, H. (2014). Isolated vascular vertigo. *Journal of Stroke, 16*, 124–130.

Lee, H., & Cho, Y. W. (2003). Auditory disturbance as a prodrome of anterior inferior cerebellar artery infarction. *Journal of Neurology, Neurosurgery, and Psychiatry, 74*, 1644–1648.

Lee, H., Kim, J. S., Chung, E. J., et al. (2009). Infarction in the territory of anterior inferior cerebellar artery: Spectrum of audiovestibular loss. *Stroke: A Journal of Cerebral Circulation, 40*, 3745–3751.

Lee, J. O., Park, S. H., Kim, H. J., Kim, M. S., Park, B. R., & Kim, J. S. (2014a). Vulnerability of the vestibular organs to transient ischemia: Implications for isolated vascular vertigo. *Neuroscience Letters, 558*, 180–185.

Lee, S. H., Park, S. H., Kim, J. S., Kim, H. J., Yunusov, F., & Zee, D. S. (2014b). Isolated unilateral infarction of the cerebellar tonsil: Ocular motor findings. *Annals of Neurology, 75*, 429–434.

Lee, S. U., Park, S. H., Park, J. J., et al. (2015). Dorsal medullary infarction: Distinct syndrome of isolated central vestibulopathy. *Stroke: A Journal of Cerebral Circulation, 46*, 3081–3087.

Lee, S. H., Kim, H. J., & Kim, J. S. (2018). Ocular motor dysfunction due to brainstem disorders. *Journal of Neuro-Ophthalmology: The Official Journal of the North American Neuro-Ophthalmology Society, 38*(3), 393–412.

Lefkowitz, D., LaBenz, M., Nudo, S. R., Steg, R. E., & Bertoni, J. M. (1999). Hyperacute ischemic stroke missed by diffusion-weighted imaging. *AJNR American Journal of Neuroradiology, 20*, 1871–1875.

Leigh, R., & Zee, D. S. (2015). *The neurology of eye movements* (5th ed.). New York: Oxford University Press.

Levy, C., Laissy, J. P., Raveau, V., et al. (1994). Carotid and vertebral artery dissections: Three-dimensional time-of-flight MR angiography and MR imaging versus conventional angiography. *Radiology, 190*, 97–103.

Mantokoudis, G., Tehrani, A. S., Wozniak, A., et al. (2015). VOR gain by head impulse video-oculography differentiates acute vestibular neuritis from stroke. *Otology & Neurotology: Official Publication of the American Otological Society, American Neurotology Society [and] European Academy of Otology and Neurotology, 36*, 457–465.

Mian, O. S., Glover, P. M., & Day, B. L. (2015). Reconciling magnetically induced vertigo and nystagmus. *Frontiers in Neurology, 6*, 201.

Moon, I. S., Kim, J. S., Choi, K. D., et al. (2009). Isolated nodular infarction. *Stroke: A Journal of Cerebral Circulation, 40*, 487–491.

Morita, S., Suzuki, M., & Iizuka, K. (2011). False-negative diffusion-weighted MRI in acute cerebellar stroke. *Auris, Nasus, Larynx, 38*, 577–582.

Morrow, M. J., & Sharpe, J. A. (1988). Torsional nystagmus in the lateral medullary syndrome. *Annals of Neurology, 24*, 390–398.

Nazerian, P., Bigiarini, S., Pecci, R., et al. (2018). Duplex sonography of vertebral arteries for evaluation of patients with acute vertigo. *Ultrasound in Medicine & Biology, 44*(3), 584–592.

Newman-Toker, D. E. (2015). Missed stroke in acute vertigo and dizziness: It is time for action, not debate. *Annals of Neurology, 79*(1), 27–31.

Newman-Toker, D. E., & Edlow, J. A. (2015). TiTrATE: A novel, evidence-based approach to diagnosing acute dizziness and vertigo. *Neurologic Clinics, 33*, 577–599. viii.

Newman-Toker, D. E., Kattah, J. C., Alvernia, J. E., & Wang, D. Z. (2008). Normal head impulse test differentiates acute cerebellar strokes from vestibular neuritis. *Neurology, 70*, 2378–2385.

Newman-Toker, D. E., Kerber, K. A., Hsieh, Y. H., et al. (2013a). HINTS outperforms ABCD2 to screen for stroke in acute continuous vertigo and dizziness. *Academic Emergency Medicine: Official Journal of the Society for Academic Emergency Medicine, 20*, 986–996.

Newman-Toker, D. E., Saber Tehrani, A. S., Mantokoudis, G., et al. (2013b). Quantitative video-oculography to help diagnose stroke in acute vertigo and dizziness: Toward an ECG for the eyes. *Stroke: A Journal of Cerebral Circulation, 44*, 1158–1161.

Norrving, B., & Cronqvist, S. (1991). Lateral medullary infarction: Prognosis in an unselected series. *Neurology, 41*, 244–248.

Oppenheim, C., Stanescu, R., Dormont, D., et al. (2000). False-negative diffusion-weighted MR findings in acute ischemic stroke. *AJNR American Journal of Neuroradiology, 21*, 1434–1440.

Park, H. K., Kim, J. S., Strupp, M., & Zee, D. S. (2013). Isolated floccular infarction: Impaired vestibular responses to horizontal head impulse. *Journal of Neurology, 260*, 1576–1582.

Park, K. M., Shin, K. J., Ha, S. Y., Park, J. S., & Kim, S. E. (2014). A case of acute vestibular neuritis visualized by three-dimensional FLAIR-VISTA magnetic resonance imaging. *Neuroophthalmology, 38*, 60–61.

Pasaoglu, L. (2017). Vertebrobasilar system computed tomographic angiography in central vertigo. *Medicine, 96*, e6297.

Provenzale, J. M., & Sorensen, A. G. (1999). Diffusion-weighted MR imaging in acute stroke: Theoretic considerations and clinical applications. *AJR American Journal of Roentgenology, 173*, 1459–1467.

Quint, D. J., & Spickler, E. M. (1990). Magnetic resonance demonstration of vertebral artery dissection. Report of two cases. *Journal of Neurosurgery, 72*, 964–967.

Rathakrishnan, R., Sharma, V. K., & Chan, B. P. (2008). Diffusion-negative MRI in acute ischemic stroke: A case report. *Cases Journal, 1*, 65.

Roberts, D. C., Marcelli, V., Gillen, J. S., Carey, J. P., Della Santina, C. C., & Zee, D. S. (2011). MRI magnetic field stimulates rotational sensors of the brain. *Current Biology, 21*, 1635–1640.

Saber Tehrani, A. S., Kattah, J. C., Mantokoudis, G., et al. (2014). Small strokes causing severe vertigo: Frequency of false-negative MRIs and nonlacunar mechanisms. *Neurology, 83*, 169–173.

Saber Tehrani, A. S., DeSanto, J. R., & Kattah, J. C. (2017). Neuroimaging "HINTS" of the lateral medullary syndrome. *Journal of Neuro-Ophthalmology: The Official Journal of the North American Neuro-Ophthalmology Society, 37*, 403–404.

Sacco, R. L., Freddo, L., Bello, J. A., Odel, J. G., Onesti, S. T., & Mohr, J. P. (1993). Wallenberg's lateral medullary syndrome. Clinical-magnetic resonance imaging correlations. *Archives of Neurology, 50*, 609–614.

Searls, D. E., Pazdera, L., Korbel, E., Vysata, O., & Caplan, L. R. (2012). Symptoms and signs of posterior circulation ischemia in the new England medical center posterior circulation registry. *Archives of Neurology, 69*, 346–351.

Shaikh, A. M. B., Sundararajan, S., & Katirji, B. (2014). Gravity-dependent nystagmus and inner-ear dysfunction suggest anterior and posterior inferior cerebellar artery infarct. *Journal of Stroke and Cerebrovascular Diseases, 23*, 788–790.

Shen, Y. Y., Dai, T. M., Liu, H. L., Wu, W., & Tu, J. L. (2015). Ramsay hunt syndrome complicated by brainstem encephalitis in varicella-zoster virus infection. *Chinese Medical Journal, 128*, 3258–3259.

Silbert, P. L., Mokri, B., & Schievink, W. I. (1995). Headache and neck pain in spontaneous internal carotid and vertebral artery dissections. *Neurology, 45*, 1517–1522.

Strupp, M., Jager, L., Muller-Lisse, U., Arbusow, V., Reiser, M., & Brandt, T. (1998). High resolution Gd-DTPA MR imaging of the inner ear in 60 patients with idiopathic vestibular neuritis: No evidence for contrast enhancement of the labyrinth or vestibular nerve. *Journal of Vestibular Research: Equilibrium & Orientation, 8*, 427–433.

Sue, D. E., Brant-Zawadzki, M. N., & Chance, J. (1992). Dissection of cranial arteries in the neck: Correlation of MRI and arteriography. *Neuroradiology, 34*, 273–278.

Sugiura, M., Naganawa, S., Teranishi, M., Sato, E., Kojima, S., & Nakashima, T. (2006). Inner ear hemorrhage in systemic lupus erythematosus. *The Laryngoscope, 116*, 826–828.

Tarnutzer, A. A., Berkowitz, A. L., Robinson, K. A., Hsieh, Y. H., & Newman-Toker, D. E. (2011). Does my dizzy patient have a stroke? A systematic review of bedside diagnosis in acute vestibular syndrome. *CMAJ: Canadian Medical Association Journal = Journal de l'Association Medicale Canadienne, 183*, E571–E592.

Tieleman, A., Casselman, J. W., Somers, T., et al. (2008). Imaging of intralabyrinthine schwannomas: A retrospective study of 52 cases with emphasis on lesion growth. *AJNR American Journal of Neuroradiology, 29*, 898–905.

Tien, R., Dillon, W. P., & Jackler, R. K. (1990). Contrast-enhanced MR imaging of the facial nerve in 11 patients with Bell's palsy. *AJNR American Journal of Neuroradiology, 11*, 735–741.

von Brevern, M., Bertholon, P., Brandt, T., et al. (2015). Benign paroxysmal positional vertigo: Diagnostic criteria. *Journal of Vestibular Research: Equilibrium & Orientation, 25*, 105–117.

Wallenberg, A. (1895). Acute Bulbaraffection (Embolie der art.cerebellar.post,inf.sinistr.?). *Arch Psychiat Nervenkr, 27*, 504–540.

Wallenberg, A. (1901). Anatomischer Befund in einemals "acute bulbar affection (Embolie der Art. cerebellar.post.inf sinistr.?)" besschriebenen Falle. *Arch Psychiat Nervenkr, 34*, 923–959.

Ward, B., & Zee, D. (2016). Dizziness and vertigo during MRI. *The New England Journal of Medicine, 375*, e44.

Whitehead, R. E., MacDonald, C. B., Melhem, E. R., & McMahon, L. (1998). Spontaneous labyrinthine hemorrhage in sickle cell disease. *AJNR American Journal of Neuroradiology, 19*, 1437–1440.

Yacovino, D. A., Akly, M. P., Luis, L., & Zee, D. S. (2018). The Floccular syndrome: Dynamic changes in eye movements and vestibulo-ocular reflex in isolated infarction of the cerebellar flocculus. *Cerebellum, 17*(2), 122–131.

Combined Central and Peripheral Degenerative Vestibular Disorders

David J. Szmulewicz

Abstract Differentiation of the aetiology and localization of balance disorders has traditionally involved differentiation between central (originating in the brain) and peripheral (originating in the vestibular labyrinth or the nerve) disorders. Given the increasing recognition of *combined* central (cerebellar) and peripheral (vestibular) pathology, this attempt at dichotomization may in fact function to obscure accurate diagnosis. This chapter, a review of combined central and peripheral degenerative disorders, emphasizes that identification of such combined phenotypes reduces the number of potential differential diagnoses, providing a greater opportunity for a definitive diagnosis and further disease discovery.

Keywords Vestibulo-ocular reflex · Balance · Cerebellum · Vertigo

1 Introduction

Clinical examination of eye movements in patients with imbalance and/or incoordination is a critical diagnostic tool in differentiating challenging balance disorders (Newman-Toker et al. 2008; Kerber 2009). Physiologically, accurate eye movements function to provide a stable image of an organism's visual world. As humans, this process is extremely important and facilitates safe and effective navigation of our environment. Two vital mechanisms in this process are the vestibulo-ocular reflex (VOR) and smooth pursuit (SP) eye movements. Recent developments in diagnostic technology have meant that objective, portable and non-invasive oculomotor measurement is readily accessible (MacDougall et al. 2009; Szmulewicz et al. 2011a).

D. J. Szmulewicz (✉)
Balance Disorders and Ataxia Service, Royal Victoria Eye and Ear Hospital, Melbourne, VIC, Australia

Cerebellar Ataxia Clinic, Department of Neuroscience, Alfred Health, Monash University, Melbourne, VIC, Australia

© Springer Nature Switzerland AG 2019

A. Shaikh, F. Ghasia (eds.), *Advances in Translational Neuroscience of Eye Movement Disorders*, Contemporary Clinical Neuroscience, https://doi.org/10.1007/978-3-030-31407-1_17

The SP system functions to maintain foveation of a *moving* target. It operates at low velocity and frequency of target movement. At greater than 60 degrees per second or 1 Hz, SP rapidly fails, and consequently eye motion falls behind target motion so that corrective saccades are needed in order to re-foveate. SP gain (i.e. the ratio of eye angular velocity to target angular velocity) is a function of target velocity and acceleration. A target with continuously changing velocity is pursued more accurately if the pattern of movement is predictable (e.g. sinusoidal) (Lisberger et al. 1981).

The VOR is an exquisitely sensitive, bilaterally coupled inner ear system for detecting head motion, which then rapidly coordinates oculomotor adjustments aimed at ensuring maximal visual clarity during fast physiological head movement. This is accomplished by converting the head movement metrics (velocity, distance and direction) into rapid, corrective eye movements (saccades) that drive the eyes in the opposite direction and in an equal amount to the head movement. In doing so, the eye movement cancels out the potential effect that head movement may have on the retina and thus prevents degradation of vision that would otherwise result from motion of the target relative to the eye. The vestibulocerebellum is the portion of the cerebellum involved in governing responses to stimuli important for motion detection, such as head motion or a visual target motion (Voogd et al. 1996). The vestibulocerebellum is required to modulate the VOR, a plastic or 'learning' reflex. Via this mechanism (and others), the cerebellum exerts a key role in vision (Ramat et al. 2001).

2   Visual-Vestibular Interactions

Eye movements accomplish two key visual goals: to be able to 'follow' a moving visual target, i.e. to fixate, and to 'look' at a new target, i.e. to refixate. The maintenance of stable vision during head motion requires compensatory eye movements, which requires the interaction of several oculomotor control systems. For VOR, the addition of visual input such as SP results in the visually enhanced vestibulo-ocular reflex (VVOR). The VVOR is the addition of VOR *and* vision (i.e. SP) and functions to stabilize the image of an earth-fixed target, while the subject is moving. VOR suppression (VORS) is a cerebellar-mediated oculomotor task and usually acts to suppress VOR, so that gaze can be changed during head movement, when, for example, one follows a moving target with eyes and head. In this way, VORS involves subtraction of the contribution of the VOR from eye movement.

While the study of VOR in isolation is a useful research metric, in reality the VOR is constantly interacting with other oculomotor systems. The processes by which the VOR is informed by visual information primarily occur within the vestibular nuclei (Jones 1985).

3 The Clinical Utility of the VVOR

Baloh et al. showed that an abnormal VVOR was only present in patients with a combined impairment of cerebellar ataxia and bilateral vestibular hypofunction (Baloh et al. 1979). An abnormality of the VVOR reflects a compound deficit of the three key compensatory eye movement systems: the vestibulo-ocular reflex (VOR), the optokinetic reflex OKR and smooth pursuit. Other researchers went on to exploit this combined central and peripheral vestibular pathology in their work on vestibular physiology (Bronstein et al. 1991; Waterston et al. 1992). Anatomically, the VOR reflects a component of vestibular function, whilst SP + OKR are cerebella functions. An abnormal VVOR can be demonstrated clinically by turning a patient's head slowly (at approximately 0.5 Hz) in the yaw or pitch axes while the patient's gaze is directed at an earth-fixed target and observing that the compensatory eye movements are saccadic rather than smooth (Fig. 1).

It was not until the work of Miglaccio et al. that cerebellar ataxia with bilateral vestibulopathy (CABV) was described as a distinct clinical syndrome with the

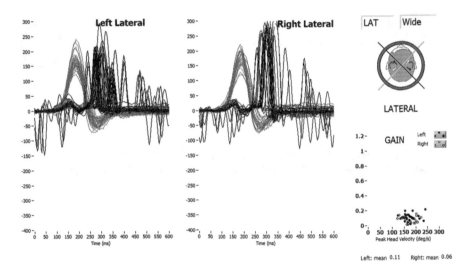

Fig. 1 Top panel: bilateral vestibulopathy shown on horizontal impulsive testing recorded using an ICS Impulse video-oculographic system (GN Otometrics, Taastrup, Denmark). The head rotation stimulus is shown in red (up to a peak angular velocity of 250°/sec and an angular acceleration of 2000°/sec); eye movement response is shown in black. Inset shows horizontal impulses in a normal subject with a VOR gain of 1, while main panel shows a patient with a bilateral vestibulopathy with a gain less than 0.1 in each direction. There are overt catch-up saccades. Middle panel: cerebellar impairment results in saccadic visual pursuit recorded using an ICS video-oculographic system (GN Otometrics, Denmark). There is no head rotation (red) as the eyes slowly pursue a slow moving target; salvos of corrective saccades are required (black). Lower panel: impaired horizontal VVOR recorded using an ICS video-oculographic system (GN Otometrics, Denmark). The head rotation stimulus is shown in red and the eye movement response is shown in black. The CABV patient makes salvos of catch-up saccades in response to a reduced VVOR gain

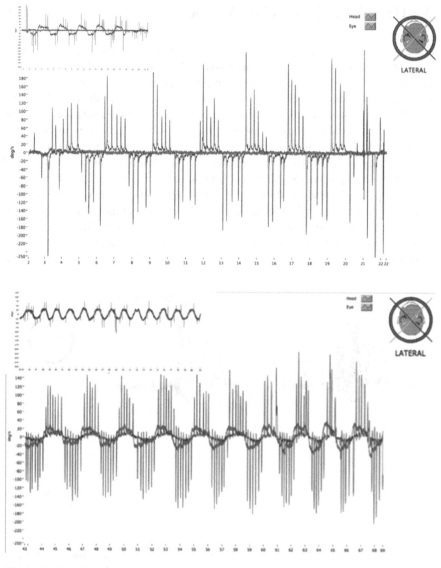

Fig. 1 (continued)

VVOR as its characteristic oculomotor deficit (Migliaccio 2004). A somatosensory deficit was identified in a subset of CABV patients that allowed the characterization of a novel clinical syndrome: cerebellar ataxia with neuronopathy and vestibular areflexia syndrome (CANVAS) (Szmulewicz et al. 2011b). In utilizing the VVOR, we were able to identify many patients with both cerebellar ataxia and a bilateral vestibulopathy who had diagnoses other than CANVAS, and so we altered the use of the term 'CABV' from a *syndrome* to a *phenotype*, which could be seen in a number of diseases including spinocerebellar ataxia (SCA) types 3 and 6 (Huh et al. 2015; Szmulewicz et al. 2014a), Friedreich's ataxia (FA)[14], multiple system atrophy of the cerebellar type (MSAc)[13], Wernicke syndrome (Kattah et al. 2013) and combined pathology of independent aetiologies (for example, a patient with a Chiari malformation who underwent a surgical posterior decompression followed by a post-surgical infection which was treated with gentamicin that was complicated by an iatrogenic bilateral vestibulopathy) (Szmulewicz et al. 2011a).

The emergence of the CABV phenotype questions the traditional anatomical delineation between *peripheral* and *central* vestibular aetiologies of imbalance. The *vestibular nuclei* are located in the medulla, i.e. located outside the cerebellum and, while not named as cerebellar nuclei, are generally considered to be the anatomical *equivalent of* the deep *cerebellar nuclei of* the vestibulocerebellum (Martin 2012). The cerebellar nuclei and the lateral vestibular nucleus constitute the efferent pathways of the cerebellum (Rahimi-Balaei et al. 2015). The lateral vestibular nucleus receives cerebellar input (in the form of the Purkinje cell axons) and is therefore often considered to be a cerebellar, rather than a vestibular nucleus (Voogd et al. 1991). Given the intimate developmental association of the cerebellar and vestibular nuclei, the differentiation of the two may be viewed as somewhat contrived and of limited functional value. This forms the basis of a challenge to the traditional *central versus peripheral vestibular aetiology* dichotomy and whether this may be a somewhat artificial delineation, particularly in the sphere of the more complex balance disorders.

Gait ataxia may be due to an impairment of any one or more of cerebellar, vestibular and somatosensory function. It then follows that (Newman-Toker et al. 2008) in any patient presenting with an ataxic gait, all three of these systems are evaluated and (Kerber 2009) where a diagnosis of an isolated bilateral vestibulopathy, a cerebellar ataxia or a peripheral neuropathy (or neuronopathy) has been made, that the patient should be assessed for the presence of one or two of the other deficits which may cause an ataxic gait. Cerebellar dysfunction may be clinically assessed by examining for the presence of cerebellar oculomotor abnormalities (e.g. saccadic visual pursuit, gaze-evoked or down-beat nystagmus), limb ataxia and cerebellar dysarthria. Cerebellar atrophy on MRI should also be sought although it may be absent in the earlier stages of a cerebellar disorder (Szmulewicz et al. 2011a). A bilateral vestibulopathy is most readily seen during the bedside examination with the head impulse test (Halmagyi and Curthoys 1988) and objectively using modalities such as the video head impulse test, bithermal caloric irrigation or the rotational chair. Somatosensory impairment may manifest as abnormal or absent perception of light touch, pin prick, vibration or joint position sense. This is best confirmed objectively with a nerve conduction study looking for reduced or absent sensory nerve action potentials (SNAPs) as the bedside assessment of sensation is far from reliable (Szmulewicz et al. 2015). CABV can be seen during the oculomotor examination as a broken-up (or saccadic) VVOR (Migliaccio 2004; Petersen et al. 2013) (Fig. 1). An abnormal VVOR is best visualized with infrared video-oculography or quantitatively with rapid video-oculography (Szmulewicz et al. 2011a). We have found the rotational chair to be a less sensitive modality.

3.1 The Video Visually Enhanced Vestibulo-ocular Reflex (VVOR)

Development of the quantitative bedside VVOR, which employs rapid video-oculographic (rVOG) technology, was the next increment in the diagnosis of vestibulo-cerebellar disease (Szmulewicz et al. 2014a). Portable rVOG is a relatively new field of diagnostic eye movement quantification, whose utility has been facilitated by the more recent development of a lightweight, minimum-slip high-speed video eye tracking system (MacDougall et al. 2009). This has allowed in-office and bedside non-invasive, objective and accurate oculomotor assessment.

4 Disease Taxonomy and the CABV Phenotype

An abnormal VVOR and, hence, combined cerebellar and bilateral vestibular impairment may be seen in an increasing number of disorders (Table 1). Its presence plays an ongoing role in the value of deep phenotyping in describing new ataxic disorders as well as further defining existing ones.

4.1 Cerebellar Ataxia with Neuronopathy and Vestibular Areflexia (CANVAS)

The cardinal triad of CANVAS are cerebellar impairment, bilateral vestibular hypofunction and a neuronopathy (ganglionopathy) (Szmulewicz et al. 2011a). The discovery of CANVAS relied on identifying patients with combined central and

Table 1 Key differential diagnoses for the CABV phenotype

Inherited
Friedreich's ataxia
Spinocerebellar ataxia (particularly SCA3, SCA6)
Cerebellar ataxia with neuronopathy and vestibular areflexia syndrome (CANVAS)[a]
Acquired
Multiple system atrophy with predominant cerebellar ataxia (MSAc)
Idiopathic cerebellar ataxia with bilateral vestibulopathy (iCABV)
Wernicke encephalopathy
[a]Dual pathology

[a]The mode of inheritance in CANVAS is yet to be established

peripheral vestibular pathology, i.e. the CABV phenotype (Szmulewicz et al. 2014b). Following this, neurophysiological testing was used to select those who had a third component of pathology – a somatosensory impairment (Szmulewicz et al. 2015). In addition to an abnormal VVOR (Szmulewicz et al. 2011a; Migliaccio 2004; Petersen et al. 2013), patients with CANVAS have impairments in peripheral sensory perception (including light touch, vibration or proprioception) (Szmulewicz et al. 2011b; 2014b). CANVAS patients primarily present with slowly progressive gait ataxia and somatosensory impairment, which generally begins in the fifth to seventh (but may present as early as the third) decades of life. There is no published clinical examination data regarding the sensory dysfunction of the geniculate or trigeminal ganglia although petrological and electrophysiological evidence of these cranial ganglionopathies exists (Szmulewicz et al. 2015). In addition to cerebellar impairment, bilateral vestibulopathy and somatosensory deficits, common accompaniments may include chronic neurogenic cough and autonomic dysfunction (e.g. orthostatic hypotension) (Szmulewicz et al. 2014b). Hearing is unaffected by CANVAS (Szmulewicz et al. 2016).

Post-mortem pathology reveals a consistent pattern of cerebellar atrophy involving the anterior and dorsal vermis (lobules VI, VIIa and VIIb) and, hemispherically, crus I (Szmulewicz et al. 2011b; 2014c). Temporal bone otopathology finds a severe vestibular (Scarpa's) neuronopathy (ganglionopathy) which underlies the bilateral vestibulopathy. The geniculate and trigeminal ganglia, but not the spiral ganglion, are similarly affected (Szmulewicz et al. 2011c). The somatosensory deficit seen in CANVAS reflects the marked neuronal loss seen in the dorsal root ganglia (DRG) (Szmulewicz et al. 2014c). The DRG neuronal loss is responsible for secondary axonal degeneration which manifests clinically as reduced perception of sensation and is electrophysiologically identifiable as reduced or absent SNAPs (Szmulewicz et al. 2015).

A genetic aetiology is strongly suggested by the pedigrees of multiple families with several affected members (in at least a subset of the cohort) and is consistent with either an autosomal recessive or an autosomal dominant pattern of inheritance with reduced penetrance (Szmulewicz et al. 2016, 2014b). For the patient who presents with the CABV phenotype in combination with electrophysiological evidence of peripheral sensory impairment, a diagnostic protocol was constructed, in part, to exclude other causes of gait ataxia which may manifest with the same clinical triad as CANVAS. These disorders include several of the autosomal dominant spinocerebellar ataxias and late-onset FA (Table 1) (Szmulewicz et al. 2016).

4.2 Idiopathic Cerebellar Ataxia with Bilateral Vestibulopathy (iCABV)

A not insignificant number of cases with the CABV phenotype are unable to be diagnosed with a specific disorder or underlying aetiology. We have labelled this 'idiopathic cerebellar ataxia with bilateral vestibulopathy (iCABV)'. We have found that the manifestation of the final component of the CANVAS diagnostic triad may take greater than 10 years. We speculate that a proportion of iCABV cases will go on to develop CANVAS, that is, *CANVAS in evolution* (Szmulewicz et al. 2014b). For this reason, we recommend that a patient with the CABV phenotype should be reviewed at regular intervals to ascertain whether they will go on to satisfy the diagnostic criteria for CANVAS (or indeed another disorder), i.e. develop a somatosensory deficit (Szmulewicz et al. 2016). It is possible that in time, other diagnostic entities which present with a compound cerebellar and bilateral vestibulopathy will be recognized.

4.3   Friedreich's Ataxia

Friedreich's ataxia (FA) is the most common inherited ataxia (Koeppen 2011), and a mutation in the gene encoding frataxin located on chromosome 9q 26 is generally responsible. Its heritability is autosomal recessive (Campuzano et al. 1996; Delatycki and Corben 2012) with prevalence varying between 2 and 4.5 per 100,000 (Barbeau 1978). Symptom onset generally occurs between 8 and 19 years of age (Friedreich 1863; Campanella et al. 2015). Late-onset FA begins after the age of 40 years (Klockgether and Chamberlain 1993). The hallmark of cerebellar pathology in FA is progressive atrophy of the dentate nucleus of the cerebellum as a result of neuronal loss (Koeppen 2011). The widespread neurodegenerative sequelae seen in FA include corticospinal, dorsal column and spinocerebellar tract degeneration. Sensorineural hearing loss is seen in approximately 20 percent (Koeppen 2011). Mild cerebellar atrophy, particularly affecting the vermis, may be seen later in the disease course (Giroud et al. 1994). Neuronal loss within Clarke's columns and a dorsal root ganglionopathy also exist (Hashida et al. 1997) and probably underlie, at least in part, the proprioceptive component of the ataxia.

Oculomotor disturbances are common, and gaze fixation is often interrupted by saccadic intrusions which are most commonly square-wave jerks but occasionally ocular flutter. VOR gain is reduced in most (Fahey et al. 2008). Vestibular ganglia cell loss with vestibular nerve atrophy has been reported, while the vestibular end organ and the facial nerve appear unaffected. There is marked loss of spiral ganglion cells with a near-normal organ of Corti (Spoendlin 1974; Oppenheimer 1979).

4.4   Spinocerebellar Ataxia Type 3 (Machado–Joseph Disease)

Spinocerebellar ataxia type 3 (SCA3) or Machado–Joseph disease is the most prevalent spinocerebellar ataxia and comprises 20 to 50% of all SCAs. Age of onset varies from childhood to late adult life, but most commonly symptoms begin between 20 and 45 years of age (Moseley et al. 1998; Schols et al. 1995).

The cerebellum is a particular target of abnormal gene expression (Hashida et al. 1997), and the MRI tends to visualize a combination of fourth ventricle enlargement with cerebellar vermal, hemispheric, superior cerebellar peduncle and pontine atrophy (Murata et al. 1998). Macroscopic pathology discloses atrophy of the cerebellum, pons and cranial nerves III–XII (Buttner et al. 1998). Gaze-evoked nystagmus, slowing of saccades, saccadic dysmetria, square-wave jerks and rebound nystagmus may all be seen (Buttner et al. 1998; Burk et al. 1999). Impairment of the VOR is a common feature (Gordon et al. 2003; Zeigelboim et al. 2013).

Hearing abnormalities have been reported in approximately a third of SCA 3 patients (Hoche et al. 2008). Although there appears to be a dearth of published otopathology in SCA 3, Hoche et al. reported widespread pathological involvement of the auditory brainstem nuclei which may account for the auditory abnormalities (Linnemann et al. 2016). A combination of reduced VOR gain and a peripheral sensory neuropathy may also accompany the cerebellar atrophy (Matsumura et al. 1997).

4.5 Spinocerebellar Ataxia Type 6

Spinocerebellar ataxia type 6 (SCA6) is a late-onset and slowly progressive cerebellar ataxia (Zhuchenko et al. 1997). A CAG repeat(polyglutamine) expansion in the CACNA1A gene coding for a voltage-dependent calcium channel is the principal genetic abnormality (Rajakulendran et al. 2011). Other mutations in this gene are responsible for episodic ataxia type 2 and familial hemiplegic migraine (Solodkin and Gomez 2012). Disease onset in most patients is in the sixth decade (Schols et al. 2000). This late onset of symptoms most likely accounts for fact that SCA6 is found in approximately 10% of patients with apparently idiopathic sporadic cerebellar ataxia (Yu-Wai-Man et al. 2009). MRI abnormalities are generally limited to cerebellar atrophy (Schols et al. 2000). Oculomotor abnormalities are comprised of saccadic pursuit, square-wave jerks, down-beat nystagmus and saccadic dysmetria (Crane et al. 2000). Vestibular function in SCA6 patients has been reportedly normal (Buttner et al. 1998), hypoactive (Crane et al. 2000; Gomez et al. 1997) or hyperactive (Yu-Wai-Man et al. 2009). More recently the VOR gain in response to high acceleration and frequency stimulation has been reported to be is increased in mild disease and decreased in more severe disease (Huh et al. 2015). Episodic vertigo is often present (Schols et al. 1997), while peripheral sensory impairment is a more variable feature (Graham and Oppenheimer 1969) with a recent study reporting a somatosensory deficit in 22% of SCA6 patients (Matsumura et al. 1997).

4.6 Multiple System Atrophy with Predominant Cerebellar Ataxia (MSAc)

Multiple system atrophy (MSA) encompasses the former diagnostic entities of olivo-pontocerebellar atrophy, striatonigral degeneration and Shy-Drager syndrome (Hara et al. 2007). MSA is primarily a sporadic disorder although familial cases have been identified (Berciano 2002). It is a neurodegenerative disease causing various degrees of extra-pyramidal, pyramidal, cerebellar and autonomic impairment (Bower et al. 1997). The annual incidence of MSA is approximately 3 per 100,000 (Schrag et al. 1999) with an estimated prevalence between 2 and 5 cases per 100,000 population (Tison et al. 2000; Kollensperger et al. 2010). The mean age of onset of 52 (range from 31 to 78) years (Tison et al. 2000; Kollensperger et al. 2010). MSA with predominant parkinsonism (MSAp) occurs two to four times as commonly as MSAc (Quinn 2005; Watanabe et al. 2002) except in Japan, where MSAc occurs approximately twice as often (Brooks and Seppi 2009). The MRI brain scan in patients with MSA may reveal atrophy of the putamen, pons and middle cerebellar peduncles (Jellinger and Lantos 2010; Anderson et al. 2008). Degeneration of transverse pontocerebellar fibres may give rise to the somewhat characteristic 'hot cross bun sign'. This sign however lacks specificity and has been seen in other neurodegenerative conditions (Jellinger and Lantos 2010).

The following oculomotor abnormalities are seen in MSA patients: excessive square-wave jerks, mild vertical supranuclear gaze palsy, gaze-evoked nystagmus, positioning downbeat nystagmus, saccadic hypometria, impaired SP and abnormal VOR suppression (Suarez et al. 1992). It appears that disruption of the cerebellum's descending inhibition on the vestibular apparatus may result in abnormalities in the VOR gain (Suarez et al. 1992; Ikeda et al. 2011). We have found that VOR gain may be reduced in a subset of MSAc patients (Szmulewicz et al. 2014a). While there do not appear to be any published reports of otopathology in MSAc patients, one study was unable to find any hearing impairment in 32 MSAc patients as compared to age-matched controls (Victor et al. 1989).

4.7 Wernicke Encephalopathy

Wernicke encephalopathy (WE) may be an acute life-threatening syndrome which may complicate thiamine deficiency (Harper 1983). It is most often associated with chronic alcoholism (78,79). Autopsy studies have found a higher incidence of WE in the general population than are recognized clinically (Vege et al. 1991; Ghez 1969). WE patients present with a range of central oculomotor abnormalities and a bilateral vestibulopathy[73]. Given the known cerebellar involvement in WE, it may present with the CABV phenotype. Kattah et al. (2013) describe a series of patients who while not encephalopathic suffered with Wernicke *disease* and a bilateral vestibulopathy. This further emphasizes the importance of recognizing the oculomotor abnormalities as a cue to commencing urgent thiamine replacement therapy in this group of ataxic patients.

4.8 CABV Due to Dual Pathology

The CABV phenotype may be the combination of two separate pathologies, for instance, a patient who suffered a cerebellar haemorrhagic stroke and then received parenteral gentamicin for sepsis which was then complicated by vestibulotoxicity. While this patient has the combination of cerebellar ataxia and a bilateral vestibulopathy, they are not due to a common aetiology. Other cases we have seen include haemorrhage into a cerebellar tumour complicated by superficial siderosis leading to bilateral vestibular hypofunction.

5 Conclusion

Identification of the CABV phenotype not only reduces the number of potential differential diagnoses for any presenting patient but also outlines a diagnostic pathway with a greater opportunity for obtaining a definitive diagnosis. In the broader research context, the VVOR has a role in phenotyping as a means to improved disease taxonomy, while at a clinical level, it aids diagnosis. Combined cerebellar and peripheral vestibular dysfunction should be sought in any chronic, particularly progressive, gait ataxia. The presence of either a bilateral vestibulopathy or cerebellar ataxia should not preclude the investigation of the other or, indeed, a somatosensory deficit. This approach enables increased diagnostic accuracy, more targeted management and further elucidation of the interplay between these three systems in an increasing range of balance disorders. The combination of more detailed phenotypes and the expansion and increased availability of genetic testing promises improved nosology, as a prelude to more definitive treatments.

References

Anderson, T., Luxon, L., Quinn, N., Daniel, S., David Marsden, C., & Bronstein, A. (2008). Oculomotor function in multiple system atrophy: Clinical and laboratory features in 30 patients. *Movement Disorders, 23*(7), 977–984. https://doi.org/10.1002/mds.21999.

Baloh, R. W., Jenkins, H. A., Honrubia, V., Yee, R. D., & Lau, C. G. Y. (1979). Visual-vestibular interaction and cerebellar atrophy. *Neurology, 29*(1), 116–116. https://doi.org/10.1212/WNL.29.1.116.

Barbeau, A. (1978). Friedreich's ataxia 1978: An overview. *The Canadian Journal of Neurological Sciences, 5*, 161–165.

Berciano, J. (2002). Multiple system atrophy and idiopathic late-onset cerebellar ataxia. In M. Manto & M. Pandolfo (Eds.), *The cerebellum and its disorders* (pp. 178–197). Cambridge: Cambridge University Press.

Bower, J. H., Maraganore, D. M., McDonnell, S. K., & Rocca, W. A. (1997). Incidence of progressive supranuclear palsy and multiple system atrophy in Olmsted County, Minnesota, 1976 to 1990. *Neurology, 49*(5), 1284–1288. https://doi.org/10.1212/WNL.49.5.1284.

Bronstein, A. M., Mossman, S., Luxon, L. M., et al. (1991). The neck-eye reflex in patients with reduced vestibular and optokinetic function. *Brain, 114*(Pt 1A), 1–11.

Brooks, D. J., & Seppi, K. (2009). Neuroimaging Working Group on MSA. Proposed neuroimaging criteria for the diagnosis of multiple system atrophy. *Movement Disorders, 24,* 949–964.

Burk, K., Fetter, M., Abele, M., et al. (1999). Autosomal dominant cerebellar ataxia type I: oculomotor abnormalities in families with SCA1, SCA2 and SCA3. *Journal of Neurology, 246,* 789–797. Gordon C. R, Zivotofsy A. Z, Caspi A. (2014). Impaired vestibulo-ocular reflex (VOR) in spinocerebellar ataxia type 3 (SCA3). *J Vestib Res, 24,* 351–355.

Buttner, N., Geschwind, D., Jen, J. C., et al. (1998). Oculomotor phenotypes in autosomal dominant ataxias. *Archives of Neurology, 55,* 1353–1357.

Campanella, G., Filla, A., De Falco, F., Mansi, D., Durivage, A., & Barbeau, A. (2015). Friedreich's ataxia in the South of Italy: A clinical and biochemical survey of 23 patients. *The Canadian Journal of Neurological Sciences, 7*(04), 351–357. https://doi.org/10.1017/S0317167100022873.

Campuzano, V., Montermini, L., Molto, M. D., et al. (1996). Friedreich's ataxia: Autosomal recessive disease caused by an intronic GAA triplet repeat expansion. *Science, 271*(5254), 1423–1427. https://doi.org/10.1126/science.271.5254.1423.

Crane, B. T., Tian, J. R., & Demer, J. L. (2000). Initial vestibulo-ocular reflex during transient angular and linear acceleration in human cerebellar dysfunction. *Experimental Brain Research, 130*(4), 486–496.

Delatycki, M. B., & Corben, L. A. (2012). Clinical features of Friedreich ataxia. *Journal of Child Neurology, 27*(9), 1133–1137. https://doi.org/10.1177/0883073812448230.

Fahey, M. C., Cremer, P. D., Aw, S. T., et al. (2008). Vestibular, saccadic and fixation abnormalities in genetically confirmed Friedreich ataxia. *Brain, 131*(4), 1035–1045. https://doi.org/10.1093/brain/awm323.

Friedreich, N. (1863). Ueber degenerative Atrophie der spinalen Hinterstränge. *Archiv für pathologische Anatomie und Physiologie und für klinische Medicin, 26,* 391–419. https://doi.org/10.1007/BF01930976.pdf.

Ghez, C. (1969). Vestibular paresis: A clinical feature of Wernicke's disease. *Journal of Neurology, 32,* 134–139.

Giroud, M., Septien, L., Pelletier, J. L., & Dueret, N. (1994). Decrease in cerebellar blood flow in patients with Friedreich's ataxia: A TC-HMPAO SPECT study of three cases. *Neurological Research, 16,* 342–344.

Gomez, C. M., Thompson, R. M., & Gammack, J. T. (1997). Spinocerebellar ataxia type 6: Gaze-evoked and vertical nystagmus, Purkinje cell degeneration, and variable age of onset. *Annals of Neurology, 42*(6), 933–950.

Gordon, C. R., Joffe, V., Vainstein, G., & Gadoth, N. (2003). Vestibulo-ocular areflexia in families with spinocerebellar ataxia type 3 (Machado-Joseph disease). *Journal of Neurology, Neurosurgery, and Psychiatry, 74*(10), 1403–1406.

Graham, J. G., & Oppenheimer, D. R. (1969). Orthostatic hypotension and nicotine sensitivity in a case of multiple system atrophy. *Journal of Neurology, Neurosurgery, and Psychiatry, 32,* 28–34.

Halmagyi, G. M., & Curthoys, I. S. (1988). A clinical sign of canal paresis. *Archives of Neurology, 45,* 737–739.

Hara, K., Momose, Y., Tokiguchi, S., et al. (2007). Multiplex families with multiple system atrophy. *Archives of Neurology, 64,* 545–551.

Harper, C. (1983). The incidence of Wernicke's encephalopathy in Australia – a neuropathological study of 131 cases. *Journal of Neurology, 46*(7), 593–598.

Hashida, H., Goto, J., Kurisaki, H., et al. (1997). Brain regional differences in the expansion of a CAG repeat in the spinocerebellar ataxias: Dentatorubral-pallidoluysian atrophy, Machado-Joseph disease, and spinocerebellar ataxia type 1. *Annals of Neurology, 41,* 505–511.

Hoche, F., Seidel, K., Brunt, E. R., et al. (2008). Involvement of the auditory brainstem system in spinocerebellar ataxia type 2 (SCA2), type 3 (SCA3) and type 7 (SCA7). *Neuropathology and Applied Neurobiology, 34*(5), 479–491. https://doi.org/10.1111/j.1365-2990.2007.00933.x.

Huh, Y. E., Kim, J. S., Kim, H.-J., et al. (2015). Vestibular performance during high-acceleration stimuli correlates with clinical decline in SCA6. *Cerebellum, 14*(3), 284–291. https://doi.org/10.1007/s12311-015-0650-3.

Ikeda, Y., Nagai, M., Kurata, T., et al. (2011). Comparisons of acoustic function in SCA31 and other forms of ataxias. *Neurological Research, 33*(4), 427–432. https://doi.org/10.1179/1743132810Y.0000000011.

Jellinger, K. A., & Lantos, P. L. (2010). Papp-Lantos inclusions and the pathogenesis of multiple system atrophy: An update. *Acta Neuropathologica, 119*, 657–667.

Jones, G. M. (1985). Adaptive modulation of VOR parameters by vision. *Reviews of Oculomotor Research, 1*, 21–50.

Kattah, J. C., Dhanani, S. S., Pula, J. H., Mantokoudis, G., Tehrani, A. S. S., & Toker, D. E. N. (2013). Vestibular signs of thiamine deficiency during the early phase of suspected Wernicke encephalopathy. *Neurology: Clinical Practice, 3*(6), 460–468. https://doi.org/10.1212/01.CPJ.0000435749.32868.91.

Kerber, K. A. (2009). Vertigo and dizziness in the emergency department. *Emergency Medicine Clinics of North America, 27*(1), 39–50. https://doi.org/10.1016/j.emc.2008.09.002.

Klockgether, T., & Chamberlain, S. (1993). Late-onset Friedreich's ataxia: Molecular genetics, clinical neurophysiology, and magnetic resonance imaging. *Archives of Neurology, 50*(8), 803–806.

Koeppen, A. H. (2011). Friedreich's ataxia: Pathology, pathogenesis, and molecular genetics. *Journal of the Neurological Sciences, 303*(1–2), 1–12. https://doi.org/10.1016/j.jns.2011.01.010.

Kollensperger, M., Geser, F., Ndayisaba, J. P., et al. (2010). Presentation, diagnosis, and management of multiple system atrophy in Europe: Final analysis of the European multiple system atrophy registry. *Movement Disorders, 25*, 2604–2612.

Linnemann, C., Tezenas du Montcel, S., Rakowicz, M., et al. (2016). Peripheral Neuropathy in Spinocerebellar Ataxia type 1, 2, 3, and 6. *Cerebellum, 15*(2), 165–173. https://doi.org/10.1007/s12311-015-0684-6.

Lisberger, S. G., Evinger, C., Johanson, G. W., & Fuchs, A. F. (1981). Relationship between eye acceleration and retinal image velocity during foveal smooth pursuit in man and monkey. *Journal of Neurophysiology, 46*(2), 229–249. https://doi.org/10.1152/jn.1981.46.2.229.

MacDougall, H. G., Weber, K. P., McGarvie, L. A., Halmagyi, G. M., & Curthoys, I. S. (2009). The video head impulse test: diagnostic accuracy in peripheral vestibulopathy. *Neurology, 73*(14), 1134–1141. https://doi.org/10.1212/WNL.0b013e3181bacf85.

Martin, J. (2012). *Neuroanatomy text and atlas* (4th ed.). New York: McGraw Hill Professional.

Matsumura, R., Futamura, N., Fujimoto, Y., et al. (1997). Spinocerebellar ataxia type 6. Molecular and clinical features of 35 Japanese patients including one homozygous for the CAG repeat expansion. *Neurology, 49*, 1238–1243.

Migliaccio, A. A. (2004). Cerebellar ataxia with bilateral vestibulopathy: description of a syndrome and its characteristic clinical sign. *Brain, 127*(2), 280–293. https://doi.org/10.1093/brain/awh030.

Moseley, M. L., Benzow, K. A., & Schut, L. J. (1998). Incidence of dominant spinocerebellar and Friedreich triplet repeats among 361 ataxia families. *Neurology, 51*(6), 1666–1671.

Murata, Y., Yamaguchi, S., Kawakami, H., et al. (1998). Characteristic mag- netic resonance imaging findings in Machado-Joseph disease. *Archives of Neurology, 55*, 33–37.

Newman-Toker, D. E., Hsieh, Y.-H., Camargo, C. A., Pelletier, A. J., Butchy, G. T., & Edlow, J. A. (2008). Spectrum of dizziness visits to US emergency departments: Cross-sectional analysis from a nationally representative sample. *Mayo Clinic Proceedings, 83*(7), 765–775. https://doi.org/10.4065/83.7.765.

Oppenheimer, D. R. (1979). Brain lesions in Friedreich's ataxia. *The Canadian Journal of Neurological Sciences, 6*, 173.

Petersen, J. A., Wichmann, W. W., & Weber, K. P. (2013). The pivotal sign of CANVAS. *Neurology, 81*(18), 1642–1643. https://doi.org/10.1212/WNL.0b013e3182a9f435.

Quinn, N. P. (2005). How to diagnose multiple system atrophy. *Movement Disorders, 20*(Suppl12), S5.

Rahimi-Balaei, M., Afsharinezhad, P., Bailey, K., Buchok, M., Yeganeh, B., & Marzban, H. (2015). Embryonic stages in cerebellar afferent development. *Cerebellum Ataxias, 2*(1), 7. https://doi.org/10.1186/s40673-015-0026-y.

Rajakulendran, S., Kaski, D., & Hanna, M. G. (2011). Neuronal P/Q-type calcium channel dysfunction in inherited disorders of the CNS. *Nature Reviews. Neurology, 17,* 86–96.

Ramat, S., Zee, D. S., & Minor, L. B. (2001). Translational Vestibulo-Ocular Reflex Evoked by a "Head Heave" Stimulus. *Annals of the New York Academy of Sciences, 942,* 95–113.

Schols, L., Amoiridis, G., & Langkafel, M. (1995). Machado-Joseph disease mutation as the genetic basis of most spinocerebellar ataxias in Germany. *Journal of Neurology, Neurosurgery, and Psychiatry, 59,* 449–450.

Schols, L., Amoiridis, G., Büttner, T., Przuntek, H., Epplen, J. T., & Riess, O. (1997). Autosomal dominant cerebellar ataxia: Phenotypic differences in genetically defined subtypes? *Annals of Neurology, 42*(6), 924–932.

Schols, L., Szymanski, S., Peters, S., & Przuntek, H. (2000). Genetic background of apparently idiopathic sporadic cerebellar ataxia. *Human Genetics, 107,* 132.

Schrag, A., Ben-Shlomo, A., & Quinn, N. P. (1999). Prevalence of progressive supranuclear palsy and multiple system atrophy: A cross-sectional study. *Lancet, 354,* 1771–1775.

Solodkin, A., & Gomez, C. M. (2012). Spinocerebellar ataxia type 6. *Handbook of Clinical Neurology, 103,* 461–473.

Spoendlin, H. (1974). Optic and Cochleovestibular degenerations in hereditary ataxias. *Brain, 97,* 41–48.

Suarez, H., Rosales, B., & Claussen, C. F. (1992). Plastic properties of the vestibulo-ocular reflex in olivo-ponto-cerebellar atrophy. *SOTO, 112*(4), 589–594.

Szmulewicz, D. J., Szmulewicz, D. J., Waterston, J. A., et al. (2011a). Cerebellar ataxia, neuropathy, vestibular areflexia syndrome (CANVAS): a review of the clinical features and video-oculographic diagnosis. *Annals of the New York Academy of Sciences, 1233,* 139–147. https://doi.org/10.1111/j.1749-6632.2011.06158.x.

Szmulewicz, D. J., Waterston, J. A., Halmagyi, G. M., et al. (2011b). Sensory neuropathy as part of the cerebellar ataxia neuropathy vestibular areflexia syndrome. *Neurology, 76*(22), 1903–1910. https://doi.org/10.1212/WNL.0b013e31821d746e.

Szmulewicz, D. J., Merchant, S. N., & Halmagyi, G. M. (2011c). Cerebellar ataxia with neuropathy and bilateral vestibular areflexia syndrome: A histopathologic case report. *Otology & Neurotology, 32,* e63–e65.

Szmulewicz, D., MacDougall, H., Storey, E., Curthoys, I., & Halmagyi, M. (2014a). A Novel Quantitative Bedside Test of Balance Function: The Video Visually Enhanced Vestibulo-ocular Reflex (VVOR) (S19.002). *Neurology, 82*(10 Supplement), S19.002.

Szmulewicz, D. J., McLean, C. A., MacDougall, H. G., Roberts, L., & Storey, E. (2014b). Halmagyi GM. CANVAS an update: Clinical presentation, investigation and management. *Journal of Vestibular Research, 24*(5–6), 465–474. https://doi.org/10.3233/VES-140536.

Szmulewicz, D. J., McLean, C. A., Rodriguez, M. L., et al. (2014c). Dorsal root ganglionopathy is responsible for the sensory impairment in CANVAS. *Neurology, 82*(16), 1410–1415. https://doi.org/10.1212/WNL.0000000000000352.

Szmulewicz, D. J., Seiderer, L., Halmagyi, G. M., Storey, E., & Roberts, L. (2015). Neurophysiological evidence for generalized sensory neuronopathy in cerebellar ataxia with neuropathy and bilateral vestibular areflexia syndrome. *Muscle & Nerve, 51*(4), 600–603. https://doi.org/10.1002/mus.24422.

Szmulewicz, D. J., Szmulewicz, D. J., Roberts, L., et al. (2016). Proposed diagnostic criteria for cerebellar ataxia with neuropathy and vestibular areflexia syndrome (CANVAS). *Neurology: Clinical Practice, 6*(1), 61–68. https://doi.org/10.1212/CPJ.0000000000000215.

Tison, F., Yekhlef, F., Chrysostome, V., et al. (2000). Prevalence of multiple system atrophy. *Lancet, 355,* 495.

Vege, A., Sund, S., & Lindboe, C. F. (1991). Wernicke's encephalopathy in an autopsy material obtained over a one-year period. *APMIS, 99*(8), 755–758.

Victor, M., Adams, R. A., & Collins, G. H. (1989). *The Wernicke-Korsakoff syndrome and related disorders due to alcoholism and malnutrition.* Philadelphia: FA Davis.

Voogd, J., Epema, A. H., & Rubertone, J. A. (1991). Cerebello-vestibular connections of the anterior vermis. A retrograde tracer study in different mammals including primates. *Archives Italiennes de Biologie, 129*(1), 3–19.

Voogd, J., Gerrits, N. M., & Ruigrok, T. J. (1996). Organization of the vestibulocerebellum. *Annals of the New York Academy of Sciences, 781*, 553–579.

Watanabe, H., Saito, Y., & Terao, S. (2002). Progression and prognosis in multiple system atrophy: an analysis of 230 Japanese patients. *Brain, 125*, 1070–1083.

Waterston, J. A., Barnes, G. R., Grealy, M. A., et al. (1992). Coordination of eye and head movements during smooth pursuit in patients with vestibular failure. *Journal of Neurology, Neurosurgery, and Psychiatry, 55*(12), 1125–1131.

Yu-Wai-Man, P., Gorman, G., Bateman, D. E., Leigh, R. J., & Chinnery, P. F. (2009). Vertigo and vestibular abnormalities in spinocerebellar ataxia type 6. *Journal of Neurology, 256*(1), 78–82.

Zeigelboim, B. S., Teive, H. A., Santos, R. S., et al. (2013). Audiological evaluation in spinocerebellar ataxia. *Codas, 25*, 351–357.

Zhuchenko, O., Bailey, J., Bonnen, P., et al. (1997). Autosomal dominant cerebellar ataxia (SCA6) associated with small polyglutamine expansions in the alpha 1A-voltage-dependent calcium channel. *Nature Genetics, 15*, 62–69.

Testing the Human Vestibulo-ocular Reflex in the Clinic: Video Head Impulses and Ocular VEMPs

Benjamin Nham, Leigh A. McGarvie, Rachael L. Taylor, and Miriam S. Welgampola

Abstract The video head impulse test (vHIT) and ocular vestibular-evoked myogenic potential are two techniques developed in recent times that enable non-invasive interrogation of the human vestibulo-ocular reflex in a clinical setting. In this chapter, we examine their testing technique, analysis and application in acute and episodic vestibular syndromes and chronic imbalance.

Keywords Myogenic potential · Vestibular system · Vestibulocolic reflex

1 Introduction

The vestibular system, our "sixth sense", is made of five peripheral receptors on either side, whose innervation carries signals to the vestibular nuclei within the brainstem and thence to effector muscles, which move the eyes, neck, trunk and limbs. The vestibulo-ocular reflex (VOR), the fastest reflex in the human body, was once described as an "ocular gyroscope that generates eye rotations to compensate for head movements, so the image of the outside world can remain still on the retina for as much time as possible" (Leigh 1996). As the head moves, the VOR generates slow-phase eye movements that are equal in magnitude, approximate in velocity but opposite in direction, thus maintaining stable vision. It is the VOR that enables us to see street signs and details of houses in our neighbourhood when we walk or jog

B. Nham · L. A. McGarvie · M. S. Welgampola (✉)
Central Clinical School, Institute of Clinical Neurosciences, Royal Prince Alfred Hospital
University of Sydney, Sydney, NSW, Australia
e-mail: miriam@icn.usyd.edu.au

R. L. Taylor
Audiology Department, Whangarei Hospital, Whangarei, New Zealand

New Zealand Dizziness and Balance Centre, Auckland, New Zealand

© Springer Nature Switzerland AG 2019   353
A. Shaikh, F. Ghasia (eds.), *Advances in Translational Neuroscience of Eye Movement Disorders*, Contemporary Clinical Neuroscience,
https://doi.org/10.1007/978-3-030-31407-1_18

down the street. Both rotational and translational head movements activate the VOR by stimulating the semicircular canals and otolith organs (Pozzo et al. 1990).

Vestibular function tests, past and present, assess the integrity of the VOR pathways in order to determine the extent and site of a vestibular lesion. The bithermal caloric test, described by Robert Barany in 1907, uses thermal stimulation of horizontal canals to selectively assess each horizontal VOR (Baloh et al. 2012; Baloh 2002). The sinusoidal and trapezoidal rotation tests simultaneously assess *both horizontal canals* during rotations of the head and body in the yaw plane (Hanson and Goebel 1998; Fife et al. 2000; Furman 2016). The gain (ratio of amplitude of eye velocity to the amplitude of head velocity), phase (timing relationship between head movement and reflexive eye response) and suppression of the VOR when visual fixation is present are useful parameters when seeking peripheral and central vestibular disorders. These tests are useful and still in use; however, they require cumbersome equipment and trained operators and can seldom be accommodated within an office practice. Two techniques developed in the past two decades, vestibular evoked potential (Colebatch et al. 1994; Rosengren et al. 2005) and video head impulse testing (MacDougall et al. 2009), have provided clinicians easy access to vestibular assessment within the clinic itself.

2   Head Impulse Testing

If the head is rapidly rotated 10–20 degrees to one side, the angular VOR will produce equal and opposite eye movement to maintain the visible world upon the retina. So swift is this movement that it is imperceptible to the naked eye. When a peripheral vestibular disorder such as vestibular neuritis results in an impaired VOR, this compensatory eye movement is no longer equal to the head movement, and a "catch-up saccade" is needed, to return the eye back to target. Halmagyi and Curthoys in 1988 first reported the horizontal head impulse test which enables the examiner to diagnose a defective horizontal VOR with their bare hands (Halmagyi and Curthoys 1988). Later work by the same group described head impulse testing in the plane of all three semicircular canal pairs (Cremer et al. 1998). To perform the HIT, the examiner delivers quick, low-amplitude (10–20°) head rotations in the plane of any canal pair as the subject fixes upon a stationery earth-fixed target. If the VOR is intact, the eyes remain on target (Fig. 1e). If VOR is impaired, when the

---

Fig. 1 (continued) green. In a normal VOR, head and eye velocity profiles are near identical. The gain (calculated as area under the eye velocity curve over the area under the head velocity) is close to 1 (**b**). In a subject with unilateral vestibular loss, the smooth eye velocity response is reduced and the gain is also reduced (**d**); corrective ("catch-up") saccades return the eye back to target. Saccades that occur during the head movement are called "covert saccades", while those occurring after the head movement are "overt saccades". (**e–h**) The bedside head impulse test. If the VOR is intact (**e**), when the head is thrust to the right, the eyes remain fixed on the midline target. In a left vestibulopathy, as the head is rotated to the left, the eyes initially move with the head and drift off the target (**g**); this is followed by a catch-up or refixation saccade as the eyes return to the target (**h**)

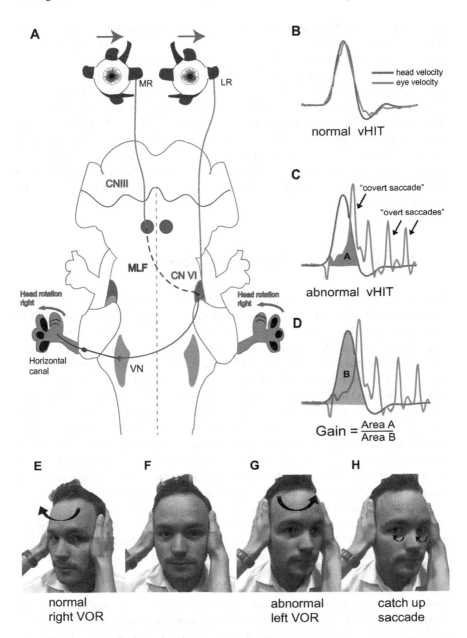

Fig. 1 (**a**) Three-neuron arc of the VOR. A head rotation to the right activates the right horizontal canal and primary vestibular afferents which project to the ipsilateral vestibular nucleus (VN) which in turn connects with the contralateral abducens nucleus and abducens interneurons which via the medial longitudinal fasciculus (MLF – dashed purple line) project to the ipsilateral oculomotor nucleus. Third-order neurons travel via the oculomotor and abducens nerves to innervate the right medial rectus (MR) and left lateral rectus (LR). (**b–d**) VOR quantification. The typical traces generated during video head impulse testing, representing head velocity in red and eye velocity in

head is rotated to the affected side, the head and eyes initially move together (Fig. 1g), followed by a rapid corrective eye movement or "catch-up saccade", which rapidly refixates the eye (Fig. 1h). Catch-up saccades that occur *after the head movement* are thought to be clinically detectable and are called "overt" saccades; those occurring *during the head movement* are undetectable to the human eye and are called "covert" saccades (Weber et al. 2009) (Fig. 1c). The quantitative head impulse (using scleral search coils or high-frequency video oculography) circumvents some drawbacks of the clinical HIT, which is subjective and relies upon the examiner's skill. In a large study of 500 patients, where quantitative video head impulse test (vHIT) was used as gold standard, the bedside HIT had a sensitivity of 66% and a low positive predictive value of 44% for detecting vestibular hypofunction (Yip et al. 2016).

3 vHIT Methods and Practical Considerations

There are several commercially available vHIT systems (Alhabib and Saliba 2017) which use high-speed digital cameras, with sampling rates ranging from 100 to 250 Hz, integrated within lightweight goggles worn by the subject. The patient sits at a ~1 metre distance from a target (fixation point on the wall at eye level). To test the horizontal canals, the examiner stands behind the patient, firmly grasps the subjects head with both hands and delivers rapid, unpredictable and small-amplitude (10–20°) fast head rotations (peak head velocity 150–300°/s) or impulses to the left or right (Halmagyi et al. 2017). It is essential that high-velocity (>150°/s) impulses are generated. The eye movement response to each impulse has a larger excitatory drive from the ipsilateral ear and a smaller excitatory contribution from silencing of the contralateral ear or "disinhibition". A high acceleration impulse effectively silences input from the contralateral ear. Thus, when the VOR is defective on one side, a high-velocity head thrust will yield an accurate low gain, whereas a low-velocity impulse allows more contributions from the intact contralateral ear and may incorrectly yield a normal gain (van den Berg et al. 2018). To minimize the subject's ability to predict the direction, the impulse should begin at the primary position. The head turn should be as abrupt as possible, in a "turn and stop" technique. Unintentional overshoot or rebound can result in stimulation of the contralateral side (Fig. 2b) (van den Berg et al. 2018; MacDougall et al. 2013). Each impulse consists of a large ipsilateral impulse followed by a much smaller contralateral impulse due to the head inertia during the stopping procedure. This has important consequences if saccade analysis is to provide complementary information to the gain analysis. This is illustrated in Fig. 2, in which the rebound during impulses towards the intact side has been varied in size. Due to the loss of the contribution from the neuritis-affected contralateral ear, the gain during the impulse towards the good side is slightly reduced, pulling the eye slightly away from the target. During the rebound, the low gain towards the affected contralateral ear means that the eye moves back in space with the head towards the centre target. The magnitude of this rebound determines whether a repositioning saccade is required and in which direction it needs to be.

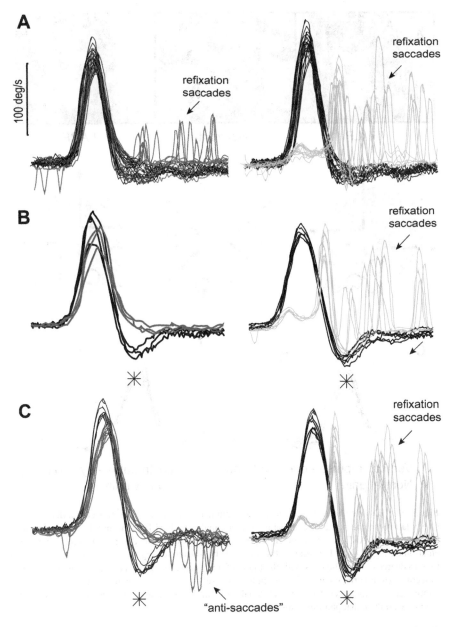

UNAFFECTED EAR **AFFECTED EAR**

A

100 deg/s

refixation saccades

refixation saccades

B

refixation saccades

C

refixation saccades

"anti-saccades"

Fig. 2 Video head impulses: the effect of "bounce back". A very small rebound will mean that the eye movement during the initial ipsilateral impulse dominates and the saccade will be in the compensatory direction for this part of the impulse (**a**). A slightly bigger rebound towards the affected side will result in the eye initially being pulled away from the target and then returned to the target during the rebound. This will require no catch-up saccades in the unaffected ear (**b**). If, however, the rebound towards the affected side is large, then the initial impulse portion will pull the eye slightly in that direction, while large rebound towards the affected ear will take the eye back past the target, requiring a repositioning saccade in the direction opposite to that of the initial impulse (**c**)

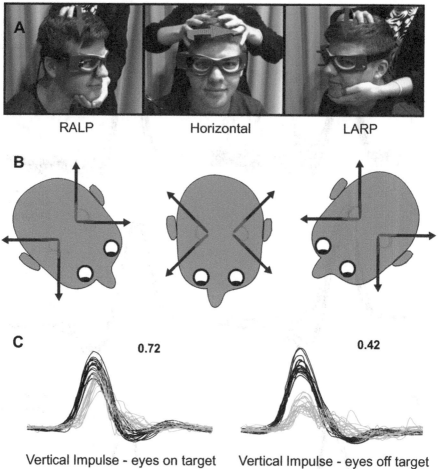

Fig. 3 The 3D video head impulse test. The examiner stands behind the subject and delivers passive rotations in the plane of each semicircular canal pair: RALP (right anterior, left posterior), horizontal and LARP (left anterior, right posterior) (**a**). A schematic diagram of head and eye orientation during 3D VHIT testing is illustrated in (**b**) – when testing vertical canals, eye deviation by ~45 degrees aligns the eyes with the plane of a vertical canal pair; thus, the compensatory eye movement (to a vertical canal plane impulse) is purely vertical, and torsional eye movements are eliminated. (**c**) illustrates how vertical canal gain can be spuriously lowered if the eyes are not aligned with the canal pair being tested

To test the vertical canals, the subject is rotated en bloc (45°) to the left or right of the mid-sagittal plane (Fig. 3), and their eyes are deviated ~ 45° in the yaw plane to in line with the mid-sagittal plane (McGarvie et al. 2015a). Turning the subject rightwards at this angle positions them for stimulation of left anterior and right posterior (LARP) canals, and leftward turning positions them for the right anterior and left posterior (RALP) pair. En bloc turning minimizes neck discomfort for the subject as compared to doing head on body rotations. The head thrusts are delivered in the LARP or RALP plane (Fig. 3), towards and away from the target. The eccentric gaze position ensures predominantly vertical eye movements when the head thrusts are performed, making for more reliable analysis, since current vHIT systems are only capable of 2D analysis and cannot measure torsion (McGarvie et al. 2015a). When gaze drifts away from the plane of stimulus, eye velocity response may show a delayed onset and spurious gain reduction without catch-up saccades. Fastening the goggles tightly around the head to minimize slippage and ensuring accidental goggle contact (with the examiners hand) does not occur during the head thrusts are important in order to avoid exaggerated gain due to artefact. Obtaining an unobstructed view of the pupil by tucking drooping eyelids under goggle rims and increasing room lighting to reduce pupil size, removing makeup, assists better pupil tracking.

4 Saccades Characteristics Could Help Track Vestibular Compensation

A sudden severe complete unilateral vestibular loss (as observed in total vestibular neuritis) initially results in a defective vHIT gain and catch-up saccades that occur after the head movement. With the passage of time, these saccades occur at shorter latencies such that they are invisible to the naked eye and begin to cluster together (MacDougall and Curthoys 2012). It has been hypothesized that clustering of saccades (both overt and covert) may signify vestibular compensation (MacDougall and Curthoys 2012; Curthoys and Manzari 2017). New tools of measuring saccade amplitude, velocity, latency and dispersion (Rey-Martinez et al. 2015) may help test these hypotheses.

5 Calculating VOR Gain

The horizontal vHIT gain is close to 1 in normal subjects (McGarvie et al. 2015b). Since the gain is also influenced by peak head velocity during impulse testing, target distance, device used and gain calculation algorithm, it is best that investigators collect their own normative data for a known range of peak head velocities rather than using a set cut-off value. All methods used to calculate the VOR gain use a ratio of eye velocity/head velocity either over the entire duration of the ipsilateral impulse (wide window method) (McGarvie et al. 2015a, 2015b) or a narrow window centred on peak head velocity (MacDougall et al. 2009; Halmagyi et al. 2017; Weber et al. 2008) or at a fixed point (at 40, 60 or 80 ms from onset) (Agrawal et al. 2014; Mossman et al. 2015). Narrow-window gain calculations are thought to be more sensitive to rapid transient signal changes including skin and goggle slip-induced artefact, while wider windows are less influenced by noise yet less sensitive to brief signal changes. A formal comparison of these methods in controls and lesions is needed, to identify the merits and shortcomings of each.

6 SHIMP

When performing the vHIT using the conventional video head impulse (HIMP) paradigm, the subject looks upon an *earth-fixed target* as the head is passively moved. A defective VOR is usually indicated by a catch-up saccade. When performing a SHIMP (suppression head impulse) paradigm, the subject fixes on a *head-fixed* target (usually from a head mounted laser) as head impulses are delivered. In the presence of a normal VOR, the eyes are involuntarily driven away from the head-fixed target and remain fixed in space leading to the healthy subject making anti-compensatory saccades to refixate on the target (MacDougall et al. 2016). The size of these saccades correlates with VOR gain. An absent VOR will result in absent saccades since there is no VOR to drive the eyes off target (Curthoys and Manzari 2017). Gain calculation using the conventional HIMP paradigm can be difficult when covert saccades are present; the SHIMP paradigm circumvents this problem (see Fig. 4) (MacDougall et al. 2016). SHIMP and HIMP parameters are complementary; the precise role of SHIMP in different vestibular disorders remains to be explored.

7 vHIT in Acute Vestibular Syndrome (AVS)

A sudden disabling attack of acute spontaneous vertigo lasting 24 hours or more ("acute vestibular syndrome") could be due to vestibular neuritis (VN) or a posterior circulations stroke (PCS). Here the vHIT provides valuable information that helps identify the cause of AVS. Vestibular neuritis is characterized by unidirectional spontaneous nystagmus, a positive head impulse and normal hearing. An abnormally decreased and asymmetrical horizontal VOR gain separates VN from PCS with high sensitivity (88%) and specificity (92%) (Mantokoudis et al. 2015). As the accelerations of head impulses increase in subjects with VN, catch-up saccade amplitudes increase, and VOR gain decreases (Weber et al. 2008; Schubert and Zee 2010). The pattern of canal hypofunction identifies the affected vestibular nerve division (Fig. 5). VN affecting both divisions is the most commonly observed subtype (55%) followed by superior (40%) and inferior VN (5%) (Magliulo et al. 2015; Taylor et al. 2016).

There are a myriad of vHIT patterns in PCS depending on the anatomical structures involved. Therefore, when approaching the subject with acute vestibular syndrome, it is best to first seek vestibular neuritis and investigate for PCS when "HINTS plus" criteria for VN are not fulfilled (Tehrani et al. 2018). Lesions involving the vestibular nucleus, nucleus prepositus hypoglossi or flocculus can have unilateral or bilateral reduced VOR gain (Kim et al. 2017). Most PCS in the posterior

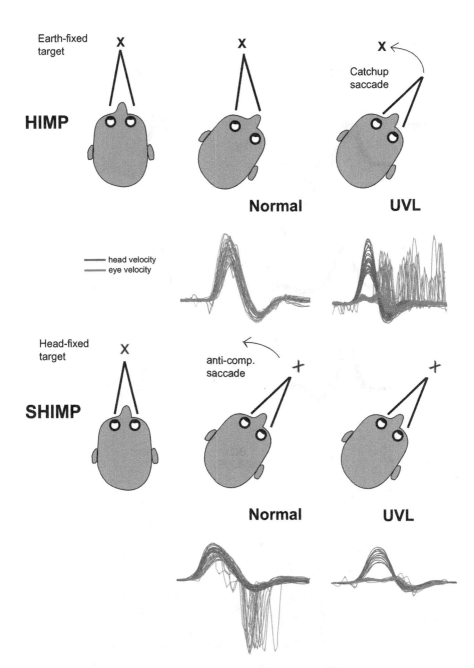

Fig. 4 The conventional head impulse (HIMP) and suppression head impulse (SHIMP) paradigms. For the HIMP, the subject focuses on an earth-fixed target. When a head impulse is delivered in the plane of an intact semicircular canal, the VOR swiftly moves the eyes in the opposite direction to remain fixed on target, generating a smooth eye velocity trace with a normal gain. In unilateral vestibular loss (UVL), when an impulse is delivered in a defective canal plane, since the VOR is defective, the head and eyes move together, the eyes move off target and a compensatory saccade is required to refixate eye upon the target. During SHIMP testing, the subject is asked to focus on a head-fixed target. If the VOR is intact, the eyes swiftly return to midline but then use an anti-compensatory saccade to redirect gaze upon the target. In UVL, since the VOR is defective, the head and eye move together, the eyes remain on the target and no saccades are required

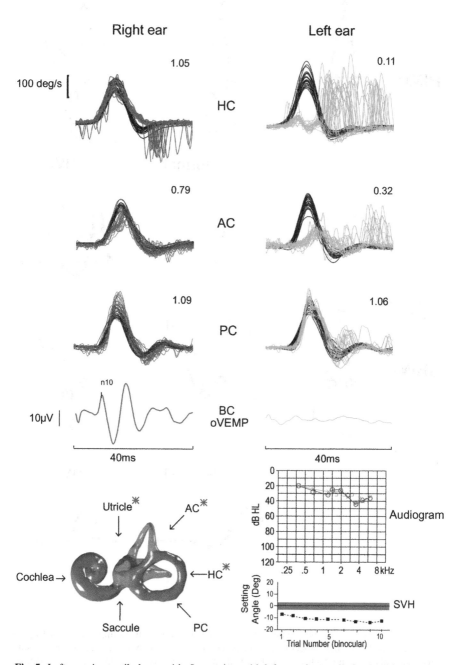

Fig. 5 Left superior vestibular neuritis. In a patient with left superior vestibular neuritis, the gains of the left horizontal (0.11) and anterior canals (0.32) are reduced significantly compared to the right side. Refixation saccades are present. There is an absent BC oVEMP on the left with preserved response on the right signifying utricular hypofunction. The audiogram is normal reflecting normal cochlear function. Subjective visual horizontal (SVH), a measure of otolith function, shows a leftward bias. Vestibular end organs innervated by the superior vestibular nerve are represented in green and marked with an asterisk (utricle, horizontal and anterior canals). Those innervated by the inferior nerve are represented in yellow (posterior canal and saccule)

inferior cerebellar artery (PICA) and superior cerebellar artery (SCA) territories will have normal vHIT results and gains. However, some PICA and SCA strokes can cause symmetrical mild reduction in VOR gain (0.7–0.8) with small amplitude saccades if projecting neurons from the nodulus/uvula or flocculus are involved. AICA strokes can have symmetric bilateral VOR gain reduction (0.3–0.4 ipsilesionally and 0.5–0.6 contralesionally) and smaller saccades compared to VN (Choi et al. 2018; Chen et al. 2014). This is thought to be secondary to involvement of the vestibular nucleus or flocculus and their contralateral connections.

8 vHIT in Episodic Vertigo Syndromes

vHIT rarely helps separate the two common causes of recurrent spontaneous vertigo: vestibular migraine and Meniere's disease. The caloric test was found to have a much higher prevalence of asymmetry in Meniere's disease (67%) than vHIT (37%), while VM was associated with 22% and 9% prevalence of asymmetry on calorics and vHIT (Cerchiai et al. 2016). The caloric-vHIT dissociation is considered a useful marker of Meniere's disease (McGarvie et al. 2015c). vHIT testing plays no role in the diagnosis of benign positional vertigo (BPV) which is identified by its history and distinctive examination findings of paroxysmal positional nystagmus in the plane of the affected semicircular canal. vHIT may however be useful in the identification of unilateral vestibular loss ipsilateral to the ear with BPV, thus implying secondary BPV.

9 vHIT in Chronic Imbalance

The patient with chronic vestibular insufficiency will complain of imbalance and gait ataxia; physical examination may reveal unilaterally or bilaterally positive head impulses, a positive matted Romberg test and impaired dynamic visual acuity. vHIT is the test of choice when seeking objective evidence of bilateral vestibular impairment (Fig. 6), since it has a well-defined normal range. Severe unilateral vestibular loss from a previous attack of VN or a vestibular schwannoma could also cause imbalance. Larger schwannomas are associated with a lower horizontal vHIT gain (Taylor et al. 2015). When using multiple indicators (low gain, covert and overt saccades), vHIT abnormalities were found in 90% patients with schwannoma compared with 62% for the caloric response (Batuecas-Caletrio et al. 2015). In cerebellar ataxia with neuropathy and vestibular areflexia (CANVAS), bilateral vestibular loss accompanies cerebellar ataxia, proprioceptive loss and a defective visually enhanced VOR (vVOR). Assessment of the vVOR can be performed and documented using the vHIT system: by performing slow sinusoidal head rotations as the subject fixes on a target. Low VOR gains and showers of compensatory saccades are typical of CANVAS.

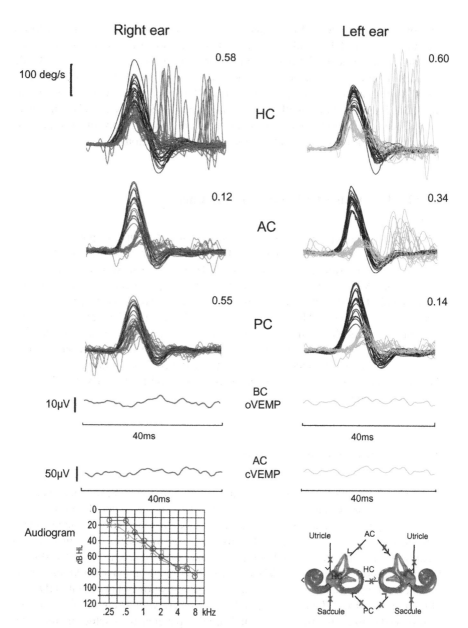

Fig. 6 Bilateral vestibular failure. In a patient with bilateral vestibular failure due to gentamicin ototoxicity, vHIT gains are reduced bilaterally, refixation saccades are present, and VEMPs are absent bilaterally. The audiogram shows bilateral age-related high-frequency sensorineural hearing loss

10 Ocular Vestibular-Evoked Myogenic Potentials: A Test of Otolith-Ocular Pathways

The otolith organs perceive linear head acceleration. Like the semicircular canals, their afferents project to extra-ocular muscles to produce compensatory eye movements during head movement and head tilt. However, otolith-evoked eye movements are extremely small, and it is impractical to use linear acceleration in a clinical setting. Ocular vestibular-evoked myogenic potentials (oVEMPs), which use non-physiological stimuli like sound and vibration, provide a more practical means of evoking and recording otolith-ocular reflexes.

Rosengren and coworkers in 2005 identified a short-latency surface potential recordable from averaged unrectified EMG recorded from infraorbital electrodes, in response to intense air- or bone-conducted sound (Rosengren et al. 2005). This reflex they found to be intact in subjects with profound sensorineural hearing loss and abolished in vestibular loss. As expected of a sound-evoked VOR, it has a short-onset latency (7 ms) and peaks at 10 ms ("n10"). Todd et al., who recorded the oVEMP surface potential and 3D eye movements in response to intense sound, found that the peak of the surface potential coincided with the onset of the sound-evoked eye movement (Todd et al. 2007). Thus the oVEMP was clearly a muscle action potential that generated an eye movement rather than an electrooculographic response (Fig. 7). The response was maximal on upgaze and abolished on downgaze (Iwasaki et al. 2007) indicating that activation of the inferior oblique muscle modulated reflex amplitude. The "oVEMP" was distinct from the blink reflex, which can in fact be triggered by bone-conducted supraorbital or glabella taps; the blink reflex occurs 4–5 ms later and is not abolished in downgaze (Fig. 8) and is still preserved despite vestibular loss (Smulders et al. 2009).

Several important attributes distinguish the oVEMP from its earlier counterpart – the cervical VEMP (Colebatch et al. 1994; Colebatch and Halmagyi 1992) or cVEMP recorded from the sternocleidomastoid muscles. While the cVEMP is an ipsilateral reflex, the oVEMP is crossed, reflecting the crossed utriculo-ocular pathways to the inferior oblique muscles. The cVEMP is an inhibitory or "relaxation" potential with a biphasic positive negative response, while the oVEMP is an excitatory potential that produces a negative-positive response. The air-conducted cVEMP, based on studies conducted on experimental animals, is thought to be a predominantly saccular response (Murofushi et al. 1995), while the oVEMP to bone-conducted sound is most likely to originate predominantly from the utricle (Manzari et al. 2010). Together these two complementary reflexes provide assessment of utriculo-ocular (oVEMP) and sacculo-collic pathways (cVEMP).

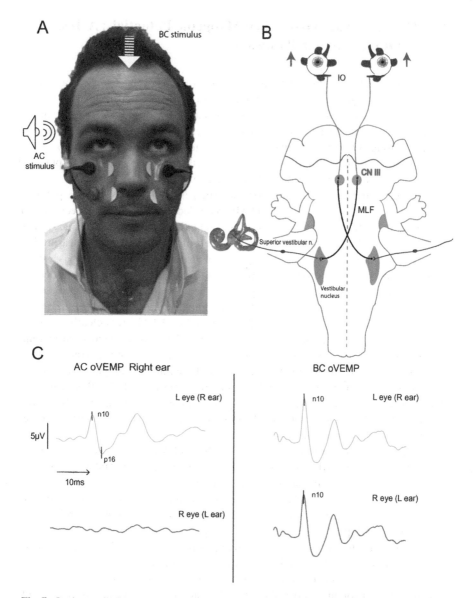

Fig. 7 Ocular vestibular-evoked myogenic potentials (oVEMP). The oVEMP montage consists of active electrodes (black) that are placed under the lower eye between the pupil and outer canthus (**a**). The reference electrodes (red) are in line with pupils approximately 2 cm below the active electrodes. The ground (purple) can be placed on the sternum or the neck. The subject looks upwards 20 degrees during the recording. The crossed otolith-ocular pathway for the oVEMP (**b**). Primary afferents from the right utricle travel in the superior division of the vestibular nerve to synapse in the vestibular nuclear complex. The second-order neurons cross via the MLF to the contralateral oculomotor nucleus to innervate the contralateral inferior oblique muscle. Thus, an air-conducted (AC) stimulus delivered over the right ear produces an oVEMP response beneath the left eye, while a bone-conducted (BC) stimulus delivered over Fz, which activates both ears simultaneously, will generate bilateral symmetric oVEMP responses (**c**)

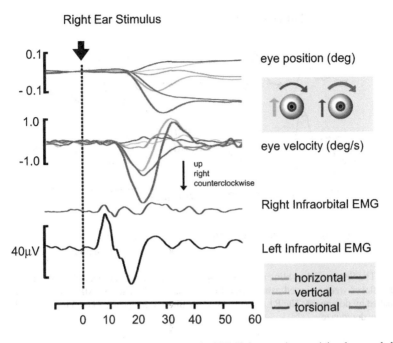

Fig. 8 The sound-evoked ocular movement and oVEMP in superior semicircular canal dehiscence. A sound delivered in the right ear generates an upward and contraversive eye movement that begins at 15 ms. The oVEMP (surface potential) recorded over the contralateral infraorbital muscles peaks just before the eye movement onset, consistent with the hypothesis that it is the compound muscle action potential generating the eye movement response

11 oVEMP Testing Methods

oVEMPs are recorded using evoked potential equipment common to most audiology or neurotology departments. Surface waveforms are produced through amplification of voltage differences from pairs of electrodes placed vertically beneath each eye, slightly lateral to the pupil (when the subject looks straight ahead). Testing is performed during upward gaze (20 degrees), to activate the inferior oblique muscles (Fig. 9) (Govender et al. 2016). Nystagmus, head tilt, alcohol intoxication, eye blinks and fatigue can attenuate but do not abolish the oVEMP response (Colebatch et al. 2016).

Bone-conducted (BC) stimuli produced by either tendon hammers, mechanical oscillators or mini-shakers placed on a midline skull location over the hairline (Fz) can elicit robust oVEMPs by simultaneously activating both otolith organs equally. VEMPs from BC stimuli bypass the middle ear ossicles and are preserved in conductive hearing loss. Widely used BC stimuli include 500 Hz tone bursts of 10–20 V peak-peak amplitude, 1 ms square wave pulses of 20 V p-p (146 dB force level) and tendon hammer taps which are near identical to a single cycle of a 125 Hz 10–20 V p-p tone burst. Air-conducted (AC) clicks or tones delivered through audiometric

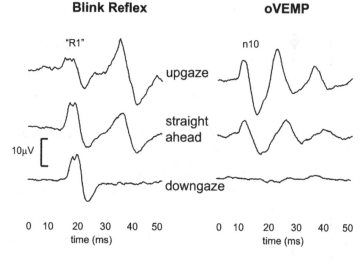

Fig. 9 The Fz tap-evoked blink reflex and supraorbital tap-evoked blink reflex: the effect of gaze. Both reflexes were recorded using infraorbital surface EMG. The oVEMP peaks at 10 ms and is abolished by downgaze. The R1 component of the blink reflex peaks about 5 ms later and is unaffected by downgaze

headphones can be used to elicit oVEMPs. AC stimuli need to be sufficiently loud (≥95 dB nHL) since oVEMPs evoked by AC stimuli are often small, have a higher threshold when compared with cVEMPs and are often absent, particularly in older patients (Welgampola and Colebatch 2001). Short (≤ 6 ms) tone bursts of 500–1000 Hz can be used to elicit oVEMPs in healthy subjects (Taylor et al. 2011). The frequency tuning of oVEMPs is similar to the tuning properties of irregular discharging otolith afferents in guinea pigs (Curthoys et al. 2016).

oVEMP parameters that are useful include response presence/absence, reflex amplitude asymmetry, reflex thresholds and peak latencies. The range of normal oVEMP amplitudes is large. Thus, the calculation of amplitude asymmetry ratios (using Jongkee's formula), rather than relying on amplitudes themselves, is recommended to control for between-subject variability. Amplitudes can be measured either from the baseline to the first negative peak (n10) or as the peak-peak (negative-positive) amplitude difference. Amplitude asymmetry ratios can be compared with the normal range of values for healthy controls. Thus, in unilateral disease, the healthy ear is the patient's own control. Ideally, clinics should collect their own normative data to account for variations in internal amplifier gains, stimulus factors and calibration settings for different equipment.

12   oVEMP Abnormalities in Superior Canal Dehiscence and Third Window Syndromes

oVEMP testing remains the most reliable non-radiological investigation, using either specific thresholds or amplitudes, to aid in SCD diagnosis. The abnormal oVEMP response in SCD is actually "hyperactive" (Fife et al. 2017). This is due to a less intense sound stimulus than normal (i.e. lower threshold) required to induce the response. The response amplitude is also abnormally increased on the affected side. (Fig. 10). While still abnormal, the threshold reductions appear to be less marked for BC oVEMP stimuli than for AC stimuli (Welgampola et al. 2008). Successful surgical treatment of SCD normalizes oVEMP thresholds and amplitudes (Fig. 10) (Welgampola et al. 2008). Other third window syndromes such as posterior canal dehiscence (Aw et al. 2011) and enlarged vestibular aqueduct syndrome (Taylor et al. 2012a) can also display large amplitudes and low reflex thresholds (Colebatch et al. 2016).

13   oVEMP in Acute Vestibular Syndrome

13.1   Vestibular Neuritis

The oVEMP is often affected in acute VN due to the frequent involvement of the superior vestibular nerve (Fig. 4). Large series of VN patients report absent or abnormal oVEMPs in ~70% of cases compared to ~40% with cVEMP abnormalities (Magliulo et al. 2015; Taylor et al. 2016). AC and BC oVEMPs demonstrate similar sensitivities for detecting VN (Govender et al. 2015). However, oVEMPs reveal fewer abnormalities than the horizontal vHIT, and cVEMP and posterior vHIT results are sometimes dissociated (Taylor et al. 2016). Consequently, reliance on VEMPs alone may result in cases involving superior or inferior ampullary afferents being missed (Taylor et al. 2016).

13.2   Posterior Circulation Stroke and Other Central Lesions

oVEMPs can aid in detecting central lesions but are dependent on the structures affected. Lesions involving the MLF, the crossed ventral tegmental tract, oculomotor nuclei and the interstitial nucleus of Cajal can impair oVEMPs (Oh et al. 2016a). Unilateral cerebellar infarctions may show abnormal oVEMPs especially when clinical ocular tilt reaction is evident (Oh et al. 2016b). Delayed oVEMP latencies have been reported in multiple sclerosis (Gabelić et al. 2013).

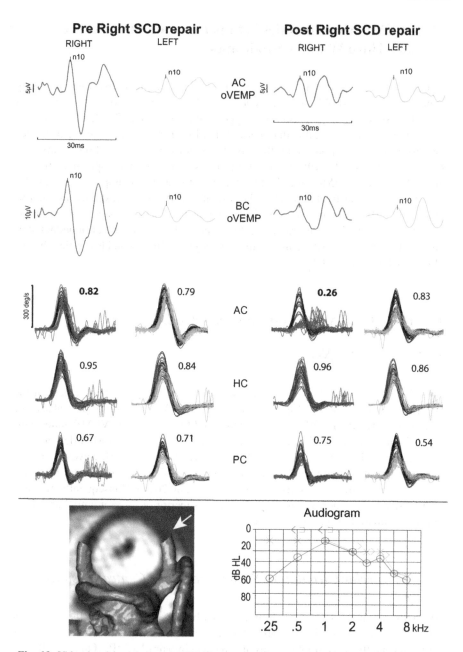

Fig. 10 Video head impulses and oVEMPs recorded from a subject with right superior canal dehiscence pre- and post-right superior canal plugging. In a patient with a right superior canal dehiscence (SCD), the amplitudes of the right AC and BC oVEMPs are pathologically enlarged compared with normal left-sided responses. Following plugging of the superior canal dehiscence, the right oVEMPs return to normal amplitudes. The vHIT gain for the right superior canal is reduced from 0.82 prior to surgery to 0.26 post-surgery. 3D reconstruction of CT imaging demonstrates a right SCD (white arrow). The pre-surgical audiogram demonstrates right-sided low-frequency hearing loss with an apparent air-bone gap

13.3   Episodic Vertigo Syndromes

Two studies report relatively high rates of oVEMP abnormalities in Meniere's disease, but only in response to AC sound (Taylor et al. 2011; Huang et al. 2011). Both studies showed a similar dissociated pattern (AC > BC) for cVEMP abnormalities. While the source of this stimulus dissociation is unclear, it could reflect subtle changes in middle ear function secondary to effects of endolymphatic hydrops on stapes footplate motion. Changes in inner ear resonance could also produce abnormalities specific to AC sound. For example, it is not uncommon for patients with MD to show an upward shift in AC VEMP tuning. However, similar tuning shifts also occur with advancing age, which complicates the clinical interpretation. The collection of age-matched normative tuning data is therefore recommended to maximize sensitivity and specificity of this outcome measure in MD. During an MD attack, contrasting with the more common finding of normal or reduced amplitudes, BC oVEMP amplitudes are sometimes enlarged (Taylor et al. 2011).

13.4   oVEMPs in Chronic Imbalance

While some investigators report normal inter-ictal oVEMPs in vestibular migraine (Taylor et al. 2012b), others report asymmetric response (Zaleski et al. 2015) and absent unilateral or bilateral AC oVEMPs (Boldingh et al. 2011; Kim et al. 2015). Differences in study design, subject variables such as disease duration, frequency of attacks, medications and stage in the migrainous cycle might account for some of these diverse findings. As expected, oVEMPs play no role in diagnosis of idiopathic BPPV but may help identify underlying unilateral vestibulopathy.

In vestibular schwannoma (VS), similar rates of asymmetrical cVEMPs and oVEMPs have been reported (Lin et al. 2014). Small schwannomas show normal responses, whereas large schwannomas are more likely to affect both cVEMPs and oVEMPs together (Lin et al. 2014; Chiarovano et al. 2014). Some patients with abnormal VEMPs in the presence of VS have symmetrical hearing (Taylor et al. 2015; Batuecas-Caletrio et al. 2015) and would thus be missed by audiometric screening protocols. Using VEMPs, vHIT and audiometry in the subject presenting with imbalance will optimize detection of an undiagnosed schwannoma. Although VEMPs are considered to be only complementary to vHIT when diagnosing bilateral vestibular failure (Strupp et al. 2017), it is highly likely that they contribute significantly to diagnosing the aetiology of BVL. VEMPs are likely to be preserved in CANVAS (Kirchner et al. 2011; Rust et al. 2017) but impaired in bilateral Meniere's disease (Huang and Young 2015; Agrawal et al. 2013), gentamicin toxicity (Fig. 6) (Agrawal et al. 2013; Ozluoglu et al. 2008) and superficial siderosis (Lee et al. 2018).

Since the prevalence of VEMP abnormalities approaches a 100% only in superior canal dehiscence, it has been proposed that the sole application of the VEMP test is the diagnosis of SCD. Clinicians who use vestibular function tests to complement their history and physical examination will find VEMPs to be useful measures of otolith function that provide valuable diagnostic information in the assessment of both peripheral and central vestibular disorders.

14   Conclusion

In the past decade, laboratory tests of both semicircular canal and otolith function have moved from specialized laboratories to a clinic and office practice. vHIT and oVEMP combined enable assessment of canal and otolith-ocular pathways. Used in conjunction with cVEMP, which assesses sacculo-collic reflexes, these tests provide a powerful, non-invasive five-test battery that enables clinicians to profile and diagnose vestibular disorders within the clinic itself.

References

Agrawal, Y., Bremova, T., Kremmyda, O., & Strupp, M. (2013). Semicircular canal, saccular and utricular function in patients with bilateral vestibulopathy: Analysis based on etiology. *Journal of Neurology, 260*(3), 876–883.

Agrawal, Y., Schubert, M. C., Migliaccio, A. A., et al. (2014). Evaluation of quantitative head impulse testing using search coils versus video-oculography in older individuals. *Otology & Neurotology, 35*, 283–288.

Alhabib, S., & Saliba, I. (2017). Video head impulse test: A review of the literature. *European Archives of Oto-Rhino-Laryngology, 274*, 1215–1222.

Aw, S. T., Aw, G. E., Todd, M. J., Bradshaw, A. P., & Halmagyi, G. M. (2011). Three-dimensional vibration-induced vestibulo-ocular reflex identifies vertical semicircular canal dehiscence. *Journal of the Association for Research in Otolaryngology, 12*(5), 549–558.

Baloh, R. W. (2002). Robert Barany and the controversy surrounding his discovery of the caloric reaction. *Neurology, 58*, 1094–1099.

Baloh, R. W., Halmagyi, G. M., & Zee, D. S. (2012). The history and future of neuro-otology. *Continuum Lifelong Learning Neurology, 18*(5), 1001–1015.

Batuecas-Caletrio, A., Santa Cruz-Ruiz, S., Muñoz-Herrera, A., & Perez-Fernandez, N. (2015). The map of dizziness in vestibular schwannoma. *Laryngoscope, 125*, 2784–2789.

Boldingh, M. I., Ljøstad, U., Mygland, A., & Monstad, P. (2011). Vestibular sensitivity in vestibular migraine: VEMPs and motion sickness susceptibility. *Cephalalgia, 31*(11), 1211–1219.

Cerchiai, N., Navari, E., Dallan, I., Sellari-Franceschini, S., & Casani, A. P. (2016). Assessment of vestibulo-oculomotor reflex in Ménière's disease: Defining an instrumental profile. *Otology & Neurotology, 37*(4), 380–384.

Chen, L., Todd, M., Halmagyi, G. M., & Aw, S. (2014). Head impulse gain and saccade analysis in pontine-cerebellar stroke and vestibular neuritis. *Neurology, 83*, 1513–1522.

Chiarovano, E., Darlington, C., Vidal, P.-P., Lamas, G., & de Waele, C. (2014). The role of cervical and ocular vestibular evoked myogenic potentials in the assessment of patients with vestibular schwannomas. *PLoS One, 9*(8), e105026. https://doi.org/10.1371/journal.pone.0105026.

Choi, J. Y., Kim, H. J., & Kim, J. S. (2018). Recent advances in head impulse test findings in central vestibular disorders. *Neurology, 90*, 602–612. https://doi.org/10.1212/WNL.0000000000005206.

Colebatch, J. G., & Halmagyi, G. M. (1992). Vestibular evoked potentials in human neck muscles before and after unilateral vestibular deafferentation. *Neurology, 42*(8), 1635–1636.

Colebatch, J. G., Halmagyi, G. M., & Skuse, N. F. (1994). Myogenic potentials generated by a click-evoked vestibulocollic reflex. *Journal of Neurology, Neurosurgery, and Psychiatry, 57,* 190–197.

Colebatch, J. G., Rosengren, S. M., & Welgampola, M. S. (2016). Vestibular-evoked myogenic potentials. *Handbook of Clinical Neurology, 137,* 133–155.

Cremer, P. D., Halmagyi, G. M., Aw, S. T., et al. (1998). Semicircular canal plane head impulses detect absent function of individual semicircular canals. *Brain, 121*(Pt 4), 699–716.

Curthoys, I. S., & Manzari, L. (2017). Clinical application of the head impulse test of semicircular canal function. *Hearing, Balance and Communication, 15*(3), 113–126.

Curthoys, I. S., Vulovic, V., Burgess, A. M., Sokolic, L., & Goonetilleke, S. C. (2016). The response of guinea pig primary utricular and saccular irregular neurons to bone-conducted vibration (BCV) and air-conducted sound (ACS). *Hearing Research, 331,* 131–143.

Fife, T. D., Tusa, R. J., Furman, J. M., et al. (2000). Assessment: vestibular testing techniques in adults and children: report of the Therapeutics and Technology Assessment Subcommittee of the American Academy of Neurology. *Neurology, 55*(10), 1431–1441.

Fife, T. D., Colebatch, J. G., Kerber, K. A., Brantberg, K., Strupp, M., Lee, H., Walker, M. F., Ashman, E., Fletcher, J., Callaghan, B., & Gloss, D. S. (2017). 2nd Practice guideline: Cervical and ocular vestibular evoked myogenic potential testing Report of the Guideline Development, Dissemination, and Implementation Subcommittee of the American Academy of Neurology. *Neurology, 89*(22), 2288–2296.

Furman, J. M. (2016). Rotational testing. *Handbook of Clinical Neurology, 137,* 177–186.

Gabelić, T., Krbot, M., Šefer, A. B., et al. (2013). Ocular and cervical vestibular evoked myogenic potentials in patients with multiple sclerosis. *Journal of Clinical Neurophysiology, 30*(1), 86–91.

Govender, S., Dennis, D. L., & Colebatch, J. G. (2015). Vestibular evoked myogenic potentials (VEMPs) evoked by air-and bone-conducted stimuli in vestibular neuritis. *Clinical Neurophysiology, 126*(10), 2004–2013.

Govender, S., Cheng, P. Y., Dennis, D. L., & Colebatch, J. G. (2016). Electrode montage and gaze effects on ocular vestibular evoked myogenic potentials (oVEMPs). *Clinical Neurophysiology, 127,* 2846–2854.

Halmagyi, G. M., & Curthoys, I. S. (1988). A clinical sign of canal paresis. *Archives of Neurology, 45,* 737–739.

Halmagyi, G. M., Chen, L., MacDougall, H. G., Weber, K. P., McGarvie, L. A., & Curthoys, I. S. (2017). The video head impulse test. *Frontiers in Neurology, 8,* 258.

Hanson, J. M., & Goebel, J. A. (1998). Comparison of manual whole-body and passive and active head-on-body rotational testing with conventional rotary chair testing. *Journal of Vestibular Research, 8*(3), 273–282.

Huang, C. H., & Young, Y. H. (2015). Bilateral Meniere's disease assessed by an inner ear test battery. *Acta Oto-laryngol, 135*(3), 233–238.

Huang, C. H., Wang, S. J., & Young, Y. H. (2011). Localization and prevalence of hydrops formation in Meniere's disease using a test battery. *Audiology & Neuro-otology, 16,* 41–48.

Iwasaki, S., McGarvie, L. A., Halmagyi, G. M., et al. (2007). Head taps evoke a crossed vestibulo-ocular reflex. *Neurology, 68,* 1227–1229.

Kim, C. H., Jang, M. U., Choi, H. C., & Sohn, J. H. (2015). Subclinical vestibular dysfunction in migraine patients: A preliminary study of ocular and rectified cervical vestibular evoked myogenic potentials. *The Journal of Headache and Pain, 16,* 93.

Kim, S. H., Kim, H. J., & Kim, J. S. (2017). Isolated vestibular syndromes due to brainstem and cerebellar lesions. *Journal of Neurology, 264*(Suppl 1), S63–S69.

Kirchner, H., Kremmyda, O., Hufner, K., Stephan, T., Zingler, V., Brandt, T., et al. (2011). Clinical, electrophysiological, and MRI findings in patients with cerebellar ataxia and a bilaterally pathological head-impulse test. *Annals of the New York Academy of Sciences, 1233,* 127–138.

Lee, S. Y., Lee, D. H., Bae, Y. J., Song, J. J., Kim, J. S., & Koo, J. W. (2018). Bilateral Vestibulopathy in superficial Siderosis. *Frontiers in Neurology, 6*(9), 422. https://doi.org/10.3389/fneur.2018.00422. eCollection 2018.

Leigh, R. (1996). What is the vestibulo-ocular reflex and why do we need it. In R. W. Baloh & G. M. Halmagyi (Eds.), *Disorders of the vestibular system* (1st ed.). New York: Oxford University Press.

Lin, K. L., Chen, C. M., Wang, S. J., & Young, Y. H. (2014). Correlating vestibular schwannoma size with vestibular-evoked myogenic potential results. *Ear and Hearing, 35*(5), 571–576.

MacDougall, H. G., & Curthoys, I. S. (2012). Plasticity during vestibular compensation: The role of saccades. *Frontiers in Neurology, 3*, 21.

MacDougall, H. G., Weber, K. P., McGarvie, L. A., Halmagyi, G. M., & Curthoys, I. S. (2009). The video head impulse test. *Neurology, 73*(14), 1134–1141.

MacDougall, H. G., McGarvie, L. A., Halmagyi, G. M., Curthoys, I. S., & Weber, K. P. (2013). Application of the video head impulse test to detect vertical semicircular canal dysfunction. *Otology & Neurotology, 34*(6), 974–979.

MacDougall, H. G., McGarvie, L. A., Halmagyi, G. M., Rogers, S. J., Manzari, L., Burgess, A. M., Curthoys, I. S., & Weber, K. P. (2016). A new saccadic indicator of peripheral vestibular function based on the video head impulse test. *Neurology, 87*(4), 410–418.

Magliulo, G., Iannella, G., Gagliardi, S., & Re, M. (2015). A 1-year follow-up study with C-VEMPs, O-VEMPs and video head impulse testing in vestibular neuritis. *European Archives of Oto-Rhino-Laryngology, 272*(11), 3277–3281.

Mantokoudis, G., Saber Tehrani, A., Wozniak, A., Eibenberger, K., Kattah, J. C., Guede, C. I., et al. (2015). VOR gain by head impulse video-oculography differentiates acute vestibular neuritis from stroke. *Otology & Neurotology, 36*, 457–465.

Manzari, L., Tedesco, A., Burgess, A. M., & Curthoys, I. S. (2010). Ocular vestibular-evoked myogenic potentials to bone-conducted vibration in superior vestibular neuritis show utricular function. *Otolaryngology and Head and Neck Surgery, 143*(2), 274–280.

McGarvie, L. A., Martinez-Lopez, M., Burgess, A. M., MacDougall, H. G., & Curthoys, I. S. (2015a). Horizontal Eye Position Affects Measured Vertical VOR Gain on the Video Head Impulse Test. *Frontiers in Neurology, 6*, 58.

McGarvie, L. A., MacDougall, H. G., Halmagyi, G. M., Burgess, A. M., Weber, K. P., & Curthoys, I. S. (2015b). The video head impulse test (vHIT) of semicircular canal function – Age-dependent normative values of VOR gain in healthy subjects. *Frontiers in Neurology, 6*, 154.

McGarvie, L. A., Curthoys, I. S., MacDougall, H. G., & Halmagyi, G. M. (2015c). What does the dissociation between the results of video head impulse versus caloric testing reveal about the vestibular dysfunction in Ménière's disease? *Acta Otolaryngol, 135*(9), 859–865.

Mossman, B., Mossman, S., Purdie, G., et al. (2015). Age dependent normal horizontal VOR gain of head impulse test as measured with video-oculography. *Journal of Otolaryngology – Head and Neck Surgery, 44*, 29. https://doi.org/10.1186/s40463-015-0081-7.

Murofushi, T., Curthoys, I. S., Topple, A. N., Colebatch, J. G., & Halmagyi, G. M. (1995). Responses of guinea pig primary vestibular neurons to clicks. *Experimental Brain Research, 103*(1), 174–178.

Oh, S. Y., Kim, J. S., Lee, J. M., et al. (2016a). Ocular vestibular evoked myogenic potentials induced by air-conducted sound in patients with acute brainstem lesions. *Journal of Neurology, 263*(2), 210–220.

Oh, S. Y., Kim, H. J., & Kim, J. S. (2016b). Vestibular-evoked myogenic potentials in central vestibular disorders. *Neurology, 87*(16), 1704–1712.

Ozluoglu, L. N., Akkuzu, G., Ozgirgin, N., & Tarhan, E. (2008). Reliability of the vestibular evoked myogenic potential test in assessing intratympanic gentamicin therapy in Meniere's disease. *Acta Oto-Laryngologica, 128*, 422.

Pozzo, T., Berthoz, A., & Lefort, L. (1990). Head stabilization during various locomotor tasks in humans. I. Normal subjects. *Experimental Brain Research, 82*, 97–106.

Rey-Martinez, J., Batuecas-Caletrio, A., Matino, E., et al. (2015). HITCal: A software tool for analysis of video head impulse test responses. *Acta Oto-Laryngologica, 135*, 886–894.

Rosengren, S. M., McAngus Todd, N. P., & Colebatch, J. (2005). Vestibular-evoked extraocular potentials produced by stimulation with bone-conducted sound. *Clinical Neurophysiology, 116*(8), 1938–1948.

Rust, H., Peters, N., Allum, J. H. J., Wagner, B., Honegger, F., & Baumann, T. (2017). VEMPs in a patient with cerebellar ataxia, neuropathy and vestibular areflexia (CANVAS). *Journal of the Neurological Sciences, 378*, 9–11.

Schubert, M. C., & Zee, D. S. (2010). Saccade and vestibular ocular motor adaptation. *Restorative Neurology and Neuroscience, 28*(1), 9–18. https://doi.org/10.3233/RNN-2010-0523.

Smulders, Y. E., Welgampola, M. S., Burgess, A. M., McGarvie, L. A., Halmagyi, G. M., & Curthoys, I. S. (2009). The n10 component of the ocular vestibular-evoked myogenic potential (oVEMP) is distinct from the R1 component of the blink reflex. *Clinical Neurophysiology, 120*(8), 1567–1576.

Strupp, M., Kim, J. S., Murofushi, T., Straumann, D., Jen, J. C., Rosengren, S. M., Santina, C. D. C., & Kingma, H. (2017). Bilateral vestibulopathy: Diagnostic criteria Consensus document of the Classification Committee of the Bárány Society. *Journal of Vestibular Research, 27*(4), 177–189.

Taylor, R. L., Wijewardene, A. A., Gibson, W. P., Black, D. A., Halmagyi, G. M., & Welgampola, M. S. (2011). The vestibular evoked-potential profile of Meniere's disease. *Clinical Neurophysiology, 122*, 1256–1263.

Taylor, R. L., Bradshaw, A. P., Magnussen, J. S., Gibson, W. P., Halmagyi, G. M., & Welgampola, M. S. (2012a). Augmented ocular vestibular evoked myogenic potentials to air-conducted sound in large vestibular aqueduct syndrome. *Ear Hear, 33*(6), 768–771.

Taylor, R. L., Zagami, A. S., Gibson, W. P., et al. (2012b). Vestibular evoked myogenic potentials to sound and vibration: Characteristics in vestibular migraine that enable separation from Meniere's disease. *Cephalalgia, 32*, 213–225.

Taylor, R. L., Kong, J., Flanagan, S., Pogson, J., Croxson, G., Pohl, D., & Welgampola, M. S. (2015). Prevalence of vestibular dysfunction in patients with vestibular schwannoma using video head-impulses and vestibular-evoked potentials. *Journal of Neurology, 262*(5), 1228–1237.

Taylor, R. L., McGarvie, L. A., Reid, N., Young, A. S., Halmagyi, G. M., & Welgampola, M. S. (2016). Vestibular neuritis affects both superior and inferior vestibular nerves. *Neurology, 87*(16), 1704–1712.

Tehrani, A. S., Kattah, J. C., Kerber, K. A., et al. (2018). Diagnosing stroke in acute dizziness and vertigo pitfalls and pearls. *Stroke, 49*, 788–795.

Todd, N. P., Rosengren, S. M., Aw, S. T., & Colebatch, J. G. (2007). Ocular vestibular evoked myogenic potentials (OVEMPs) produced by air- and bone-conducted sound. *Clinical neurophysiology, 118*(2), 381–390.

van den Berg, R., Rosengren, S., & Kingma, H. (2018). Laboratory examinations for the vestibular system. *Current Opinion in Neurobiology, 31*(1), 111–116.

Weber, K. P., Aw, S. T., Todd, M. J., McGarvie, L. A., Curthoys, I. S., & Halmagyi, G. M. (2008). Head impulse test in unilateral vestibular loss – vestibulo-ocular reflex and catch-up saccades. *Neurology, 70*(454–63), 63.

Weber, K., MacDougall, H., Halmagyi, G. M., & Curthoys, I. (2009). Impulsive testing of semicircular-canal function using video-oculography. *Annals of the New York Academy of Sciences, 1164*, 486–491.

Welgampola, M. S., & Colebatch, J. G. (2001). Characteristics of tone burst-evoked myogenic potentials in the sternocleidomastoid muscles. *Otology & Neurotology, 22*, 796–802.

Welgampola, M. S., Myrie, O. A., Minor, L. B., et al. (2008). Vestibular-evoked myogenic potential thresholds normalize on plugging superior canal dehiscence. *Neurology, 70*, 464–472.

Yip, C. W., Glaser, M., Frenzel, C., Bayer, O., & Strupp, M. (2016). Comparison of the bedside head-impulse test with the video head-impulse test in a clinical practice setting: A prospective study of 500 outpatients. *Frontiers in Neurology, 7*, 58.

Zaleski, A., Bogle, J., Starling, A., et al. (2015). Vestibular evoked myogenic potentials in patients with vestibular migraine. *Otology & Neurotology, 36*(2), 295–302.

The Influence of Deep Brain Stimulation on Eye Movements

Salil Patel, Maksymilian A. Brzezicki, James J. FitzGerald, and Chrystalina A. Antoniades

Abstract Over the past half-century, electrical stimulation of specific areas of the brain has revolutionised the treatment of movement disorders. The insertion of electrodes into the brain for therapeutic purposes also provides a unique opportunity for research into human brain function and pathophysiology. Deep brain stimulation (DBS) of parts of the basal ganglia has helped illuminate the importance of the deep nuclei in oculomotor control (Fitzgerald and Antoniades, Curr Opin Neurol, 29(1), 69–73, 2016). The reverse is also true: changes in eye movements are yielding clues to the mechanism of action of DBS, which remains poorly understood. This chapter provides an overview of the relationship between deep brain stimulation (DBS) and eye movements.

Keywords Saccade · Deep brain stimulation · Basal ganglia · Eye movement

Authors "James J. FitzGerald and Chrystalina A. Antoniades" are contributed equally to this work.

S. Patel · M. A. Brzezicki · C. A. Antoniades (✉)
NeuroMetrology Lab, Nuffield Department of Clinical Neurosciences, University of Oxford, Oxford, UK
e-mail: chrystalina.antoniades@ndcn.ox.ac.uk

J. J. FitzGerald (✉)
NeuroMetrology Lab, Nuffield Department of Clinical Neurosciences, University of Oxford, Oxford, UK

Nuffield Department of Surgical Sciences, University of Oxford, Oxford, UK
e-mail: James.Fitzgerald@nds.ox.ac.uk

© Springer Nature Switzerland AG 2019
A. Shaikh, F. Ghasia (eds.), *Advances in Translational Neuroscience of Eye Movement Disorders*, Contemporary Clinical Neuroscience,
https://doi.org/10.1007/978-3-030-31407-1_19

1 Deep Brain Stimulation

The surgical treatment of neurological conditions (FitzGerald et al. 2018) has a long history (Aminoff and Daroff n.d.) The frequency of lesional, destructive surgeries ebbed and flowed over the last half-century – greatly reduced by the introduction of levodopa for Parkinson's disease (PD), but then seeing a resurgence years later as the long-term complications of medication became clear. Deep brain stimulation (DBS) (Aziz et al. 1991; Okun 2014; Limousin-Dowsey et al. 1999; Sandvik et al. 2012) entered mainstream practice in the 1990s, although it was first used for Parkinson's-induced motor fluctuations in the 1960s (Sironi 2011; Bekthereva et al. 1963). DBS is now the dominant surgical treatment for PD. By stimulating specific deep brain structures, motor symptoms of PD can be reduced dramatically. The efficacy of DBS is supported by high-quality clinical trial evidence (Deuschl et al. 2006; Weaver et al. 2009).

DBS systems consist of leads, which are passed into the brain through small holes in the skull, connected to an implantable pulse generator (IPG) via subcutaneously tunnelled extension wires. The IPG is much like a cardiac pacemaker and contains a battery together with electronics to generate a stream of small current pulses for delivery to the brain. The brain leads are implanted using a stereotactic frame, allowing precise submillimetric targeting, which is necessary to maximise the therapeutic effect while minimising the chances of side effects due to unwanted stimulation of other nearby structures. Unlike lesional procedures DBS is reversible, and there are several stimulation parameters that can be titrated wirelessly, including pulse frequencies, durations and amplitudes. The precise location of stimulation can also be varied: leads typically have several longitudinally spaced electrical contacts at their tips (Yousif and Liu 2007), allowing the activated region of brain tissue to be moved up and down the lead by a few millimetres. In recent years, more complex electrode configurations have been available which allow the stimulated region to also be moved in the plane perpendicular to the lead.

The most commonly used stimulation targets for treating akinetic-rigid PD are the subthalamic nucleus (STN) and globus pallidus interna (GPi). These nuclei are key components of basal ganglia networks that are also involved in oculomotor function. By monitoring changes in ocular movement, investigations using DBS as an independent variable can uncover clues regarding the mechanism behind stimulation, the nature of oculomotor circuitry and the function of deep brain nuclei.

The rapid wash-in and wash-out periods, and precise knowledge of the geographical area of brain being stimulated, means that using DBS as a modulator to investigate changes in physiology is in many respects easier than using pharmaceutical interventions.

2 Saccades

Saccades are rapid eye movements that shift the point of fixation. The circuitry behind such movement has been mapped in detail, thanks to lesional, imaging and increasingly computational studies. A large number of areas have been implicated including cortical structures, the basal ganglia (BG), the thalamus and the superior colliculus (SC). The cortico-basal ganglia circuitry controlling movement has been described in terms of two main pathways known as the direct and indirect pathways (Parent and Hazrati 1995). Cortical neurons project to striatal medium spiny neurons (MSNs), which themselves either project straight to the GPi/SNr (direct pathway) or reach the GPi/SNr via the GPe and STN (indirect pathway). In this model the direct pathway promotes movement, while the indirect pathway inhibits movement, and together these reciprocal effects mediate action selection. This is undoubtedly a rather simplistic view – for example, many striatal neurons project to both parts of the pallidum – but it is still the prevalent model. The SC receives convergent inputs from both pathways – and most of the time is under tonic inhibition by the SNr, suppressing saccadic movement (Hikosaka et al. 2000).

The frontal eye fields (FEFs) are cortical areas that specialise in the induction of saccadic movement. They project both to the BG (caudate nucleus) and also directly to the SC. The caudate nucleus helps select the most relevant action via direct pathway inhibitory projections to the SNr, which interrupt the tonic inhibition of the SC, generating a saccade towards a specific target. Targets of saccadic movement are retinotopically mapped on the FEFs, and increased activity in a particular area of the map correlates with a saccadic movement towards the corresponding geographical location (Schall 2004).

Given that multiple areas of the BG are involved in the control of saccades, it is no surprise that PD causes disruption of saccadic eye movements (Antoniades et al. 2015a; Shields et al. 2007; Pretegiani and Optican 2017). Dopaminergic projections from the SNc to the striatum modulate activity in MSNs, and the effect of DA is usually described as increasing activity in the direct pathway while decreasing activity in the indirect pathway. As described above, in a non-diseased brain, the SNr plays an inhibitory role during periods of non-movement, and this is suppressed by direct pathway activity when saccadic movement is required. A lack of dopamine, and therefore reduced direct pathway activity, will reduce the efficiency of saccade generation. At the same time, increased indirect pathway activity leads to overactivity of the STN, leading to an increase in inhibition of the SC (Nambu et al. 2015; Obeso et al. 2009).

The simplest type of saccadic movement is the prosaccade, a natural movement which shifts gaze towards a visual stimulus (see Fig. 1). Other types of saccadic movements have been described, which are largely artificial but may be very useful as experimental tasks. Antisaccades are saccades directed *away* from a visual stimulus. These are complex movements, requiring both an inhibition of the normal

Fig. 1 Antisaccadic movement (**a**) away from the light source juxtaposed with prosaccadic movement toward the light source (**b**)

prosaccadic reflex and subsequent volitional generation of a saccade in the other direction. Memory-guided saccades are saccades that are executed towards the remembered position of a target that is no longer present.

Saccadic movements may be measured using electronic eye-tracking equipment of which there are now several types available (Antoniades and FitzGerald 2016). Multiple parameters may be precisely quantified, including latency (the time between stimulus presentation and onset of the saccade), saccade amplitude and saccade velocity. More complex measures may bring out subtler deficits, for example, the antisaccadic error rate (AER; the proportion of trials in the antisaccade task where the subject makes an erroneous prosaccade) (Everling and Fischer 1998) can demonstrate cognitive issues in very early PD patients, well before they would be apparent with standard clinical measures (Antoniades et al. 2015a).

Prosaccadic latency (PSL) has been shown to be prolonged in patients with advanced PD (Chan et al. 2005). Bilateral DBS targeting the STN has been shown in many studies to reduce PSL (Temel et al. 2008; Yugeta et al. 2010; Fawcett et al. 2009; Sauleau et al. 2008). A study by Antoniades et al. (2012) found the latency to be *increased* immediately following implantation of STN electrodes (before switching on the stimulation), but then *reduced* to a level below baseline once stimulation was switched on. The mechanism of action of DBS is not well understood, but the diametrically opposite nature of these effects is evidence at least that it does not simply create a lesional effect.

A further clue to the mechanism behind DBS may lie in differences in effect between the two most commonly stimulated locations – the GPi and STN (see Fig. 2). Bilateral DBS targeting the GPi has been shown to reduce the AER, whereas STN stimulation does not (Antoniades et al. 2015b). The authors proposed a theory focusing on striatal medium spiny neurons (MSNs): these normally fire at low rates, but in PD the rate of firing is substantially increased (by up to 15-fold in primate models). It is proposed this high rate of MSN firing constitutes 'neural noise' that impairs information flow through the striatum, reducing the ability of higher prefrontal circuits to exercise control over the lower-level motor and oculomotor circuits, leading to the increase in AER. This was in agreement with a computational study by Guthrie et al. exploring the role of MSNs in striatal information flow (Guthrie et al. 2013), which found that when the firing rate of MSNs is increased, this transfer of information is impaired.

It was proposed that stimulation of the GPi may induce antidromic action potentials that travel up MSN axons to collateral branch points, and then propagate orthodromically into their very extensive collateral network, leading to widespread release of GABA at MSN-MSN synapses (Kang and Lowery 2014). Despite each individual stimulation-induced MSN synaptic potential being small, due to the multitude of MSNs, the effect could be sufficient to reduce the baseline firing rate of MSNs, damping the neural noise and improving the AER. STN stimulation does not affect MSNs and hence such stimulation has no effect on the AER. In comparison, levodopa does influence MSNs and has been shown to improve the AER (Hood et al. 2007).

In the study undertaken above, all stimulation was bilateral, targeting either GPi or STN. Goelz et al. (2016) looked at the effect of unilateral STN stimulation on sac-

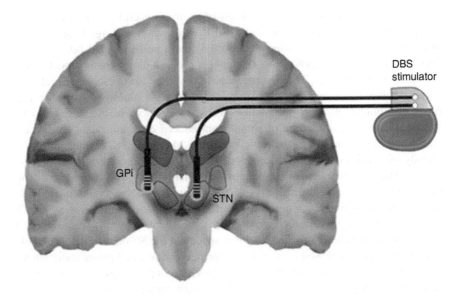

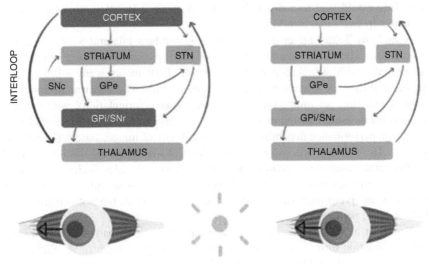

Fig. 2 A comparison between GPi and STN stimulation, highlighting the proposed effect on basal ganglia circuitry and resultant antisaccadic error rate (Antoniades et al. 2015b)

cadic movement compared to both bilateral stimulation and a no-stimulation control. Bilateral stimulation was shown to reduce prosaccadic latency to a significant degree compared to unilateral and no-stimulation paradigms. A novel measurement, prosaccadic gain, was defined as the ratio of primary saccade amplitude to target displacement. Gain was also significantly increased in the bilateral group compared to unilateral and no-stimulation paradigms. AER was found to be significantly higher in the bilateral stimulation group. Interestingly, both unilateral and bilateral paradigms increased antisaccadic gain. The authors propose three possible mechanisms to explain these findings. The first mechanism proposed the dorsolateral prefrontal cortex (DLPFC) – an area implicated in voluntary saccade inhibition – was disrupted by STN stimulation. STN stimulation is known to increase blood flow to the DLPFC. Therefore, the nature of stimulation-induced disruption may be direct in nature or indirect, via basal ganglia-thalamo-cortical projections. As mentioned previously, bilateral stimulation induced more prosaccadic errors compared to unilateral lesions. The authors suggest that unilateral stimulation may well cause less disruption to the DLPFC and any change is easier to counteract via the non-stimulated side. The second mechanism of action is predicated on the relationship between the STN and SC. A higher prosaccadic error rate in the bilateral group, it was suggested, stems from a reduction of the modulatory capacity of the SC on the DLPFC. A relationship between increased prosaccadic error rate and reduced prosaccadic latency was found in patients with schizophrenia. This same coupling was noted in the bilateral stimulation group hinting at DLPFC disruption (an area commonly implicated in the neuropathology of schizophrenia). A third suggested mechanism was that bilateral stimulation acts on the DLPFC and/or SC to interfere with proactive inhibition. This would lead to a disruption-induced increase in automatic sensorimotor responses (prosaccadic errors).

A study by Tokushige et al. (2017) investigated how bilateral STN stimulation affected saccadic visual scanning. Images of varying complexity were shown to 20 participants, with stimulation either on or off. Participants were asked to scan images and then try to memorise them for 10 seconds. The aim was to search for single target ring per image (highlighted by colour or orientation), with saccadic movements measured during this 10-second search period. The authors found a difference in performance depending on the complexity of image, suggesting intrinsic saccades (involuntary actions) were more plentiful when viewing simple images. A similar relationship was apparent between extrinsic saccades (induced by visual cues) and complex images. The conclusion was that STN stimulation disrupts the pathways involving intrinsic saccades (and hence the BG), but not extrinsic, non-BG, saccadic pathways.

Yugeta et al. (2017) measured saccade amplitude in 32 patients with bilateral STN stimulation turned on and off. A significant improvement of amplitude was seen with stimulation. Deviation angle – the angle between target location and saccade location – was also significantly reduced in the stimulation cohort. Interestingly, the authors changed the eccentricity (conical shape) of stimulation using 5-, 10-, 20- and 30-degree variations. The saccadic deviation angle was found not to differ regardless of eccentricity, with stimulation improving eye movement in all variations to similar degrees.

DBS of the STN has been shown to not only reduce saccadic latency but also reduce the reaction time when moving towards low-probability targets (Antoniades et al. 2014). Non-stimulated patients and healthy controls show shorter reaction times towards high-probability targets and longer reaction times towards low-probability targets. During STN stimulation, the reaction time towards high-probability targets decreases as expected, but the reaction time towards low-probability targets does not increase as before. This finding mirrored computational studies in which the STN helped normalise the neural representation of prior probabilities. Electrophysiological studies suggest reaction times are strongly influenced by pre-stimulus activity and the probability of an action. The computational model used presumed such probabilities are encoded within the activity before activity onset. An increase in reaction time towards low-probability targets is thought to involve feedback from the STN. Stimulation, it is theorised, interferes with this feedback and, hence, modulates the mechanism responsible for slowing reaction time but not the mechanism responsible for decreasing reaction time.

3   Other Types of Eye Movements

Four distinct types of eye movement exist (Kennard and Leigh 2011). In addition to saccades, these include smooth-pursuit movements, the vestibulo-ocular reflex and vergence movements. Compared to the large number of studies that have looked at the effects of DBS on saccades, there is a paucity of literature investigating the effect of stimulation on these other types of eye movement.

The vestibulo-ocular reflex (VOR) moves the eyes in the opposite direction to a movement of the head, to stabilise the image on the retina. A study by Shaikh et al. (2017) examined the effect of STN stimulation on the VOR. Stimulation of the most distal electrode, in the most medical aspect of the STN, was found to induce symptoms of angular rotation. In addition, a positive correlation was found between the velocity of rotation and amount of current delivered. Electrode displacement towards the dorsal STN induced a sensation akin to riding on a swing. The authors attributed this effect to unintended stimulation of a vestibulo-thalamic pathway adjacent to the STN.

A study by Nilsson et al. (2013) looked at the effect of bilateral STN stimulation on smooth-pursuit by tracking eye movements as a horizontal target moved from side to side. The latency of movement was calculated as the difference between the point at which the target starts to move and the point at which the velocity of the eye was quicker than 5°/s (therefore indicating the movement was not smooth-pursuit but saccadic in nature). The study found that stimulation improved smooth-pursuit gains and movement accuracy to a significant degree. Historical studies have shown patients with PD tend to have abnormal smooth-pursuits but only recently has this begun to be investigated anatomically. Recent studies suggest the involvement of the globus pallidus in the mechanism inducing smooth movement. Such evidence provides a possible route as to how STN stimulation would affect smooth-pursuits – via BG and SC circuitry.

4   Stimulation and Visuospatial Attention

Visuospatial attention is the ability to direct attention towards a part of the visual field, in order to selectively process specific information (Bartolomeo et al. 2012). Spatial neglect – a common feature of cerebrovascular events – is characterised by the lack of ability to explore a specific area. For example, hemispatial neglect is defined as neglect affecting the hemifield contralateral to the side of a lesion. Studies describing spatial neglect in patients with BG lesions suggest that the deep nuclei are involved in visuospatial attention (Karnath and Rorden 2012).

A study by Schmalbach et al. (2014) investigated whether unilateral right, unilateral left or bilateral STN stimulation would affect visuospatial attention. Thirteen participants, with DBS in situ for PD, were tested in all three conditions. A series of symmetrical pictures were shown on a computer screen, with the participant given free rein to look anywhere on the screen at the beginning of each picture change. Both saccadic movement and fixed duration (the time between saccades) were measured from the left eye. Participants in the unilateral left stimulation condition had significantly shorter periods of fixation on the left hemispheric side and significantly longer periods of fixation on the right hemispheric side. No significant changes in saccadic velocity, amplitude or acceleration occurred between the parameters tested. Due to the significant unilateral nature of effect, the authors proposed the unilateral left stimulation setting was akin to a right-sided nigrostriatal lesion. Efferent connections between the STN, putamen and caudate nucleus (CN) were highlighted as a potential location for the mechanism of action. Interestingly both the putamen and CN have been implicated in human spatial orientation lending some credence to the proposal that a modulation of these areas would reduce neglect and increase spatial attention (Karnath et al. 2002).

5   Considerations

It is important to note that prosaccadic latency times increase gradually with age, from their minimum between 20 and 30 years of age (Munoz et al. 1998). As PD is a disease of the elderly, the measured saccadic variables often overlap with naturally prolonged reaction times. It is important to compare saccadic variables in PD patients with those of age-appropriate healthy controls.

DBS is undertaken in patients with diseased brains. Whether damaged by dopaminergic loss as in Parkinson's disease, or by other disorders, it is important to be aware of the multiple variables at play when assessing the effects of DBS on eye movements. The interplay between disease-related changes, stimulation and eye movement circuitry is complex. Care must be taken not to assume the pathways being studied function in an identical manner in non-pathological brains. Nevertheless, an increasing body of evidence predicated on the relationship between brain stimulation and oculomotor function illustrates the promising nature of this intersection.

Search Strategy
1. Deep brain stimulation∗ AND (eye OR ocular OR eye movement)
2. 1 AND (STN)
3. 1 AND (GPi)
4. 1 AND (saccade)
5. 1 AND (vestibulo-ocular)
6. 1 AND (pursuit)
7. 1 AND (smooth-pursuit)
8. 1 AND (Parkinson's)
9. 1 AND (antisaccade)
10. 1 AND (prosaccade)
11. 1 AND (visual attention)
12. 1 OR 2 OR 3 OR 4 OR 5 OR 6 OR 7 OR 8 OR 9 OR 10 OR 11

References

Aminoff, M. and Daroff, R. (n.d.). Encyclopedia of the neurological sciences. Elsevier

Antoniades, C. A., & Fitzgerald, J. J. (2016). Using saccadometry with deep brain stimulation to study normal and pathological brain function. *Journal of Visualized Experiments*, 113, e53640.

Antoniades, C., Buttery, P., FitzGerald, J., Barker, R., Carpenter, R., & Watts, C. (2012). Deep brain stimulation: Eye movements reveal anomalous effects of electrode placement and stimulation. *PLoS One, 7*(3), e32830.

Antoniades, C., Bogacz, R., Kennard, C., FitzGerald, J., Aziz, T., & Green, A. (2014). Deep brain stimulation abolishes slowing of reactions to unlikely stimuli. *Journal of Neuroscience, 34*(33), 10844–10852.

Antoniades, C., Demeyere, N., Kennard, C., Humphreys, G., & Hu, M. (2015a). Antisaccades and executive dysfunction in early drug-naive Parkinson's disease: The discovery study. *Movement Disorders, 30*(6), 843–847.

Antoniades, C., Rebelo, P., Kennard, C., Aziz, T., Green, A., & FitzGerald, J. (2015b). Pallidal deep brain stimulation improves higher control of the oculomotor system in Parkinson's disease. *Journal of Neuroscience, 35*(38), 13043–13052.

Aziz, T., Peggs, D., Sambrook, M., & Crossman, A. (1991). Lesion of the subthalamic nucleus for the alleviation of 1-methyl-4-phenyl-1,2,3,6-tetrahydropyridine (MPTP)-induced parkinsonism in the primate. *Movement Disorders, 6*(4), 288–292.

Bartolomeo, P., Thiebaut de Schotten, M., & Chica, A. (2012). Brain networks of visuospatial attention and their disruption in visual neglect. *Frontiers in Human Neuroscience, 6*, 110.

Bekthereva, N., Grachev, K., Orlova, A., & Iatsuk, L. (1963). Utilization of multiple electrodes implanted in the subcortical structure of the human brain for the treatment of hyperkinesis. *Zhurnal Nevropatologii i Psikhiatrii Imeni S.S. Korsakova, 63*, 3–8.

Chan, F., Armstrong, I., Pari, G., Riopelle, R., & Munoz, D. (2005). Deficits in saccadic eye-movement control in Parkinson's disease. *Neuropsychologia, 43*(5), 784–796.

Deuschl, G., Deutschländer, A., Volkmann, J., Schäfer, H., Bötzel, K., Daniels, C., Deutschländer, A., Dillmann, U., Eisner, W., Gruber, D., & Hamel, W. (2006). A randomized trial of deep-brain stimulation for Parkinson's disease. *New England Journal of Medicine, 355*, 896–908.

Everling, S., & Fischer, B. (1998). The antisaccade: A review of basic research and clinical studies. *Neuropsychologia, 36*(9), 885–899.

Fawcett, A., González, E., Moro, E., Steinbach, M., Lozano, A., & Hutchison, W. (2009). Subthalamic nucleus deep brain stimulation improves saccades in Parkinson's disease. *Neuromodulation: Technology at the Neural Interface, 13*(1), 17–25.

FitzGerald, J., & Antoniades, C. (2016). Eye movements and deep brain stimulation. *Current Opinion in Neurology, 29*(1), 69–73.

FitzGerald, J., Lu, Z., Jareonsettasin, P., & Antoniades, C. (2018). Quantifying motor impairment in movement disorders. *Frontiers in Neuroscience, 12*, 202.

Goelz, L., David, F., Sweeney, J., Vaillancourt, D., Poizner, H., Metman, L., & Corcos, D. (2016). The effects of unilateral versus bilateral subthalamic nucleus deep brain stimulation on prosaccades and antisaccades in Parkinson's disease. *Experimental Brain Research, 235*(2), 615–626.

Guthrie, M., Leblois, A., Garenne, A., & Boraud, T. (2013). Interaction between cognitive and motor cortico-basal ganglia loops during decision making: A computational study. *Journal of Neurophysiology, 109*(12), 3025–3040.

Hikosaka, O., Takikawa, Y., & Kawagoe, R. (2000). Role of the basal ganglia in the control of purposive saccadic eye movements. *Physiological Reviews, 80*(3), 953–978.

Hood, A., Amador, S., Cain, A., Briand, K., Al-Refai, A., Schiess, M., & Sereno, A. (2007). Levodopa slows prosaccades and improves antisaccades: An eye movement study in Parkinson's disease. *Journal of Neurology, Neurosurgery & Psychiatry, 78*(6), 565–570.

Kang, G., & Lowery, M. (2014). Effects of antidromic and orthodromic activation of STN afferent axons during DBS in Parkinson's disease: A simulation study. *Frontiers in Computational Neuroscience, 8*, 32.

Karnath, H., & Rorden, C. (2012). The anatomy of spatial neglect. *Neuropsychologia, 50*(6), 1010–1017.

Karnath, H., Himmelbach, M., & Rorden, C. (2002). The subcortical anatomy of human spatial neglect: Putamen, caudate nucleus and pulvinar. *Brain, 125*(2), 350–360.

Kennard, C., & Leigh, R. (2011). *Neuro-ophthalmology*. Edinburgh: Elsevier.

Limousin-Dowsey, P., Pollak, P., Blercom, N., Krack, P., Benazzouz, A., & Benabid, A. (1999). Thalamic, subthalamic nucleus and internal pallidum stimulation in Parkinson's disease. *Journal of Neurology, 246*(S2), II42–II45.

Munoz, D., Broughton, J., Goldring, J., & Armstrong, I. (1998). Age-related performance of human subjects on saccadic eye movement tasks. *Experimental Brain Research, 121*(4), 391–400.

Nambu, A., Tachibana, Y., & Chiken, S. (2015). Cause of parkinsonian symptoms: Firing rate, firing pattern or dynamic activity changes? *Basal Ganglia, 5*(1), 1–6.

Nilsson, M., Patel, M., Rehncrona, S., Magnusson, M., & Fransson, P. (2013). Subthalamic deep brain stimulation improves smooth pursuit and saccade performance in patients with Parkinson's disease. *Journal of Neuroengineering and Rehabilitation, 10*(1), 33.

Obeso, J., Marin, C., Rodriguez-Oroz, C., Blesa, J., Benitez-Temiño, B., Mena-Segovia, J., Rodríguez, M., & Olanow, C. (2009). The basal ganglia in Parkinson's disease: Current concepts and unexplained observations. *Annals of Neurology, 64*(S2), S30–S46.

Okun, M. (2014). Deep-brain stimulation—Entering the era of human neural-network modulation. *New England Journal of Medicine, 371*(15), 1369–1373.

Parent, A., & Hazrati, L. (1995). Functional anatomy of the basal ganglia. II. The place of subthalamic nucleus and external pallidum in basal ganglia circuitry. *Brain Research Reviews, 20*(1), 128–154.

Pretegiani, E., & Optican, L. (2017). Eye movements in Parkinson's disease and inherited parkinsonian syndromes. *Frontiers in Neurology, 8*, 592.

Sandvik, U., Koskinen, L., Lundquist, A., & Blomstedt, P. (2012). Thalamic and subthalamic deep brain stimulation for essential tremor. *Neurosurgery, 70*(4), 840–846.

Sauleau, P., Pollak, P., Krack, P., Courjon, J., Vighetto, A., Benabid, A., Pélisson, D., & Tilikete, C. (2008). Subthalamic stimulation improves orienting gaze movements in Parkinson's disease. *Clinical Neurophysiology, 119*(8), 1857–1863.

Schall, J. (2004). On the role of frontal eye field in guiding attention and saccades. *Vision Research, 44*(12), 1453–1467.

Schmalbach, B., Günther, V., Raethjen, J., Wailke, S., Falk, D., Deuschl, G., & Witt, K. (2014). The subthalamic nucleus influences visuospatial attention in humans. *Journal of Cognitive Neuroscience, 26*(3), 543–550.

Shaikh, A., Straumann, D., & Palla, A. (2017). Motion illusion—Evidence towards human vestibulo-thalamic projections. *The Cerebellum, 16*(3), 656–663.

Shields, D., Gorgulho, A., Behnke, E., Malkasian, D., & Desalles, A. (2007). Contralateral conjugate eye deviation during deep brain stimulation of the subthalamic nucleus. *Journal of Neurosurgery, 107*(1), 37–42.

Sironi, V. (2011). Origin and evolution of deep brain stimulation. *Frontiers in Integrative Neuroscience, 5*, 42.

Temel, Y., Visser-Vandewalle, V., & Carpenter, R. (2008). Saccadic latency during electrical stimulation of the human subthalamic nucleus. *Current Biology, 18*(10), R412–R414.

Tokushige, S., Matsuda, S., Oyama, G., Shimo, Y., Umemura, A., Sekimoto, S., Sasaki, T., Inomata-Terada, S., Yugeta, A., Hamada, M., Ugawa, Y., Hattori, N., Tsuji, S., & Terao, Y. (2017). How deep brain stimulation affects saccades in visual scanning in Parkinson's disease patients. *Journal of the Neurological Sciences, 381*, 1040.

Weaver, F., Follett, K., Stern, M., Hur, K., Harris, C., Marks, W., Rothlind, J., Sagher, O., & Reda, D. (2009). Bilateral deep brain stimulation vs best medical therapy for patients with advanced Parkinson disease: A randomized controlled trial. *JAMA, 301*, 63–73.

Yousif, N., & Liu, X. (2007). Modeling the current distribution across the depth electrode–brain interface in deep brain stimulation. *Expert Review of Medical Devices, 4*(5), 623–631.

Yugeta, A., Terao, Y., Fukuda, H., Hikosaka, O., Yokochi, F., Okiyama, R., Taniguchi, M., Takahashi, H., Hamada, I., Hanajima, R., & Ugawa, Y. (2010). Effects of STN stimulation on the initiation and inhibition of saccade in Parkinson disease. *Neurology, 74*(9), 743–748.

Yugeta, A., Terao, Y., & Ugawa, Y. (2017). Improvement of saccade amplitude by the STN DBS and target eccentricity. *Journal of the Neurological Sciences, 381*, 1053.

Eyelid Dysfunction in Neurodegenerative, Neurogenetic, and Neurometabolic Disease

Ali G. Hamedani and Daniel R. Gold

Abstract Eyelid movement is neuroanatomically linked to eye movement, and thus eyelid abnormalities frequently accompany and sometimes precede or overshadow eye movement abnormalities in neurodegenerative disease. In this chapter, we summarize the various eyelid abnormalities that can occur in inherited and acquired neurodegenerative disorders in the context of the neuroanatomic pathways that are affected. The epidemiology and clinical approach to diagnosis and treatment of ptosis, eyelid retraction, abnormal spontaneous and reflexive blinking, blepharospasm, and eyelid apraxia will also be reviewed.

Keywords Eyelid · Neuro-ophthalmology · Brainstem

1 Ptosis

1.1 Overview of Eyelid Elevation

During wakefulness, eye opening is maintained by tonic contraction of the levator palpebrae superioris (LPS). The degree of LPS activity depends on a number of factors, especially vertical eye position, and momentary pauses in this activity allow for blinking, both of which are discussed later in this chapter.

The LPS, which is innervated by the oculomotor nerve, originates from the lesser wing of the sphenoid bone at the orbital apex. It courses through the orbit, where it lies superior to the superior rectus muscle, and inserts on the superior tarsal plate as well as directly on the skin of the upper eyelid, forming the lid crease. A secondary muscle (the superior tarsal muscle, also known as Müller's muscle), which is

A. G. Hamedani (✉)
Department of Neurology, Hospital of the University of Pennsylvania, Philadelphia, PA, USA
e-mail: Ali.hamedani@pennmedicine.upenn.edu

D. R. Gold
Departments of Neurology, Ophthalmology, Neurosurgery, and Otolaryngology – Head & Neck Surgery, Johns Hopkins Hospital, Baltimore, MD, USA

© Springer Nature Switzerland AG 2019
A. Shaikh, F. Ghasia (eds.), *Advances in Translational Neuroscience of Eye Movement Disorders*, Contemporary Clinical Neuroscience,
https://doi.org/10.1007/978-3-030-31407-1_20

innervated by oculosympathetic nerve fibers arising from the superior cervical ganglion, originates from the distal aponeurosis of the LPS and inserts on the superior tarsal plate. Muscles extrinsic to the eyelids themselves, such as the frontalis and other facial nerve-innervated muscles, can indirectly affect eyelid position as well.

Inadequate eyelid elevation in primary gaze results in ptosis. However, ptosis is not synonymous with LPS weakness. Levator function can be assessed clinically by measuring the difference in position of the upper eyelid margin in downgaze and then in upgaze; it is normally greater than or equal to 12 mm (Liu et al. 2010). The examiner must also control for frontalis contraction to ensure that levator function is measured accurately. Ptosis with reduced levator function implies a lesion of the LPS or its motor control. In contrast, ptosis of mechanical origin will have normal levator function.

1.2   Ptosis due to Levator Weakness

The LPS is preferentially affected in mitochondrial myopathies due to its enriched supply of mitochondria (Porter et al. 1989). This is illustrated by the prominent ptosis that accompanies the *chronic progressive external ophthalmoplegia (CPEO)* phenotype, where patients often maintain chronic frontalis overaction in an effort to overcome their ptosis. CPEO can occur in isolation or in association with other mitochondrial disorders such as the Kearns–Sayre syndrome, sensory ataxic neuropathy with dysarthria and ophthalmoplegia (SANDO), Leigh syndrome, and mitochondrial neurogastrointestinal encephalopathy (MNGIE), among others (McClelland et al. 2016). CPEO can be caused by mutations in the mitochondrial or nuclear genome (Milone and Wong 2013). Mitochondrial mutations are characterized by large deletions, rearrangements, or point mutations involving genes encoding for tRNA synthetases and are typically somatic rather than germline, resulting in a sporadic rather than maternal pattern of inheritance. Nuclear mutations may be inherited in an autosomal dominant or recessive pattern. Many of these genes (e.g., OPA1) are responsible for mitochondrial homeostasis, and thus patients can acquire secondary somatic mitochondrial DNA mutations over time, accounting for some of the variable expressivity in these disorders (Taylor and Turnbull 2005). Secondary mitochondrial mutations also occur in SCA28 (Gorman et al. 2015), which is caused by mutations in the *AFG3L2* gene (Zuhlke et al. 2015). This may explain why ptosis is more prominent in this disease compared to other autosomal-dominant spinocerebellar ataxias (Paulson 2009). Mitochondrial disorders also have a predilection for the basal ganglia due to their high metabolic activity. In fact, a CPEO-like syndrome accompanied by symmetric parkinsonism has been reported in families with *c10orf2* (Twinkle) and *POLG1* mutations (Kiferle et al. 2013). Ptosis has been described in cases of early-onset Parkinson's disease (PD) due to *PARK2* mutations (Amboni et al. 2009), but is not a typical manifestation of idiopathic PD or other acquired neurodegenerative disorders.

Most other inherited myopathies spare the LPS. Notable exceptions include *oculopharyngeal muscular dystrophy* (*OPMD*) and *myotonic dystrophy*. For this reason, both were thought at one point to be mitochondrially mediated: aggregates of dysmorphic mitochondria have been observed in muscle biopsies of OPMD (Wong et al. 1996), and myotonic dystrophy has a more severe clinical phenotype when it is inherited maternally rather than paternally, with congenital presentations seen exclusively in the children of affected mothers (Yum et al. 2017). However, mitochondrial DNA sequencing studies have failed to detect any variants associated with phenotype severity (Poulton et al. 1995). Congenital myasthenic syndromes, which are caused by mutations in any number of genes involved in presynaptic or postsynaptic neuromuscular transmission, also frequently result in ptosis (Engel et al. 2015).

1.3 Congenital Ptosis

When the LPS does not form properly in utero, the result is myogenic ptosis that is present at birth. Because there is actual dysgenesis and hypoplasia of the LPS, *congenital ptosis* is characterized by a reduced or absent lid crease and lid lag in downgaze in addition to reduced levator excursion. It typically occurs sporadically, though mutations in several genes have been identified in familial cases (Pavone et al. 2005). Due to its shared embryologic origin with the superior rectus, a congenital elevator palsy may be seen (SooHoo et al. 2014). There may also be signs of aberrant reinnervation. For example, in the Marcus Gunn jaw winking phenomenon, the ptotic eyelid retracts with lateral jaw movement as in sucking or chewing, indicating that it is innervated in part by the motor portion of the trigeminal nerve. While histologic examination revealing reduced muscle fiber number and fibrosis led many to suspect that the disease is primarily myopathic in pathogenesis (Clark et al. 1995), some experts now group congenital ptosis with the congenital cranial dysinnervation disorders (CCDD) (Mendes et al. 2015). *Congenital fibrosis of the extraocular muscles* is another CCDD characterized by prominent ptosis; it resembles CPEO except that it is present at birth and the ophthalmoplegia is restrictive in origin rather than paretic (Heidary et al. 2008). Aberrant reinnervation can also produce eyelid abnormalities in the absence of true ptosis. Take the Duane retraction syndrome type 1, for example, where there is an abduction deficit from impaired development of the abducens nerve. In adduction, the palpebral fissure narrows, not because of reduced eyelid opening or increased eyelid closure, but because of simultaneous contraction of the medial and lateral recti due to dual innervation by the oculomotor nerve, resulting in retraction of the globe into the orbit (Yuksel et al. 2010).

1.4 Mechanical Ptosis

Ptosis with preserved levator function is usually caused by a defect of the aponeurotic insertion of the LPS onto the upper eyelid. This can be assessed clinically by measuring the lid crease height, which is the distance between the lid crease and upper eyelid margin as measured in downgaze and is normally less than 10 mm. *Levator dehiscence-disinsertion syndrome* is the most common cause of acquired ptosis in adulthood. It occurs when the LPS loses its insertion site on the superior tarsal plate and then reinserts on a more proximal portion of the tarsal plate or eyelid skin, resulting in an abnormally increased lid crease height. However, since the muscle belly itself is unaffected, levator function is preserved (Kersten et al. 1995). Levator dehiscence is commonly seen with advancing age but can be accelerated by eyelid manipulation during regular contact lens use, frequent rubbing of the eyes, botulinum injection of the orbicularis oculi (OO) for the treatment of blepharospasm (Verhulst et al. 1994), or ocular surgery. Mechanical ptosis can also be inherited, either in relative isolation (e.g., autosomal-dominant blepharophimosis-ptosis-epicanthus inversus syndrome) or in the setting of other craniofacial abnormalities (e.g., trisomy 13, Turner syndrome, Noonan syndrome, Cornelia de Lange syndrome, congenital arthrogryposes (Allen 2013)) (Fig. 1).

While the LPS is more active in upgaze and less active in downgaze, the superior tarsal muscle remains equally active in all directions of gaze, as it is not a primary eyelid elevator but instead is modulated by level of arousal and sympathetic tone. Thus, a lesion of the superior tarsal muscle or its oculosympathetic innervation (as part of Horner's syndrome with ipsilateral miosis) results in relatively mild ptosis with preserved levator function rather than the more severe ptosis with reduced levator function that is seen in true LPS weakness of neurogenic (e.g., oculomotor nerve palsy) or myogenic origin. Reduced sympathetic tone of the superior tarsal muscle is responsible for the ptosis seen in disorders of neurotransmitter synthesis

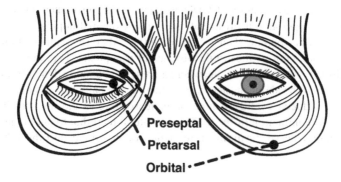

Fig. 1 Anatomy of the eyelids. The orbicularis oculi (OO) is innervated by the facial nerve and consists of two segments: the palpebral portion (comprised of preseptal and pretarsal components), which is contained within the eyelid itself, and the orbital portion, which is located outside the eyelid surrounding the orbit (Modified with permission from reference (Skarf 2004), Fig. 24.5)

such as tyrosine hydroxylase deficiency, aromatic L-amino acid decarboxylase deficiency, dopamine beta-hydroxylase deficiency, and brain dopamine-serotonin vesicular transport disease (Ng et al. 2015).

1.5   Treatment of Ptosis

Treatment options for ptosis include conservative measures (e.g., temporary taping of the upper eyelids, eyelid crutches attached to eyeglasses) and surgical intervention such as shortening of the LPS, resection of the superior tarsal muscle, and frontalis suspension to elevate the entire upper eyelid complex. Both are generally reserved for cases where ptosis obstructs the visual axis or results in cosmetic distress. The primary risk associated with ptosis surgery is incomplete eyelid closure during normal blinking and sleep (lagophthalmos) causing exposure keratopathy. Surgery should therefore be approached with caution if the muscles of eyelid closure are already weak or Bell's phenomenon (upward movement of the eyes during eye closure) is reduced, as is frequently the case in CPEO and other myopathies.

2   Eyelid Retraction

2.1   Overview of Vertical Eye and Eyelid Position Coordination

Under normal conditions, the eyelids elevate in upgaze and depress in downgaze. This is important for maximizing protection of the cornea and tear film while avoiding obscuration of the visual axis. This coordination of vertical eye and eyelid position is exquisitely sensitive, as the velocity and gain of eyelid movement roughly matches that of the corresponding eye movement, be it a saccade or smooth pursuit (Becker and Fuchs 1988). A single nucleus (the central caudal nucleus [CCN] of the midbrain) is shared by both the left and right oculomotor nucleus complexes and innervates both LPS muscles, allowing the eyelids to elevate and depress equally (Rucker 2011). In electrophysiologic studies of primates, the CCN has a basal firing rate in primary gaze. In upward saccades, it experiences a burst of increased firing (and in downward saccades, it experiences a pause in firing) after which the basal firing rate resumes (Becker and Fuchs 1988; Fuchs et al. 1992; Evinger et al. 1984). This is similar to what happens in the superior rectus subnucleus during vertical saccades, suggesting shared supranuclear control with the CCN (Evinger et al. 1984; Robinson 1970). In primates, this connection is mediated by a population of neurons called the M-group, which lies adjacent to the rostral interstitial nucleus of the medial longitudinal fasciculus (riMLF). The M-group receives excitatory input from the riMLF (which is responsible for generating vertical and torsional saccades) and superior colliculus (SC) during upgaze. The M-group provides some reinforcement to the oculomotor subnuclei responsible for supraduction (namely, the superior rectus and inferior oblique), but its primary output is to the CCN (Horn et al. 2000). The M-group is inhibited by the interstitial nucleus of Cajal (iNC) and nucleus of the posterior commissure (nPC) during downgaze (Horn and Buttner-Ennever 2008), allowing the eyelids to follow the downward eye movements (Fig. 2).

2.2 Eyelid Retraction in Midbrain Dysfunction

Dysfunction of the M-group and its afferent and efferent connections is the mechanism by which central nervous system disease causes eyelid retraction. Because of its proximity to the riMLF, it is often accompanied by a vertical saccadic or gaze palsy, as in the dorsal midbrain syndrome (also known as the *pretectal* or *Parinaud syndrome*, where eyelid retraction is referred to as Collier's sign) (Keane 1990) and *progressive supranuclear palsy (PSP)*. Eyelid retraction in these disorders reflects a dissociation between eye position and eyelid position such that the CCN is overactive relative to vertical eye position. This may be related to overstimulation of the M-group in an effort to overcome a vertical gaze palsy, as eyelid retraction is often more prominent during attempted upgaze. Alternatively, retraction could result from underinhibition of the M-group by the iNC and nPC during attempted downgaze, which manifests clinically as lid lag (a failure of the eyelids to lower sufficiently during attempted downgaze).

2.3 Neurodegenerative Diseases Associated with Eyelid Retraction

Eyelid retraction is present in virtually all patients with PSP (Friedman et al. 1992) (in contrast, it has only been rarely reported in PD (Onofrj et al. 2011)), and together with frontalis and procerus contraction often results in a characteristic surprised appearance or "stare". This finding coupled with the typical vertical supranuclear gaze palsy is consistent with the radiographic finding of midbrain atrophy (hummingbird sign). Eyelid retraction is also a distinguishing feature of *spinocerebellar ataxia 3* (also known as *Machado–Joseph disease*); in one study, 65% of patients with SCA3 had eyelid retraction resulting in a "bulging eyes" appearance compared to less than 5% of patients with other autosomal-dominant spinocerebellar ataxias (Moro et al. 2013). The mechanism for eyelid retraction in SCA3 is unclear, as midbrain atrophy is rarely seen on imaging in this disorder (Table 1).

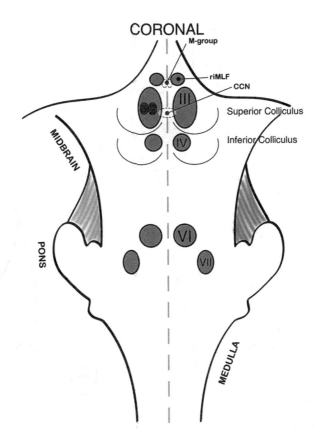

Fig. 2 Supranuclear control of eyelid movement. (**a**) The central caudal nucleus (CCN) of the midbrain is a midline structure that contributes fibers to both oculomotor nerves, thereby innervating bilateral levator palpebrae (LP) muscles. The CCN maintains a tonic level of activity to keep the eyes open during wakefulness, and this resting tone transiently increases with upward eye movements (so the visual target is not occluded by the eyelids) and decreases with downward eye movements. The rostral interstitial nucleus of the median longitudinal fasciculus (riMLF) initiates the upward saccade by providing excitatory input to the oculomotor subnuclei responsible for elevation, i.e., superior rectus (SR) and inferior oblique (IO). (**b**) The riMLF moves the eyes upward, but also influences eyelid position via excitation of the nearby M-group (M). The M-group has strong excitatory projections to the CCN, resulting in activation of the LP muscles bilaterally and eyelid elevation that occurs simultaneously with upward eye movements. The M-group also has weak excitatory influence over the SR and IO subnuclei to assist in coordination of eyelid movements and excites the frontalis muscles via excitatory projections to the facial nucleus. A normal blink leads to cessation of LP firing and contraction of the orbicularis oculi (OO) muscles, which are innervated by the facial nerve. It is thought that the superior colliculus (SC) plays a significant role in the coordination of LP and OO activity, via projections to the CCN and facial nuclei, respectively. The SC is inhibited by the substantia nigra pars reticulata (SNr), which explains abnormalities in blinking in parkinsonian disorders. In parkinsonism, due to increased SNr activity, there is greater SC inhibition and a decreased blink rate. Afferents from trigeminal and pretectal nuclei also project to the SC, and these inputs modulate reflexive blinking due to corneal stimulation and bright light, respectively

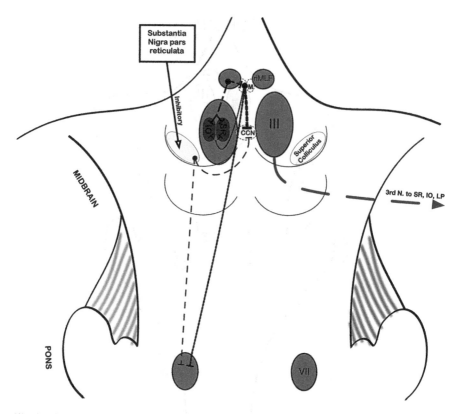

Fig. 2 (Continued)

Table 1 Relative frequency of eyelid disorders in inherited and acquired neurodegenerative disease

Eyelid Disorder	PD	PSP	MSA	SCA	HD
Ptosis	−	−	−	± (esp. SCA 28)	−
Eyelid retraction	±	+++	−	± (esp. SCA 3)	±
Decreased blinking	++	+++	±	±	−
Increased blinking	±	−	−	−	+++
Blepharospasm	±	+++	+	±	±
Apraxia of eyelid opening	±	+++	+	±	±

PD Parkinson's disease, *PSP* progressive supranuclear palsy, *MSA* multiple systems atrophy, *SCA* spinocerebellar ataxia, *HD* Huntington's disease

2.4 Lid Nystagmus

The association between vertical eye and eyelid position is maintained even when eye movement is involuntary, such as in upbeat nystagmus (UBN). Occasionally, rhythmic movements of the eyelids can be seen without visible UBN, resulting in so-called eyelid nystagmus or lid flutter (Milivojevic et al. 2013). This dissociation between eye and eyelid movement probably occurs similarly to eyelid retraction, as eyelid nystagmus is often associated with midbrain ischemic and compressive lesions (Howard 1986; Safran et al. 1982). Eyelid nystagmus can be enhanced by attempted convergence (also known as Pick's sign) (Sanders et al. 1968). This reflects the fact that under normal conditions, convergence increases the basal firing rate of the LPS, resulting in a small degree of eyelid retraction.

3 Decreased Blinking

3.1 Overview of Eyelid Closure

The orbicularis oculi (OO), which is innervated by the facial nerve, is the primary muscle of eyelid closure. It originates from multiple bony and connective tissue structures surrounding the medial canthus. The palpebral orbicularis is contained within the upper and lower eyelids themselves and inserts on connective tissue structures surrounding the lateral canthus. It can be further subdivided into pretarsal and preseptal components, which is important for understanding eyelid apraxias (discussed later in this chapter). The orbital portion of the orbicularis oculi lies outside the eyelids and forms a muscular ellipse encircling the orbit (Liu et al. 2010). The distinction between the palpebral and orbital portions of the orbicularis oculi is not only anatomic but also physiologic, as they are innervated by separate populations of motor neurons within the facial nucleus (VanderWerf et al. 2003). Other muscles of facial expression, such as the corrugator, can draw the eyelids downward or narrow the palpebral fissure as well.

3.2 Supranuclear Control of Spontaneous Blinking

An average person blinks 15–20 blinks times per minute; this frequency varies considerably between individuals and with age and is somewhat higher in women than men (Sforza et al. 2008). During a blink, LPS activity abruptly ceases, and the palpebral portion of the OO (but not the orbital portion) contracts. Note that this represents active eyelid depression, in contrast to the downward eyelid movement during downgaze, which occurs passively via relaxation of the LPS but without active contraction of the OO. As soon as the upper and lower eyelids appose, the OO abruptly stops firing, and basal activity of the LPS resumes, resulting in eye opening (Esteban and Salinero 1979). Blinking occurs rapidly so as to not disrupt visual input. In addition to occurring spontaneously, blinking can also occur reflexively in response to visual threat, bright light, tactile stimulation of the cornea or eyelids, or loud noise. These are all mediated by brainstem reflex arcs except for blinking to threat, which requires cortical input. Unlike spontaneous blinks, reflexive blinks may involve the orbital portion of the OO, which serves to provide extra protection against a potentially harmful stimulus. Blink reflexes must also be modulated so that they are not overly sensitive and do not result in unwanted blinks in response to non-threatening stimuli.

Empiric evidence suggests that the SC plays a major role in coordinating LPS and OO activity during blinks. It receives afferent projections from the trigeminal sensory nucleus and the dorsal midbrain, which are stimulated in response to tactile corneal stimulation and bright light, respectively, and it sends efferent projections to both the facial motor nucleus and the supraoculomotor area, which lies directly above the CCN (Schmidtke and Buttner-Ennever 1992). Inhibitory microstimulation of the SC in primates has been shown to both suppress spontaneous blinking and increase sensitivity to blink reflexes (Gnadt et al. 1997; Basso et al. 1996).

The SC is inhibited by the pars reticulata of the substantia nigra (SNr), and signaling through these projections in the nigrocollicular pathway is mediated by dopamine. In animals, the administration of apomorphine and other dopamine agonists increases spontaneous blinking, and this effect is abolished by pre-treatment with sulpiride, a dopamine-receptor antagonist (Karson 1983). Dopamine level in the caudate nucleus correlates with blink rate in animal models of MPTP-induced parkinsonism (Taylor et al. 1999). Anticholinergic drugs, which induce a relative excess of dopamine owing to the importance of dopamine–acetylcholine balance in the basal ganglia, also increase spontaneous blinking in experimental models (Basso et al. 1996).

3.3 Reduced Spontaneous Blinking in Parkinsonism

The spontaneous blink rate is roughly 30% lower in *PD* patients compared to healthy controls (Biousse et al. 2004; Golbe et al. 1989). Both levodopa therapy (Agostino et al. 2008) and deep brain stimulation (DBS) of the subthalamic nucleus (Bologna et al. 2012) increase the blink rate. Rarely, PD patients may have an increased spontaneous blink rate that paradoxically decreases with levodopa therapy. This is typically seen in more advanced PD, and given the increase in spontaneous blink rate that characterizes blepharospasm (a focal dystonia discussed later in this chapter), it has been postulated by some to represent a type of "off-dystonia" (Kimber and Thompson 2000). Blink rate is dramatically reduced in PSP to as low as three blinks per minute, making this a feature that can help distinguish it from PD. "Slow blinks" have also been observed in PSP (Friedman et al. 1992), whereas in PD, the blink rate is reduced but the eyelid movements themselves are of normal amplitude and velocity.

Since a major function of spontaneous blinking is to evenly distribute the tear film, signs and symptoms of dry eye are extremely common in parkinsonism. In one study, 63% of PD patients complained of dry eye and related symptoms and roughly 50% had objective evidence of xerophthalmia as measured by the Schirmer or tear film build-up time tests (Taylor et al. 1999). Chronically dry eyes can also lead to blepharitis and meibomian gland disease, both of which are increased in PD (Nowacka et al. 2014). Autonomic dysfunction may further compound dry eye symptoms in PD mainly due to inadequate tear production. Artificial tears are often recommended, but their efficacy has not been specifically studied in this population. A trial of LipiFlow (a pulsating thermal eyepiece) compared to warm compresses for the treatment of meibomian gland dysfunction in PD is currently underway (NCT02894658)   (https://clinicaltrials.gov/ct2/show/NCT02894658?term=lipiflo w&rank=4).

3.4   Reflexive Blinking in Parkinsonism

Given the inverse relationship between spontaneous blinking and blink reflexes, a decrease in dopamine in the SNr would be expected to enhance reflexive blinking. In PD, this manifests clinically as the glabellar reflex (blinking in response to tapping the nasion or forehead that fails to habituate, also known as Myerson's sign) (Brodsky et al. 2004), though this may be seen in many other structural, metabolic, and degenerative disorders as well. Reflexive blinking to other stimuli such as bright light is also increased in parkinsonism, particularly in PSP. Like eyelid retraction, the fact that reflexive blinking to light is more pronounced in PSP compared to PD is explained by the prominent midbrain involvement in the former (Kuniyoshi et al. 2002).

Trigeminally mediated blink reflexes can be studied electrophysiologically by stimulating the supraorbital nerve and recording OO activity using surface or needle electrodes (Aramideh and Ongerboer de Visser 2002). This elicits two responses: R1, a brief ipsilateral response with a latency of about 10 ms, and R2, a more sustained bilateral response with a latency of about 30 ms. When the LPS is recorded, there are two corresponding periods of electromyographic silence (SP1 and SP2). This means that the LPS and OO are never co-activated under normal conditions. In addition, repetitive stimulation can be performed to assess reflex excitability. Normally, through a combination of membrane refractoriness after hyperpolarization and activation of negative feedback circuits, a second stimulus elicits a weaker response compared to the first stimulus. Specifically, R2 is absent when the interval between two stimuli is less than 200 ms, reduced by 50–60% at an interval of 500 ms, and reduced by 10–30% at an interval of 1500 ms.

The R2 latency is mildly prolonged in PD, consistent with intrinsic brainstem pathology in the early Braak stages of the disease. R2 prolongation has been shown to be greater in PD patients with dyskinesias compared to those without and in dementia with Lewy bodies (DLB) compared to PD, both of which are likely a reflection of a greater burden of Lewy bodies in the brainstem (Iriarte et al. 1989; Bonanni et al. 2007). The blink reflex is also hyperexcitable in PD—that is, when the supraorbital nerve is stimulated twice in rapid succession, the second R2 exhibits less of a decrease in PD patients compared to healthy controls (Gnadt et al. 1997). Dopaminergic therapy and STN DBS in both humans (Costa et al. 2006) and animals (Kaminer et al. 2015) reduces blink reflex excitability to normal. The degree of blink reflex hyperexcitability also correlates with multiple measures of disease severity including bradykinesia, rigidity, gait impairment, dysarthria, and quality of life (Iriarte et al. 1988; Matsumoto et al. 1992).

4 Increased Blinking and Blink-Assisted Saccades

4.1 Spontaneous and Reflexive Blinking in Hyperkinetic Movement Disorders

Since a relative deficiency of dopamine in the basal ganglia reduces spontaneous blinking and increases reflexive blinking, a relative excess of dopamine would be expected to increase spontaneous blinking and reduce reflexive blinking. This is indeed seen in hyperkinetic movement disorders such as *Huntington's disease (HD)*, where at approximately 36 blinks per minute (Karson et al. 1984), the average blink rate is nearly double the normal rate, and up to 75% of HD patients have subjectively increased blinking (Fekete and Jankovic 2014). A case of juvenile HD was characterized by excessive blinking (40 blinks per minute) that preceded other disease manifestations by over 2 years (Xing et al. 2008). Increased spontaneous blinking is the first clinical manifestation of blepharospasm (see below) and has also been described in Wilson's disease (Verma et al. 2012). Tourette syndrome (Tharp et al. 2015) and schizophrenia, which are also thought to be mediated by a relative excess of dopamine and are treated with dopamine antagonists, are characterized by increased spontaneous blinking as well. In HD, increased spontaneous blinking is accompanied by reduced reflexive blinking, as the electrophysiologic blink reflex has been shown to be underexcitable compared to normal in both symptomatic (Esteban and Gimenez-Roldan 1975) and pre-symptomatic (De Tommaso et al. 2001) individuals.

4.2 Relationship Between Blinking and Saccades

In normal individuals, spontaneous blinks are partially inhibited during voluntary saccades. This is done to avoid disrupting visual input during a visually guided task and also because saccades are slower and less accurate when they are interrupted by blinks (Rottach et al. 1998). In parkinsonism, however, patients fail to suppress blinks during voluntary saccades. For example, while normal individuals blinked an average of 15.7 times per minute when fixating in primary gaze and 9.1 times per minute when asked to alternate looking left and right every five seconds in one study, PD patients experienced a slight increase in blink rate (from 12.5 to 14.8 per minute), and PSP patients a substantial increase in blink rate (from 3.0 to 5.3 per minute) during horizontal eye movements (Golbe et al. 1989).

Paradoxically, while blinks reduce the speed and accuracy of saccades in normal individuals, in certain patients (saccadic or ocular motor apraxia) blinks can help to initiate saccades. In HD, difficulty with saccade initiation and prolonged saccadic latency are among the earliest clinical manifestations (Lasker and Zee 1997; Golding et al. 2006). Patients frequently employ head thrusts and blinks when performing saccades, and as the disease progresses, patients may be unable to initiate voluntary saccades without an obligatory blink (35% in one study) (Leigh et al. 1983). Blink-assisted saccades (also termed blink-saccade synkinesis (Zee et al. 1983)) can occur in any disease where saccade initiation is impaired. Examples include the autosomal-dominant spinocerebellar ataxias, the autosomal recessive ataxias with ocular motor apraxia, ataxia-telangiectasia, congenital ocular motor apraxia, Gaucher disease, Niemann-Pick disease type C, Joubert syndrome, and others (Cogan et al. 1981). Preceding a normal saccade, omnipause neurons in the dorsal pons cease firing. This enables activation of excitatory burst neurons in the parapontine reticular formation and riMLF (which are normally inhibited by omnipause cells), allowing them to generate a horizontal or vertical/torsional saccade, respectively. For reasons that are not entirely clear, omnipause neurons also stop firing during blinks (Schultz et al. 2010). If ocular motor apraxia is caused by a failure to inhibit omnipause neurons through normal supranuclear pathways during saccade initiation, then blink-saccade synkinesis may represent a collateral pathway to inhibit omnipause neurons and generate saccades.

5 Blepharospasm

5.1 Introduction to Blepharospasm

Blepharospasm is characterized by periods of involuntary, sustained, active eyelid closure. As it involves the co-contraction of agonist (OO) and antagonist (LPS) muscles, blepharospasm meets the definition of a focal dystonia. Not surprisingly, more than half of patients have a *geste antagoniste*, that is, a sensory trick that can temporarily relieve symptoms (Martino et al. 2010). Blepharospasm presents between the fourth and sixth decades of life and is more common in women than men (Steeves et al. 2012). The initial clinical manifestation of blepharospasm is an increase in spontaneous blink rate (Bentivoglio et al. 2006). Over time, blinks become increasingly forceful and prolonged, involving both the orbital and palpebral portions of the OO. These have been termed "dystonic blinks" and are different from normal spontaneous blinks, which only involve the palpebral OO. Eventually these blinks coalesce into periods of sustained eyelid closure whose frequency and duration can be so severe as to produce functional blindness (Valls-Sole and Defazio 2016).

Co-contraction of the LPS and OO also occurs in eyelid myotonia, but for different reasons. In myotonia, neuronal activation of the OO during voluntary or reflexive (e.g., sneezing) eyelid closure appropriately ceases, but the muscle itself fails to relax appropriately. This is seen in myotonic dystrophy as well as the non-dystrophic myotonias (e.g., myotonia congenita, paramyotonia congenita), which are caused by mutations in voltage-gated chloride and sodium channel genes. Myotonia can be treated with sodium channel-blocking antiepileptic drugs and the antiarrhythmic drug mexilitene (Matthews et al. 2010).

5.2 Pathophysiology of Blepharospasm

Blepharospasm is thought to represent a disorder of overactive reflex blinking, specifically to light. Several clinical and electrophysiologic observations support this hypothesis:

1. Patients with blepharospasm frequently complain of photophobia and that bright light triggers spasms of eyelid closure (Katz and Digre 2016). Because the ratio of blepharospasm to cervical dystonia patients in movement disorders cohorts varies by season and latitude (Molloy et al. 2016), some have even suggested that sun exposure is a risk factor for the disorder, though this theory is not universally accepted. Polarized sunglasses have been shown to be a useful adjunctive treatment for blepharospasm (Blackburn et al. 2009). Ocular surface symptoms and sensitivity to tactile stimulation of the cornea and eyelids have also been reported (Martino et al. 2005).

2. A feature that distinguishes reflexive from spontaneous blinks is the activation of the orbital portion of the OO in the former but not the latter. Given that the dystonic blinks of blepharospasm involve the orbital portion of the OO, it is suggested that they are generated via reflexive rather than spontaneous blinking circuits.

3. While spontaneous and most reflexive blinks are characterized by reciprocal coordination of LPS and OO activity such that they never co-activate, in normal reflexive blinking to light, subclinical overlap in LPS and OO activity does occur and can be seen at the electromyographic level (Manning and Evinger 1986). The co-contraction of these muscles in blepharospasm is therefore thought to represent an exaggeration of the normal blink reflex to light.

4. The electrophysiologic blink reflex has been found to be hyperexcitable in patients with blepharospasm, that is, it does not appropriately habituate or downregulate with repetitive stimulation (Tisch et al. 2006; Schwingenschuh

et al. 2011). Similar findings have been reported in other focal dystonias as well (e.g., cervical dystonia, spasmodic dysphonia), raising the possibility of a shared pathophysiology (Tolosa et al. 1988; Cohen et al. 1989). During a *geste antagoniste*, the R2 duration shortens, but the degree of excitability does not change (Gómez-Wong et al. 1998). High-frequency supraorbital nerve stimulation has been studied as a method of inducing long-term potentiation of the blink reflex to treat blepharospasm, but with disappointing results (Kranz et al. 2013).

The localization of blepharospasm remains unknown. The vast majority of patients have normal neuroimaging; in a single case series of 1114 patients, only 18 (1.6%) had abnormal brain MRIs, and lesions localized to a variety of areas including the basal ganglia, thalami, cerebellum, midbrain, and even cortex (Khooshnoodi et al. 2013). Structural imaging studies have yielded inconsistent results, and it is unclear whether any changes that are seen reflect primary pathology or secondary adaptations in response to changes occurring elsewhere in the brain (Hallett et al. 2008). Functional neuroimaging studies have shown hypermetabolism of a variety of cortical and deep gray matter foci, both at rest (Zhou et al. 2013) and during tasks such as voluntary blinking (Baker et al. 2003). Using dopamine-receptor SPECT imaging, decreased striatal dopamine binding has been found in up to one-third of patients of blepharospasm (Horie et al. 2009). In one unique case report of craniocervical dystonia treated with deep brain stimulation, electrical activity in the globus pallidus interna was recorded prior to electrode implantation and was found to be increased (Foote et al. 2005).

5.3   Epidemiology and Natural History of Blepharospasm

Blepharospasm may remain limited to the OO or may spread to adjacent muscles of the face, jaw, and neck, which is known as craniocervical dystonia or Meige syndrome. This spread usually occurs within 5 years of onset (Abbruzzese et al. 2008; Svetel et al. 2015) and is said to occur in up to 60% of patients, though this may be an overestimate due to referral bias in tertiary movement disorders centers (Martino et al. 2012; Weiss et al. 2006). Risk factors for generalization include greater age of onset, female sex, and a prior history of minor head trauma (Defazio and Livrea 2002). In two separate cohorts, a single nucleotide polymorphism in the 3′ untranslated region of the TOR1A gene, which is the causative gene in DYT1, was also associated with a twofold increase in risk (Defazio et al. 2009). Spontaneous remission may occur in up to 12% of patients (Grandas et al. 1998). Blepharospasm can occur in patients with inherited focal (e.g., autosomal-dominant focal dystonia) and generalized (e.g., DYT1) dystonias. There are reports of blepharospasm in other inherited neurodegenerative disorders as well (e.g., autosomal-dominant spinocerebellar ataxias (Gorman et al. 2015), neurodegeneration with brain iron accumulation (Egan et al. 2005)). It also presents in acquired parkinsonian disorders, especially PSP (20–70% of patients affected) but occasionally multiple system atrophy and rarely PD (Friedman et al. 1992; Yoon et al. 2005; Barclay and Lang 1997). Comorbid blepharospasm and apraxia of eyelid opening may also occur (Defazio et al. 2001), especially in PSP.

5.4   Treatment of Blepharospasm

Botulinum toxin chemodenervation of the OO is the treatment of choice for blepharospasm. In addition to weakening the OO, it lowers the spontaneous blink rate (Conte et al. 2016) and reduces blink reflex hyperexcitability (Quartarone et al. 2006). In mild cases where photophobia and increased blink frequency are the chief complaints, polarized lenses may be helpful. Some patients report symptomatic improvement by wearing a tight band around the forehead, which serves to provide a constant *geste antagoniste*. Other studies of noninvasive treatment include behavioral therapy to encourage eyelid relaxation, biofeedback using EMG recording of the frontalis muscle (Surwit and Rotberg 1984), and transcranial magnetic stimulation (Kranz et al. 2010). Medications (e.g., anticholinergics, baclofen, benzodiazepines, levodopa) that are used to treat other dystonias, especially generalized dystonias where botulinum chemodenervation is not feasible, have been attempted in blepharospasm with mixed results. Surgical myectomy was routinely performed in the past but is now mainly reserved for patients who are refractory or intolerant to botulinum toxin (Pariseau et al. 2013). Deep brain stimulation has also been successfully performed in a few refractory cases (Yamada et al. 2016).

6   Apraxia of Eyelid Opening and Closure

6.1   Overview of Voluntary Eyelid Control

In addition to spontaneous and reflexive blinking, eyelid closure can be initiated and sustained at will. The supranuclear pathways responsible for voluntary eyelid control are poorly understood. Electrical stimulation of various frontal, temporal, parietal, and occipital foci can elicit eye opening or closure, and the cortical control of eyelid position is thought to have a right hemispheric predominance (Liu et al. 2010). A syndrome of difficulty initiating and maintaining voluntary eye opening during a normal waking state was first described by Goldstein and Cogan in 1965, who called it "apraxia of lid opening" (Goldstein and Cogan 1965). The use of the term apraxia has since been criticized, and other names such as blepharocolysis, akinesia of lid opening and function, eyelid freezing, and involuntary levator palpebrae inhibition of supranuclear origin have been proposed. Clinically and electrophysiologically, apraxia of eyelid opening (AEO) is characterized by difficulty initiating eyelid elevation with a lack of appropriate LPS activation following sustained voluntary eyelid closure in the absence of inappropriate orbicularis oculi activity. The frontalis muscle is often tonically activated in an attempt to secondarily elevate the upper eyelids. Spontaneous blinking and reflexive eye opening are normal, confirming that the neuromuscular apparatus of the levator palpebrae is intact and that the disorder is one of supranuclear control. PET studies have found hypometabolism in the anterior cingulate and supplemental motor areas of the frontal lobes (Smith et al. 1995; Suzuki et al. 2003).

Some of the clinical features of AEO resemble a focal dystonia. As many as one-third of patients learn to adopt a *geste antagoniste*, typically a light touch of the eyelids, that allows them to initiate eye opening (Defazio et al. 1988). Blepharospasm may co-exist, and AEO is occasionally unmasked by chemodenervation of the OO to treat blepharospasm. This should not be mistaken as treatment failure or ptosis due to the spread of botulinum toxin into the LPS, as the treatment is actually to provide additional botulinum toxin to the pretarsal OO (Aramideh et al. 1995; Rana et al. 2012; Esposito et al. 2014). In fact, while AEO is defined by clinical and electromyographic OO silence, selective recordings of the pretarsal OO have shown persistent activity during attempted eyelid opening in some patients (Tozlovanu et al. 2001). This finding has been difficult to replicate on a larger scale because of its technical barriers, so it remains unclear if these patients truly have AEO, a subtle variant of blepharospasm, or a distinct entity altogether that some have termed "OO motor persistence".

6.2   Neurodegenerative Diseases Associated with Apraxia of Eyelid Opening

AEO may occur in isolation or in association with an underlying neurodegenerative disorder. Of 32 patients with AEO seen at a regional referral center in Puglia, Italy, over a 10-year period, 10 were healthy, 10 had blepharospasm, 6 had PSP, and 3 had idiopathic PD (Lamberti et al. 2002). AEO is seen in anywhere from 30% to 45% of patients with PSP (Nath et al. 2003). Furthermore, AEO typically coincides with or precedes the onset of parkinsonism in PSP, whereas it is a much later manifestation of PD. AEO is also seen in MSA and corticobasal syndrome. In the latter, however, the underlying pathology at autopsy is one of PSP (with neuronal tau-positive inclusions) rather than true corticobasal ganglionic degeneration (which is characterized by astrocytic tau-positive inclusions and balloon neurons) (Ouchi et al. 2014). There are also reports of AEO in amyotrophic lateral sclerosis (ALS) with or without frontotemporal disease, HD, SCA2 (Kanazawa et al. 2007), SCA3 (Cardoso et al. 2000), Wilson's disease, and chorea-acanthocytosis among others.

Apraxia of eyelid closure has also been described but is much less common. These patients constrict the corrugator and procerus muscles during attempted voluntary eyelid closure but not the OO (Aramideh et al. 1997); the LPS remains active, and thus this disorder represents a failure to both inhibit the LPS and activate the

OO. However, patients are able to close their eyes normally during spontaneous and reflexive blinking. It has been reported in PSP (Friedman et al. 1992), HD (Bonelli and Niederwieser 2002), Creutzfeldt–Jakob disease (CJD), ALS (Fukishima et al. 2007), and acquired frontal and parietal lobe disease.

6.3   Treatment of Apraxia of Eyelid Opening

A number of modalities can be used to treat AEO. Conservative measures include wearing goggles (Hirayama et al. 2000) or eyelid crutches (Ramasamy et al. 2007). While these provide some direct mechanical elevation of the upper eyelid, their primary effect is probably as a *geste antagoniste*. Levodopa may be effective in isolated (Dewey Jr and Maraganore 1994) or PD-associated AEO (Yamada et al. 2004; Lee et al. 2004), but it appears to worsen PSP-associated AEO. Anticholinergics (Krack and Marion 1994), atypical antipsychotics (Tokisato et al. 2015), and methylphenidate (Eftekhari et al. 2015) have been tried in individual case reports. Given the finding of abnormal EMG activity in the pretarsal OO in some patients with AEO, botulinum injection of the pretarsal OO is frequently performed with success (Piccione et al. 1997). In cases of comorbid blepharospasm and AEO, surgical myectomy (Georgescu et al. 2008) or frontalis suspension (Dressler et al. 2017; De Groot et al. 2000) can treat both disorders simultaneously but is generally reserved as a last resort.

PD-associated AEO is especially complicated as it can be confounded by DBS (Baizabal-Carvallo and Jankovic 2016). While AEO may be present in untreated PD and improve after STN DBS (Fuss et al. 2004; Weiss et al. 2010a), it more frequently emerges or worsens after STN DBS (anywhere from 2% to 31% of patients). This is thought to occur via the spread of current into the adjacent corticobulbar tract, particularly when higher voltages are applied to more caudal contact points. This effect seems to be specific to the STN, as AEO does not seem to occur (and may even improve (Goto et al. 2000)) following the more rostral GPi DBS, and experimental low-frequency stimulation of the STN at certain voltage thresholds can even induce myoclonus in the pretarsal OO (Tommasi et al. 2012). The weaning of levodopa following DBS could also theoretically unmask a pre-existing, levodopa-responsive AEO (Umemura et al. 2008). AEO usually occurs within a year of DBS implantation, and an increase in spontaneous blink rate can be a harbinger of AEO during programming sessions (Weiss et al. 2010b). Treatment is challenging and consists of reducing voltage, increasing frequency, and administering levodopa in addition to conventional AEO therapies.

References

Abbruzzese, G., Berardelli, A., Girlanda, P., Marchese, R., Martino, D., Morgante, F., Avanzino, L., Colosimo, C., & Defazio, G. (2008). Long-term assessment of the risk of spread in primary late-onset focal dystonia. *JNNP, 79*, 392–396.

Agostino, R., Bologna, M., Dinapoli, L., Gregori, B., Fabbrini, G., Accornero, N., & Berardelli, A. (2008). Voluntary spontaneous, and reflex blinking in Parkinson's disease. *Movement Disorders, 23*(5), 669–675.

Allen, R. C. (2013). Genetic diseases affecting the eyelids: What should a clinician know? *Current Opinion in Ophthalmology, 24*, 463–477.

Amboni, M., Pellecchia, M. T., Cozzolino, A., Picillo, M., Vitale, C., Barone, P., Varrone, A., Garavaglia, B., Gambelli, S., & Federico, A. (2009). Cerebellar and pyramidal dysfunctions, palpebral ptosis and weakness as the presenting symptoms of PARK-2. *Movement Disorders, 24*(2), 303–305.

Aramideh, M., & Ongerboer de Visser, B. W. (2002). Brainstem reflexes: Electrodiagnostic techniques, physiology, normative data, and clinical applications. *Muscle & Nerve, 26*(1), 14–30.

Aramideh, M., Eekhof, J. L., Bour, L. J., Koelman, J. H., & Speelman, J. D. (1995). Ongerboer de Visser BW. Electromyography and recovery of the blink reflex in involuntary eyelid closure: A comparative study. *JNNP, 58*(6), 692–698.

Aramideh, M., Kwa, I. H., Brans, J. W., Speelman, J. D., & Verbeeten, B., Jr. (1997). Apraxia of eyelid closure accompanied by denial of eye opening. *Movement Disorders, 12*(6), 1105–1108.

Baizabal-Carvallo, J. F., & Jankovic, J. (2016). Movement disorders induced by deep brain stimulation. *Parkinsonism & Related Disorders, 25*, 1–9.

Baker, R. S., Andersen, A. H., Morecraft, R. J., & Smith, C. D. (2003). A functional magnetic resonance imaging study in patients with benign essential blepharospasm. *Journal of Neuro-Ophthalmology, 23*(1), 11–15.

Barclay, C. L., & Lang, A. E. (1997). Dystonia in progressive supranuclear palsy. *JNNP, 62*(4), 352–356.

Basso, M. A., Powers, A. S., & Evinger, C. (1996). An explanation for reflex blink hyperexcitability in Parkinson's disease. I. Superior colliculus. *The Journal of Neuroscience, 16*(22), 7308–7317.

Becker, W., & Fuchs, A. F. (1988). Lid-eye coordination during vertical gaze changes in man and monkey. *Journal of Neurophysiology, 60*, 1227–1252.

Bentivoglio, A. R., Daniele, A., Albanese, A., Tonali, P. A., & Fasano, A. (2006). Analysis of blink rate in patients with blepharospasm. *Movement Disorders, 21*(8), 1225–1229.

Biousse, V., Skibell, B. C., Watts, R. L., Loupe, D. N., Drews-Botsch, C., & Newman, N. J. (2004). Ophthalmologic features of Parkinson's disease. *Neurology, 62*(2), 177–180.

Blackburn, M. K., Lamb, R. D., Digre, K. B., Smith, A. G., Warner, J. E., McClane, R. W., Nandedkar, S. D., Langeberg, W. J., Holubkov, R., & Katz, B. J. (2009). FL-41 tint improves blink frequency, light sensitivity, and functional limitations in patients with benign essential blepharospasm. *Ophthalmology, 116*(5), 997–1001.

Bologna, M., Fasano, A., Modugno, N., Fabbrini, G., & Berardelli, A. (2012). Effects of subthalamic nucleus deep brain stimulation and L-DOPA on blinking in Parkinson's disease. *Experimental Neurology, 235*(1), 265–272.

Bonanni, L., Anzellotti, F., Varanese, S., Thomas, A., Manzoli, L., & Onofrj, M. (2007). Delayed blink reflex in dementia with Lewy bodies. *JNNP, 78*(10), 1137–1139.

Bonelli, R. M., & Niederwieser, G. (2002). Apraxia of eyelid closure in Huntington's disease. *Journal of Neural Transmission, 109*(2), 197–201.

Brodsky, H., Dat Vuong, K., Thomas, M., & Jankovic, J. (2004). Glabellar and palmomental reflexes in parkinsonian disorders. *Neurology, 63*(6), 1096–1098.

Cardoso, F., de Oliveira, J. T., Puccioni-Sohler, M., Fernandes, A. R., de Mattos, J. P., & Lopes-Cendes, I. (2000). Eyelid dystonia in Machado-Joseph disease. *Movement Disorders, 15*(5), 1028–1030.

Clark, B. J., Kemp, E. G., Behan, W. M. H., et al. (1995). Abnormal extracellular material in the levator palpebrae superioris complex in congenital ptosis. *Archives of Ophthalmology, 113*, 1414–1419.

Cogan, D. G., Chu, F. C., Reingold, D., & Barranger, J. (1981). Ocular motor signs in some metabolic diseases. *Archives of Ophthalmology, 99*(10), 1802–1808.

Cohen, L. G., Ludlow, C. L., Warden, M., Estegui, M., Agostino, R., Sedory, S. E., Holloway, E., Dambrosia, J., & Hallett, M. (1989). Blink reflex excitability recovery curves in patients with spasmodic dysphonia. *Neurology, 39*(4), 572–577.

Conte, A., Berardelli, I., Ferrazzano, G., Pasquini, M., Berardelli, A., & Fabbrini, G. (2016). Non-motor symptoms in patients with adult-onset focal dystonia: Sensory and psychiatric disturbances. *Parkinsonism & Related Disorders, 22*, S111–S114.

Costa, J., Valls-Sole, J., Valldeoriola, F., Pech, C., & Rumia, J. (2006). Single subthalamic nucleus deep brain stimuli inhibit the blink reflex in Parkinson's disease patients. *Brain, 129*, 1758–1767.

De Groot, V., De Wilde, F., Smet, L., & Tassignon, M. J. (2000). Frontalis suspension combined with blepharoplasty as an effective treatment for blepharospasm associated with apraxia of eyelid opening. *Ophthalmic Plastic and Reconstructive Surgery, 16*(1), 34–38.

De Tommaso, M., Sciruicchio, V., Spinelli, A., Specchio, N., Difruscolo, O., Puca, F., & Specchio, L. M. (2001). Features of the blink reflex in individuals at risk for Huntington's disease. *Muscle & Nerve, 24*(11), 1520–1525.

Defazio, G., & Livrea, P. (2002). Epidemiology of primary blepharosmasm. *Movement Disorders, 17*(1), 7–12.

Defazio, G., Livrea, P., Lamberti, P., De Salvia, R., Laddomada, G., Giorelli, M., & Ferrari, E. (1988). Isolated so-called apraxia of eyelid opening: Report of 10 cases and a review of the literature. *European Neurology, 39*, 204–210.

Defazio, G., Livrea, P., De Salvia, R., Manobianca, G., Coviello, V., Anaclerio, D., Guerra, V., Martino, D., Valluzzi, F., Liguori, R., & Logroscino, G. (2001). Prevalence of primary blepharospasm in a community of Puglia region, Southern Italy. *Neurology, 56*(11), 1579–1581.

Defazio, G., Matarin, M., Peckham, E. L., Martino, D., Valente, E. M., Singleton, A., Crawley, A., Aniello, M. S., Brancati, F., Abbruzzese, G., Girlanda, P., Livrea, P., Hallett, M., & Berardelli, A. (2009). The TOR1A polymorphism rs1182 and the risk of spread in primary blepharospasm. *Movement Disorders, 24*(4), 613–616.

Dewey, R. B., Jr., & Maraganore, D. M. (1994). Isolated eyelid-opening apraxia: Report of a new levodopa-responsive syndrome. *Neurology, 44*(9), 1752–1754.

Dressler, D., Karapantzou, C., Rohrbach, S., Schneider, S., & Laskawi, R. (2017). Frontalis suspension surgery to treat patients with blepharospasm and eyelid opening apraxia: Long term results. *Journal of Neural Transmission, 124*(2), 253–257.

Eftekhari, K., Choe, C. H., Vagefi, M. R., Gausas, R. E., & Eckstein, L. A. (2015). Oral methylphenidate for the treatment of refractory facial dystonias. *Ophthalmic Plastic and Reconstructive Surgery, 31*(3), e65–e66.

Egan, R. A., Weleber, R. G., Hogarth, P., Gregory, A., Coryell, J., Westaway, S. K., Gitschier, J., Das, S., & Hayflick, S. J. (2005). Neuro-ophthalmologic and electroretinographic findings in pantothenate kinase-associated neurodegeneration (formerly Hallervorden-Spatz disease). *American Journal of Ophthalmology, 140*(2), 267–274.

Engel, A. G., Shen, X. M., Selcen, D., & Sine, S. M. (2015). Congenital myasthenic syndromes: Pathogenesis, diagnosis, and treatment. *Lancet Neurology, 14*(4), 420–434.

Esposito, M., Fasano, A., Crisci, C., Dubbioso, R., Iodice, R., & Santoro, L. (2014). The combined treatment with orbital and pretarsal bnotulinum toxin injections in the management of poorly responsive blepharospasm. *Neurological Sciences, 35*(3), 397–400.

Esteban, A., & Gimenez-Roldan, S. (1975). Blink reflex in Huntington's chorea and Parkinson's disease. *Acta Neurologica Scandinavica, 52*(2), 145–157.

Esteban, A., & Salinero, E. (1979). Reciprocal reflex activity in ocular muscles: Implications in spontaneous blinking and Bell's phenomenon. *European Neurology, 18*, 157–165.

Evinger, C., Shaw, M. D., Peck, C. K., Manning, K. A., & Baker, R. (1984). Blinking and associated eye movements in humans, guinea pigs, and rabbits. *Journal of Neurophysiology, 52*(2), 323–339.

Fekete, R., & Jankovic, J. (2014). Upper facial chorea in Huntington disease. *Journal of Clinical Movement Disorders, 1*, 7.

Foote, K. D., Sanchez, J. C., & Okun, M. S. (2005). Staged deep brain stimulation for refractory craniofacial dystonia with blepharospasm: Case report and physiology. *Neurosurgery, 56*, E415.

Friedman, D. I., Jankovic, J., & McCrary, J. A. (1992). Neuro-ophthalmic findings in progressive supranuclear palsy. *Journal of Clinical Neuro-ophthalmology, 12*(2), 104–109.

Fuchs, A. F., Becker, W., Ling, L., Langer, T. P., & Kaneko, C. R. (1992). Discharge patterns of levator palpebrae superioris motoneurons during vertical lid and eye movements in the monkey. *Journal of Neurophysiology, 68*(1), 233–243.

Fukishima, T., Hasegawa, A., Matsubara, N., & Kolke, R. (2007). An apraxia of eyelid closure in association with frontal lobe atrophy in a patient with amyotrophic lateral sclerosis. *Rinshō Shinkeigaku, 47*(5), 226–230.

Fuss, G., Spiegel, J., Magnus, T., Moringlane, J. R., Becker, G., & Dillmann, U. (2004). Improvement of apraxia of lid opening by STN-stimulation in a 70 year-old with Parkinson's disease: A case report. *Minimally Invasive Neurosurgery, 47*(1), 58–60.

Georgescu, D., Vagefi, M. R., McMullan, T. F., McCann, J. D., & Anderson, R. L. (2008). Upper eyelid myectomy in blepharospasm with associated apraxia of lid opening. *American Journal of Ophthalmology, 145*(3), 541–547.

Gnadt, J. W., Lu, S. M., Breznen, B., Basso, M. A., Henriquez, V. M., & Evinger, C. (1997). Influence of the superior colliculus on the primate blink reflex. *Experimental Brain Research, 116*, 389–398.

Golbe, L. I., Davis, P. H., & Lepore, F. E. (1989). Eyelid movement abnormalities in progressive supranuclear palsy. *Movement Disorders, 4*(4), 297–302.

Golding, C. V., Danchaivijitr, C., Hodgson, T. L., Tabrizi, S. J., & Kennard, C. (2006). Identification of an oculomotor biomarker of preclinical Huntington disease. *Neurology, 67*, 485–487.

Goldstein, J. E., & Cogan, D. G. (1965). Apraxia of lid opening. *Archives of Ophthalmology, 73*, 155–159.

Gómez-Wong, E., Martí, M. J., Cossu, G., Fabregat, N., Tolosa, E. S., & Valls-Solé, J. (1998). The 'geste antagonistique' induces transient modulation of the blink reflex in human patients with blepharospasm. *Neuroscience Letters, 251*(2), 125–128.

Gorman, G. S., Pfeffer, G., Griffin, H., Blakely, E. L., Kurzawa-Akanbi, M., Gabriel, J., Sitarz, K., Roberts, M., Schoser, B., Pyle, A., Schaefer, A. M., McFarland, R., Turnbull, D. M., Horvath, R., Chinnery, P. F., & Taylor, R. W. (2015). Clonal expansion of secondary mitochondrial DNA deletions associated with spinocerebellar ataxia type 28. *JAMA Neurology, 72*(1), 106–111.

Goto, S., Kihara, K., Hamasaki, T., Nishikawa, S., Hirata, Y., & Ushio, Y. (2000). Apraxia of lid opening is alleviated by pallidal stimulation in a patient with Parkinson's disease. *European Journal of Neurology, 7*(3), 337–340.

Grandas, F., Traba, A., Alonso, F., & Esteban, A. (1998). Blink reflex recovery cycle in patients with blepharospasm unilaterally treated with botulinum toxin. *Clinical Neuropharmacology, 21*(5), 307–311.

Hallett, M., Evinger, C., Jankovic, J., & Stacy, M. (2008). Update on blepharospasm: Report from the BEBRF international workshop. *Neurology, 71*, 1275–1282.

Heidary, G., Engle, E. C., & Hunter, D. G. (2008). Congenital fibrosis of the extraocular muscles. *Seminars in Ophthalmology, 23*(1), 3–8.

Hirayama, M., Kumano, T., Aita, T., Nakagawa, H., & Kuriyama, M. (2000). Improvement of apraxia of eyelid opening by wearing goggles. *Lancet, 356*(9239), 1413.

Horie, C., Suzuki, Y., Kiyosawa, M., Mochizuki, M., Wakakura, M., Oda, K., Ishiwata, K., & Ishii, K. (2009). Decreased dopamine D receptor binding in essential blepharospasm. *Acta Neurologica Scandinavica, 119*(1), 49–54.

Horn, A. K., & Buttner-Ennever, J. A. (2008). Brainstem circuits controlling lid-eye coordination in monkeys. *Progress in Brain Research, 171*, 87–95.

Horn, A. K., Buttner-Ennever, J. A., Gayde, M., & Messoudi, A. (2000). Neuroanatomical identification of mesencephalic premotor neurons coordinating eyelid with upgaze in the monkey and man. *The Journal of Comparative Neurology, 420*, 19–34.

Howard, R. S. (1986). A case of convergence evoked eyelid nystagmus. *Journal of Clinical Neuroophthalogy, 6*(3), 169–171.

https://clinicaltrials.gov/ct2/show/NCT02894658?term=lipiflow&rank=4

Iriarte, L. M., Chacon, J., Madrazo, J., Chaparro, P., & Vadillo, J. (1988). Blink reflex in 57 parkinsonian patients with correlation between the clinical and electrophysiological parameters. *Functional Neurology, 3*(2), 147–156.

Iriarte, L. M., Chacon, J., Madrazo, J., Chaparro, P., & Vadillo, J. (1989). Blink reflex in dyskinetic and nondyskinetic patients with Parkinson's disease. *European Neurology, 29*(2), 67–70.

Kaminer, J., Thakur, P., & Evinger, C. (2015). Effects of subthalamic deep brain stimulation on blink abnormalities of 6-OHDA lesioned rats. *Journal of Neurophysiology, 1113*(9), 3038–3046.

Kanazawa, M., Thimohata, T., Sato, M., Onodera, O., Tanaka, K., & Nishizawa, M. (2007). Botulinum toxin A injections improve apraxia of eyelid opening without overt blepharospasm associated with neurodegenerative diseases. *Movement Disorders, 22*(4), 597–598.

Karson, C. N. (1983). Spontaneous eye-blink rates and dopaminergic systems. *Brain, 106*, 643–653.

Karson, C. N., Burns, R. S., LeWitt, P. A., Foster, N. L., & Newman, R. P. (1984). Blink rates and disorders of movement. *Neurology, 34*(5), 677–678.

Katz, B. J., & Digre, K. B. (2016). Diagnosis, pathophysiology, and treatment of photophobia. *Survey of Ophthalmology, 61*(4), 466–477.

Keane, J. R. (1990). The pretectal syndrome: 206 patients. *Neurology, 40*, 684–690.

Kersten, R. C., de Conciliis, C., & Kulwin, D. R. (1995). Acquired ptosis in the young and middle-aged adult population. *Ophthalmology, 102*, 924–928.

Khooshnoodi, M. A., Factor, S. A., & Jinnah, H. A. (2013). Secondary blepharospasm associated with structural lesions of the brain. *Journal of the Neurological Sciences, 331*(102), 98–101.

Kiferle, L., Orsucci, D., Mancuso, M., Lo Gerfo, A., Petrozzi, L., Siciliano, G., Ceravolo, R., & Bonnucelli, U. (2013). Twinkle mutation in an Italian family with external progressive ophthalmoplegia and parkinsonism: A case report and an update on the state of the art. *Neuroscience Letters, 556*, 1–4.

Kimber, T. E., & Thompson, P. D. (2000). Increased blink rate in advanced Parkinson's disease: A form of 'off'-period dystonia? *Movement Disorders, 15*(5), 982–985.

Krack, P., & Marion, M. H. (1994). "Apraxia of lid opening," a focal eyelid dystonia: Clinical study of 32 patients. *Movement Disorders, 9*, 610.

Kranz, G., SHamim, E. A., Lin, P. T., Kranz, G. S., & Hallett, M. (2010). Transcranial magnetic brain stimulation modulates blepharospasm: A randomized controlled study. *Neurology, 75*(16), 1465–1471.

Kranz, G., Shamim, E. A., Lin, P. T., Kranz, G. S., & Hallet, M. (2013). Long term depression-like plasticity of the blink reflex for the treatment of blepharospasm. *Movement Disorders, 28*(4), 498–503.

Kuniyoshi, S., Riley, D. E., Zee, D. S., Reich, S. G., Whitney, C., & Leigh, R. J. (2002). Distinguishing progressive supranuclear palsy from other forms of Parkinson's disease: Evaluation of new signs. *Annals of the New York Academy of Sciences, 956*, 484–486.

Lamberti, P., De Mari, M., Zenzola, A., Aniello, M. S., & Defazio, G. (2002). Frequency of apraxia of eyelid opening in the general population and in patients with extrapyramidal disorders. *Neurological Sciences, 23*, S81–S82.

Lasker, A. G., & Zee, D. S. (1997). Ocular motor abnormalities in Huntington's disease. *Vision Research, 37*, 3639–3645.

Lee, K. C., Finley, R., & Miller, B. (2004). Apraxia of lid opening: dose-dependent response to carbidopa-levodopa. *Pharmacotherapy, 24*(3), 401–403.

Leigh, R. J., Newman, S. A., Folstein, S. E., Lasker, A. G., & Jensen, B. A. (1983). Abnormal ocular motor control in Huntington's disease. *Neurology, 33*(10), 1268–1275.

Liu, G. T., Volpe, N. J., & Galetta, S. L. (2010). *Neuro-ophthalmology: Diagnosis and management* (2nd ed.). Philadelphia: Elsevier.

Manning, K. A., & Evinger, C. (1986). Different forms of blinks and their two-stage control. *Experimental Brain Research, 64*(3), 579–588.

Martino, D., Defazio, G., Alessio, G., et al. (2005). Relationship between eye symptoms and blepharospasm: A multicenter case-control study. *Movement Disorders, 20*, 1564–1570.

Martino, D., Liuzzi, D., Macerollo, A., Aniello, M. S., Livrea, P., & Defazio, G. (2010). The phenomenology of the *geste antagoniste* in primary blepharospasm and cervical dystonia. *Movement Disorders, 25*(4), 407–412.

Martino, D., Berardelli, A., Abbruzzese, G., Bentivoglio, A. R., Esposito, M., Fabbrini, G., Guidubaldi, A., Girlanda, P., Liguori, R., Marinelli, L., Morgante, F., Santoro, L., & Defazio, G. (2012). Age at onset and symptom spread in primary focal dystonia. *Movement Disorders, 27*(11), 1447–1450.

Matsumoto, H., Noro, H., Kaneshige, Y., Chiba, S., Miyano, N., Motoi, Y., & Yanada, Y. (1992). A correlation study between blink reflex habituation and clinical state in patients with Parkinson's disease. *Journal of the Neurological Sciences, 107*(2), 155–159.

Matthews, E., Fialho, D., Tan, S. V., Venance, S. L., Cannon, S. C., Sternberg, D., Fontaine, B., Amato, A. A., Barohn, R. J., Griggs, R. C., Hanna, M. G., & The CINCH investigators. (2010). The non-dystrophic myotonias: Molecular pathogenesis, diagnosis, and treatment. *Brain, 133*(1), 9–22.

McClelland, C., Manousakis, G., & Lee, M. S. (2016). Progressive external ophthalmoplegia. *Current Neurology and Neuroscience Reports, 16*(6), 53.

Mendes, S., Beselga, D., Campos, S., Neves, A., Campos, J., Carvalho, S., Silvo, E., & Castro Sousa, J. P. (2015). Possible rare congenital dysinnervation disorder: Congenital ptosis associated with adduction. *Strabismus, 23*(1), 33–35.

Milivojevic, I., Bakran, Z., Adamec, I., Miletic Grskovic, S., & Habek, M. (2013). Eyelid nystagmus and primary position upbeat nystagmus. *Neurological Sciences, 34*(8), 1463–1464.

Milone, M., & Wong, L. J. (2013). Diagnosis of mitochondrial myopathies. *Molecular Genetics and Metabolism, 110*(1–2), 35–41.

Molloy, A., Williams, L., Kimmich, O., Butler, J. S., Beiser, I., McGovern, E., O'Riordan, S., Reilly, R. B., Walsh, C., & Hutchinson, M. (2016). Sun exposure is an environmental risk factor for the development of blepharospasm. *JNNP, 87*(4), 420–424.

Moro, A., Munhoz, R. P., Arruda, W. O., Raskin, S., & Teive, H. A. (2013). Clinical relevance of "bulging eyes" for the differential diagnosis of spinocerebellar ataxias. *Arquivos de Neuro-Psiquiatria, 71*(7), 428–430.

Nath, U., Ben-Shiomo, Y., Thomson, R. G., Lees, A. J., & Burn, D. J. (2003). Clinical features oand natural history of progressive supranuclear palsy: A clinical cohort study. *Neurology, 60*(6), 910–916.

Ng, J., Papandreou, A., Jeales, S. J., & Kurian, M. A. (2015). Monoamine neurotransmitter disorders – clinical advances and future directions. *Nature Reviews. Neurology, 11*, 567–584.

Nowacka, B., Lubinski, W., Honczarenko, K., Potemkowski, A., & Safranow, K. (2014). Ophthalmological features of Parkinson's disease. Ophthalmological features of Parkinson's disease. *Medical Science Monitor, 20*, 2243–2249.

Onofrj, M., Monaco, D., Bonani, L., Onofrj, V., Bifolchetti, S., Manzoli, L., & Thomas, A. (2011). Eyelid retraction in dementia with Lewy bodies and Parkinson's disease. *Journal of Neurology, 258*(8), 1542–1544.

Ouchi, H., Toyoshima, Y., Tada, M., Oyake, M., Aida, I., Tomita, I., Satoh, A., Tsujihata, M., Takahashi, H., Nishizawa, M., & Shimohata, T. (2014). Pathology and sensitivity of current clinical criteria in corticobasal syndrome. *Movement Disorders, 29*(2), 238–244.

Pariseau, B., Worley, M. W., & Anderson, R. L. (2013). Myectomy for blepharospasm. *Current Opinion in Ophthalmology, 24*(5), 488–493.

Paulson, H. L. (2009). The spinocerebellar ataxias. *Journal of Neuro-Ophthalmology, 29*(3), 227–237.

Pavone, P., Barbagallo, M., Parano, E., et al. (2005). Clinical heterogeneity in familial congenital ptosis: Analysis of fourteen cases in one family over five generations. *Pediatric Neurology, 33*, 251–254.

Piccione, F., Mancini, E., Tonin, P., & Bizzarini, M. (1997). Botulinum toxin treatment of apraxia of eyelid opening in progressive supranuclear palsy: Report of two cases. *Archives of Physical Medicine and Rehabilitation, 78*(5), 525–529.

Porter, J. D., Burns, L. A., & May, P. J. (1989). Morphological substrate for eyelid movements: Innervation and structure of primary levator palpebrae superioris and orbicularis oculi muscles. *The Journal of Comparative Neurology, 287*(1), 64–81.

Poulton, J., Harley, H. G., Dasmahapatra, J., Brown, G. K., Potter, C. G., & Sykes, B. (1995). Mitochondrial DNA does not appear to influence the congenital onset type of myotonic dystrophy. *Journal of Medical Genetics, 32*(9), 732–735.

Quartarone, A., Sant'Angelo, A., Battaglia, F., Bagnato, S., Rizzo, V., Morgante, F., Rothwell, J. C., Siebner, H. R., & Girlanda, P. (2006). Enhanced long-term potentiation-like plasticity of the trigeminal blink reflex circuit in blepharospasm. *The Journal of Neuroscience, 26*(2), 716–721.

Ramasamy, B., Rowe, F., Freeman, G., Owen, M., & Noonan, C. (2007). Modified Lundie loops improve apraxia of eyelid opening. *Journal of Neuro-Ophthalmology, 27*(1), 32–35.

Rana, A. Q., Kabir, A., Dogu, O., Patel, A., & Khondker, S. (2012). Prevalence of blepharospasm and apraxia of eyelid opening in patients with parkinsonism, cervical dystonia, and essential tremor. *European Neurology, 68*(5), 318–321.

Robinson, D. A. (1970). Oculomotor unit behavior in the monkey. *Journal of Neurophysiology, 33*(3), 393–403.

Rottach, K. G., Das, V. E., Wohlgemuth, W., Zivotofsky, A. Z., & Leigh, R. J. (1998). Properties of horizontal saccades accompanied by blinks. *Journal of Neurophysiology, 79*(6), 2895–2902.

Rucker, J. C. (2011). Normal and abnormal lid function. *Handbook of Clinical Neurology, 102*(3), 403–424.

Safran, A. B., Berney, J., & Safran, E. (1982). Convergence-evoked eyelid nystagmus. *AJO, 93*(1), 48–51.

Sanders, M. D., Hoyt, W. F., & Daroff, R. B. (1968). Lid nystagmus evoked by ocular convergence: An ocular electromyographic study. *JNNP, 31*(4), 368–371.

Schmidtke, N., & Buttner-Ennever, J. A. (1992). Nervous control of eyelid function: A review of clinical, experimental, and pathological data. *Brain, 115*, 227–247.

Schultz, K. P., Williams, C. R., & Busettini, C. (2010). Macaque pontine omnipause neurons play no direct role in the generation of eye blinks. *Journal of Neurophysiology, 103*(4), 2255–2274.

Schwingenschuh, P., Katschnig, P., Edwards, M. J., Teo, J. T., Korlipara, L. V., Rothwell, J. C., & Bhatia, K. P. (2011). The blink reflex recovery cycle differs between essential and presumed psychogenic blepharospasm. *Neurology, 76*(7), 610–614.

Sforza, C., Rango, M., Galante, D., Bresolin, N., & Ferrario, V. F. (2008). Spontaneous blinking in healthy persons: An optoelectric study of eyelid motion. *Ophthalmic & Physiological Optics, 28*(4), 345–353.

Skarf, B. (2004). Normal and abnormal eyelid function. In N. R. Miller et al. (Eds.), *Walsh & Hoyt's clinical neuro-ophthalmology* (pp. 1177–1229). Philadelphia: Lippincott, Williams, & Wilkins.

Smith, D., Ishikawa, T., Dhawan, V., Winterkorn, J. S., & Eidelberg, D. (1995). Lid-opening apraxia is associated with medial frontal hypometabolism. *Movement Disorders, 10*(3), 341–344.

SooHoo, J. R., Davies, B. W., Allard, F. D., & Durairaj, V. D. (2014). Congenital ptosis. *Survey of Ophthalmology, 59*, 483–492.

Steeves, T. D., Day, L., Dykeman, J., Jette, N., & Pringsheim, T. (2012). The prevalence of primary dystonia: A systematic review and meta-analysis. *Movement Disorders, 27*(14), 1789–1796.

Surwit, R. S., & Rotberg, M. (1984). Biofeedback therapy of essential blepharospasm. *American Journal of Ophthalmology, 98*(1), 28–31.

Suzuki, Y., Kiyosawa, M., Ohno, N., Michizuki, M., Inaba, A., Mizusawa, H., Ishii, K., & Senda, M. (2003). Gluocse hypometablism in medial frontal cortex of patients with apraxia of lid opening. *Graefe's Archive for Clinical and Experimental Ophthalmology, 241*(7), 529–534.

Svetel, M., Pekmezovic, T., Tomic, A., Kresojevic, N., & Kostic, V. S. (2015). The spread of primary late-onset focal dystonia in a long-term follow up study. *Clinical Neurology and Neurosurgery, 132*, 41–43.

Taylor, R. W., & Turnbull, D. M. (2005). Mitochondrial DNA mutations in human disease. *Nature Reviews. Genetics, 6*(5), 389–402.

Taylor, J. R., Elsworth, J. D., Lawrence, M. S., Sladek, J. R., Jr., Roth, R. H., & Redmond, D. E., Jr. (1999). Spontaneous blink rates correlate with dopamine levels in the caudate nucleus of MPTP-treated monkeys. *Experimental Neurology, 158*(1), 214–220.

Tharp, J. A., Wendelken, C., Mathews, C. A., Marco, E. J., Schreier, H., & Bunge, S. A. (2015). Tourette syndrome: Complementary insights from measures of cognitive control, eyeblink rate, and pupil diameter. *Frontiers in Psychiatry, 6*, 95.

Tisch, S., Limousin, P., Rothwell, J. C., Asselman, P., Quinn, N., Jahanshahi, M., Bhatia, K. P., & Hariz, M. (2006). Changes in blink reflex excitability after globus pallidus internus stimulation for dystonia. *Movement Disorders, 21*(10), 1650–1655.

Tokisato, K., Fukunaga, K., Tokunaga, M., Watanabe, S., Nakanishi, R., & Yamanaga, H. (2015). Aripiprazole can improve apraxia of eyelid opening in Parkinson's disease. *Internal Medicine, 54*(23), 3061–3064.

Tolosa, E., Montserrat, L., & Bayes, A. (1988). Blink reflex studies in patients with focal dystonias. *Advances in Neurology, 50*, 517–524.

Tommasi, G., Krack, P., Fraix, V., & Pollak, P. (2012). Effects of varying subthalamic nucleus stimulation on apraxia of lid opening in Parkinson's disease. *Journal of Neurology, 259*(1944), 2592.

Tozlovanu, V., Forget, R., Iancu, A., & Boghen, D. (2001). Prolonged orbicularis oculi activity: A major factor in apraxia of lid opening. *Neurology, 57*(6), 1013–1018.

Umemura, A., Toyoda, T., Yamamoto, K., Oka, Y., Ishii, F., & Yamada, K. (2008). Apraxia of eyelid opening after subthalamic deep brain stimulation may be caused by reduction of levodopa. *Parkinsonism & Related Disorders, 14*(8), 655–657.

Valls-Sole, J., & Defazio, G. (2016). Blepharospasm: Update on epidemiology, clinical aspects, and pathophysiology. *Frontiers in Neurology, 7*(45), 1–8.

VanderWerf, F., Brassinga, P., Reits, D., et al. (2003). Eyelid movements: Behavioral studies of blinking in humans under different stimulus conditions. *Journal of Neurophysiology, 89*, 2784–2796.

Verhulst, S., Smet, H., De Wilde, F., & Tassignon, M. J. (1994). Levator aponeurosis dehiscence in a patient treated with botulinum toxin for blepharospasms and eyelid apraxia. *Bull Soc Belge Ophthalmol, 252*, 51–53.

Verma, R., Lalla, R., & Patil, T. B. (2012). Is blinking of the eyes affected in extrapyramidal disorders? An interesting observation in a patient with Wilson disease. *BML Case Reports*. https://doi.org/10.1136/bcr-2012-007367.

Weiss, E. M., Hershey, T., Karimi, M., Racette, B., Tabbal, S. D., Mink, J. W., Paniello, R. C., & Perlmutter, J. S. (2006). Relative risk of spread of symptoms among the focal onset dystonias. *Movement Disorders, 21*(8), 1175–1181.

Weiss, D., Wächter, T., Breit, S., Jacob, S. N., Pomper, J. K., Asmus, F., Valls-Solé, J., Plewnia, C., Gasser, T., Gharabaghi, A., & Krüger, R. (2010a). Involuntary eyelid closure after STN-DBS: Evidence for different pathophysiological entities. *JNNP, 81*(9), 1002–1007.

Weiss, D., Wächter, T., Breit, S., et al. (2010b). Involuntary eyelid closure after STN-DBS: Evidence for different pathophysiological entities. *Journal of Neurology, Neurosurgery, and Psychiatry, 81*, 1002–1007.

Wong, K. T., Dick, D., & Anderson, J. R. (1996). Mitochondrial abnormalities in oculopharyngeal muscular dystrophy. *Neuromuscular Disorders, 6*(3), 163–166.

Xing, S., Chen, L., Chen, X., Pei, Z., Zeng, J., & Li, J. (2008). Excessive blinking as an initial manifestation of juvenile Huntington's disease. *Neurological Sciences, 29*(4), 275–277.

Yamada, S., Matsuo, K., Hirayama, M., & Sobue, G. (2004). The effects of levodopa on apraxia of lid opening: A case report. *Neurology, 62*(5), 830–831.

Yamada, K., Shinojima, N., Hamasaki, T., & Kuratsu, J. (2016). Pallidal stimulation for medically intractable blepharospasm. *BML Case Reports.* https://doi.org/10.1136/bcr-2015-214241.

Yoon, W. T., Chung, E. J., Lee, S. H., Kim, B. J., & Lee, W. Y. (2005). Clinical analysis of blepharospasm and apraxia of eyelid opening in patients with parkinsonism. *Journal of Clinical Neurology, 1*(2), 159–165.

Yuksel, D., Orban de Xivry, J. J., & Lefevre, P. (2010). Review of the major findings about Duane retraction syndrome (DRS) leading to an updated form of classification. *Vision Research, 50*(23), 2334–2347.

Yum, K., Wang, E. T., & Kalsotra, A. (2017). Myotonic dystrophy: Disease repeat range, penetrance, age of onset, and relationship between repeat size and phenotypes. *Current Opinion in Genetics & Development, 44*, 30–37.

Zee, D. S., Chu, F. C., Leigh, R. J., Savino, P. J., Schatz, N. J., Reingold, D. B., & Cogan, D. G. (1983). Blink-saccade dynkinesis. *Neurology, 33*(9), 1233–1236.

Zhou, B., Wang, J., Huang, Y., Yang, Y., Gong, Q., & Zhou, D. (2013). A resting state functional magnetic resonance imaging study of patients with benign essential blepharospasm. *Journal of Neuro-Ophthalmology, 33*(3), 235–240.

Zuhlke, C., Mikat, B., Timmann, D., Wieczorek, D., Gillessen-Kaesbach, G., & Burk, K. (2015). Spinocerebellar ataxia 28: Novel AFG3L2 mutation in a German family with young onset, slow progression and saccadic slowing. *Cerebellum Ataxias, 2*, 19.

Eye Movements in Autosomal Dominant Spinocerebellar Ataxias

Alessandra Rufa and Francesca Rosini

Abstract Spinocerebellar ataxias (SCAs) are clinically heterogeneous forms of degenerative disorders with autosomal dominant pattern of inheritance. Among more than 40 forms of SCAs, each may have peculiar neurological constellation. Historically, the study of eye movements has played a fundamental role in the identification of various forms of neurological disorders, and utilizing the eye movements to differentiate various forms of SCAs remains most critical. In this chapter, our goal is to elucidate the eye movement abnormalities in autosomal-dominant spinocerebellar ataxias through a systematic review of the literature. Clinical, genetic, and neuropathological/neuroimaging aspects of the SCAs are also briefly discussed.

Keywords Ataxia · Apraxia · Neurodegeneration · Eye movement · Cerebellum

Abbreviations

GEN	gaze-evoked nystagmus
PAN	periodic alternating nystagmus
SWJs	square wave jerks
SWOs	square wave oscillations
VOR	vestibulo-ocular reflex

A. Rufa (✉) · F. Rosini
Eye Tracking & Visual Application Lab EVALAB, Department of Medicine Surgery and Neuroscience, Neurology and Neurometabolic Unit, University of Siena, Siena, Italy
e-mail: rufa@unisi.it

© Springer Nature Switzerland AG 2019   415
A. Shaikh, F. Ghasia (eds.), *Advances in Translational Neuroscience of Eye Movement Disorders*, Contemporary Clinical Neuroscience,
https://doi.org/10.1007/978-3-030-31407-1_21

1 Introduction

Spinocerebellar ataxias (SCAs) are a clinically heterogeneous group of hereditary late onset, neurodegenerative disorders, with an autosomal dominant pattern of inheritance. In Europe, SCAs have estimated prevalence of 0.8–3.0/1,00,000 population (van de Warrenburg et al. 2003; Klockgether 2008). The prevalence worldwide, however, shows differences among geographical areas due to a founder effect of the gene. In this respect, SCA2 is higher in Cuba, SCA3 in Azores, and DRPLA in Japan. SCA1, 2, 3, and 6 are the most frequently diagnosed SCAs, all over the world, among these, SCA 3 is the commonest type worldwide. There are approximately 42 clinically different types of SCA identified so far. To date, 22 different genes (SCA1, 2, 3, 5, 6, 7, 8, 10, 11, 12, 13, 14, 15, 17, 22, 23, 27, 28, 31, 35, 36, 37, 38, 40, 41, 42, 43, 44, 45, 46, 47 and dentatorubropallidoluysian atrophy [DRPLA]) and additionally 10 different gene loci (SCA 4, 18, 19, 20, 21, 25,26, 29, 30, and 32) are identified. SCA1, 2, 3, 6, 7, and 17 are translated CAG repeat expansions coding for an elongated polyglutamine tract within the respective proteins; they belong to the group of polyglutaminopathies, also including Huntington's disease, DRPLA, and spinobulbar muscular atrophy (Bettencourt et al. 2016; Jones et al. 2017). SCA8, 10, and 12 are due to untranslated repeat expansions in noncoding regions of the genes. SCA5, SCA13, SCA14, SCA27, and 16q22-linked SCA are associated with point mutations. SCA4 and 16q22-linked SCA (SCA31) have been mapped to the same chromosomal region, but no mutations have been found in the original SCA4 family (Manto 2010; Klockgether 2008) (Figs. 1 and 2).

The absence of a familiar history does not rule out a genetic ataxia. Patient may present with a de novo mutation, or affected relatives may have died before the onset of symptoms. Due to the anticipation phenomenon, very common in the dominant ataxias, the elderly relatives carrying the mutation might not have fully developed the symptoms or may be clinically unaffected. Moreover, false attribution of paternity should be taken into account.

Clinical presentation of dominant ataxia is extremely heterogeneous. Common clinical findings usually include a late onset presentation, typically around the third or fourth decade (McIntosh et al. 2017). In addition, all patients exhibit a slowly progressive cerebellar syndrome with various combinations of oculomotor disorders, dysarthria, dysmetria/kinetic tremor, and/or ataxic gait. Diagnosis is often complicated by the possible concomitance of pigmentary retinopathy, extra-pyramidal disorders and various movement disorders (parkinsonism, dyskinesias, dystonia, chorea), pyramidal signs, cortical symptoms (seizures, cognitive impairment/ behavioral symptoms), and peripheral neuropathy (Manto 2010). Some clinical signs/symptoms may provide a clue to address the diagnosis. In this regard, pure cerebella ataxia is typical of SCA6, ophthalmoplegia is quite specific of SCA28, saccade slowing is an hallmark of SCA2; pigmentary retinopathy is usually seen in SCA7. Spasticity can be encountered in SCA3. Dyskinesias and behavioral changes are typical of SCA17 and DRPLA; seizures are often seen in SCA10, SCA17, and DRPLA (Manto 2010).

Fig. 1 MRI of a 42-year-old patient with SCA2, showing cerebellar atrophy

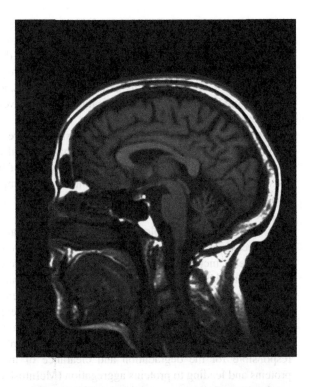

Fig. 2 MRI of a 60-year-old patient with SCA8

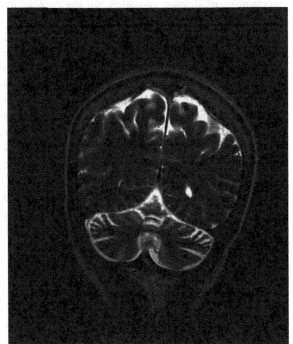

Overall, eye movement abnormalities are encountered in almost all SCAs, though they are, with some important exceptions, not specific. Usually, they are related to the cerebellar impairment, such as saccadic dysmetria, changes of pursuit gain, and various types of nystagmus. However, progressive atrophy of pons is responsible for extreme slowing of saccades in SCA2, and, to a much lesser extent, of SCA1 and SCA3. Analogously, frontal cortex involvement is thought to be linked to increased latency in SCA2.

In this respect, brain MRI of SCAs can show a pure cerebellar atrophy, a pattern of olivopontocerebellar atrophy and a global cerebral atrophy. In most cases, the involvement of extracerebellar structures is typical of the later stage of disease of most of the SCAs, with some rare exceptions (for example, SCA6, whose degeneration is usually confined to cerebellum). SCA1, SCA2, and SCA7 are characterized by olivopontocerebellar atrophy (Table 1).

A hallmark neuropathological feature of polyglutamine ataxias (and polyglutamine diseases in general) is the presence of disease proteins aggregation and neuronal nuclear inclusions, also containing other proteins such as chaperones, proteasomal subunits, and transcription factors (Schmidt et al. 2002).

Over recent years, it was observed that many genetic ataxias share a similar mechanism of disease causation. Particularly, the expansions of polyglutamine tracts are associated with transcriptional dysregulation, due to compromission of the ubiquitin-proteasome system and macroautophagy (autophagy) process, responsible for the degradation and clearance of misfolded and/or superfluous proteins and leading to proteins aggregation (McIntosh et al. 2017). Other potential mechanism of disease causation are the protein misfolding resulting in altered function, the formation of toxic oligomeric complexes, mitochondrial dysfunction, impaired axonal transport, aberrant neuronal signaling including excitotoxicity, cellular protein homeostasis impairment, and RNA toxicity (Shakkottai and Fogel 2013; Williams and Paulson 2008). Either direct ion-channel mutations or secondary ion-channel dysfunction has been implicated in the pathogenesis of SCA5, SCA6, SCA13, SCA15/16, SCA19/22, and SCA27 (Shakkottai and Paulson 2009; Chen et al. 2003), while mutations in signal transduction molecules are the direct cause of disease in SCA11, SCA12, SCA14, and SCA23 (Shakkottai and Fogel 2013). The putative mechanism of disease in SCA8, SCA10, SCA31, and SCA36 includes transcriptional alterations and the generation of antisense transcripts, sequestration of mRNA-associated protein complexes that lead to aberrant mRNA splicing, and processing and alterations in cellular processes, including activation of abnormal signaling cascades and failure of protein quality control pathways (Todd and Paulson 2010).

2 Spinocerebellar Ataxia Type 1

SCA1 is observed in Italian, South African, Australian, German, Chinese, Korean, Japanese, and Spanish populations (Sun et al. 2016).

Table 1 Eye movement change in SCAs

Subtype	Genetic Change	Eye Movement Abnormalities
SCA1	*Ataxin 1*	Saccades hypo/hypermetria, slow/normal speed Impaired fixation with saccadic intrusions, or oscillations, GEN impaired VOR
SCA2	*Ataxin 2*	Slow saccades, hypometric saccades, Ny, increased latency, and high direction errors of antisaccades
SCA3	*Ataxin 3*	Slow and dysmetric saccades, ophthalmoplegia, impaired VOR, ny, and saccadic intrusion or oscillations
SCA4	*16q32*	Slow saccades, ophthalmoparesis, impaired VOR, ny
SCA5	*βIII-Spectrin*	Slow saccades, fragmented pursuit, ny
SCA6	*CACNA1A*	GEN, rebound ny, PAN, impaired pursuit and VOR, fast saccades
SCA7	*Ataxin 7*	Slow saccades, impaired pursuit and VOR, saccadic intrusions
SCA8	*Ataxin 8*	Slow saccades, ophthalmoparesis and diplopia, fast downward saccades, ny, saccadic pursuit, SWJ
SCA10	*Ataxin 10*	Ny, SWJ and other ocular oscillations, abnormal pursuit, hypometric saccades
SCA11	*Tubulin kinase-2*	GEN, downbeat ny, fragmented pursuit, and saccades
SCA12	*PPP2R2B*	Ny, fast saccades, abnormal VOR
SCA13	*KCNC3*	GEN
SCA14	*PRKCG*	GEN and SWJ, slow and hypometric saccades
SCA15/16	*ITPR1*	GEN, VOR changes
SCA17	*TATA-box binding protein*	Hypometric saccades, errors and increased latency in antisaccades, abnormal pursuit, GEN and rebound ny, abnormal VOR
SCA 18	*IFRD1*	GEN
SCA19/22	*KCND3*	Normal speed saccades, dysmetric saccades, abnormal pursuit, GEN
SCA 20	*11q22.2-11q12.3*	Fast and hypermetric saccades, GEN, downbeat ny, SWJ
SCA 21	*TMEM240*	SWJ and pursuit changes, normal sacacdes
SCA 23	*PDYN on 20p13-12.3*	Slow saccades and GEN
SCA 25	*2p15-p21*	GEN, downbeat ny, hypermetric saccades, saccadic intrusions
SCA 26	*19p13.3*	Dysmetric saccades, abnormal VOR gain and pursuit
SCA27	*Fibroblast growth factor-14*	Impired saccades and pursuit, ny
SCA 28	*AFG3L2*	Slow saccades, ophthalmoparesis, palpebral ptosis, pursuit and OKN impairment
SCA 29	*ITPR1*	GEN and fixation instability
SCA 30	*4q34.3-q35.1.*	Hypermetric saccades, ny
SCA 31	*ODZ3*	GEN
SCA 34	*ELOVL4*	GEN, supranuclear ophthalmoplegia, impaired smooth pursuit
SCA 35	*20p13-12.2.*	Slow saccades
SCA 36	*NOP56*	Impaired pursuit, GEN, supranuclear palsy
SCA 37	*DAB1*	Vertical gaze impairment, GEN, SWJ

(continued)

Table 1 (continued)

Subtype	Genetic Change	Eye Movement Abnormalities
SCA 38	*ELOVL5*	Slow saccades, GEN
SCA 40	*CCDC88C*	Vertical gaze restriction, saccade hypometria
SCA 42	*CACNA1G*	Saccadic pursuit, diplopia, GEN
SCA 43	*MME*	Ny
SCA 44	*GRM1*	Saccadic hypermetria, abnormal pursuit
SCA 45	*FAT2*	Ny
SCA 46	*PLD3*	Slow saccades, ny, saccadic pursuit, SWJ
SCA 47	*PUM1*	Ny
DPRLA	*ATN1*	Saccadic apraxia, slow saccades, ophthalmoplegia, saccadic pursuit, ny, SWJ

VOR Vestibulo-ocular reflex, *GEN* Gaze-evoked nystagmus, *NY* Nystagmus, *PAN* Periodic alternating nystagmus, *SWJ* Square wave jerks, *OKN* Optokinetic nystagmus

SCA1 gene maps at the locus 6p23; the number of CAG repeats ranges from 6 to 44 in the general population, being highly polymorphic (Manto 2010). SCA1 patients typically have expansions between 39 and 82 repeats (Orr et al. 1993). The length of the repeat, the age of onset, and the severity of the disease are strictly correlated: For longer repeats, the onset occurs earlier and the phenotype is more severe. This is due to the dynamic aspect of the mutation during transmission and is especially evident for paternal transmission (Goldfarb et al. 1996). The CAG repeat in the SCA1 gene encodes a polyglutamine tract. Both SCA1 mRNA and the protein ataxin-1 are ubiquitously expressed. This protein is localized mainly in the nuclei of Purkinje and brainstem neurons.

The onset of the disease usually occurs during the third or fourth decade of life; anticipation phenomenon is very common, successive generations tend to present with an earlier onset and more severe manifestations (Orr et al. 1993). Cerebellum and brainstem-related symptoms are the most common clinical presentation, but patients may exhibit also extrapyramidal signs and peripheral neuropathy. Disease progression takes 10–15 years, with a gradual worsening of the ataxia and of the bulbar involvement. Neuropathology studies have demonstrated a progressive severe neuronal loss in the cerebellum (cerebellar cortx and dentate nuclei) and pons. Neuroimaging confirms the pontocerebellar atrophy, and sometimes the presence of a midline hyperintensity of the pons on T2-weighted sequences has been observed (Adachi et al. 2006).

Oculomotor abnormalities, even though not specific, have often been described in SCA1, particularly, saccadic dysmetria (Kim et al. 2013a, b; Teive et al. 2012; Bürk et al. 1999), resulting in both hypometria (Alexandre et al. 2013) or hypermetria (Rivaud-Pechoux et al. 1998; Vale et al. 2010). Reduced saccadic speed has been inconstantly observed (Alexandre et al. 2013, Buttner et al. 1998, Klostermann et al. 1997), generally with latency preservation (Alexandre et al. 2013). As a consequence of the brainstem involvement, both horizontal and vertical ophthalmoparesis may be present (Klostermann et al. 1997), as well as gaze-evoked nystagmus, expression of the gaze holding impairment (Kim et al. 2013b).

Gaze steady fixation may be interrupted by saccadic intrusions, both SWJs and SO (Kim et al. 2013a). Impaired vestibulocular reflex and optokinetic nystagmus have also been reported (Manto 2010).

3   Spinocerebellar Ataxia Type 2

Epidemiologically, SCA2 is more often seen in Italy, India, Mexico, and Cuba.

It is caused by an expansion of an unstable CAG repeat in the ataxin-2 gene (12q24.1). The CAG trinucleotide repeat is not highly polymorphic in normal individuals, ranging from 15 to 32 repeats. A number of 33 CAG repeats is sufficient to cause the disease and, as for SCA1, there is a clear correlation between the length of the repeat and the age of onset (McIntosh et al. 2017). The polyQ protein affected in SCA2, ATXN2, is implicated in a range of biological functions, including cell specification, actin filament formation, receptor-mediated signaling, and secretion, probably due to its role in cytoplasmic RNA-related functions, in particular the regulation of translation (Orr 2012).

Neuropathological characteristic of the disease is the extensive atrophy of cerebellum and pons, more severely than in SCA1, with dilation of the fourth ventricle. Often, atrophy progression may involve the frontotemporal lobes. Neuronal loss is localized principally in the cerebellar Purkinje cells, dentate nuclei, and substantia nigra. Demyelination may occur in the posterior columns of the spinal cord and spinocerebellar tract (Scherzed et al. 2012; Seidel et al. 2012).

Given this progressive widespread involvement, clinical symptoms may vary, but all patients manifest with cerebellar ataxia. Often, a subclinical peripheral neuropathy may occur; seldom, a hyperreflexia due to pyramidal tract involvement may be observed. Some families present with parkinsonism or amyotrophic lateral sclerosis phenotypes (Pulst 2016). In other individuals, a cognitive decline may develop. Rufa et al (2002) reported pigmented retinopathy in a SCA2 family.

Besides cerebellar ataxia, the hallmark of the disease is the severe slowing of ocular movements, which has been observed in up to 90% of patients, mostly due to the impairment of the excitatory burst neurons in PPRF (Magaña et al. 2013; Velázquez-Pérez et al. 2004; Wadia and Swami 1971; Zee et al. 1976a, b; Bürk et al. 1999; Federighi et al. 2011; Vale et al. 2010). In a 5-year prospective follow-up study, Rodriguez-Labrada et al. (Rodríguez-Labrada et al. 2016) observed that the progression rate of saccade slowing was influenced by expansion size. Few studies also reported hypometric saccades, nystagmus, and SWJs intrusions in SCA2 patients (Teive et al. 2012; Kim et al. 2013a, b; Moscovich et al. 2015), though at a lower extent than other cerebellar patients. In this regard, (Federighi et al. 2011; Rufa and Federighi 2011) hypothesized that SCA2 patients could more easily maintain the accuracy of the movement due to their slow velocity and longer duration, allowing a visual reference during the execution of movement that may guide the eye landing to the target. Increased saccadic latency has been correlated with executive functions but not with demographic, clinical, or molecular variables, in Cuban population (Rodríguez-Labrada et al. 2011). An executive deficit for SCA2 patients was also demonstrated in the antisaccade task by the same authors (Rodríguez-Labrada et al. 2014), reporting a significant increase of antisaccadic errors with low correction rate and prolonged latency in respect to the normal population. Antisaccadic data were correlated with the number of CAG repeat expansion in this study. The authors hypothesized that antisaccadic changes in these patients are due to the spreading of atrophy in frontal/dorsolateral prefrontal cortex. In this respect, our group recently reported that in the antisaccade task, SCA2 patients had prolonged and variable latency and increased directional error rate than normal, but this behavior was not different from that of another cerebellar population with isolated cerebellar atrophy (LOCA, late onset cerebellar ataxia). Authors suggested a role for the cerebellum in the execution of voluntary movements by modulating its initiation and reducing reflexive responses that would perturb goal-directed actions (Pretegiani et al. 2018)

4 Spinocerebellar Ataxia Type 3

The frequency of SCA3, or Machado–Joseph disease (MJD), is high in France, USA, Japan, Portugal, Brazil, and China. In several studies, it is reported as the most common dominant ataxia worldwide (Coutinho and Andrade 1978; Sequeiros and Coutinho 1993).

The MJD1 gene of 1776 bp maps to the 14q32.1 region (Kawaguchi et al. 1994). The normal range of CAG repeats in the MJD1 gene is up to 47. The expanded repeat size is more than 44 (Padiath et al. 2005), with 86 being the largest expanded repeat reported. There is an overlap of normal repeat sizes (up to 47), with the smallest expanded repeat of 45 CAGs. Individuals with a repeat length in the overlapping region do not always present with the disease (reduced penetrance) (Riess et al. 2008). The instability of CAG expansion, especially for paternal transmission, causes anticipation in children from affected fathers. The age of onset is influenced not only by the triplet repeat length but also by additional factors such as homozygosity for the expanded CAG repeat, which predisposes to earlier age at onset (Lang et al. 1994), sex (the symptoms appear later in females), and probably lifestyle. The length of the repeats is associated with different phenotypes, being a length of CAG motifs of less than 73 related to peripheral neuropathy, and more than 73 to pyramidal symptoms. The product is the protein ataxin-3, an ubiquitin-binding protein, whose definite functions are still to be clarified, probably participating in proteasomal protein degradation (Wang et al. 2000).

Neuropathology of SCA3 is characterized by atrophy of cerebellum, pons and medulla oblongata, and cranial nerves from III to XII, as well as depigmentation of substantia nigra. The areas initially involved by degenerative process are the cerebellum, substantia nigra, and a subset of vestibular, oculomotor and precerebellar nuclei, while thalamus and the other interested areas become involved later (Scherzed et al. 2012).

The age of onset can be extremely variable, but often it ranges from 20 to 40 years of life. Clinical presentation is often heterogeneous, with a mix of cerebellar and extra-cerebellar deficits. Usually, first symptoms include slurred speech, gait ataxia, and visual blurring. On the basis of neurological onset and clinical presentation, a classification into four subtypes has been proposed:

- Type I: Early onset (10–30 years), extrapyramidal (dystonia), and pyramidal defects
- Type II: Intermediate onset (20–50 years), with pyramidal, and cerebellar deficits
- Type III: Late onset (40–75 years), cerebellar deficits, and peripheral neuropathy
- Type IV: Variable onset, peripheral neuropathy, and parkinsonism

Ocular movements are very frequently abnormal in SCA3 patients. One finding is represented by slowing of ocular movements (Teive et al. 2012), though usually not as severe as in SCA2; the neuroanatomical substrate is thought to be the involvement of the network including excitatory and inhibitory burst neurons and omnipause neurons in the brainstem (Rüb et al. 2003). Complete ophthalmoplegia and/or gaze paralysis have often been reported (Teive et al. 2012; Moscovich et al. 2015; Vale et al. 2010).

Alexandre et al. (2013) found normal saccade velocity in a cohort of SCA3 patients, except in one case with increased mean velocity and an increase of intra-subject variability of amplitude. Saccadic dysmetria were found also by Kim et al. (Kim et al. 2013a, b) and Moscovich et al. (2015).

Impairment of vestibulocular reflex or complete vestibular areflexia (Gordon et al. 2003; Buttner et al. 1998) is also described in SCA3.Very commonly reported are the gaze holding abnormalities, GEN (Moscovich et al. 2015; Kim et al. 2013a, b; Teive et al. 2012; Vale et al. 2010), and the impairment of steady fixation with saccadic intrusions, including SWJs and SWOs (Kim et al. 2013a, b).

5 Spinocerebellar Ataxia Type 4

SCA4 has been firstly reported in a pedigree from USA of Scandinavian origin (Flanigan et al. 1996). Age of onset is variable, with a peak in fourth or fifth decade.

SCA4 maps to chromosome 16q22.1; this locus overlaps with that of a possibly related endemic cerebellar syndrome described in Japan, the 16q-ADCA or SCA31, which constitutes the third most common dominant ataxia form in Japan after MJD and SCA6. However, Edener et al. (2011) excluded a pathogenic pentanucleotide repeat in the BEAN gene, which causes SCA31, as a cause of SCA4 in the family reported by Hellenbroich et al. (2003), indicating that SCA4 and SCA31 are not allelic disorders. Cerebellar and brainstem neurodegeneration with marked neuronal loss, especially in the Purkinje cells layers and cerebellar nuclei (Hellenbroich et al. 2006), is the neuropathological characteristic of the disease. Degeneration is also observed in the posterior columns and roots of spinal cord.

Clinically, SCA4 is characterized by ataxia with prominent axonal sensory neuropathy. Full tendon areflexia is sometimes observed; more rarely a Babinski sign is present. The 16q-ADCA of Japan is often a pure ataxia syndrome, while rarely dementia and hearing impairment have been reported.

Ocular movement abnormalities are represented by saccade slowing and ophthalmoparesis, reduced VOR response, and GEN (Maschke et al. 2005).

6 Spinocerebellar Ataxia Type 5

SCA5 has been described in a population descending from US President Abraham Lincoln. Though the onset could be variable, symptoms in most cases begin during the third or fourth decades. Anticipation is common and early cases are usually maternally inherited (Ranum et al. 1994).

Deletions and missense mutations in the beta-III spectrin gene (STBN2) linked to chromosome 11q13 cause SCA5. Spectrin mutations affect glutamate signaling.

SCA5 is mainly a cerebellar degeneration. In this respect, clinical symptoms are constituted by gait and stance ataxia, and, to a lesser extent, by dysarthria and kinetic tremor.

Oculomotor abnormalities are especially characterized by downbeat nystagmus or GEN (Bürk et al. 2004). More rarely, saccadic dysmetria and saccade slowing have been reported.

7 Spinocerebellar Ataxia Type 6

SCA6 was commonly seen in Korean (15–23%), Japanese (6–23%), Dutch (11–23%), Australian (17%), and German populations (10–22%), and less frequently seen in the UK (5%), India (0–4%), and China (0–3%) (Sun et al. 2016). It is the only polyglutamate disease caused by mutation in a membrane protein (Voltage-gated calcium channel alpha 1 a subunit) (Zhuchenko et al. 1997; Mondal et al. 2013). It is caused by CAG triplet expansion in *CACNA1A* gene, which maps to 19p13. Normal alleles range from 4 to 18 units, with expanding alleles ranging from 20 to 30 repeats. The expanded CAG repeat number is inversely correlated with age at onset; homozygous patients show an earlier onset and a more rapid progression (Manto 2010). The size of SCA6 expanded alleles is generally stable during transmission, and anticipation is rare. A dosage effect was observed. Age of onset varies from 19 to 73 years, with a peak around 45 years.

Neuropathological studies have shown a severe loss of Purkinje cells and macroscopically a cerebellar atrophy. Atrophy of cerebral cortex, thalamus, midbrain, pons, medulla oblongata is occasionally reported (Seidel et al. 2012). In neuroimaging studies, however, atrophy prevails in the vermis, while it is less pronounced in the cerebellar hemispheres and not present in the middle cerebellar peduncles, pons, or other posterior fossa structures.

Usually, SCA6 is a pure cerebellar ataxia, with gait and limb ataxia, cerebellar dysarthria as the commonest symptoms complained by patients. The involvement of other structures, leading to pyramidal or extrapyramidal signs, is very rare.

Ocular movements speed is generally preserved. Saccades may show variable dysmetria, likely due to an impairment of cerebellar fastigial nucleus (Zhang et al. 2016), while ophthalmoparesis is rare (Moscovich et al. 2015). Diverse kinds of nystamus can be encountered: downbeat, upbeat, GEN, PAN, pendular (Gomez et al. 1997; Kim et al. 2013a, b; Zee et al. 1976a, b; Moscovich et al. 2015; Liang et al. 2016). Moreover, stability of gaze could be affected by intrusions of SWJs subtype (Kim et al. 2013a, b; Moscovich et al. 2015).

The VOR is usually hyperactive (Gomez et al. 1997), and the pursuit may be impaired.

8   Spinocerebellar Ataxia Type 7

SCA7 is rare. It is found in Swedish, Finnish, and South African (22.2%) populations as well as in Brazilian (6%), Spanish (1%), Portuguese (1%), and Australian (2%) populations (Sun et al. 2016). SCA7 is caused by a CAG triplet repeat expansion in gene *ATXN7*, mapping in chromosome 3p12-21.1. The gene product is the protein ataxin-7. Healthy individuals have a range of 7 to 34 CAGs triplets. Repeats over 37 cause the phenotype. The length of the repeats is correlated with an earlier onset, a more rapid progression and severity; a correlation exists between number of repeats and phenotype, as for 59 or under the onset is purely cerebellar, while above 59 the diseases starts with visual impairment (Johansson et al. 1998). Anticipation phenomenon is common, especially in male transmission.

Neuropathology of the disease is macroscopically characterized by atrophy of the cerebellar cortex, dentate nuclei, inferior olivary complex and olivocerebellar tract, spinocerebellar tract, subthalamic nucleus, and pyramidal tracts. Neuronal loss is found extensively in Purkinje cells, inferior olives, motor nuclei of cranial nerves, spinal motor neurons, and substantia nigra (Michalik et al. 2004). In addition, severe degeneration is observed in the retina (Seidel et al. 2012).

At neuroimaging, a marked atrophy of cerebellum and pons is evident.

SCA7 shows a variable clinical severity, ranging from infantile death (Benton et al. 1998) to slow progression in elderly. Cerebellar ataxia is the most common symptoms. Patients may complain about dysarthria, dysphagia, hypoacusia. Involvement of the visual system with pigmentary retinal dystrophy is frequent (Vale et al. 2010; Michalik et al. 2004).

As for the efferent system, saccades may show slowing (Schöls et al. 2008) and increased latency (Oh et al. 2001; Maschke et al. 2005), progressing to ophthalmoparesis (Teive et al. 2012). Usually, gaze holding and fixation are preserved, as well as VOR, though nystagmus has been reported in few SCA7 patients (Vale et al. 2010; Teive et al. 2012). Pursuit may be impaired.

9   Spinocerebellar Ataxia Type 8

SCA8 is usually an adult onset (fourth or fifth decade) neurodegenerative disease, however the age of onset may vary (Sun et al. 2016). Given the presence of extra-cerebellar features such as parkinsonism and migraine, its phenotype is more complex in comparison to other SCAs.

SCA8 is due to CTA and CTG triplet repeat expansion in *ATXN8OS/ATXN8/* Kelch-like (KLHL1) genes, located on chromosome 13q21. Normal allele range from 15 to 434 repeats, while expanded range from 89 to 155. The expansion is expressed in both direction, in coding and non-coding regions, suggesting a toxic gain-of-function pathogenesis of the disease (Manto 2010). Relationship between the expanded repeats and the age of onset, disease progression or disease severity is debated. SCA8 also show reduced penetrance, and the expansion carriers may remain asymptomatic for ataxia. Therefore, genetic counseling is critical to explain the results in asymptomatic individuals or family members (Sun et al. 2016).

Neuropathology of SCA8 macroscopically show a severe cerebellar atrophy, with loss of Purkinje cells and neuronal loss at the inferior olives, but relative sparing of dentate nuclei (Ito et al. 2006). Pancerebellar atrophy is observed at brain MRI.

Symptoms at onset are often dysarthria and gait instability. Patients may then complain of cognitive decline, rapidly progressive parkinsonism resembling cortico-basal syndrome, limb spasticity, hyperreflexia, myoclonic jerks at arms and fingers and athetotic movements, and rarely of an amyotrophic lateral sclerosis-like phenotype.

Ocular motor examination can point out a saccadic dysmetria (downward hypermetria). Rarely saccades are slow or there is ophthalmoparesis. GEN nystamus may be encountered and fixation is usually disturbed by PAN or SWJs. Diplopia is often complained by patients (Zee and Leigh 2015; Maschke et al. 2005; Ikeda et al. 2000).

10   Spinocerebellar Ataxia Type 10

SCA10 occurs in Latino-American populations, being the second most common SCA in Mexico (after SCA2) and Brazil (after SCA3).

It is caused by a repeat expansion primarily composed of ATTCT pentanucleo-tide repeats. It is localized in intron 9 of *ATXN10* gene on chromosome 22q13.3 (previously known as *E46L* gene). SCA10 is the only known human disease caused by an expansion of pentanucleotide repeats. The number of repeats is polymorphic and ranges from 10 to 29 in normal population, while the expanded allele range from 800 to 4500. The length of repeat expansion is inversely correlated with age at onset (Matsuura et al. 2000). The repeats show instability during paternal, but not maternal, transmission. The gene product is a protein called ataxin-10.

Clinically, cardinal features of SCA10 are represented by cerebellar syndrome, including gait ataxia, dysarthria, slurred speech, kinetic tremor, and oculomotor abnormalities described below, and epilepsy, which is reported only in pedigree of Mexican origin and comprehends generalized or partial motor seizures (Rasmussen et al. 2001). Rarely, mild peripheral neuropathy or hyperreflexia have been reported (Rasmussen et al. 2001). Neuroimaging shows pancerebellar atrophy.

Ocular motor involvement is wide. Saccades are generally hypometric (Teive et al. 2010, 2012; Rasmussen et al. 2001; Lin and Ashizawa 2005) but of normal speed. Gaze-evoked nystagmus is described (Teive et al. 2010). Fixation may be disturbed by ocular oscillations. Pursuits are often impaired.

11   Spinocerebellar Ataxia Type 11

SCA11 is caused by mutations of the gene encoding tau tubulin kinase-2 protein (TTK2) located at chromosome 15q14-21.3. Insertions or deletions have been reported.

Onset occurs usually during the second decade, progression is slow and life expectancy is normal. SCA11 is a relatively pure cerebellar disease, confirmed neuropathologically by an extensive cerebellar atrophy with severe loss of Purkinje cells. At brain MRI, besides cerebellar vermis atrophy, a slight atrophy of pons and medulla has been observed.

Clinical features are those of cerebellar involvement, with gait and truncal ataxia, dysarthria.

Oculomotor examination shows pursuit impairment, GEN or downbeat nystagmus are common as well as saccadic abnormalities (Manto 2010; Worth et al. 1999; Johnson et al. 2008; Zee and Leigh 2015).

12   Spinocerebellar Ataxia Type 12

SCA12 is very rare worldwide. It has been detected in Indian families, American families of German origin, Chinese, Singaporean, and Italian families. It is caused by CAG repeat expansion in gene *PPP2R2B*, mapped to 5q31-q33. Repeats ranges from 55 to 78 in affected, from 9 to 28 in normal alleles (Holmes et al. 1999). The age of onset varies from 9 to 55 years. Typical clinical feature is action tremor, beginning in the fourth decade, variably associated with head tremor, limb ataxia, dysmetria, hyperreflexia, and dementia in advanced cases. Some families showed psychiatric symptoms (O'Hearn et al. 2001). MRI of the brain showed cortical or cerebellar atrophy and white matter hyperintensities. Typical oculomotor abnormalities include dysmetric saccades, impaired pursuit, and cerebellar nystagmus (O'Hearn et al. 2001).

13   Spinocerebellar Aataxia Type 13

SCA13 is due to conventional mutations in the voltage-gated potassium channel *KCNC3* (chromosome 19q13.3-q13.4) (Herman-Bert et al. 2000). The p.R420H (c.G1260A) mutation in *KCNC3* is associated with late-onset ataxia, whereas the p.F448 L (c.

C1344A) and p.R423H (c.G1268A) mutations are related to early-onset ataxia, delayed motor milestones, mental retardation, and epilepsy (Sun et al. 2016; Subramony et al. 2013; Figueroa et al. 2010, 2011). KCNC3 is expressed in Purkinje cells, cerebellar granule cells, and cerebellar neurons. It is supposed that SCA13 mutations impair the firing properties of fast-spiking cerebellar neurons and influence neuronal function, and also interfere with the capacity of cerebellar neurons to handle oxidative stress (Waters et al. 2006). Neuroimaging show usually cerebellar hypoplasia for the early-onset subtypes and progressive cerebellar atrophy for the adult-onset.

Among ocular motor abnormalities, no saccades or pursuit changes have so far been reported, while GEN has been described in the R448L and R423H phenotypes (Herman-Bert et al. 2000; Khare et al. 2017).

14 Spinocerebellar Aataxia Type 14

SCA14 has been described in Japan and European families (Hiramoto et al. 2006; van de Warrenburg et al. 2003; Chelban et al. 2018). It is caused by conventional, usually missense, mutations in the protein kinase C gamma *(PRKCG)* gene in chromosome 19q13.4-qter.

The age of onset may vary, but usually the disease manifests in adulthood, around the third decade. (Yamashita et al. 2000; Chen et al. 2003). Clinical characteristics are extremely heterogeneous even among the same family. Generally, the onset symptom is instability of gait, with slow progression. Other features as myoclonus, dystonia, parkinsonian syndrome, reduced tendon reflexes, cognitive deficits, epilepsy have been reported (Stevanin et al. 2004a, b). Brain MRI shows pure cerebellar atrophy.

Ocular motor examination frequently shows gaze-evoked nystagmus and/or saccadic intrusions of SWJs type (Hiramoto et al. 2006; van de Warrenburg et al. 2003; Chen et al. 2005a). Occasionally, slow and hypometric saccades have been observed.

15 Spinocerebellar Ataxia Type 15/16

SCA15 is a very rare dominant cerebellar ataxia, originally described in an Australian kindred (Gardner et al. 2005) and later observed in European and Japan families. In the latter, it was firstly identified as SCA16 (Iwaki et al. 2008). SCA15 and SCA16 are associated with deletions of the inositol 1,4,5-triphosphate receptor (ITPR1) gene.

Age of onset may vary from seven to 72 years. Major clinical finding is a very slow progressive cerebellar ataxia, with associated postural and kinetic tremor. Rarely, orolingual dyskinesias or mioclonus have been described. Japanese pedigree showed pyramidal signs (Hara et al. 2008). Cognitive impairment may also be present. However, the SCA15 is a relatively pure cerebellar ataxia, confirmed at neuroimaging (cerebellar atrophy, mainly of the vermis) (Storey and Gardner 2012).

Saccades may sometimes be mildly hypometric. Pursuit can be found impaired, as well as gain of vestibulocular reflex. In approximately 80% of affected individuals, a GEN nystagmus has been described (Dudding et al. 2004; Gardner et al. 2005; Miyoshi et al. 2001; Storey and Gardner 2012; Zee and Leigh 2015).

16   Spinocerebellar Ataxia Type 17

SCA17 has been reported in Japanese, Germans, French, Chinese, Koreans, Italian Mexicans, Greeks, and Indians, albeit with a frequency much lower than SCA1–3 (Sun et al. 2016; Nakamura et al. 2001; Rolfs et al. 2003). SCA17 is due to an expanded CAG repeat in the TATA-box binding protein (TBP), which is a transcription factor expressed from a single gene on chromosome 6q27. The TBP protein is the DNA-binding subunit of the RNA polymerase II transcription factor D (TFIID) implicated in mRNA transcription. A complete loss of normal TBP function is not compatible with life. Normal repeat range is between 25 and 42 residues, with most alleles containing 32–39. The threshold for pathological expansion varies from 43 to 45; this variability is partly related to incomplete penetrance. Repeat expansions are quite stable during transmission, and anticipation is very rarely documented. Moreover, the inverse relationship between age at onset and size of repeats is not as strong as in other dynamic disease mutations. Clinical variability is high, even in the same families or in the same size of expansions (Koide et al. 1999).

Neuropathology demonstrates the presence of a diffuse cerebral atrophy, predominantly in the cerebellum. Loss of Purkinje cells is severe. Caudate nuclei can show marked atrophy. The pathological hallmarks of this disease are the presence of neuronal intranuclear inclusions containing the pathological proteins, as well as heat shock proteins and ubiquitin, although with a lower frequency (Stevanin and Brice 2008).

Age of onset ranges between 3 and 75 years, with a mean of 35 years. Clinical presentation can be really heterogeneous. Main features are represented by cerebellar deficits (gait ataxia, limb dysmetria, dysarthria), dementia (up to 75% of patients), extrapyramidal signs. Most patients, indeed, was diagnosed as having Huntington's disease prior to molecular analysis. SCA17 is also known as Huntington's disease-like (HDL4). Occasionally, hypogonadotropic hypogonadism has been described. Parkinson's disease–like, Creutzfeldt-Jakob disease–like, and Alzheimer's disease–like phenotypes have also been reported with small SCA17 expansions, while pyramidal signs and dystonia are common in larger-sized amplification (>50) (Sun et al. 2016; Chen et al. 2010).

At the beginning of symptoms, brain neuroimaging can be normal or show a moderate atrophy. With the progression of symptoms, the atrophy become more pronounced in the cerebellum but the cerebellar cortex and caudate nuclei are impaired as well; the brainstem is relatively spared.

Ocular motor abnormalities have often been reported in SCA17 patients, albeit their presence is occasional in the clinical picture. Hubner et al. (Hübner et al.

2007) found normal saccade velocity, with saccadic hypometria in SCA17 patients compared to healthy controls. Smooth pursuit initiation and maintenance were affected, with increase of latency and decrease of acceleration in a step-ramp paradigm. Steady fixation and gaze-holding were interrupted by GEN, rebound and downbeat nystagmus. A study on antisaccades and memory-guided tasks demonstrated a pathological increase of errors. These oculomotor abnormalities were correlated with disease duration, but not with repeat length. Similar findings were described by Mariotti et al. (2007), though in their patients nystagmus was not reported while they presented with hyperreflexia of vestibule-ocular reflexes. Lin et al. (2007) reported a case of SCA17 with a phenotype resembling that of PSP/CBD, including supranuclear palsy.

17 Spinocerebellar Ataxia Type 18

SCA18 is a rare clinical entity, only described in a five generation American family of Irish ancestry. It is also known as sensorimotor neuropathy with ataxia (Brkanac et al. 2002, 2009). Brkanac et al. (Brkanac et al. 2002) identified a pathological mutation in the *IFRD1* gene, mapped to locus 7q22-q32, which segregated perfectly with the phenotype of the affected family. However, they concluded that mutation analysis of *IFRD1* in additional patients with similar phenotypes was needed for demonstration of causality and further evaluation of its importance in neurologic diseases (Brkanac et al. 2009).

Age of onset was in the second or third decade. Main clinical symptom was a gait ataxia, but patients also complained of dysmetria, tendon hyporeflexia/areflexia, muscular weakness and atrophy, pes cavus. At brain MRI, cerebellar atrophy was pointed out.

Ocular motor examination showed GEN, while no other changes were reported (Brkanac et al. 2002).

18 Spinocerebellar Ataxia Type 19/22

SCA19 was firstly reported in a Dutch family (Schelhaas et al. 2001). In 2003, a family with similar characteristics and a linkage at the same locus was described in China and called SCA22 (Chung et al. 2003), but these disorders could represent the same SCA subtype. The SCA19 locus is located on chromosome 1p21-q21 (Dutch family) and 1p21-q23 (Chinese family), and the gene *KCND3* is affected by conventional mutations.

In the Dutch family, age of onset ranges from 10 to 45 years. Anticipation was empirically noted, but not yet demonstrated. Clinical features mostly include a cerebellar syndrome (ataxia, dysarthria, oculomotor abnormalities) with very slow progression. Additionally, mood disturbances, impulsive behavior, irregular postural head tremor, myoclonic jerks, and tendon hypo/areflexia have been reported, but not in the Chinese family. Brain MRI showed cerebellar atrophy of vermis and hemispheres; occasionally, also atrophy of cerebral cortex have been described.

Oculomotor abnormalities are typical of cerebellar impairment: saccades are dysmetric but with normal speed; pursuit are impaired and GEN may occur during gaze holding (Schelhaas et al. 2001; Chung et al. 2003; Zee and Leigh 2015).

19 Spinocerebellar Ataxia Type 20

SCA20 has been reported in an Australian pedigree of Anglo-Celtic origin (Knight et al. 2004) and is associated with a duplication at chromosome 11q22.2–11q12.3 (Knight et al. 2008). The gene locus overlaps that of SCA5, but the phenotypes are different.

Age of onset varies from 19 to 64 years, with a peak around the fourth decade. Anticipation phenomenon is described. The clinical pictures are characterized by a very slow progression, usually dysarthria without ataxia as first symptom. Cerebellar ataxia usually follows later in the course of the disease, sometimes with head and upper limbs tremor and spasmodic dysphonia. Majority of patients exhibit palatal tremor. Occasionally, pyramidal features without spasticity may be present. Neuroimaging of SCA20 is peculiar: brain CT scans show dentate nuclei calcifications and occasionally concomitant pallidal calcifications. At brain MRI, cerebellar atrophy is usually seen and, in the cases of palatal tremor, a pseudohypertrophy of inferior olives. SCA20 needs to be differentiated by familiar progressive ataxia and palatal tremor: in the latter, no olivary hypertrophy is seen.

Saccades appear hypermetric, pursuit are impaired. Gaze holding is affected by GEN or downbeat nystagmus, while steady fixation may be disturbed by SWJs (Storey et al. 2005; Knight et al. 2004).

20 Spinocerebellar Ataxia Type 21

SCA21 was firstly observed in a French family (Devos et al. 2001). The gene *TMEM240* on chromosome 1p36.33–p36.32 (previously mapped on 7p21.3-p15.1) was found to be affected by conventional mutations (Vuillaume et al. 2002; Delplanque et al. 2014).

The age of onset ranges from 1 to 30 years.

Clinical picture is characterized by slowly progressive cerebellar syndrome associated with mild cognitive impairment and extrapyramidal features, poorly responsive to L-Dopa. An isolated cerebellar atrophy has been observed at brain MRI (Delplanque et al. 2008). Neuropathological study evidenced a slight loss of Purkinje cells, with normal macroscopic appearance of cerebellum and brain.

Neurophthalmological examination pointed out, in the affected, the impairment of smooth pursuit and the presence of SWJs on steady fixation. No saccades abnormalities were described.

21 Spinocerebellar Ataxia Type 23

SCA23 was described in a two-generation Dutch family (Verbeek et al. 2004) with slowly progressive ataxia. The candidate gene identified is *PDYN* on chromosome 20p13–12.3; in the affected, three heterozygous mutations were identified (Bakalkin et al. 2010). At neuropathological examination in one SCA23 patient, atrophy of frontotemporal regions, cerebellar vermis, pons, and spinal cord was identified (Manto 2010; Verbeek et al. 2004)

The mean age of onset is 50.4 years. Clinical features include, besides slowly progressive cerebellar syndrome, rare presence of pyramidal signs or decreased vibration sense below the knees.

At neurophthalmological examination, saccades were reported to be dysmetric and with mild reduction of speed. Rarely, a GEN has been observed (Jezierska et al. 2013; Verbeek et al. 2004).

22 Spinocerebellar Ataxia Type 25

Stevanin et al. (Stevanin et al. 2004a, b) reported a large family of Southeastern France with cerebellar ataxia with sensory involvement. SCA25 maps on chromosome 2p15–p21 and is genetically characterized by reduced penetrance without evidence of anticipation.

The age of onset ranged from 17 months to 39 years. Clinical symptomatology was characterized by cerebellar ataxia and severe sensory neuropathy, confirmed at EMG examination. Patients also complained of scoliosis and pes cavus. Global cerebellar atrophy was observed at brain MRI.

Saccades were found hypermetric of normal speed. Fixation and gaze-holding stability were affected by the presence of SWJs, GEN, and downbeat nystagmus.

23 Spinocerebellar Ataxia Type 26

Yu et al. (2005) reported a new form of SCA in a six-generation family of Norwegian ancestry, immigrated to the USA. The locus SCA26 mapped on chromosome 19p13.3, adjacent to the gene of SCA6.

Age of onset ranged from 26 to 60 years, without anticipation. Patients present with pure cerebellar syndrome, with ataxia, dysarthria, normal intelligence. Pure cerebellar atrophy is seen at MRI.

Ocular movement examination pointed out dysmetric saccades, impairment of horizontal and vertical pursuit, sporadic GEN (Yu et al. 2005).

24 Spinocerebellar Ataxia Type 27

SCA27 was firstly reported in a Dutch family (van Swieten et al. 2003). SCA27 is caused by a missense mutation (F145S) in the gene encoding fibroblast growth factor-14 (FGF14) on chromosome 13q34 (Brusse et al. 2006; Misceo et al. 2009).

The first appearing symptom, starting in childhood, is a tremor at both hands of high frequency and small amplitude. Gait ataxia appears later in the course of the disease. Patients also complain of orofacial dyskinesias, depression and aggressive behavior, occasional mental retardation, pes cavus (Dalski et al. 2005). Moderate cerebellar atrophy is present at brain MRI.

Saccades are dysmetric, pursuit are impaired. Gaze-holding is impaired by GEN and fixation stability by SWJs (Dalski et al. 2005; Coebergh et al. 2014).

25 Spinocerebellar Ataxia Type 28

SCA28 was initially described in a four-generation Italian family (Cagnoli et al. 2006) with slowly progressive ataxia. The candidate disease locus, named SCA28, was identified on chromosome 18p11.22–q11.2; several conventional mutations were later identified for gene *AFG3L2* (Cagnoli et al. 2010; Di Bella et al. 2010), probably affecting the mithocondrial and protheolytic functions of the product protein by dominant-negative and loss of function effects (Zuhlke et al. 2015).

The reported age at onset varied from 12 to 36 years, without evidence of anticipation. Patients showed a similar phenotype. First symptom was usually the gait ataxia and unbalanced standing. Other symptoms included pyramidal syndrome (Cagnoli et al. 2010) or dystonia. MRI showed isolated cerebellar atrophy.

Two patterns of ocular movement deficits were observed. Patients with shorter duration of disease prevalently showed cerebellar nystagmus (Cagnoli et al. 2006). Those with longer disease duration presented with dysmetric saccades, slowing of saccades to ophthalmoparesis, palpebral ptosis, pursuit and OKN impairment (Mariotti et al. 2008; Politi et al. 2016), most likely due to a specific impairment of mitochondrial energy metabolism.

26 Spinocerebellar Ataxia Type 29

SCA29 is an autosomal dominant neurologic disorder characterized by onset in infancy of delayed motor development and mild cognitive delay. It is caused by mutations in the *ITPR1* gene in the 3p24.2-3pter chromosome; heterozigous mutations also cause SCA15, which is different for later age at onset and normal cognition.

The disease usually appears in infancy. Clinical features are those of a slowly progressive cerebellar syndrome (gait and limb ataxia, intentional tremor, dysarthria), though some patients exhibited partial seizures (Huang et al. 2012). Brain MRI showed cerebellar vermis atrophy.

Oculomotor examination showed prevalently GEN and fixation instability (Huang et al. 2012; Dudding et al. 2004).

27 Spinocerebellar Ataxia Type 30

SCA30 has been reported in an Australian family of Anglo-Celtic ethnicity (Storey et al. 2009). Disease locus was identified on chromosome 4q34.3–q35.1.

Mean age of onset was 52 years, but with very insidious onset. The phenotype was characterized by relatively pure, slowly evolving gait and appendicular ataxia with mild to moderate dysarthria. In some cases, hyperreflexia at lower limbs was described. No neuropathy was observed. Brain MRI of two patients showed pure cerebellar atrophy with preservation of the brainstem.

At neuro-ophthalmic examination, all patients had hypermetric saccades with normal gain of VOR. One patient showed GEN.

28 Spinocerebellar Ataxia Type 31

For a further description, see SCA4. This ataxia has been described in a Japanese population. The candidate gene ODZ3 might be involved.

Clinical features are those of a pure cerebellar ataxia, though some patients present with dementia and hearing impairment; differently from SCA4, no peripheral neuropathy was observed.

GEN was the most common eye movement abnormality seen (Nagaoka et al. 2000; Ouyang et al. 2006; Hirano et al. 2009).

29 Spinocerebellar Aataxia Type 34

SCA34 was firstly described in a five-generation French–Canadian family (Giroux and Barbeau 1972; Cadieux-Dion et al. 2014); later, it was identified in another family from Japan (Ozaki et al. 2015).

Disease locus was identified on chromosome 6q14.1. Various missense mutations were identified in affected patients in gene *ELOVL4*.

Besides slowly progressive cerebellar ataxia, the French–Canadian family presented with papulosquamous erythematous ichthyosiform plaques appearing soon after birth, with disappearance with age. Cerebellar syndrome then became the predominant feature, and many patients become wheelchair-bound in later life. Japan variant differed for the absence of dermatological involvement and the minor severity of ataxia. Conversely, patients might present pyramidal tract signs and autonomic symptoms. Brain imaging showed cerebellar and pontine atrophy. In addition, four patients had cruciform hyperintensities ("hot cross bun sign") and two other patients had pontine midline linear hyperintensity, both of which often appear in patients with multiple system atrophy.

In the French–Canadian family, nystagmus was mainly seen. In the Japan variant, many patients had GEN (78%), while less commonly a supranuclear ophthalmoplegia (33%) and impaired smooth pursuit (56%) were observed.

30 Spinocerebellar Ataxia Type 35

SCA35 was prevalently described in Chinese families (Wang et al. 2010; Li et al. 2013; Guo et al. 2014). The age of onset varied from teenage to late adulthood, and the disorder was slowly progressive. Missense mutations were identified in TGM6 gene on chromosome 20p13–12.2.

Clinical pictures included limb and gait ataxia, dysarthria, hand tremor, hyperreflexia. Brain imaging showed cerebellar atrophy. Patients also exhibited dysmetric saccades, sometimes slow, and saccadic pursuit. Nystagmus has not been reported.

31 Spinocerebellar Ataxia Type 36

SCA36 has been described in Japanese families and in another family from Costa de Morte, in Galicia, Spain (Kobayashi et al. 2011; Ikeda et al. 2012; García-Murias et al. 2012). SCA36 is caused by heterozygous expansion of an intronic GGCCTG hexanucleotide repeat in the NOP56 gene on chromosome 20p13. Unaffected individuals carry 3–14 repeats, whereas affected individuals carry 650–2500 repeats.

Mean age of onset was in the fifth decade. Clinical features include gait ataxia, truncal instability, dysarthria. Later in the course of the disease, patients developed tongue atrophy with fasciculation or skeletal muscle atrophy at lower limbs and trunk and signs of lower motor neuron involvement at EMG. Brain MRI showed pancerebellar atrophy, later evolving in olivopontocerebellar atrophy.

Eye movement examination pointed out impaired smooth pursuit and GEN (Ikeda et al. 2012); rarely, vertical or horizontal gaze limitation has been observed (García-Murias et al. 2012).

32 Spinocerebellar Aataxia Type 37

SCA37, described in Spanish (Serrano-munuera et al. 2013) and Portuguese (Seixas et al. 2017) families, is caused by heterozygous ATTTC insertion, ranging from 31 to 75 repeats, in the *DAB1* gene (603448) on chromosome 1p32. There was an inverse correlation between ATTTC insertion size and age of onset. In addition, there was instability upon transmission of the pathogenic repeat, with an increase in length particularly when the father was the transmitting parent. Onset was usually in the fourth decade, with gait instability and dysarthria. Later on, patients had truncal ataxia, dysmetria, and dysphagia. At brain MRI, cerebellar atrophy with sparing of brainstem was observed. All patients had ocular movement abnormalities. Particularly, saccades were dysmetric, prominently in the vertical plane. Analogously, pursuit and OKN were impaired, particularly vertically. Few patients had GEN and saccadic intrusions of SWJs subtype.

33 Spinocerebellar Ataxia Type 38

SCA38, reported in four unrelated families, three from Italy and one from France, is caused by heterozygous missense mutations in the *ELOVL5* gene on chromosome 6p12. The onset was between the third and fourth decade. All patients had slowly progressive gait ataxia. Some exhibited peripheral axonal neuropathy. All patients had ocular movement abnormalities such as slow saccades and GEN (Di Gregorio et al. 2014).

34 Spinocerebellar Ataxia Type 40

SCA 40 has been reported in a family from Hong Kong with five affected individuals and is caused by heterozygous mutation in the CCDC88C gene on chromosome 14q32 (Tsoi et al. 2014)

Clinical features included gait ataxia, wide-base gait, intentional tremor, scanning speech, hyperreflexia, spastic paraparesis. Patients were wheelchair-bound. Brain MRI showed pontocerebellar atrophy.

Ocular movement abnormalities comprehended saccadic dysmetria and impaired vertical gaze.

35 Spinocerebellar Aataxia Type 42

SCA 42 was described in French and Japanese families (Coutelier et al. 2015; Morino et al. 2015). It is caused by heterozygous missense mutations in the CACNA1G gene on chromosome 17q21.

Age of onset was highly variable. Clinical features included gait ataxia, dysarthria and, less commonly, urinary disturbances and decreased distal vibratory sense, pyramidal signs.

Patients exhibited saccadic pursuit, diplopia, GEN.

36 Spinocerebellar Ataxia Type 43

SCA43 was reported in a Belgian family. It is caused by heterozygous missense mutations in the MME gene on chromosome 3q25. Main clinical features were cerebellar ataxia and severe peripheral neuropathy. Cerebellar atrophy was observed at brain MRI.

Inconstantly, GEN was observed (Depondt et al. 2016).

37 Spinocerebellar Ataxia Type 44

SCA44, caused by heterozygous missense mutation in the GRM1 gene on chromosome 6q24, has been reported in two unrelated families with slowly progressive spinocerebellar ataxia (onset range 20–50 years) (Watson et al. 2017).

Clinical phenotype included gait ataxia, though no patient was wheelchairbound, dysphagia, dysarthria, dydiadocokinesia, and rarely spasticity. At brain MRI, a cerebellar atrophy was seen.

Ocular movement changes, mainly saccadic hypermetria and impaired smooth pursuit, were inconstantly described.

38 Spinocerebellar Ataxia Type 45

SCA45 is caused by heterozygous missense mutations in the *FAT2* gene on chromosome 5q33. It has recently been described (Nibbeling et al. 2017) in a family whose affected members had pure cerebellar syndrome. Brain MRI showed cerebellar vermis atrophy and hemosiderin deposit on meencephalon.

Patients related to the family presented with nystagmus, which was absent in an unrelated, affected patient.

39 Spinocerebellar Aataxia Type 46

SCA 46 is caused by heterozygous mutation in the PLD3 gene on chromosome 19q13. One Dutch family has been reported by van Dijk et al. (1995). Affected patients had adult-onset sensory ataxic neuropathy with cerebellar signs and variable sensory neuropathy at lower limbs. No cerebellar atrophy was evident at brain MRI. At follow-up (Nibbeling et al. 2017), patients had variable combination of sensory polineuropathy and cerebellar ataxia, the former being prevalent. Only in one patient there was mild cerebellar atrophy.

Ocular movement abnormalities included nystagmus, saccadic slowness and dysmetria, saccadic pursuit, presence of SWJs at steady fixation.

40 Spinocerebellar Ataxia Type 47

SCA47 has been recently identified (Gennarino et al. 2018) and is related to heterozygous missense mutations in the *PUM1* gene on chromosome 1p35.

Gennarino et al. (2018) reported a family, affected by adult-onset cerebellar ataxia, with dysmetria, dysarthria, gait ataxia, and mild cerebellar vermis atrophy at brain MRI. Some patients exhibited diplopia. Moreover, they reported an early-onset phenotype in two unrelated girls with delayed motor development, early-onset ataxia, and short stature. The less severely affected girl showed chorea, ataxia, dysarthria, spasticity, ballismus, and fine-motor incoordination, with normal brain imaging. The second girl was more severely affected with an epileptic encephalopathy, progressive ataxia, hypotonicity of the lower limbs, global developmental delay with cortical visual impairment, and stereotypic hand movements. She also showed hypotonia, scoliosis, facial dysmorphism, and low bone mineral density. At brain MRI, enlargement of fourth ventricle and shortening of cerebellar vermis were noticed. The two distinct phenotypes were related to a diverse reduction of PUM1 levels.

41 Dentatorubral-Pallidoluysian Atrophy (DPRLA)

DPRLA term was firstly used in a sporadic case (Smith et al. 1958). Familiar cases were then reported in Japan (Naito and Oyanagi 1982; Sasaki et al. 2003), with a prevalence ranging from 0.2 to 0.7 per 100.000. Other rare cases have been described in European and North American families (Farmer et al. 1989; Connarty et al. 1996).

It is caused by CAG repeat expansions in gene *ATN1* on chromosome 12p13.31. In normal individuals, size of expansion ranges from 6 to 35 (with repeats larger than 17 prevalently seen in Japanese population), while in affected ranges from 48 to 93. Anticipation is severe and more prominent for paternal transmission. Neuropathological changes are eponym with the name and consist of combined dentatorubral and pallidoluysian systems degeneration. Neuronal inclusions have been described, as well as accumulation of the mutant DRPLA protein (atrophin-1).

The onset varied from childhood to late adulthood. Clinical heterogeneity is marked and patients exhibit various combinations of cerebellar ataxia, choreoathetosis, epilepsy, myoclonus, dementia, and psychiatric symptoms. The juvenile form (<20 years) is mainly characterized by progressive myoclonic epilepsy and intellectual deterioration. With presentation after 20 years, the occurrence of seizures decrease and the main clinical features are cerebellar ataxia, choreoathetosis, and dementia, mimicking Huntington's disease. Typical MRI findings are the atrophy in the cerebellum and brainstem, particularly pontomesencephalic tegmentum (Manto 2010). Occasional white matter hyperintensities on T2-weighted images are seen. Rarely, the appearance of the pons resembles a central pontine myelinolysis.

At neurophthalmological examination, some patients exhibit increase of saccadic latency and slow saccades, progressing to palsy. Impairment of smooth pursuit is common, as well as GEN and SWJs on steady fixation (Muñoz et al. 1999; Vinton et al. 2005).

References

Adachi, M., Kawanami, T., Ohshima, H., & Hosoya, T. (2006). Characteristic signal changes in the pontine base on T2- and multishot diffusion-weighted images in spinocerebellar ataxia type 1. *Neuroradiology, 48*(1), 8–13.

Alexandre, M. F., Rivaud-Péchoux, S., Challe, G., Durr, A., & Gaymard, B. (2013). Functional consequences of oculomotor disorders in hereditary cerebellar ataxias. *Cerebellum, 12*(3), 396–405.

Bakalkin, G., Watanabe, H., Jezierska, J., Depoorter, C., Verschuuren-Bemelmans, C., Bazov, I., Artemenko, K. A., Yakovleva, T., Dooijes, D., Van de Warrenburg, B. P. C., Zubarev, R. A., Kremer, B., Knapp, P. E., Hauser, K. F., Wijmenga, C., Nyberg, F., Sinke, R. J., & Verbeek, D. S. (2010). Prodynorphin mutations cause the neurodegenerative disorder spinocerebellar ataxia type 23. *American Journal of Human Genetics, 87*, 593–603.

Benton, C. S., de Silva, R., Rutledge, S. L., Bohlega, S., Ashizawa, T., & Zoghbi, H. Y. (1998). Molecular and clinical studies in SCA-7 define a broad clinical spectrum and the infantile phenotype. *Neurology, 51*(4), 1081–1086.

Bettencourt, C., Hensman-Moss, D., Flower, M., Wiethoff, S., Brice, A., Goizet, C., Stevanin, G., Koutsis, G., Karadima, G., Panas, M., Yescas-Gómez, P., García-Velázquez, L. E., Alonso-Vilatela, M. E., Lima, M., Raposo, M., Traynor, B., Sweeney, M., Wood, N., Giunti, P., SPATAX Network, D. A., Holmans, P., Houlden, H., Tabrizi, S. J., & Jones, L. (2016). DNA repair pathways underlie a common genetic mechanism modulating onset in polyglutamine diseases. *Annals of Neurology, 79*(6), 983–990.

Brkanac, Z., Fernandez, M., Matsushita, M., Lipe, H., Wolff, J., Bird, T. D., & Raskind, W. H. (2002). Autosomal dominant sensory/motor neuropathy with Ataxia (SMNA): Linkage to chromosome 7q22-q32. *American Journal of Medical Genetics, 114*(4), 450–457.

Brkanac, Z., Spencer, D., Shendure, J., Robertson, P. D., Matsushita, M., Vu, T., Bird, T. D., Olson, M. V., & Raskind, W. H. (2009). IFRD1 is a candidate gene for SMNA on chromosome 7q22-q23. *American Journal of Human Genetics, 84*(5), 692–697.

Brusse, E., de Koning, I., Maat-Kievit, A., Oostra, B. A., Heutink, P., & van Swieten, J. C. (2006). Spinocerebellar ataxia associated with a mutation in the fibroblast growth factor 14 gene (SCA27): A new phenotype. *Movement Disorders, 21*(3), 396–401.

Bürk, K., Fetter, M., Abele, M., Laccone, F., Brice, A., Dichgans, J., & Klockgether, T. (1999). Autosomal dominant cerebellar ataxia type I: Oculomotor abnormalities in families with SCA1, SCA2, and SCA3. *Journal of Neurology, 246*(9), 789–797.

Bürk, K., Zühlke, C., König, I. R., Ziegler, A., Schwinger, E., Globas, C., Dichgans, J., & Hellenbroich, Y. (2004). Spinocerebellar ataxia type 5: Clinical and molecular genetic features of a German kindred. *Neurology, 62*(2), 327–329. Review.

Buttner, N., Geschwind, D., Jen, J. C., Perlman, S., Pulst, S. M., & Baloh, R. W. (1998). Oculomotor phenotypes in autosomal dominant ataxias. *Archives of Neurology, 55*(10), 1353–1357.

Cadieux-Dion, M., Turcotte-Gauthier, M., Noreau, A., Martin, C., Meloche, C., Gravel, M., Drouin, C. A., Rouleau, G. A., Nguyen, D. K., & Cossette, P. (2014). Expanding the clinical phenotype associated with ELOVL4 mutation: Study of a large French-Canadian family with autosomal dominant spinocerebellar ataxia and erythrokeratodermia. *JAMA Neurology, 71*(4), 470–475.

Cagnoli, C., Mariotti, C., Taroni, F., Seri, M., Brussino, A., Michielotto, C., Grisoli, M., Di Bella, D., Migone, N., Gellera, C., Di Donato, S., & Brusco, A. (2006). SCA28, a novel form of autosomal dominant cerebellar ataxia on chromosome 18p11.22-q11.2. *Brain, 129*(Pt 1), 235–242.

Cagnoli, C., Stevanin, G., Brussino, A., Barberis, M., Mancini, C., Margolis, RL., Holmes, SE., Nobili, M., Forlani, S., Padovan, S., Pappi, P., Zaros, C., Leber, I., Ribai, P., Pugliese, L., Assalto, C., Brice, A., Migone, N., Dürr, A., & Brusco A. (2010). Missense mutations in the AFG3L2 proteolytic domain account for ~1.5% of European autosomal dominant cerebellar ataxias. *Human Mutation 31*(10), 1117–1124. https://doi.org/10.1002/humu.21342.

Chelban, V., Wiethoff, S., Fabian-Jessing, B. K., Haridy, N. A., Khan, A., Efthymiou, S., Becker, E. B. E., O'Connor, E., Hersheson, J., Newland, K., Hojland, A. T., Gregersen, P. A., Lindquist, S. G., Petersen, M. B., Nielsen, J. E., Nielsen, M., Wood, N. W., Giunti, P., & Houlden, H. (2018). Genotype-phenotype correlations, dystonia and disease progression in spinocerebellar ataxia type 14. *Movement Disorders*. https://doi.org/10.1002/mds.27334.

Chen, D-H., Brkanac, Z., Verlinde, C. L. M. J., Tan, X-J., Bylenok, L., Nochlin, D., Matsushita, M., Lipe, H., Wolff, J., Fernandez, M., Cimino, P. J., Bird, T. D., Raskind,W. H. (2003). Missense mutations in the regulatory domain of PKCγ: A new mechanism for dominant nonepisodic cerebellar ataxia. *The American Journal of Human Genetics, 72*(4), 839–849.

Chen, D. H., Cimino, P. J., Ranum, L. P., Zoghbi, H. Y., Yabe, I., Schut, L., Margolis, R. L., Lipe, H. P., Feleke, A., Matsushita, M., Wolff, J., Morgan, C., Lau, D., Fernandez, M., Sasaki, H., Raskind, W. H., & Bird, T. D. (2005a). The clinical and genetic spectrum of spinocerebellar ataxia 14. *Neurology, 64*(7), 1258–1260.

Chen, D. H., Bird, T. D., & Raskind, W. H. (2005b). Spinocerebellar Ataxia Type 14. In: Adam MP, Ardinger HH, Pagon RA, Wallace SE, Bean LJH, Stephens K, Amemiya A, editors. GeneReviews® [Internet]. Seattle (WA): University of Washington, Seattle; 1993–2019. [updated 2013 Apr 18].

Chen, C. M., Lee, L. C., Soong, B. W., Fung, H. C., Hsu, W. C., Lin, P. Y., Huang, H. J., Chen, F. L., Lin, C. Y., Lee-Chen, G. J., & Wu, Y. R. (2010). SCA17 repeat expansion: Mildly expanded CAG/CAA repeat alleles in neurological disorders and the functional implications. *Clinica Chimica Acta, 411*(5–6), 375–380.

Chung, M. Y., Lu, Y. C., Cheng, N. C., & Soong, B. W. (2003). A novel autosomal dominant spinocerebellar ataxia (SCA22) linked to chromosome 1p21-q23. *Brain, 126*(Pt 6), 1293–1299.

Coebergh, J. A., Fransen van de Putte, D. E., Snoeck, I. N., Ruivenkamp, C., van Haeringen, A., & Smit, L. M. (2014). A new variable phenotype in spinocerebellar ataxia 27 (SCA 27) caused by a deletion in the FGF14 gene. *European Journal of Paediatric Neurology, 18*(3), 413–415.

Connarty, M., Dennis, N. R., Patch, C., Macpherson, J. N., & Harvey, J. F. (1996). Molecular re-investigation of patients with Huntington's disease in Wessex reveals a family with dentatorubral and pallidoluysian atrophy. *Human Genetics, 97*(1), 76–78.

Coutelier, M., Blesneac, I., Monteil, A., Monin, M. L., Ando, K., Mundwiller, E., Brusco, A., Le Ber, I., Anheim, M., Castrioto, A., Duyckaerts, C., Brice, A., Durr, A., Lory, P., & Stevanin, G. (2015). A recurrent mutation in CACNA1G alters Cav3.1 T-type calcium-channel conduction and causes autosomal-dominant cerebellar ataxia. *American Journal of Human Genetics, 97*(5), 726–737.

Coutinho, P., & Andrade, C. (1978). Autosomal dominant system degeneration in Portuguese families of the Azores Islands. A new genetic disorder involving cerebellar, pyramidal, extrapyramidal and spinal cord motor functions. *Neurology, 28*, 703–709.

Dalski, A., Atici, J., Kreuz, F. R., Hellenbroich, Y., Schwinger, E., & Zühlke, C. (2005). Mutation analysis in the fibroblast growth factor 14 gene: Frameshift mutation and polymorphisms in patients with inherited ataxias. *European Journal of Human Genetics, 13*(1), 118–120.

Delplanque, J., Devos, D., Vuillaume, I., De Becdelievre, A., Vangelder, E., Maurage, C. A., Dujardin, K., Destée, A., & Sablonnière, B. (2008). Slowly progressive spinocerebellar ataxia with extrapyramidal signs and mild cognitive impairment (SCA21). *Cerebellum, 7*(2), 179–183.

Delplanque, J., Devos, D., Huin, V., Genet, A., Sand, O., Moreau, C., Goizet, C., Charles, P., Anheim, M., Monin, M. L., Buee, L., Destee, A., Grolez, G., Delmaire, C., Dujardin, K., Dellacherie, D., Brice, A., Stevanin, G., Strubi-Vuillaume, I., Dürr, A., & Sablonnière, B. (2014). TMEM240 mutations cause spinocerebellar ataxia 21 with mental retardation and severe cognitive impairment. *Brain, 137*, 2657–2663.

Depondt, C., Donatello, S., Rai, M., Wang, F. C., Manto, M., Simonis, N., & Pandolfo, M. (2016). MME mutation in dominant spinocerebellar ataxia with neuropathy (SCA43). *Neurological Genetics, 2*(5), e94.

Devos, D., Schraen-Maschke, S., Vuillaume, I., Dujardin, K., Nazé, P., Willoteaux, C., Destée, A., & Sablonnière, B. (2001). Clinical features and genetic analysis of a new form of spinocerebellar ataxia. *Neurology, 56*(2), 234–238.

Di Bella, D., Lazzaro, F., Brusco, A., Plumari, M., Battaglia, G., Pastore, A., Finardi, A., Cagnoli, C., Tempia, F., Frontali, M., Veneziano, L., Sacco, T., Boda, E., Brussino, A., Bonn, F., Castellotti, B., Baratta, S., Mariotti, C., Gellera, C., Fracasso, V., Magri, S., Langer, T., Plevani, P., Di Donato, S., Muzi-Falconi, M., & Taroni, F. (2010). Mutations in the mitochondrial protease gene AFG3L2 cause dominant hereditary ataxia SCA28. *Nature Genetics, 42*(4), 313–321.

Di Gregorio, E., Borroni, B., Giorgio, E., Lacerenza, D., Ferrero, M., Lo Buono, N., Ragusa, N., Mancini, C., Gaussen, M., Calcia, A., Mitro, N., Hoxha, E., Mura, I., Coviello, D. A., Moon, Y. A., Tesson, C., Vaula, G., Couarch, P., Orsi, L., Duregon, E., Papotti, M. G., Deleuze, J. F., Imbert, J., Costanzi, C., Padovani, A., Giunti, P., Maillet-Vioud, M., Durr, A., Brice, A., Tempia, F., Funaro, A., Boccone, L., Caruso, D., Stevanin, G., & Brusco, A. (2014). ELOVL5 mutations cause spinocerebellar ataxia 38. *American Journal of Human Genetics, 95*(2), 209–217.

Dudding, T. E., Friend, K., Schofield, P. W., Lee, S., Wilkinson, I. A., & Richards, R. I. (2004). Autosomal dominant congenital non-progressive ataxia overlaps with the SCA15 locus. *Neurology, 63*(12), 2288–2292.

Edener, U., Bernard, V., Hellenbroich, Y., Gillessen-Kaesbach, G., & Zühlke, C. (2011). Two dominantly inherited ataxias linked to chromosome 16q22.1: SCA4 and SCA31 are not allelic. *Journal of Neurology, 258*(7), 1223–1227.

Farmer, T. W., Wingfield, M. S., Lynch, S. A., Vogel, F. S., Hulette, C., Katchinoff, B., & Jacobson, P. L. (1989). Ataxia, chorea, seizures, and dementia. Pathologic features of a newly defined familial disorder. *Archives of Neurology, 46*(7), 774–779.

Federighi, P., Cevenini, G., Dotti, M. T., Rosini, F., Pretegiani, E., Federico, A., & Rufa, A. (2011). Differences in saccade dynamics between spinocerebellar ataxia 2 and late-onset cerebellar ataxias. *Brain, 134*(Pt 3), 879–891.

Figueroa, K. P., Minassian, N. A., Stevanin, G., Waters, M., Garibyan, V., Forlani, S., Strzelczyk, A., Bürk, K., Brice, A., Dürr, A., Papazian, D. M., & Pulst, S. M. (2010). KCNC3: Phenotype, mutations, channel biophysics-a study of 260 familial ataxia patients. *Human Mutation, 31*(2), 191–196.

Figueroa, K. P., Waters, M. F., Garibyan, V., Bird, T. D., Gomez, C. M., Ranum, L. P., Minassian, N. A., Papazian, D. M., & Pulst, S. M. (2011). Frequency of KCNC3 DNA variants as causes of spinocerebellar ataxia 13 (SCA13). *PLoS One, 6*(3), e17811.

Flanigan, K., Gardner, K., Alderson, K., Galster, B., Otterud, B., Leppert, M. F., Kaplan, C., & Ptácek, L. J. (1996). Autosomal dominant spinocerebellar ataxia with sensory axonal neuropathy (SCA4): Clinical description and genetic localization to chromosome 16q22.1. *American Journal of Human Genetics, 59*(2), 392–399.

García-Murias, M., Quintáns, B., Arias, M., Seixas, A. I., Cacheiro, P., Tarrío, R., Pardo, J., Millán, M. J., Arias-Rivas, S., Blanco-Arias, P., Dapena, D., Moreira, R., Rodríguez-Trelles,

F., Sequeiros, J., Carracedo, A., Silveira, I., & Sobrido, M. J. (2012). 'Costa da Morte' ataxia is spinocerebellar ataxia 36: Clinical and genetic characterization. *Brain, 135*(Pt 5), 1423–1435.

Gardner, R. J., Knight, M. A., Hara, K., Tsuji, S., Forrest, S. M., & Storey, E. (2005). Spinocerebellar ataxia type 15. *Cerebellum, 4*(1), 47–50.

Gennarino, V. A., Palmer, E. E., McDonell, L. M., Wang, L., Adamski, C. J., Koire, A., See, L., Chen, C. A., Schaaf, C. P., Rosenfeld, J. A., Panzer, J. A., Moog, U., Hao, S., Bye, A., Kirk, E. P., Stankiewicz, P., Breman, A. M., McBride, A., Kandula, T., Dubbs, H. A., Macintosh, R., Cardamone, M., Zhu, Y., Ying, K., Dias, K. R., Cho, M. T., Henderson, L. B., Baskin, B., Morris, P., Tao, J., Cowley, M. J., Dinger, M. E., Roscioli, T., Caluseriu, O., Suchowersky, O., Sachdev, R. K., Lichtarge, O., Tang, J., Boycott, K. M., Holder, J. L., Jr., & Zoghbi, H. Y. (2018). A mild PUM1 mutation is associated with adult-onset ataxia, whereas haploinsufficiency causes developmental delay and seizures. *Cell, 172*(5), 924–936.e11.

Giroux, J. M., & Barbeau, A. (1972). Erythrokeratodermia with ataxia. *Archives of Dermatology, 106*(2), 183–188.

Goldfarb, L. G., Vasconcelos, O., Platonov, F. A., Lunkes, A., Kipnis, V., Kononova, S., Chabrashvili, T., Vladimirtsev, V. A., Alexeev, V. P., & Gajdusek, D. C. (1996). Unstable triplet repeat and phenotypic variability of spinocerebellar ataxia type 1. *Annals of Neurology, 39*(4), 500–506.

Gomez, C. M., Thompson, R. M., Gammack, J. T., Perlman, S. L., Dobyns, W. B., Truwit, C. L., Zee, D. S., Clark, H. B., & Anderson, J. H. (1997). Spinocerebellar ataxia type 6: Gaze-evoked and vertical nystagmus, Purkinje cell degeneration, and variable age of onset. *Annals of Neurology, 42*(6), 933–950.

Gordon, C. R., Joffe, V., Vainstein, G., & Gadoth, N. (2003). Vestibulo-ocular arreflexia in families with spinocerebellar ataxia type 3 (Machado-Joseph disease). *Journal of Neurology, Neurosurgery, and Psychiatry, 74*(10), 1403–1406.

Guo, Y. C., Lin, J. J., Liao, Y. C., Tsai, P. C., Lee, Y. C., & Soong, B. W. (2014). Spinocerebellar ataxia 35: Novel mutations in TGM6 with clinical and genetic characterization. *Neurology, 83*(17), 1554–1561.

Hara, K., Shiga, A., Nozaki, H., Mitsui, J., Takahashi, Y., Ishiguro, H., Yomono, H., Kurisaki, H., Goto, J., Ikeuchi, T., Tsuji, S., Nishizawa, M., & Onodera, O. (2008). Total deletion and a missense mutation of ITPR1 in Japanese SCA15 families. *Neurology, 71*(8), 547–551.

Hellenbroich, Y., Bubel, S., Pawlack, H., Opitz, S., Vieregge, P., Schwinger, E., & Zühlke, C. (2003). Refinement of the spinocerebellar ataxia type 4 locus in a large German family and exclusion of CAG repeat expansions in this region. *Journal of Neurology, 250*(6), 668–671.

Hellenbroich, Y., Gierga, K., Reusche, E., Schwinger, E., Deller, T., de Vos, R. A., Zühlke, C., & Rüb, U. (2006). Spinocerebellar ataxia type 4 (SCA4): Initial pathoanatomical study reveals widespread cerebellar and brainstem degeneration. *Journal of Neural Transmission (Vienna), 113*(7), 829–843.

Herman-Bert, A., Stevanin, G., Netter, J. C., Rascol, O., Brassat, D., Calvas, P., Camuzat, A., Yuan, Q., Schalling, M., Dürr, A., & Brice, A. (2000). Mapping of spinocerebellar ataxia 13 to chromosome 19q13.3-q13.4 in a family with autosomal dominant cerebellar ataxia and mental retardation. *American Journal of Human Genetics, 67*(1), 229–235.

Hiramoto, K., Kawakami, H., Inoue, K., Seki, T., Maruyama, H., Morino, H., Matsumoto, M., Kurisu, K., & Sakai, N. (2006). Identification of a new family of spinocerebellar ataxia type 14 in the Japanese spinocerebellar ataxia population by the screening of PRKCG exon 4. *Movement Disorders, 21*(9), 1355–1360.

Hirano, R., Takashima, H., Okubo, R., Okamoto, Y., Maki, Y., Ishida, S., Suehara, M., Hokezu, Y., & Arimura, K. (2009). Clinical and genetic characterization of 16q-linked autosomal dominant spinocerebellar ataxia in South Kyushu, Japan. *Journal of Human Genetics, 54*(7), 377–381.

Holmes, S. E., O'Hearn, E. E., McInnis, M. G., Gorelick-Feldman, D. A., Kleiderlein, J. J., Callahan, C., Kwak, N. G., Ingersoll-Ashworth, R. G., Sherr, M., Sumner, A. J., Sharp, A. H., Ananth, U., Seltzer, W. K., Boss, M. A., Vieria-Saecker, A. M., Epplen, J. T., Riess, O., Ross,

C. A., & Margolis, R. L. (1999). Expansion of a novel CAG trinucleotide repeat in the 5′ region of PPP2R2B is associated with SCA12. *Nature Genetics, 23*(4), 391–392.

Huang, L., Chardon, J. W., Carter, M. T., Friend, K. L., Dudding, T. E., Schwartzentruber, J., Zou, R., Schofield, P. W., Douglas, S., Bulman, D. E., & Boycott, K. M. (2012). Missense mutations in ITPR1 cause autosomal dominant congenital nonprogressive spinocerebellar ataxia. *Orphanet Journal of Rare Diseases, 7,* 67.

Hübner, J., Sprenger, A., Klein, C., Hagenah, J., Rambold, H., Zühlke, C., Kömpf, D., Rolfs, A., Kimmig, H., & Helmchen, C. (2007). Eye movement abnormalities in spinocerebellar ataxia type 17 (SCA17). *Neurology, 69*(11), 1160–1168.

Ikeda, Y., Shizuka, M., Watanabe, M., Okamoto, K., & Shoji, M. (2000). Molecular and clinical analyses of spinocerebellar ataxia type 8 in Japan. *Neurology, 54*(4), 950–955.

Ikeda, Y., Ohta, Y., Kobayashi, H., Okamoto, M., Takamatsu, K., Ota, T., Manabe, Y., Okamoto, K., Koizumi, A., & Abe, K. (2012). Clinical features of SCA36: A novel spinocerebellar ataxia with motor neuron involvement (Asidan). *Neurology, 79*(4), 333–341.

Ito, H., Kawakami, H., Wate, R., Matsumoto, S., Imai, T., Hirano, A., & Kusaka, H. (2006). Clinicopathologic investigation of a family with expanded SCA8 CTA/CTG repeats. *Neurology, 67*(8), 1479–1481.

Iwaki, A., Kawano, Y., Miura, S., Shibata, H., Matsuse, D., Li, W., Furuya, H., Ohyagi, Y., Taniwaki, T., Kira, J., & Fukumaki, Y. (2008). Heterozygous deletion of ITPR1, but not SUMF1, in spinocerebellar ataxia type 16. *Journal of Medical Genetics, 45*(1), 32–35. Epub 2007 Oct 11.

Jezierska, J., Stevanin, G., Watanabe, H., Fokkens, M. R., Zagnoli, F., Kok, J., Goas, J. Y., Bertrand, P., Robin, C., Brice, A., Bakalkin, G., & Durr, A. (2013). Verbeek DS Identification and characterization of novel PDYN mutations in dominant cerebellar ataxia cases. *Journal of Neurology, 260*(7), 1807–1812.

Johansson, J., Forsgren, L., Sandgren, O., Brice, A., Holmgren, G., & Holmberg, M. (1998). Expanded CAG repeats in Swedish spinocerebellar ataxia type 7 (SCA7) patients: Effect of CAG repeat length on the clinical manifestation. *Human Molecular Genetics, 7*(2), 171–176.

Johnson, J., Wood, N., Giunti, P., & Houlden, H. (2008). Clinical and genetic analysis of spinocerebellar ataxia type 11. *Cerebellum, 7*(2), 159–164.

Jones, L., Houlden, H., & Tabrizi, S. J. (2017). DNA repair in the trinucleotide repeat disorders. *Lancet Neurology, 16*(1), 88–96.

Kawaguchi, Y., Okamoto, T., Taniwaki, M., Aizawa, M., Inoue, M., Katayama, S., Kawakami, H., Nakamura, S., Nishimura, M., Akiguchi, I., et al. (1994). CAG expansions in a novel gene for Machado-Joseph disease at chromosome 14q32.1. *Nature Genetics, 8*(3), 221–228.

Khare, S., Nick, J. A., Zhang, Y., Galeano, K., Butler, B., Khoshbouei, H., Rayaprolu, S., Hathorn, T., Ranum, L. P. W., Smithson, L., Golde, T. E., Paucar, M., Morse, R., Raff, M., Simon, J., Nordenskjöld, M., Wirdefeldt, K., Rincon-Limas, D. E., Lewis, J., Kaczmarek, L. K., Fernandez-Funez, P., Nick, H. S., & Waters, M. F. (2017). A KCNC3 mutation causes a neurodevelopmental, non-progressive SCA13 subtype associated with dominant negative effects and aberrant EGFR trafficking. *PLoS One, 12*(5), e0173565.

Kim, J. S., Kim, J. S., Youn, J., Seo, D. W., Jeong, Y., Kang, J. H., Park, J. H., & Cho, J. W. (2013a). Ocular motor characteristics of different subtypes of spinocerebellar ataxia: Distinguishing features. *Movement Disorders, 28*(9), 1271–1277.

Kim, J. S., Son, T. O., Youn, J., Ki, C. S., & Cho, J. W. (2013b). Non-ataxic phenotypes of SCA8 mimicking amyotrophic lateral sclerosis and parkinson disease. *Journal of Clinical Neurology, 9*(4), 274–279. https://doi.org/10.3988/jcn.2013.9.4.274. Epub 2013 Oct 31.

Klockgether, T. (2008). The clinical diagnosis of autosomal dominant spinocerebellar ataxias. *The Cerebellum, 7*(2), 101–105.

Klostermann, W., Zühlke, C., Heide, W., Kömpf, D., & Wessel, K. (1997). Slow saccades and other eye movement disorders in spinocerebellar atrophy type 1. *Journal of Neurology, 244*(2), 105–111.

Knight, M. A., Gardner, R. J., Bahlo, M., Matsuura, T., Dixon, J. A., Forrest, S. M., & Storey, E. (2004). Dominantly inherited ataxia and dysphonia with dentate calcification: Spinocerebellar ataxia type 20. *Brain, 127*(Pt 5), 1172–1181.

Knight, M. A., Hernandez, D., Diede, S. J., Dauwerse, H. G., Rafferty, I., van de Leemput, J., Forrest, S. M., Gardner, R. J., Storey, E., van Ommen, G. J., Tapscott, S. J., Fischbeck, K. H., & Singleton, A. B. (2008). A duplication at chromosome 11q12.2-11q12.3 is associated with spinocerebellar ataxia type 20. *Human Molecular Genetics, 17*(24), 3847–3853.

Kobayashi, H., Abe, K., Matsuura, T., Ikeda, Y., Hitomi, T., Akechi, Y., Habu, T., Liu, W., Okuda, H., & Koizumi, A. (2011). Expansion of intronic GGCCTG hexanucleotide repeat in NOP56 causes SCA36, a type of spinocerebellar ataxia accompanied by motor neuron involvement. *American Journal of Human Genetics, 89*(1), 121–130.

Koide, R., Kobayashi, S., Shimohata, T., Ikeuchi, T., Maruyama, M., Saito, M., Yamada, M., Takahashi, H., & Tsuji, S. (1999). A neurological disease caused by an expanded CAG tri-nucleotide repeat in the TATA-binding protein gene: A new polyglutamine disease? *Human Molecular Genetics, 8*(11), 2047–2053.

Lang, A. E., Rogaeva, E. A., Tsuda, T., Hutterer, J., & George-Hyslop, P. (1994). Homozygous inheritance of the Machado-Joseph disease gene. *Annals of Neurology, 36*, 443–447.

Li, M., Pang, S. Y., Song, Y., Kung, M. H., Ho, S. L., & Sham, P. C. (2013). Whole exome sequencing identifies a novel mutation in the transglutaminase 6 gene for spinocerebellar ataxia in a Chinese family. *Clinical Genetics, 83*(3), 269–273.

Liang, L., Chen, T., & Wu, Y. (2016). The electrophysiology of spinocerebellar ataxias. *Neurophysiologie Clinique, 46*(1), 27–34.

Lin, X., & Ashizawa, T. (2005). Recent progress in spinocerebellar ataxia type-10 (SCA10). *Cerebellum, 4*(1), 37–42.

Lin, I. S., Wu, R. M., Lee-Chen, G. J., Shan, D. E., & Gwinn-Hardy, K. (2007). The SCA17 phenotype can include features of MSA-C, PSP and cognitive impairment. *Parkinsonism & Related Disorders, 13*(4), 246–249.

Magaña, J. J., Velázquez-Pérez, L., & Cisneros, B. (2013). Spinocerebellar ataxia type 2: Clinical presentation, molecular mechanisms, and therapeutic perspectives. *Molecular Neurobiology, 47*(1), 90–104.

Manto, M. (2010). *Cerebellar disorders. A practical approach to diagnosis and management.* New York: Cambridge University Press.

Mariotti, C., Alpini, D., Fancellu, R., Soliveri, P., Grisoli, M., Ravaglia, S., Lovati, C., Fetoni, V., Giaccone, G., Castucci, A., Taroni, F., Gellera, C., & Di Donato, S. (2007). Spinocerebellar ataxia type 17 (SCA17): Oculomotor phenotype and clinical characterization of 15 Italian patients. *Journal of Neurology, 254*(11), 1538–1546.

Mariotti, C., Brusco, A., Di Bella, D., Cagnoli, C., Seri, M., Gellera, C., Di Donato, S., & Taroni, F. (2008). Spinocerebellar ataxia type 28: A novel autosomal dominant cerebellar ataxia characterized by slow progression and ophthalmoparesis. *Cerebellum, 7*(2), 184–188.

Maschke, M., Oehlert, G., Xie, T. D., Perlman, S., Subramony, S. H., Kumar, N., Ptacek, L. J., & Gomez, C. M. (2005). Clinical feature profile of spinocerebellar ataxia type 1-8 predicts genetically defined subtypes. *Movement Disorders, 20*(11), 1405–1412.

Matsuura, T., Yamagata, T., Burgess, D. L., Rasmussen, A., Grewal, R. P., Watase, K., Khajavi, M., McCall, A. E., Davis, C. F., Zu, L., Achari, M., Pulst, S. M., Alonso, E., Noebels, J. L., Nelson, D. L., Zoghbi, H. Y., & Ashizawa, T. (2000). Large expansion of the ATTCT pentanucleotide repeat in spinocerebellar ataxia type 10. *Nature Genetics, 26*(2), 191–194.

McIntosh, C. S., Aung-Htut, M. T., Fletcher, S., & Wilton, S. D. (2017). Polyglutamine ataxias: From clinical and molecular features to current therapeutic strategies. *J Genet Syndr Gene Ther, 8*, 2.

Michalik, A., Martin, J. J., & Van Broeckhoven, C. (2004). Spinocerebellar ataxia type 7 associated with pigmentary retinal dystrophy. *European Journal of Human Genetics, 12*(1), 2–15.

Misceo, D., Fannemel, M., Barøy, T., Roberto, R., Tvedt, B., Jaeger, T., Bryn, V., Strømme, P., & Frengen, E. (2009). SCA27 caused by a chromosome translocation: Further delineation of the phenotype. *Neurogenetics, 10*(4), 371–374.

Miyoshi, Y., Yamada, T., Tanimura, M., Taniwaki, T., Arakawa, K., Ohyagi, Y., Furuya, H., Yamamoto, K., Sakai, K., Sasazuki, T., & Kira, J. (2001). A novel autosomal dominant spinocerebellar ataxia (SCA16) linked to chromosome 8q22.1-24.1. *Neurology, 57*(1), 96–100.

Mondal, B., Paul, P., Paul, M., & Kumar, H. (2013). An update on Spino-cerebellar ataxias. *Annals of Indian Academy of Neurology, 16*(3), 295–303.

Morino, H., Matsuda, Y., Muguruma, K., Miyamoto, R., Ohsawa, R., Ohtake, T., Otobe, R., Watanabe, M., Maruyama, H., Hashimoto, K., & Kawakami, H. (2015). A mutation in the low voltage-gated calcium channel CACNA1G alters the physiological properties of the channel, causing spinocerebellar ataxia. *Molecular Brain, 8,* 89.

Moscovich, M., Okun, M. S., Favilla, C., Figueroa, K. P., Pulst, S. M., Perlman, S., Wilmot, G., Gomez, C., Schmahmann, J., Paulson, H., Shakkottai, V., Ying, S., Zesiewicz, T., Kuo, S. H., Mazzoni, P., Bushara, K., Xia, G., Ashizawa, T., & Subramony, S. H. (2015). Clinical evaluation of eye movements in spinocerebellar ataxias: A prospective multicenter study. *Journal of Neuro-Ophthalmology, 35*(1), 16–21.

Muñoz, E., Milà, M., Sánchez, A., Latorre, P., Ariza, A., Codina, M., Ballesta, F., & Tolosa, E. (1999). Dentatorubropallidoluysian atrophy in a Spanish family: A clinical, radiological, pathological, and genetic study. *Journal of Neurology, Neurosurgery, and Psychiatry, 67*(6), 811–814.

Nagaoka, U., Takashima, M., Ishikawa, K., Yoshizawa, K., Yoshizawa, T., Ishikawa, M., Yamawaki, T., Shoji, S., & Mizusawa, H. (2000). A gene on SCA4 locus causes dominantly inherited pure cerebellar ataxia. *Neurology, 54*(10), 1971–1975.

Naito, H., & Oyanagi, S. (1982). Familial myoclonus epilepsy and choreoathetosis: Hereditary dentatorubral-pallidoluysian atrophy. *Neurology, 32*(8), 798–807.

Nakamura, K., Jeong, S. Y., Uchihara, T., Anno, M., Nagashima, K., Nagashima, T., Ikeda, S., Tsuji, S., & Kanazawa, I. (2001). SCA17, a novel autosomal dominant cerebellar ataxia caused by an expanded polyglutamine in TATA-binding protein. *Human Molecular Genetics, 10*(14), 1441–1448.

Nibbeling, E. A. R., Duarri, A., Verschuuren-Bemelmans, C. C., Fokkens, M. R., Karjalainen, J. M., Smeets, C. J. L. M., de Boer-Bergsma, J. J., van der Vries, G., Dooijes, D., Bampi, G. B., van Diemen, C., Brunt, E., Ippel, E., Kremer, B., Vlak, M., Adir, N., Wijmenga, C., van de BPC, W., Franke, L., Sinke, R. J., & Verbeek, D. S. (2017). Exome sequencing and network analysis identifies shared mechanisms underlying spinocerebellar ataxia. *Brain, 140*(11), 2860–2878.

O'Hearn, E., Holmes, S. E., Calvert, P. C., Ross, C. A., & Margolis, R. L. (2001). SCA-12: Tremor with cerebellar and cortical atrophy is associated with a CAG repeat expansion. *Neurology, 56,* 299–303.

Oh, A. K., Jacobson, K. M., Jen, J. C., & Baloh, R. W. (2001). Slowing of voluntary and involuntary saccades: An early sign in spinocerebellar ataxia type 7. *Annals of Neurology, 49*(6), 801–804.

Orr, H. T. (2012). Cell biology of spinocerebellar ataxia. *The Journal of Cell Biology, 197*(2), 167–177.

Orr, H. T., Chung, M. Y., Banfi, S., Kwiatkowski, T. J., Jr., Servadio, A., Beaudet, A. L., McCall, A. E., Duvick, L. A., Ranum, L. P., & Zoghbi, H. Y. (1993). Expansion of an unstable trinucleotide CAG repeat in spinocerebellar ataxia type 1. *Nature Genetics, 4*(3), 221–226.

Ouyang, Y., Sakoe, K., Shimazaki, H., Namekawa, M., Ogawa, T., Ando, Y., Kawakami, T., Kaneko, J., Hasegawa, Y., Yoshizawa, K., Amino, T., Ishikawa, K., Mizusawa, H., Nakano, I., & Takiyama, Y. (2006). 16q-linked autosomal dominant cerebellar ataxia: A clinical and genetic study. *Journal of the Neurological Sciences, 247*(2), 180–186.

Ozaki, K., Doi, H., Mitsui, J., Sato, N., Iikuni, Y., Majima, T., Yamane, K., Irioka, T., Ishiura, H., Doi, K., Morishita, S., Higashi, M., Sekiguchi, T., Koyama, K., Ueda, N., Miura, Y., Miyatake, S., Matsumoto, N., Yokota, T., Tanaka, F., Tsuji, S., Mizusawa, H., & Ishikawa, K. (2015). A novel mutation in ELOVL4 leading to spinocerebellar ataxia (SCA) with the hot cross bun sign but lacking erythrokeratodermia: A broadened spectrum of SCA34. *JAMA Neurology, 72*(7), 797–805.

Padiath, Q. S., Srivastava, A. K., Roy, S., Jain, S., & Brahmachari, S. K. (2005). Identification of a novel 45 repeat unstable allele associated with a disease phenotype at the MJD1/SCA3 locus. *American Journal of Medical Genetics. Part B, Neuropsychiatric Genetics, 133B*(1), 124–126.

Politi, L. S., Bianchi Marzoli, S., Godi, C., Panzeri, M., Ciasca, P., Brugnara, G., Castaldo, A., Di Bella, D., Taroni, F., Nanetti, L., & Mariotti, C. M. R. I. (2016). Evidence of cerebellar and extraocular muscle atrophy differently contributing to eye movement abnormalities in SCA2 and SCA28 diseases. *Investigative Ophthalmology & Visual Science, 57*(6), 2714–2720.

Pretegiani, E., Piu, P., Rosini, F., Federighi, P., Serchi, V., Tumminelli, G., Dotti, M. T., Federico, A., & Rufa, A. (2018). Anti-saccades in cerebellar ataxias reveal a contribution of the cerebellum in executive functions. *Frontiers in Neurology, 9*, 274.

Pulst, S. M. (2016). Degenerative ataxias, from genes to therapies. The 2015 Cotzias Lecture. *Neurology, 86*(24), 2284–2290.

Ranum, L. P. W., Schut, L. J., Lundgren, J. K., Orr, H. T., & Livingston, D. M. (1994). Spinocerebellar ataxia type 5 in a family descended from the grandparents of President Lincoln maps to chromosome 11. *Nature Genetics, 8*, 280–284.

Rasmussen, A., Matsuura, T., Ruano, L., Yescas, P., Ochoa, A., Ashizawa, T., & Alonso, E. (2001). Clinical and genetic analysis of four Mexican families with spinocerebellar ataxia type 10. *Annals of Neurology, 50*(2), 234–239.

Riess, O., Rüb, U., Pastore, A., Bauer, P., & Schöls, L. (2008). SCA3: Neurological features, pathogenesis and animal models. *Cerebellum, 7*(2), 125–137.

Rivaud-Pechoux, S., Dürr, A., Gaymard, B., Cancel, G., Ploner, C. J., Agid, Y., Brice, A., & Pierrot-Deseilligny, C. (1998). Eye movement abnormalities correlate with genotype in autosomal dominant cerebellar ataxia type I. *Annals of Neurology, 43*(3), 297–302.

Rodríguez-Labrada, R., Velázquez-Pérez, L., Seigfried, C., Canales-Ochoa, N., Auburger, G., Medrano-Montero, J., Sánchez-Cruz, G., Aguilera-Rodríguez, R., Laffita-Mesa, J., Vázquez-Mojena, Y., Verdecia-Ramirez, M., Motta, M., & Quevedo-Batista, Y. (2011). Saccadic latency is prolonged in Spinocerebellar Ataxia type 2 and correlates with the frontal-executive dysfunctions. *Journal of the Neurological Sciences, 306*(1–2), 103–107.

Rodríguez-Labrada, R., Velázquez-Pérez, L., Aguilera-Rodríguez, R., Seifried-Oberschmidt, C., Peña-Acosta, A., Canales-Ochoa, N., Medrano-Montero, J., Estupiñan-Rodríguez, A., Vázquez-Mojena, Y., González-Zaldivar, Y., & Laffita Mesa, J. M. (2014). Executive deficit in spinocerebellar ataxia type 2 is related to expanded CAG repeats: Evidence from antisaccadic eye movements. *Brain and Cognition, 91*, 28–34.

Rodríguez-Labrada, R., Velázquez-Pérez, L., Auburger, G., Ziemann, U., Canales-Ochoa, N., Medrano-Montero, J., Vázquez-Mojena, Y., & González-Zaldivar, Y. (2016). Spinocerebellar ataxia type 2: Measures of saccade changes improve power for clinical trials. *Movement Disorders, 31*(4), 570–578. https://doi.org/10.1002/mds.26532. Epub 2016 Feb 5.

Rolfs, A., Koeppen, A. H., Bauer, I., Bauer, P., Buhlmann, S., Topka, H., Schöls, L., & Riess, O. (2003). Clinical features and neuropathology of autosomal dominant spinocerebellar ataxia (SCA17). *Annals of Neurology, 54*(3), 367–375.

Rüb, U., Brunt, E. R., Gierga, K., Schultz, C., Paulson, H., de Vos, R. A., & Braak, H. (2003). The nucleus raphe interpositus in spinocerebellar ataxia type 3 (Machado-Joseph disease). *Journal of Chemical Neuroanatomy, 25*(2), 115–127.

Rufa, A., & Federighi, P. (2011). Fast versus slow: Different saccadic behavior in cerebellar ataxias. *Annals of the New York Academy of Sciences, 1233*, 148–154.

Sasaki, H., Yabe, I., & Tashiro, K. (2003). The hereditary spinocerebellar ataxias in Japan. *Cytogenetic and Genome Research, 100*(1–4), 198–205.

Schelhaas, H. J., Ippel, P. F., Hageman, G., Sinke, R. J., van der Laan, E. N., & Beemer, F. A. (2001). Clinical and genetic analysis of a four-generation family with a distinct autosomal dominant cerebellar ataxia. *Journal of Neurology, 248*(2), 113–120.

Scherzed, W., Brunt, E. R., Heinsen, H., de Vos, R. A., Seidel, K., Bürk, K., Schöls, L., Auburger, G., Del Turco, D., Deller, T., Korf, H. W., den Dunnen, W. F., & Rüb, U. (2012). Pathoanatomy of cerebellar degeneration in spinocerebellar ataxia type 2 (SCA2) and type 3 (SCA3). *Cerebellum, 11*(3), 749–760.

Schmidt, T., Lindenberg, K. S., Krebs, A., Schöls, L., Laccone, F., Herms, J., Rechsteiner, M., Riess, O., & Landwehrmeyer, G. B. (2002). Protein surveillance machinery in brains with spinocerebellar ataxia type 3: Redistribution and differential recruitment of 26S proteasome subunits and chaperones to neuronal intranuclear inclusions. *Annals of Neurology, 51*, 302–310.

Schöls, L., Linnemann, C., & Globas, C. (2008). Electrophysiology in spinocerebellar ataxias: Spread of disease and characteristic findings. *Cerebellum, 7*(2), 198–203.

Seidel, K., Siswanto, S., Brunt, E. R., den Dunnen, W., Korf, H. W., & Rüb, U. (2012). Brain pathology of spinocerebellar ataxias. *Acta Neuropathologica, 124*(1), 1–21.

Seixas, A. I., Loureiro, J. R., Costa, C., Ordóñez-Ugalde, A., Marcelino, H., Oliveira, C. L., Loureiro, J. L., Dhingra, A., Brandão, E., Cruz, V. T., Timóteo, A., Quintáns, B., Rouleau, G. A., Rizzu, P., Carracedo, Á., Bessa, J., Heutink, P., Sequeiros, J., Sobrido, M. J., Coutinho, P., & Silveira, I. (2017). A pentanucleotide ATTTC repeat insertion in the non-coding region of DAB1, mapping to SCA37, causes spinocerebellar ataxia. *American Journal of Human Genetics, 101*(1), 87–103.

Sequeiros, J., & Coutinho, P. (1993). Epidemiology and clinical aspects of Machado-Joseph disease. *Advances in Neurology, 61*, 139–153.

Serrano-Munuera, C., Corral-Juan, M., Stevanin, G., San Nicolás, H., Roig, C., Corral, J., Campos, B., de Jorge, L., Morcillo-Suárez, C., Navarro, A., Forlani, S., Durr, A., Kulisevsky, J., Brice, A., Sánchez, I., Volpini, V., & Matilla-Dueñas, A. (2013). New subtype of spinocerebellar ataxia with altered vertical eye movements mapping to chromosome 1p32. *JAMA Neurology, 70*(6), 764–771.

Shakkottai, V. G., & Fogel, B. L. (2013). Clinical neurogenetics: Autosomal dominant spinocerebellar ataxias. *Neurologic Clinics, 31*(4), 487–1007.

Shakkottai, V. G., & Paulson, H. L. (2009). Physiologic alterations in ataxia: Channeling changes into novel therapies. *Archives of Neurology, 66*(10), 1196–1201.

Smith, J. K., Gonda, V. E., & Malamud, N. (1958). Unusual form of cerebellar ataxia; combined dentato-rubral and pallido-Luysian degeneration. *Neurology, 8*(3), 205–209.

Spinocerebellar Ataxia Type 14. Chen DH, Bird TD, Raskind WH. In: Adam MP, Ardinger HH, Pagon RA, Wallace SE, Bean LJH, Stephens K, Amemiya A, editors. GeneReviews® [Internet]. Seattle (WA): University of Washington, Seattle; 1993–2019. 2005 Jan 28 [updated 2013 Apr 18].

Stevanin, G., & Brice, A. (2008). Spinocerebellar ataxia 17 (SCA17) and Huntington's disease-like 4 (HDL4). *Cerebellum, 7*(2), 170–178.

Stevanin, G., Hahn, V., Lohmann, E., Bouslam, N., Gouttard, M., Soumphonphakdy, C., Welter, M. L., Ollagnon-Roman, E., Lemainque, A., Ruberg, M., Brice, A., & Durr, A. (2004a). Mutation in the catalytic domain of protein kinase C gamma and extension of the phenotype associated with spinocerebellar ataxia type 14. *Archives of Neurology, 61*(8), 1242–1248.

Stevanin, G., Bouslam, N., Thobois, S., Azzedine, H., Ravaux, L., Boland, A., Schalling, M., Broussolle, E., Dürr, A., & Brice, A. (2004b). Spinocerebellar ataxia with sensory neuropathy (SCA25) maps to chromosome 2p. *Annals of Neurology, 55*(1), 97–104.

Storey, E., & Gardner, R. J. (2012). Spinocerebellar ataxia type 15. *Handbook of Clinical Neurology, 103*, 561–565.

Storey, E., Knight, M. A., Forrest, S. M., & Gardner, R. J. (2005). Spinocerebellar ataxia type 20. *Cerebellum, 4*(1), 55–57.

Storey, E., Bahlo, M., Fahey, M., Sisson, O., Lueck, C. J., & Gardner, R. J. (2009). A new dominantly inherited pure cerebellar ataxia, SCA 30. *Journal of Neurology, Neurosurgery, and Psychiatry, 80*(4), 408–411.

Subramony, S. H., Advincula, J., Perlman, S., Rosales, R. L., Lee, L. V., Ashizawa, T., & Waters, M. F. (2013). Comprehensive phenotype of the p.Arg420his allelic form of spinocerebellar ataxia type 13. *Cerebellum, 12*(6), 932–936.

Sun, Y. M., Lu, C., & Wu, Z. Y. (2016). Spinocerebellar ataxia: Relationship between phenotype and genotype – a review. *Clinical Genetics, 90*(4), 305–314.

Takahashi, H., Ishikawa, K., Tsutsumi, T., Fujigasaki, H., Kawata, A., Okiyama, R., Fujita, T., Yoshizawa, K., Yamaguchi, S., Tomiyasu, H., Yoshii, F., Mitani, K., Shimizu, N., Yamazaki, M., Miyamoto, T., Orimo, T., Shoji, S., Kitamura, K., & Mizusawa, H. (2004). A clinical and genetic study in a large cohort of patients with spinocerebellar ataxia type 6. *Journal of Human Genetics, 49*(5), 256–264.

Teive, H. A., Munhoz, R. P., Raskin, S., Arruda, W. O., de Paola, L., Werneck, L. C., & Ashizawa, T. (2010). Spinocerebellar ataxia type 10: Frequency of epilepsy in a large sample of Brazilian patients. *Movement Disorders, 25*(16), 2875–2878.

Teive, H. A., Munhoz, R. P., Arruda, W. O., Lopes-Cendes, I., Raskin, S., Werneck, L. C., & Ashizawa, T. (2012). Spinocerebellar ataxias: Genotype-phenotype correlations in 104 Brazilian families. *Clinics (São Paulo, Brazil), 67*(5), 443–449.

Todd, P. K., & Paulson, H. L. (2010). RNA-mediated neurodegeneration in repeat expansion disorders. *Annals of Neurology, 67*(3), 291–300.

Tsoi, H., Yu, A. C., Chen, Z. S., Ng, N. K., Chan, A. Y., Yuen, L. Y., Abrigo, J. M., Tsang, S. Y., Tsui, S. K., Tong, T. M., Lo, I. F., Lam, S. T., Mok, V. C., Wong, L. K., Ngo, J. C., Lau, K. F., Chan, T. F., & Chan, H. Y. (2014). A novel missense mutation in CCDC88C activates the JNK pathway and causes a dominant form of spinocerebellar ataxia. *Journal of Medical Genetics, 51*(9), 590–595.

Vale, J., Bugalho, P., Silveira, I., Sequeiros, J., Guimarães, J., & Coutinho, P. (2010). Autosomal dominant cerebellar ataxia: Frequency analysis and clinical characterization of 45 families from Portugal. *European Journal of Neurology, 17*(1), 124–128.

van de Warrenburg, B. P., Verbeek, D. S., Piersma, S. J., Hennekam, F. A., Pearson, P. L., Knoers, N. V., Kremer, H. P., & Sinke, R. J. (2003). Identification of a novel SCA14 mutation in a Dutch autosomal dominant cerebellar ataxia family. *Neurology, 61*(12), 1760–1765.

van Dijk, G. W., Wokke, J. H., Oey, P. L., Franssen, H., Ippel, P. F., & Veldman, H. (1995). A new variant of sensory ataxic neuropathy with autosomal dominant inheritance. *Brain, 118*(Pt 6), 1557–1563.

van Swieten, J. C., Brusse, E., de Graaf, B. M., Krieger, E., van de Graaf, R., de Koning, I., Maat-Kievit, A., Leegwater, P., Dooijes, D., Oostra, B. A., & Heutink, P. (2003). A mutation in the fibroblast growth factor 14 gene is associated with autosomal dominant cerebellar ataxia. *American Journal of Human Genetics, 72*(1), 191–199.

Velázquez-Pérez, L., Seifried, C., Santos-Falcón, N., Abele, M., Ziemann, U., Almaguer, L. E., Martínez-Góngora, E., Sánchez-Cruz, G., Canales, N., Pérez-González, R., Velázquez-Manresa, M., Viebahn, B., von Stuckrad-Barre, S., Fetter, M., Klockgether, T., & Auburger, G. (2004). Saccade velocity is controlled by polyglutamine size in spinocerebellar ataxia 2. *Annals of Neurology, 56*(3), 444–447.

Verbeek, D. S., van de Warrenburg, B. P., Wesseling, P., Pearson, P. L., Kremer, H. P., & Sinke, R. J. (2004). Mapping of the SCA23 locus involved in autosomal dominant cerebellar ataxia to chromosome region 20p13-12.3. *Brain, 127*, 2551–2557.

Vinton, A., Fahey, M. C., O'Brien, T. J., Shaw, J., Storey, E., Gardner, R. J., Mitchell, P. J., Du Sart, D., & King, J. O. (2005). Dentatorubral-pallidoluysian atrophy in three generations, with clinical courses from nearly asymptomatic elderly to severe juvenile, in an Australian family of Macedonian descent. *American Journal of Medical Genetics. Part A, 136*(2), 201–204.

Vuillaume, I., Devos, D., Schraen-Maschke, S., Dina, C., Lemainque, A., Vasseur, F., Bocquillon, G., Devos, P., Kocinski, C., Marzys, C., Destée, A., & Sablonnière, B. (2002). A new locus for spinocerebellar ataxia (SCA21) maps to chromosome 7p21.3-p15.1. *Annals of Neurology, 52*(5), 666–670.

Wadia, N. H., & Swami, R. K. (1971). A new form of heredo-familial spinocerebellar degeneration with slow eye movements (nine families). *Brain, 94*(2), 359–374.

Wang, G., Sawai, N., Kotliarova, S., Kanazawa, I., & Nukina, N. (2000). Ataxin-3, the MJD1 gene product, interacts with the two human homologs of yeastDNA repair protein RAD23, HHR23A and HHR23B. *Human Molecular Genetics, 9*, 1795–1803.

Wang, J. L., Yang, X., Xia, K., Hu, Z. M., Weng, L., Jin, X., Jiang, H., Zhang, P., Shen, L., Guo, J. F., Li, N., Li, Y. R., Lei, L. F., Zhou, J., Du, J., Zhou, Y. F., Pan, Q., Wang, J., Wang, J., Li, R. Q., & Tang, B. S. (2010). TGM6 identified as a novel causative gene of spinocerebellar ataxias using exome sequencing. *Brain, 133*(Pt 12), 3510–3518.

Waters, M. F., Minassian, N. A., Stevanin, G., Figueroa, K. P., Bannister, J. P., Nolte, D., Mock, A. F., Evidente, V. G., Fee, D. B., Müller, U., Dürr, A., Brice, A., Papazian, D. M., & Pulst, S. M. (2006). Mutations in voltage-gated potassium channel KCNC3 cause degenerative and developmental central nervous system phenotypes. *Nature Genetics, 38*(4), 447–451. Epub 2006 Feb 26.

Watson, L. M., Bamber, E., Schnekenberg, R. P., Williams, J., Bettencourt, C., Lickiss, J., Jayawant, S., Fawcett, K., Clokie, S., Wallis, Y., Clouston, P., Sims, D., Houlden, H., Becker, E. B. E., & Németh, A. H. (2017). Dominant mutations in GRM1 cause spinocerebellar ataxia type 44. *American Journal of Human Genetics, 101*(3), 451–458.

Williams, A. J., & Paulson, H. L. (2008). Polyglutamine neurodegeneration: Protein misfolding revisited. *Trends in Neurosciences, 31*(10), 521–528.

Worth, P. F., Giunti, P., Gardner-Thorpe, C., Dixon, P. H., Davis, M. B., & Wood, N. W. (1999). Autosomal dominant cerebellar ataxia type III: Linkage in a large British family to a 7.6-cM region on chromosome 15q14-21.3. *American Journal of Human Genetics, 65*(2), 420–426.

Yamashita, I., Sasaki, H., Yabe, I., Fukazawa, T., Nogoshi, S., Komeichi, K., Takada, A., Shiraishi, K., Takiyama, Y., Nishizawa, M., Kaneko, J., Tanaka, H., Tsuji, S., & Tashiro, K. (2000). A novel locus for dominant cerebellar ataxia (SCA14) maps to a 10.2-cM interval flanked by D19S206 and D19S605 on chromosome 19q13.4-qter. *Annals of Neurology, 48*(2), 156–163.

Yu, G. Y., Howell, M. J., Roller, M. J., Xie, T. D., & Gomez, C. M. (2005). Spinocerebellar ataxia type 26 maps to chromosome 19p13.3 adjacent to SCA6. *Annals of Neurology, 57*(3), 349–354.

Zee, D. S., & Leigh, R. J. (2015). *The neurology of eye movements* (5th ed.). New York: Oxford University Press.

Zee, D. S., Optican, L. M., Cook, J. D., Robinson, D. A., & Engel, W. K. (1976a). Slow saccades in spinocerebellar degeneration. *Archives of Neurology, 33*(4), 243–251.

Zee, D. S., Yee, R. D., Cogan, D. G., Robinson, D. A., & Engel, W. K. (1976b). Ocular motor abnormalities in hereditary cerebellar ataxia. *Brain, 99*(2), 207–234.

Zhang, X. Y., Wang, J. J., & Zhu, J. N. (2016). Cerebellar fastigial nucleus: From anatomic construction to physiological functions. *Cerebellum Ataxias, 3*, 9.

Zhuchenko, O., Bailey, J., Bonnen, P., Ashizawa, T., Stockton, D. W., Amos, C., Dobyns, W. B., Subramony, S. H., Zoghbi, H. Y., & Lee, C. C. (1997). Autosomal dominant cerebellar ataxia (SCA6) associated with small polyglutamine expansions in the alpha 1A-voltage-dependent calcium channel. *Nature Genetics, 15*(1), 62–69.

Zühlke, C., Mikat, B., Timmann, D., Wieczorek, D., Gillessen-Kaesbach, G., & Bürk, K. (2015). Spinocerebellar ataxia 28: A novel AFG3L2 mutation in a German family with young onset, slow progression and saccadic slowing. *Cerebellum Ataxias, 2*, 19.

Ocular Motor Apraxia

Caroline Tilikete and Matthieu P. Robert

Abstract Ocular motor apraxia is a syndrome of gaze shifting failure, mainly saccades, in which patients show absent or highly delayed voluntary eye movements, although other eye movements can be preserved. One phenotype is characterized by absent or great disability to perform horizontal and vertical voluntary gaze shifting, with preservation of slow and quick phases of vestibular nystagmus. It is observed in acute brain lesions and adult-onset neurodegenerative diseases and results from dysfunction of cortical (and basal ganglia) control of voluntary eye movements. The congenital form, renamed "infantile-onset saccade initiation delay," is characterized by head thrust, highly hypometric staircase saccades, increased saccade latency, and impaired quick phases of nystagmus. It may result from involvement of superior collicular, cerebellar, and/or cerebrocerebellar circuits of conjugate gaze shifting. This last phenotype is close to the one associated with Joubert syndrome, some Gaucher disease patients, ataxia-telangiectasia, and ataxia with oculomotor apraxia.

Keywords Saccades · Eye movements · Balint's syndrome · Frontal eye field · Parietal eye field · Posterior cortical atrophy · Joubert syndrome · Infantile-onset saccade initiation delay · Ataxia-telangiectasia · Ataxia with oculomotor apraxia

C. Tilikete (✉)
Hospices Civils de Lyon, Neuro-Ophthalmology, Hôpital Neurologique Pierre Wertheimer, Bron, France

Lyon I University, Lyon, France

CRNL INSERM U1028 CNRS UMR5292, ImpAct Team, Bron, France
e-mail: caroline.tilikete@inserm.fr

M. P. Robert
Ophthalmology Department, Necker-Enfants Malades University Hospital, Assistance Publique-Hôpitaux de Paris, Paris, France

COGNAC-G, UMR 8257 CNRS-IRBA-Université Paris Descartes, Paris, France

© Springer Nature Switzerland AG 2019
A. Shaikh, F. Ghasia (eds.), *Advances in Translational Neuroscience of Eye Movement Disorders*, Contemporary Clinical Neuroscience,
https://doi.org/10.1007/978-3-030-31407-1_22

1 Introduction

According to Cogan, ocular motor apraxia should refer specifically to a form of paralysis of conjugate gaze, in which gaze shifting eye movements, mainly saccades, are impaired when they are called for in willed or purposeful action, although individual random eye movements can be fully executed (Cogan 1952). This relative specificity of the deficits indicates that the brain regions responsible for ocular motor apraxia are located upstream from the saccadic brainstem generator (Leigh and Zee 2015) and possibly include the frontoparietal cortical areas, the basal ganglia, the cerebellum, and/or the superior colliculus (Zee et al. 1977). Ocular motor apraxia is a syndrome that has been mainly described in acute brain lesions, adult-onset neurodegenerative diseases, malformative disorders such as Joubert syndrome, and inherited cerebellar ataxia or as a proper entity: the congenital ocular motor apraxia recently renamed "infantile-onset saccade initiation delay" (Salman 2015).

2 The Nosology of Ocular Motor Apraxia

The first difficulty to tackle this syndrome comes from the fact that the suitability of the term has been debated for decades. Cogan, who was one of the pioneers of this syndrome description in congenital and acquired forms, already noted that "The term ocular motor apraxia may or may not be well chosen" (Cogan and Adams 1953). The term apraxia is defined by an inability to properly execute a learned skilled movement, while no weakness, sensory deficit, language, or intellectual deficit can explain it (Etcharry-Bouyx et al. 2017). While this definition can account for acquired forms of ocular motor apraxia, the debate comes from the fact that specifically in the congenital form the entity includes less specific cases of conjugate gaze failure. Furthermore, compensatory head thrusts subtending change in gaze position are so characteristic of congenital ocular motor apraxia, "as to become diagnostic of the entity" (Cogan 1952). With this assertion, a lot of publications defining ocular motor apraxia by head thrust led to include patients without any given description of the underlying gaze palsy or to include improperly patients with acquired supranuclear brainstem saccadic gaze palsy (Yee and Purvin 2007; Zackon and Noel 1991). On the other hand, ocular motor apraxia can occur without head thrust and may be underestimated if the diagnosis is made on the presence of head thrust. Some authors suggested using the terms "intermittent horizontal saccade failure" (Harris et al. 1996) or more recently "saccade initiation delay" (Salman and Ikeda 2013) instead of "ocular motor apraxia" to better describe the eye movement disorder in this infantile form.

The second confusing element is the fact that the term congenital ocular motor apraxia has not been limited to symptoms occurring at or soon after birth, but in infancy, and includes a broad range of developmental, malformative, inherited, and degenerative diseases, overlapping acquired degenerative disease of adult-onset forms. Salman therefore suggested using the term "infantile-onset" saccade initiation delay instead of "congenital" ocular motor apraxia (Salman 2015; Salman and Ikeda 2010). Infantile-onset ocular motor apraxia is often associated with head thrust and to a wider deficit in eye movements, while adult-onset descriptions are more limited to isolated ocular motor apraxia. Whether or not both phenotypes encounter for the same anatomo-clinical entity remains to be answered.

3 Shifting Gaze Network and Ocular Motor Apraxia

Ocular motor apraxia relates to disturbances of a repertoire of gaze-shifting eye movements, allowing redirecting the line of sight to a new object of interest. These eye movements are saccades, smooth pursuit, and vergence. The specificity of this syndrome is to preserve the repertoire of gaze-shifting eye movements that rely on brainstem commands only, such as quick phases of nystagmus. A basic knowledge of the neural network subtending saccadic eye movement is necessary to better understand ocular motor apraxia in its different forms (Kennard 2011).

The first level of saccades in this repertoire is represented by quick phases of nystagmus. Quick phases enable a preview of the oncoming visual scene by redirecting gaze in direction opposite to compensatory slow eye movements that are observed during sustained vestibular or optokinetic stimulation (Leigh and Zee 2015). As for voluntary saccades, the anatomic substrate controlling the dynamic of these eye movements is in the paramedian reticular formation of the pons and mesencephalon. As for voluntary saccades, they are paralyzed or slowed in cases of brainstem lesion involving these nuclei, a syndrome called supranuclear brainstem saccadic gaze palsy (Lloyd-Smith Sequeira et al. 2017; Solomon et al. 2008). But contrary to the voluntary saccades, quick phases do not relate to upstream cortical control and should theoretically not be paralyzed in ocular motor apraxia.

The second level of eye movements in this repertoire is represented by voluntary saccades. Voluntary saccades are generated under a broad range of conditions. What can be called reactive (or reflexive) saccades are generated by the appearance of unexpected novel objects seen or heard. Volitional saccades on the other hand are made as part of purposeful behavior, such as visual searching, reading, or simply on command. This voluntary control of gaze depends upon multiple cortical areas that work in network. The main ocular motor areas in the frontal lobe are the frontal eye field (FEF), the supplementary eye field, and the dorsolateral prefrontal cortex and in the parietal lobe the parietal eye field (PEF) (Kennard 2011). Both parietal and frontal cortical areas are involved in saccade triggering, yet with a gradient between more reactive saccades mainly generated by the PEF, and more volitional saccades mainly generated by the FEF (Gaymard 2012). Descending pathways from these cortical areas will connect directly or via the basal ganglia from the frontal lobe, to superior colliculus and the brainstem neurons. Finally, the cerebellar ocular motor areas (mainly the dorsal vermis and fastigial nucleus) ensure an online side-loop control of saccade amplitude system (Manto et al. 2012). Ocular motor apraxia would result from a deficit of the circuit controlling voluntary saccades, from cortical areas to descending information in the basal ganglia, superior colliculus, and cerebellum before reaching the brainstem saccade generator (Zee et al. 1977). The deficit results in great difficulties in shifting gaze, ranging from total absence of voluntary saccades to saccades performed with high latencies or highly hypometric saccades with a staircase aspect. The nosology of ocular motor apraxia does not encomprize increased saccade latency that are infraclinical and only observed on eye movement recording. When saccades are triggered, they show normal saccade dynamics, meaning normal velocity relative to amplitude, a second feature that distinguishes ocular motor apraxia from supranuclear gaze palsy.

4 Clinical Typical Presentation of Ocular Motor Apraxia

The main common manifestation of ocular motor apraxia in infant or adult-onset is abolition of eye movements on command, while random and apparently purposeless or involuntary movements of the eyes are retained (Cogan and Adams 1953).

4.1 Acquired Form of "Pure" Ocular Motor Apraxia

In the clinical observations of "pure" and complete acquired forms of adult, ocular motor apraxia manifests as a loss or great disability to perform horizontal and vertical cortically controlled eye movements (voluntary saccades, pursuit, and vergence), with conservation of brainstem-controlled eye movements such as slow and quick phases of vestibular ocular reflex (VOR) (Dehaene and Lammens 1991; Genc et al. 2004; Monaco et al. 1980; Pierrot-Deseilligny et al. 1988). Loss of initiation of both volitional and reactive saccades has been described in cases with extended frontoparietal lesions (Dehaene and Lammens 1991; Genc et al. 2004; Pierrot-Deseilligny et al. 1988) or disruption of the descending pathways (Chung et al. 2006). Loss of initiation of voluntary saccades with dissociated preservation of reactive saccades was observed in some cases with prominent involvement of the frontal lobe (Chen and Thurtell 2012; Desestret et al. 2013; Sharpe et al. 1979). As a mirror model, loss of visually guided reactive saccades (also reported as gaze apraxia or psychic paralysis of gaze in Balint's syndrome) with relative preservation of voluntary triggered saccades was observed in patients with bilateral posterior parietal lobe lesions (Biotti and Pisella 2012; Cogan 1965; Hecaen and De Ajuriaguerra 1954; Pierrot-Deseilligny et al. 1986). Acquired ocular motor apraxia is most often associated with a disturbance of the ability to shift the direction of attention, more specifically in the parietal lobe presentation (Nyffeler et al. 2005). In all cases, it appears that the harder the patient tries to turn his eyes in the desired direction, the more he is unable to do so (Cogan and Adams 1953). Some patients can better trigger some saccades with small head movements and/or blinks (Pierrot-Deseilligny et al. 1988; Rambold et al. 2006). Large head thrusts are barely observed in acquired adult-onset ocular motor apraxia, while it has been observed in patients with supranuclear brainstem saccadic gaze palsy (Yee and Purvin 2007; Lee et al. 2017). The following ocular responses during smooth pursuit or optokinetic stimulation in the horizontal and vertical plane are absent or greatly disturbed (Genc et al. 2004; Pierrot-Deseilligny et al. 1988; Chung et al. 2006; Pierrot-Deseilligny et al. 1986). Convergence has not been tested systematically but seems to be impaired (Pierrot-Deseilligny et al. 1988). Since following responses during optokinetic stimulation is impaired, the only way to check quick phase is to test the vestibulo-ocular reflex: both slow phases and quick phases of VOR must be preserved.

4.2   Congenital Ocular Motor Apraxia

In the original description of congenital ocular motor apraxia by Cogan, patients showed an inability to move their eyes voluntarily in a horizontal direction despite the fact that random eye movements and vertical ocular motility were perfectly normal (Cogan 1952; Altrocchi and Menkes 1960). It manifests by abnormal ocular following movements noticed by the parents a few weeks after birth, and infants are often thought to be blind. Around 4–6 months, while developing head control, the typical head thrusts become obvious (Salman 2015). Patients use overshooting head movement to trigger intact VOR, which drives their eyes into an extreme contraversive position in the orbit until they become aligned with the target. Then the head rotates backward, while the eyes maintain fixation on the target using again VOR. The patients show great difficulties to perform horizontal saccades in both directions on command or on purpose, but random saccades can be elicited (Cogan 1952). Smooth pursuit is not systematically impaired, but absent quick phases of horizontal optokinetic nystagmus (OKN) and VOR are noticed in the initial cases (Cogan 1952). Vertical saccades, slow phases of OKN, and VOR remain normal in most of the cases.

Ocular motor recording with the head immobilized demonstrate a number of horizontal ocular motor abnormalities, but delayed initiation (increased latency) and hypometria of voluntary saccades are the most prominent (Zee et al. 1977; Harris et al. 1996). When saccades are hypometric, multiple small saccades are sometimes required to reach the target, a pattern called staircase saccades. Saccadic velocity relative to amplitude is preserved, allowing for a clear differentiation from supranuclear brainstem saccadic gaze palsy. However, unlike acquired ocular motor apraxia, initiation of quick phases of vestibular and optokinetic nystagmus may show some failures (Zee et al. 1977; Harris et al. 1996). Vertical saccades appear normal clinically, but eye movement recording can show impairment (Orssaud et al. 2009).

5 Etiologies and Specific Clinical Phenotypes

Ocular motor apraxia is mainly a syndrome occurring in different entities such as acute brain lesions, adult-onset neurodegenerative diseases, malformative disorders such as Joubert syndrome, metabolic diseases such as Gaucher disease, and inherited cerebellar ataxia or as a proper entity named infantile-onset saccade initiation delay (congenital ocular motor apraxia).

5.1 Acquired Ocular Motor Apraxia due to Acute Brain Lesion

Acquired ocular motor apraxia is a rare entity that has been described in nonspecific brain lesions following cerebral venous thrombosis, frontoparietal tumor, meningoencephalitis, head trauma, angioma, ischemic or hemorrhagic stroke, or hypoxia (Cogan and Adams 1953; Dehaene and Lammens 1991; Genc et al. 2004; Monaco et al. 1980; Pierrot-Deseilligny et al. 1988; Chen and Thurtell 2012; Cogan 1965; Pierrot-Deseilligny et al. 1986).

The lesions are always bilateral; their precise topography has been described on brain neuropathology (Cogan and Adams 1953; Dehaene and Lammens 1991; Hecaen and De Ajuriaguerra 1954; Pierrot-Deseilligny et al. 1986; Michel and Jeannerod 2005), CT scan (Monaco et al. 1980), or MRI (Genc et al. 2004; Pierrot-Deseilligny et al. 1988; Chen and Thurtell 2012; Nyffeler et al. 2005; Rambold et al. 2006). The majority of cases involve bilaterally the frontoparietal lobes, and some patients disclose lesions limited to the frontal or to the parietal lobes. Only one patient had lesions involving bilaterally the basal ganglia (Chung et al. 2006). Precise anatomical analysis of lobar lesions showed involvement of the cortical areas involved in eye movement, either FEF (sometimes with supplementary eye fields) or PEF (Dehaene and Lammens 1991; Pierrot-Deseilligny et al. 1988; Chen and Thurtell 2012; Pierrot-Deseilligny et al. 1986).

The clinical presentation is typical of what has been described previously in complete acute acquired forms of adult. In patients with pure frontal lobe lesions, ocular motor apraxia may be isolated or associated mostly with bilateral motor deficits, apraxia of lid closure (Monaco et al. 1980), or flattened affect (Genc et al. 2004). In patients with parietal lesions, the syndrome of ocular motor apraxia is always associated with symptoms of Balint's syndrome such as optic ataxia and/or simultagnosia (Hecaen and De Ajuriaguerra 1954). Finally, the patient with bilateral basal ganglia lesions showed dysphasia, dysarthria, facial diplegia, and apraxia of lid opening (Chung et al. 2006).

5.2 Acquired Ocular Motor Apraxia due to Adult-Onset Neurodegenerative Disorders

Ocular motor apraxia may be part of symptoms observed in focal onset degenerative disorders. It has been reported regularly in the syndrome of posterior cortical atrophy (PCA). Posterior cortical atrophy was first used by Benson et al. (Benson et al. 1988). It is described as an insidious onset and progressive cognitive deficit in which the onset is characterized by early, higher-order visual deficits, whereas anterograde memory, speech and nonvisual language function, executive functions, and behavior and personality are preserved or only mildly impaired until late in the clinical course. Patients develop features such as space perception deficit, simultagnosia, object perception deficit, constructional dyspraxia, environmental agnosia, ocular motor apraxia, dressing apraxia, optic ataxia, alexia, left/right disorientation, acalculia, limb apraxia (not limb-kinetic), aperceptive prosopagnosia, agraphia, homonymous visual field defect, and finger agnosia (Crutch et al. 2017). Some symptoms are features of Balint's syndrome (ocular motor apraxia, optic ataxia, and simultagnosia) and Gerstmann's syndrome (acalculia, agraphia, finger agnosia, and left–right disorientation). Progression of PCA ultimately leads to a more diffuse pattern of cognitive dysfunction. Predominant parieto-occipital or occipito-temporal atrophy/hypometabolism/hypoperfusion is demonstrated on magnetic resonance imaging (Benson et al. 1988), FDG-PET studies (Bokde et al. 2001), and SPECT or combined PET-MRI (Moodley et al. 2015). Within PCA, Balint's syndrome and ocular motor apraxia are more correlated with damage to the bilateral dorsal parieto-occipital regions (Kas et al. 2011). In these cases frontal and mesiotemporal regions are relatively spared (Bokde et al. 2001). Neuropathological studies showed that the majority of PCA are Alzheimer disease (Formaglio et al. 2011), more rarely corticobasal degeneration, Lewy body disease, or prion disease (Crutch et al. 2017; Renner et al. 2004; Tang-Wai et al. 2004).

Ocular motor apraxia has also been described as a predominant clinical syndrome associated with frontotemporal lobar degeneration (FTLD). Clinically, progressive behavioral and language dysfunctions are the dominant manifestations of FTLD, but we described a patient with progressive reading difficulties revealing an acquired horizontal form of ocular motor apraxia associated with vertical supranuclear ophthalmoplegia, in whom neuropathological analysis diagnosed frontotemporal lobar degeneration (Desestret et al. 2013) (Fig. 1).

5.3 Ocular Motor Apraxia in Joubert Syndrome

Joubert syndrome (JS) is an inherited mid-hindbrain malformation that manifests as congenital cerebellar ataxia and can be associated with variable organ involvement (Romani et al. 2013). JS is part of the expanding group of disorders collectively

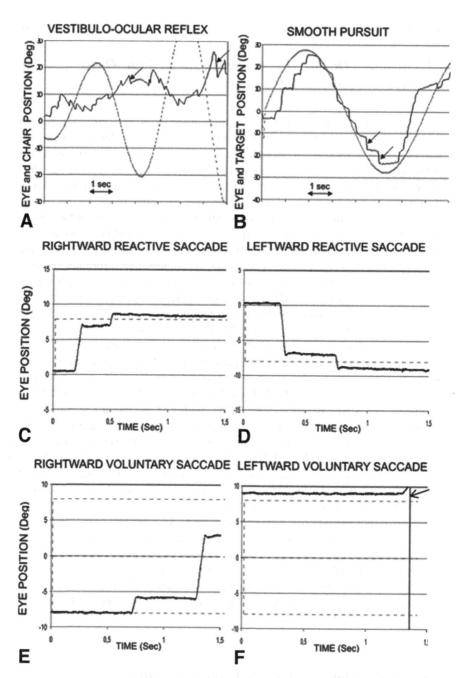

Fig. 1 Eye-movement recordings in a patient showing ocular motor apraxia in the context of fronto-temporal lobe degeneration syndrome (Desestret et al. 2013). In all six graphs, horizontal eye position (full line) and target position (dashed line) in degrees (deg) are presented relative to time in seconds (sec). Rightward and leftward reactive saccades appear normal in latency and amplitude. Rightward scanning voluntary saccades present abnormal latency and amplitude, while leftward voluntary saccades are absent and interrupted by a blink (arrow). Vestibulo-ocular reflex during pendular chair stimulation is normal and leads to rightward and leftward (arrows) quick phases (reflexive saccades). Smooth pursuit shows saccadic (arrows) following of the target

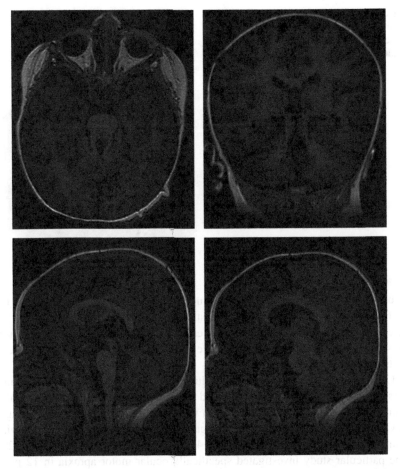

Fig. 2 The molar tooth sign (MTS) on T1-weighted MRI in a 9-month-old infant. Top left: axial images with MTS appearance, deep interpeduncular fossa, thickened superior cerebellar peduncles. Top right: coronal image showing the thickened superior cerebellar peduncles. Bottom left: sagittal image showing dysplasia of the superior vermis and enlargement of the fourth ventricle. Bottom right: parasagittal image showing thickened, elongated and horizontalized superior cerebellar peduncles

termed "ciliopathies," related to the dysfunction of the primary cilium, hence the large spectrum of associated diseases, either congenital or progressive. It typically begins during infancy by hypotonia, abnormal ocular movements (mainly ocular motor apraxia, nystagmus and strabismus), and occasionally alterations of the respiratory pattern (Joubert et al. 1969). Later, children are delayed in their acquisition of developmental steps, show intellectual disability and speech disorders due to oromotor apraxia. The molar tooth sign (MTS) seen on brain MRI is characteristic of the disease (Fig. 2). The MTS corresponds to absent or hypoplastic posterior cerebellar vermis, horizontalized, thickened and elongated superior cerebellar peduncles, deep interpeduncular fossa, showing on axial neuroimaging a unique

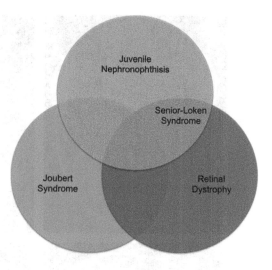

"molar tooth" appearance of these structures (Maria et al. 1997). Other organ dysfunction can be observed, the most frequent being retinal dystrophy, renal defects, polydactyly, and congenital liver fibrosis (Fig. 3). To date, more than 20 causative genes have been identified, all encoding for proteins of the primary cilium or its apparatus (Romani et al. 2013).

Ocular motor abnormalities in JS include ocular motor apraxia, defect in smooth pursuit, VOR cancellation deficit (related to cerebellar dysfunction), horizontal strabismus, pendular nystagmus (related to visual acuity defect), see-saw nystagmus, elevation of the abducting eye in lateral gaze and periodic alternating gaze deviation (Maria et al. 1997; Papanagnu et al. 2014; Tusa and Hove 1999; Weiss et al. 2009). One particular study investigated specifically ocular motor apraxia in 12 patients with JS (Tusa and Hove 1999). In this study, subjects showed partial to complete ocular motor apraxia in which initiation of saccades was prolonged or impaired. The dysfunction involved both volitional saccades and quick phases of nystagmus in the horizontal and vertical directions. When obtained, increased saccade latency and hypometric staircase saccades were commonly seen (Sturm et al. 2010). Head thrust are common.

Ocular motor apraxia in JS has been said to differ from the "idiopathic" forms of infantile-onset saccade initiation delay syndrome in the fact that both horizontal and vertical eye saccades could be involved, that it would not improve with aging and that associated cerebellar abnormal eye movements are commonly observed. However, the current view is that all infantile saccade initiation delay syndromes would lie on the same large spectrum, the most sever forms only having been formerly recognized in the past as JS.

5.4   Ocular Motor Apraxia in Gaucher Disease

Gaucher disease is a recessively inherited metabolic sphingolipid storage disease caused by deficiency in glucocerebrosidase. Besides splenomegaly and bone involvement, neurological manifestations are observed in type II and III. The major findings are seizures, myoclonus and dementia. Among these 2 forms, disturbances of horizontal gaze seem to be mainly observed in type III, where neurologic involvement progresses more slowly. Two different types of disturbances of horizontal gaze are observed: there is a predominance of slowing of horizontal saccades suggestive of supranuclear brainstem saccadic gaze palsy (Benko et al. 2011; Bohlega et al. 2000; Cogan et al. 1981) but some patients show aspects of ocular motor apraxia that resembles infantile-onset saccade initiation delay syndrome. In the latter form, the patients show no voluntary saccades or random horizontal eye movements, normal pursuit, preserved vertical eye movements and compensatory head thrusts (Cogan et al. 1981; Gross-Tsur et al. 1989; Nagappa et al. 2015).

5.5   Ocular Motor Apraxia Associated with Inherited
          Cerebellar Ataxia

Ocular motor apraxia is a dominant symptom in a group of autosomal recessive cerebellar ataxia (ARCA), including ataxia-telangiectasia (A-T), A-T like disorders and ataxia with ocular motor apraxia type 1, type 2, and type 4. Putative underlying molecular dysfunction of this group of ARCA concerns the control of polynucleotides metabolism and of cell cycle (Fogel and Perlman 2007).

Ocular Motor Apraxia in Ataxia-Telangectasia and the Like Disorders

Ataxia-telangiectasia (A-T) is an autosomal recessive inherited progressive infant-onset cerebellar ataxia characterized by associated ocular motor apraxia, choreoathetosis, dystonia, myoclonus, telangiectasias of the conjunctivae, immunodeficiency and frequent infections, and an increased risk of malignancy (Lohmann et al. 2015; Meneret et al. 2014). It is linked to the ataxia-telangiectasia mutated gene.

Besides ocular motor findings that are suggestive of cerebellar dysfunction, such as impaired smooth pursuit, optokinetic slow phases, vestibular slow phases, cancellation of vestibular slow phases, nystagmus and saccadic intrusions, the disease is characterized by ocular motor apraxia (Lewis 2001; Lewis et al. 1999; Shaikh et al. 2009, 2011). Ocular motor apraxia has been investigated in details in A-T (Lewis et al. 1999) and is characterized by reflexive and voluntary prolonged latency, hypometric amplitude, and the use of head movements to initiate gaze. This is observed in most patients. A stair-case pattern of hypometric saccades is the rule, specifically when the head is immobilized (Lewis et al. 1999). It is noteworthy that saccade peak velocities are found normal for the intended saccade amplitude and even abnormally high for the hypometric generated saccades. Head thrust is found in nearly all patients, with the particularity of not turning past the target contrary to the head thrust observed in Cogan's ocular motor apraxia.

A-T- like disorder is a rare autosomal recessive inherited progressive infant-onset cerebellar ataxia close to the A-T phenotype except for a slower progression. It is linked to hMRE11 gene mutation and has been published in 6 families (Delia et al. 2004; Fernet et al. 2005). The ocular motor apraxia is observed but has not been detailed.

Ocular Motor Apraxia in Ataxia with Ocular Motor Apraxia Types 1, 2, and 4

Ataxia with oculomotor apraxia type 1 (AOA1) is related to mutations in the aprataxin (APTX) gene (Date et al. 2001; Le Ber et al. 2003; Moreira et al. 2001; Tranchant et al. 2003) with early initial signs at the mean age of 7 years (Le Ber et al. 2003). Clinical phenotype associates usually ocular motor apraxia, progressive cerebellar ataxia, chorea, axonal sensorimotor neuropathy and deep sensory loss along with hypoalbuminemia, high cholesterol and LDL and normal alpha-fetoprotein (AFP) serum level.

AOA type 2 is caused by the mutations in the senataxin (SETX) gene (Moreira et al. 2004). The onset of the disease usually occurs later than AOA1, classically between the age of 12 and 20 years (Le Ber et al. 2004). It is one of the most frequent types of autosomal degenerative cerebellar ataxia with a relative frequency estimated at approximately 8% of non-Friedreich's ARCA (Le Ber et al. 2004). Clinical presentation is less severe than in AOA1, with polyneuropathy, ocular motor apraxia, and, less frequently pyramidal signs and movement disorder, with increased AFP serum level.

AOA4 was more recently described in a Portuguese family with mutations in the PNKP (polynucleotide kinase 30-phosphatase) gene (Bras et al. 2015). This disease develops at a mean age of 4 years with dystonia, ataxia, ocular motor apraxia, severe disabling polyneuropathy, hypoalbuminemia, high cholesterol and LDL with a normal AFP (Bras et al. 2015). The ocular motor apraxia has not been detailed in this study.

Recently, a patient with cerebellar ataxia, ocular motor apraxia, and peripheral neuropathy was found to have XRCC1 gene mutation (Hoch et al. 2017). Interestingly XRCC1 is a protein that assembles single-strand break repair multi-protein complexes that are involved in AOA1, 2 and 4 mutations. The aspect of ocular motor apraxia was not detailed.

Ocular motor apraxia as defined by horizontal gaze failure and compensatory head thrust is not a permanent symptom of AOA1 and 2. Indeed previous studies evaluated its occurrence in 86% of the AOA1 and 51% of the AOA2 patients (Le Ber et al. 2003; Anheim et al. 2009). However, ocular motor apraxia can be most

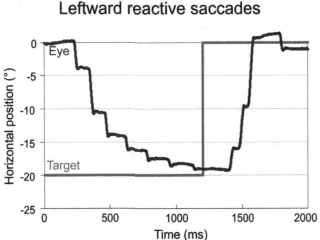

Fig. 4 Examples of saccadic recordings for a patient with AOA2. The horizontal eye position (black for control and dark gray for patient) during a reactive saccade trial is presented as a function of time and is aligned on target appearance (light gray). In these recordings, the target was presented at −20°. The primary saccade is initiated after a normal latency; however, this saccade is small (<5°) and is followed by six other saccades ("staircase") directed toward the target

frequently observed without compensatory head thrust (Panouilleres et al. 2013). In a detailed analysis of eye and head movement in AOA2 in patients without head thrust, our team found that latencies were normal for reactive saccades, scanning saccades and anti-saccades while they were longer for memory-guided saccades. Patients made hypometric primary saccades followed by staircase corrective saccades (Fig. 4). Saccade dynamics were slightly affected in patients and characterized by slower reactive and scanning saccades compared to controls. In AOA1, eye movement recording showed normal horizontal saccade latencies, with hypometric staircase aspect, with relative normal velocity (Le Ber et al. 2003). Clinical examination and eye movement recording in both AOA1 and 2 revealed less specific cerebellar oculomotor dysfunction such as fixation instability or squarewave jerks, saccadic pursuit and gaze-evoked nystagmus in all patients (Le Ber et al. 2003; Anheim et al. 2009). The few AOA4 clinically reported patients showed saccadic intrusions and difficulty initiating changes in gaze and a hypometria (undershooting) of saccades with normal speed with compensatory head thrust (Schiess et al. 2017).

Patients with AOA1 and 2 show low frontal scores on neuropsychological assessments (Le Ber et al. 2003; Klivenyi et al. 2012). However, performance on neuropsychological tests, especially those that require rapid performance and eye or hand-eye control, must be analyzed with respect to the oculomotor deficit (Clausi et al. 2013).

5.6 Idiopathic Infantile-Onset Saccade Initiation Delay (Congenital Ocular Motor Apraxia)

The syndrome of congenital ocular motor apraxia has been renamed as infantile-onset saccade initiation delay by Salman and Ikeda in 2010 (Salman and Ikeda 2010). It mainly corresponds to the original description of Cogan (1952) that has been developed previously.

An exhaustive review of all published articles recruiting 288 patients quantified the ocular motor abnormalities (Salman and Ikeda 2013). A lot of publications does not report all eye movements and only results after excluding unreported cases are presented below. Horizontal head thrusts were reported in 85% and blinks without head thrusts in 41%. Spontaneous or random saccades are observed in 100% of patients. The fast phases of the optokinetic response and vestibulo-ocular reflex were abnormal or absent in 99% and 94% respectively. Smooth ocular pursuit was abnormal or absent in 69%, especially on eye movement recording and vestibulo-ocular reflex cancellation was defective in 93% of patients. Optokinetic slow phase was found impaired or absent in 19% of reported cases. Patients may also present with nystagmus or strabismus (esophoria). Besides the ocular motor abnormalities, infantile-onset saccade initiation delay can be associated with ataxia or clumsiness, hypotonia, developmental delay, cognitive delay, speech and language delay, reading difficulties and motor delay in around 50% of cases (Salman and Ikeda 2013). The same quantifying review disclosed normal imaging in about 40% of patients. Other patients showed different pattern of abnormalities with involvement of the cerebellum (25%), cerebrum (16%) or corpus callosum (6%) being the most constant (Salman and Ikeda 2013). In a study correlating clinical manifestations and brain MRI findings, the same team demonstrated that horizontal head thrusts typically occurred in patients with normal brain MRI or infratentorial abnormalities, whereas vertical head thrusts were a more common feature in patients with supratentorial abnormalities (Salman and Ikeda 2014). The slow phases of the optokinetic response were typically impaired in patients with MRI abnormalities than in patients with a normal brain MRI (Salman and Ikeda 2014).

Another review (Wente et al. 2016) showed that most infants with "idiopathic" saccade initiation delay actually had Joubert syndrome. The imaging signs of Joubert syndrome can easily be missed in young infants. Careful analysis of brain MRI in saccade initiation delay is mandatory. The entity of "idiopathic" saccade initiation delay is shrinking over time and might disappear. It may be that such cases are ciliopathies with no sign or minimal signs on conventional brain MRI.

The clinical manifestations of ocular motor apraxia usually improve with age (Prasad and Nair 1994). Head thrust become smaller, and patients compensate impaired saccade initiation by blink and small head jerk as observed in acquired form of ocular motor apraxia (Zee et al. 1977). Some adults are even unaware that they had a congenital ocular motor apraxia during their childhood (Orssaud et al. 2009).

6   A Single Entity?

The clinical aspect of what has been called ocular motor apraxia can be divided in two different phenotypes. However, overlapping manifestations can be demonstrated.

In patients with acute or degenerative brain lesions involving the parietal or frontal eye fields, patients present with paralysis of conjugate gaze, although random eye movements can be fully executed with quite normal saccade amplitude and velocity. The absence of dysfunction of brainstem triggered quick phases of nystagmus allows distinguishing it easily from acquired supranuclear brainstem saccadic gaze palsy. In this phenotype, a deficit in the cortical circuit for the control of conjugate gaze disrupting descending information to basal ganglia, superior colliculus, cerebellum and/or brainstem saccade generator is the most likely.

In patients with infantile-onset saccade initiation failure, Joubert syndrome, inherited cerebellar ataxia associated with ocular motor apraxia and in some scare Gaucher disease, the ocular motor phenotype is dominated by head thrust, deficit in gaze shifting due to highly hypometric staircase saccades with relative appropriate saccade velocity and impaired quick phases of nystagmus and less constantly the presence of spontaneous or random saccades and increased saccade latency. Even if the clinical picture is different, manifestations such as observation of random saccades, normal saccade velocity and increased saccade latencies helps to distinguish it from acquired supranuclear brainstem saccadic gaze palsy and overlap the phenotype observed in acute or degenerative brain lesions. This eye movement deficit suggests a predominant involvement of superior collicular and/or cerebellar circuit of conjugate gaze shifting but the presence of concomitant deficit of frontal executive functions can also suggest associated cerebrocerebellar circuit involvement.

Infantile-onset saccade initiation failure is also unique among the different diseases in presenting with unique (or dominant) involvement of horizontal saccades and the syndrome improving with age suggestive of a delay in maturation of the cerebrocerebellar ocular motor circuits more than a lesional pathogenesis. All other clinical phenotypes in adults or infants share an involvement of both horizontal and vertical gaze shifting process.

Finally overlapping presentation of associated head thrust is also observed in the different phenotypes. The compensatory overshooting head thrust is a quite constant symptom of infantile-onset ocular motor apraxia phenotype, like infantile-onset saccade initiation delay, Joubert syndrome and ataxia-telangiectasia. It is less frequent in ataxia with ocular motor apraxia type 1 or 2, and not reported in patients with ocular motor apraxia due to acute or degenerative brain lesions. However, this syndrome is not specific of infantile-onset gaze palsy since it has also been described in supranuclear brainstem saccadic gaze palsy due to acute lesions in adults (Yee and Purvin 2007; Lee et al. 2017). In ataxia-telangiectasia only, head thrust not turning past the target has been observed (Leigh and Zee 2015). Finally small head thrusts and/or blinks performed to better trigger saccades has been described in both acquired ocular motor apraxia in adulthood and with aging in infantile-onset saccadic initiation failure (Zee et al. 1977; Pierrot-Deseilligny et al. 1988; Rambold et al. 2006).

7 Conclusion

The two major phenotypes of ocular motor apraxia share some common features which are random saccades, normal saccadic velocity, increased saccadic latency suggesting that both are linked to a dysfunction of the circuits controlling conjugate gaze shifting. The acquired acute or degenerative form is mostly due to a dysfunction of cortical control of this circuit through descending pathways, while the infantile-onset, malformative, inherited forms are more linked to a dysfunction of the superior colliculus/cerebellum and cerebrocerebellar pathways within this circuit.

References

Altrocchi, P. H., & Menkes, J. H. (1960). Congenital ocular motor apraxia. *Brain, 83*, 579–588.

Anheim, M., Monga, B., Fleury, M., Charles, P., Barbot, C., Salih, M., et al. (2009). Ataxia with oculomotor apraxia type 2: Clinical, biological and genotype/phenotype correlation study of a cohort of 90 patients. *Brain, 132*(Pt 10), 2688–2698.

Benko, W., Ries, M., Wiggs, E. A., Brady, R. O., Schiffmann, R., & Fitzgibbon, E. J. (2011). The saccadic and neurological deficits in type 3 Gaucher disease. *PLoS One, 6*(7), e22410.

Benson, D. F., Davis, R. J., & Snyder, B. D. (1988). Posterior cortical atrophy. *Archives of Neurology, 45*(7), 789–793.

Biotti, D., & Pisella, L. (2012). Vighetto A. [Balint syndrome and spatial functions of the parietal lobe]. *Revue Neurologique (Paris), 168*(10), 741–753.

Bohlega, S., Kambouris, M., Shahid, M., Al Homsi, M., & Al Sous, W. (2000). Gaucher disease with oculomotor apraxia and cardiovascular calcification (Gaucher type IIIC). *Neurology, 54*(1), 261–263.

Bokde, A. L., Pietrini, P., Ibanez, V., Furey, M. L., Alexander, G. E., Graff-Radford, N. R., et al. (2001). The effect of brain atrophy on cerebral hypometabolism in the visual variant of Alzheimer disease. *Archives of Neurology, 58*(3), 480–486.

Bras, J., Alonso, I., Barbot, C., Costa, M. M., Darwent, L., Orme, T., et al. (2015). Mutations in PNKP cause recessive ataxia with oculomotor apraxia type 4. *American Journal of Human Genetics, 96*(3), 474–479.

Chen, J. J., & Thurtell, M. J. (2012). Neurological picture. Acquired ocular motor apraxia due to bifrontal haemorrhages. *Journal of Neurology, Neurosurgery, and Psychiatry, 83*(11), 1117–1118.

Chung, P. W., Moon, H. S., Song, H. S., & Kim, Y. B. (2006). Ocular motor apraxia after sequential bilateral striatal infarctions. *Journal of Clinical Neurology, 2*(2), 134–136.

Clausi, S., De Luca, M., Chiricozzi, F. R., Tedesco, A. M., Casali, C., Molinari, M., et al. (2013). Oculomotor deficits affect neuropsychological performance in oculomotor apraxia type 2. *Cortex, 49*(3), 691–701.

Cogan, D. G. (1952). A type of congenital ocular motor apraxia presenting jerky head movements. *Transactions – American Academy of Ophthalmology and Otolaryngology, 56*(6), 853–862.

Cogan, D. G. (1965). Ophthalmic manifestations of bilateral non-occipital cerebral lesions. *The British Journal of Ophthalmology, 49,* 281–297.

Cogan, D. G., & Adams, R. D. (1953). A type of paralysis of conjugate gaze (ocular motor apraxia). *A.M.A. Archives of Ophthalmology, 50*(4), 434–442.

Cogan, D. G., Chu, F. C., Reingold, D., & Barranger, J. (1981). Ocular motor signs in some metabolic diseases. *Archives of Ophthalmology, 99*(10), 1802–1808.

Crutch, S. J., Schott, J. M., Rabinovici, G. D., Murray, M., Snowden, J. S., van der Flier, W. M., et al. (2017). Consensus classification of posterior cortical atrophy. *Alzheimer's & Dementia, 13*(8), 870–884.

Date, H., Onodera, O., Tanaka, H., Iwabuchi, K., Uekawa, K., Igarashi, S., et al. (2001). Early-onset ataxia with ocular motor apraxia and hypoalbuminemia is caused by mutations in a new HIT superfamily gene. *Nature Genetics, 29*(2), 184–188.

Dehaene, I., & Lammens, M. (1991). Paralysis of saccades and pursuit: Clinicopathologic study. *Neurology, 41*(3), 414–415.

Delia, D., Piane, M., Buscemi, G., Savio, C., Palmeri, S., Lulli, P., et al. (2004). MRE11 mutations and impaired ATM-dependent responses in an Italian family with ataxia-telangiectasia-like disorder. *Human Molecular Genetics, 13*(18), 2155–2163.

Desestret, V., Streichenberger, N., Panouilleres, M., Pelisson, D., Plus, B., Duyckaerts, C., et al. (2013). An elderly woman with difficulty reading and abnormal eye movements. *Journal of Neuro-Ophthalmology, 33*(3), 296–301.

Etcharry-Bouyx, F., Le Gall, D., Jarry, C., & Osiurak, F. (2017). Gestural apraxia. *Revue Neurologique (Paris), 173*(7–8), 430–439.

Fernet, M., Gribaa, M., Salih, M. A., Seidahmed, M. Z., Hall, J., & Koenig, M. (2005). Identification and functional consequences of a novel MRE11 mutation affecting 10 Saudi Arabian patients with the ataxia telangiectasia-like disorder. *Human Molecular Genetics, 14*(2), 307–318.

Fogel, B. L., & Perlman, S. (2007). Clinical features and molecular genetics of autosomal recessive cerebellar ataxias. *Lancet Neurology, 6*(3), 245–257.

Formaglio, M., Costes, N., Seguin, J., Tholance, Y., Le Bars, D., Roullet-Solignac, I., et al. (2011). In vivo demonstration of amyloid burden in posterior cortical atrophy: A case series with PET and CSF findings. *Journal of Neurology, 258*(10), 1841–1851.

Gaymard, B. (2012). Cortical and sub-cortical control of saccades and clinical application. *Revue Neurologique (Paris), 168,* 734–740.

Genc, B. O., Genc, E., Acik, L., Ilhan, S., & Paksoy, Y. (2004). Acquired ocular motor apraxia from bilateral frontoparietal infarcts associated with Takayasu arteritis. *Journal of Neurology, Neurosurgery, and Psychiatry, 75*(11), 1651–1652.

Gross-Tsur, V., Har-Even, Y., Gutman, I., & Amir, N. (1989). Oculomotor apraxia: The presenting sign of Gaucher disease. *Pediatric Neurology, 5*(2), 128–129.

Harris, C. M., Shawkat, F., Russell-Eggitt, I., Wilson, J., & Taylor, D. (1996). Intermittent horizontal saccade failure ('ocular motor apraxia') in children. *The British Journal of Ophthalmology, 80*(2), 151–158.

Hecaen, H., & De Ajuriaguerra, J. (1954). Balint's syndrome (psychic paralysis of visual fixation) and its minor forms. *Brain, 77*(3), 373–400.

Hoch, N. C., Hanzlikova, H., Rulten, S. L., Tetreault, M., Komulainen, E., Ju, L., et al. (2017). XRCC1 mutation is associated with PARP1 hyperactivation and cerebellar ataxia. *Nature, 541*(7635), 87–91.

Joubert, M., Eisenring, J. J., Robb, J. P., & Andermann, F. (1969). Familial agenesis of the cerebellar vermis. A syndrome of episodic hyperpnea, abnormal eye movements, ataxia, and retardation. *Neurology, 19*(9), 813–825.

Kas, A., de Souza, L. C., Samri, D., Bartolomeo, P., Lacomblez, L., Kalafat, M., et al. (2011). Neural correlates of cognitive impairment in posterior cortical atrophy. *Brain, 134*(Pt 5), 1464–1478.

Kennard, C. (2011). Disorders of higher gaze control. *Handbook of Clinical Neurology, 102,* 379–402.

Klivenyi, P., Nemeth, D., Sefcsik, T., Janacsek, K., Hoffmann, I., Haden, G. P., et al. (2012). Cognitive functions in ataxia with oculomotor apraxia type 2. *Frontiers in Neurology, 3,* 125.

Le Ber, I., Moreira, M. C., Rivaud-Pechoux, S., Chamayou, C., Ochsner, F., Kuntzer, T., et al. (2003). Cerebellar ataxia with oculomotor apraxia type 1: Clinical and genetic studies. *Brain, 126*(Pt 12), 2761–2772.

Le Ber, I., Bouslam, N., Rivaud-Pechoux, S., Guimaraes, J., Benomar, A., Chamayou, C., et al. (2004). Frequency and phenotypic spectrum of ataxia with oculomotor apraxia 2: A clinical and genetic study in 18 patients. *Brain, 127*(Pt 4), 759–767.

Lee, Y. H., Jeong, S. H., Kim, H. J., Hwang, J. M., Lee, W., & Kim, J. S. (2017). Vertical head thrusting in acquired supranuclear vertical ophthalmoplegia. *Journal of Neuro-Ophthalmology, 37*(4), 386–389.

Leigh, R. J., & Zee, D. S. (2015). *The neurology of eye movements* (4th ed.). Oxford: University Press.

Lewis, R. F. (2001). Ocular motor apraxia and ataxia-telangiectasia. *Archives of Neurology, 58*(8), 1312.

Lewis, R. F., Lederman, H. M., & Crawford, T. O. (1999). Ocular motor abnormalities in ataxia telangiectasia. *Annals of Neurology, 46*(3), 287–295.

Lloyd-Smith Sequeira, A., Rizzo, J. R., & Rucker, J. C. (2017). Clinical approach to supranuclear brainstem saccadic gaze palsies. *Frontiers in Neurology, 8,* 429.

Lohmann, E., Kruger, S., Hauser, A. K., Hanagasi, H., Guven, G., Erginel-Unaltuna, N., et al. (2015). Clinical variability in ataxia-telangiectasia. *Journal of Neurology, 262*(7), 1724–1727.

Manto, M., Bower, J. M., Conforto, A. B., Delgado-Garcia, J. M., da Guarda, S. N., Gerwig, M., et al. (2012). Consensus paper: Roles of the cerebellum in motor control – The diversity of ideas on cerebellar involvement in movement. *Cerebellum, 11*(2), 457–487.

Maria, B. L., Hoang, K. B., Tusa, R. J., Mancuso, A. A., Hamed, L. M., Quisling, R. G., et al. (1997). "Joubert syndrome" revisited: Key ocular motor signs with magnetic resonance imaging correlation. *Journal of Child Neurology, 12*(7), 423–430.

Meneret, A., Ahmar-Beaugendre, Y., Rieunier, G., Mahlaoui, N., Gaymard, B., Apartis, E., et al. (2014). The pleiotropic movement disorders phenotype of adult ataxia-telangiectasia. *Neurology, 83*(12), 1087–1095.

Michel, F., & Jeannerod, M. (2005). Our paper on "Visual orientation impairment in the three dimensions of space. An anatomical case". *Cortex, 41*(2), 245–247.

Monaco, F., Pirisi, A., Sechi, G. P., & Cossu, G. (1980). Acquired ocular-motor apraxia and right-sided cortical angioma. *Cortex, 16*(1), 159–167.

Moodley, K. K., Perani, D., Minati, L., Della Rosa, P. A., Pennycook, F., Dickson, J. C., et al. (2015). Simultaneous PET-MRI studies of the concordance of atrophy and hypometabolism in syndromic variants of Alzheimer's disease and frontotemporal dementia: An extended case series. *Journal of Alzheimer's Disease, 46*(3), 639–653.

Moreira, M. C., Barbot, C., Tachi, N., Kozuka, N., Uchida, E., Gibson, T., et al. (2001). The gene mutated in ataxia-ocular apraxia 1 encodes the new HIT/Zn-finger protein aprataxin. *Nature Genetics, 29*(2), 189–193.

Moreira, M. C., Klur, S., Watanabe, M., Nemeth, A. H., Le Ber, I., Moniz, J. C., et al. (2004). Senataxin, the ortholog of a yeast RNA helicase, is mutant in ataxia-ocular apraxia 2. *Nature Genetics, 36*(3), 225–227.

Nagappa, M., Bindu, P. S., Taly, A. B., & Sinha, S. (2015). Oculomotor apraxia in Gaucher disease. *Pediatric Neurology, 52*(4), 468–469.

Nyffeler, T., Pflugshaupt, T., Hofer, H., Baas, U., Gutbrod, K., von Wartburg, R., et al. (2005). Oculomotor behaviour in simultanagnosia: A longitudinal case study. *Neuropsychologia, 43*(11), 1591–1597.

Orssaud, C., Ingster-Moati, I., Roche, O., Bui Quoc, E., & Dufier, J. L. (2009). Familial congenital oculomotor apraxia: Clinical and electro-oculographic features. *European Journal of Paediatric Neurology, 13*(4), 370–372.

Panouilleres, M., Frismand, S., Sillan, O., Urquizar, C., Vighetto, A., Pelisson, D., et al. (2013). Saccades and eye-head coordination in ataxia with oculomotor apraxia type 2. *Cerebellum, 12*(4), 557–567.

Papanagnu, E., Klaehn, L. D., Bang, G. M., Ghadban, R., Mohney, B. G., & Brodsky, M. C. (2014). Congenital ocular motor apraxia with wheel-rolling ocular torsion-a neurodiagnostic phenotype of Joubert syndrome. *Journal of AAPOS, 18*(4), 404–407.

Pierrot-Deseilligny, C., Gray, F., & Brunet, P. (1986). Infarcts of both inferior parietal lobules with impairment of visually guided eye movements, peripheral visual inattention and optic ataxia. *Brain, 109*(Pt 1), 81–97.

Pierrot-Deseilligny, C., Gautier, J. C., & Loron, P. (1988). Acquired ocular motor apraxia due to bilateral frontoparietal infarcts. *Annals of Neurology, 23*(2), 199–202.

Prasad, P., & Nair, S. (1994). Congenital ocular motor apraxia: Sporadic and familial. Support for natural resolution. *Journal of Neuro-Ophthalmology, 14*(2), 102–104.

Rambold, H., Moser, A., Zurowski, B., Gbadamosi, J., Kompf, D., Sprenger, A., et al. (2006). Saccade initiation in ocular motor apraxia. *Journal of Neurology, 253*(7), 950–952.

Renner, J. A., Burns, J. M., Hou, C. E., McKeel, D. W., Jr., Storandt, M., & Morris, J. C. (2004). Progressive posterior cortical dysfunction: A clinicopathologic series. *Neurology, 63*(7), 1175–1180.

Romani, M., Micalizzi, A., & Valente, E. M. (2013). Joubert syndrome: Congenital cerebellar ataxia with the molar tooth. *Lancet Neurology, 12*(9), 894–905.

Salman, M. S. (2015). Infantile-onset saccade initiation delay (congenital ocular motor apraxia). *Current Neurology and Neuroscience Reports, 15*(5), 24.

Salman, M. S., & Ikeda, K. M. (2010). Disconnections in infantile-onset saccade initiation delay: A hypothesis. *The Canadian Journal of Neurological Sciences, 37*(6), 779–782.

Salman, M. S., & Ikeda, K. M. (2013). The syndrome of infantile-onset saccade initiation delay. *The Canadian Journal of Neurological Sciences, 40*(2), 235–240.

Salman, M. S., & Ikeda, K. M. (2014). Do the clinical features in infantile-onset saccade initiation delay (congenital ocular motor apraxia) correlate with brain magnetic resonance imaging findings? *Journal of Neuro-Ophthalmology, 34*(3), 246–250.

Schiess, N., Zee, D. S., Siddiqui, K. A., Szolics, M., & El-Hattab, A. W. (2017). Novel PNKP mutation in siblings with ataxia-oculomotor apraxia type 4. *Journal of Neurogenetics, 31*(1–2), 23–25.

Shaikh, A. G., Marti, S., Tarnutzer, A. A., Palla, A., Crawford, T. O., Straumann, D., et al. (2009). Gaze fixation deficits and their implication in ataxia-telangiectasia. *Journal of Neurology, Neurosurgery, and Psychiatry, 80*(8), 858–864.

Shaikh, A. G., Marti, S., Tarnutzer, A. A., Palla, A., Crawford, T. O., Straumann, D., et al. (2011). Ataxia telangiectasia: A "disease model" to understand the cerebellar control of vestibular reflexes. *Journal of Neurophysiology, 105*(6), 3034–3041.

Sharpe, J. A., Lo, A. W., & Rabinovitch, H. E. (1979). Control of the saccadic and smooth pursuit systems after cerebral hemidecortication. *Brain, 102*(2), 387–403.

Solomon, D., Ramat, S., Tomsak, R. L., Reich, S. G., Shin, R. K., Zee, D. S., et al. (2008). Saccadic palsy after cardiac surgery: Characteristics and pathogenesis. *Annals of Neurology, 63*(3), 355–365.

Sturm, V., Leiba, H., Menke, M. N., Valente, E. M., Poretti, A., Landau, K., et al. (2010). Ophthalmological findings in Joubert syndrome. *Eye (London, England), 24*(2), 222–225.

Tang-Wai, D. F., Graff-Radford, N. R., Boeve, B. F., Dickson, D. W., Parisi, J. E., Crook, R., et al. (2004). Clinical, genetic, and neuropathologic characteristics of posterior cortical atrophy. *Neurology, 63*(7), 1168–1174.

Tranchant, C., Fleury, M., Moreira, M. C., Koenig, M., & Warter, J. M. (2003). Phenotypic variability of aprataxin gene mutations. *Neurology, 60*(5), 868–870.

Tusa, R. J., & Hove, M. T. (1999). Ocular and oculomotor signs in Joubert syndrome. *Journal of Child Neurology, 14*(10), 621–627.

Weiss, A. H., Doherty, D., Parisi, M., Shaw, D., Glass, I., & Phillips, J. O. (2009). Eye movement abnormalities in Joubert syndrome. *Investigative Ophthalmology & Visual Science, 50*(10), 4669–4677.

Wente, S., Schroder, S., Buckard, J., Buttel, H. M., von Deimling, F., Diener, W., et al. (2016). Nosological delineation of congenital ocular motor apraxia type Cogan: An observational study. *Orphanet Journal of Rare Diseases, 11*(1), 104.

Yee, R. D., & Purvin, V. A. (2007). Acquired ocular motor apraxia after aortic surgery. *Transactions of the American Ophthalmological Society, 105*, 152–158; discussion 8–9.

Zackon, D. H., & Noel, L. P. (1991). Ocular motor apraxia following cardiac surgery. *Canadian Journal of Ophthalmology, 26*(6), 316–320.

Zee, D. S., Yee, R. D., & Singer, H. S. (1977). Congenital ocular motor apraxia. *Brain, 100*(3), 581–599.

Opsoclonus Myoclonus Syndrome

Lauren Cameron and Camilla Kilbane

Abstract Opsoclonus-myoclonus syndrome (OMS) refers to the combination of irregular, multidirectional, and chaotic eye movements that occur in all planes of gaze and is exacerbated with smooth pursuit. In addition, myoclonus, ataxia, or behavioral changes occur. Although often associated with neuroblastoma, OMS was first described in cases of encephalitis, but can occur due to various etiologies such as infections, a paraneoplastic phenomenon, or an idiopathic syndrome. OMS is rare. The pathophysiology of OMS remains unknown; however, several hypotheses have been proposed and will be discussed in this chapter. Patients usually present subacutely and require prompt workup which includes blood and cerebrospinal fluid for infections and paraneoplastic antibodies as well as imaging for occult malignancies. Treatment options include various immunomodulatory medications; however, chronic sequelae are common.

Keywords Opsoclonus · Myoclonus · Ataxia · Eye movements · Ocular flutter · Burst neurons · Brainstem · Oscillopsia

1 Definition

OMS is also known as opsoclonus-myoclonus-ataxia syndrome. "Opsoclonus" was first described by a Polish neurologist Orzechowski, in 1913, to describe rapid, chaotic, but conjugate, eye movements that he discovered in several cases of encephalitis. The term was derived from the Greek language for vision (*opsis*) and turmoil (*klonos*) (Smith and Walsh 1960). In 1962, an Australian pediatric neurologist named Marcel Kinsbourne then associated the syndrome with neuroblastoma in

L. Cameron · C. Kilbane (✉)
Department of Neurology, University Hospitals Cleveland Medical Center, Case Western Reserve University, Cleveland, OH, USA
e-mail: Camilla.kilbane@uhhospitals.org

© Springer Nature Switzerland AG 2019
A. Shaikh, F. Ghasia (eds.), *Advances in Translational Neuroscience of Eye Movement Disorders*, Contemporary Clinical Neuroscience,
https://doi.org/10.1007/978-3-030-31407-1_23

encephalopathic infants (Kinsbourne 1962). Other historical names for the same clinical phenomenon include "infantile myoclonic encephalopathy" and "dancing eyes-dancing feet" syndrome; however, "opsoclonus-myoclonus syndrome" has been the most widely accepted term to date.

Opsoclonus is defined as irregular, arrhythmic, and chaotic eye movements that occur in all directions and planes of gaze and exacerbates with pursuit (Leigh and Zee 2015). The saccadic movements of opsoclonus can be horizontal, vertical, and torsional. Opsoclonus can also be observed with closed eyelids. Smooth pursuit and optokinetic nystagmus are typically normal but may be difficult to interpret due to superimposed opsoclonus (Shawkat et al. 1993). It should be noted that ocular flutter and opsoclonus fall along a spectrum, and it is possible for ocular flutter to progress to opsoclonus. Opsoclonus can be differentiated from ocular flutter because of its multidirectional nature, whereas ocular flutter purely consists of horizontal oscillations (Leigh and Zee 2015).

Opsoclonus-myoclonus syndrome is frequently manifested by opsoclonus, myoclonus, ataxia, and behavioral changes. Myoclonus is defined as a sudden, brief, spontaneous, usually stimulus sensitive jerks of synchronized muscles typically lasting 10–50 milliseconds (Eberhardt and Topka 2017). The face, eyelids, limbs, fingers, head, and trunk are involved. However, up to one third of cases have an atypical presentation. In addition to the ocular findings, there are several other common symptoms such as tremor, dysarthria, mutism, aphasia, ataxia, and psychiatric symptoms. The syndrome typically presents itself subacutely and tends to progress quickly. There is chronic neurological sequelae in about 50–70% of patients, most notably in the cognitive and behavioral domains (Pranzatelli et al. 2017a).

2 Epidemiology and Diagnostic Criteria

OMS is a rare disorder. The incidence of OMS was 0.18 cases per million population per year in a prospective survey of the UK pediatric neurology centers. Males and females are equally affected (Pranzatelli et al. 2017a). Given the rarity of the condition, many providers will only see a few cases over their careers, and atypical presentations with, for example, delayed opsoclonus are not uncommon, occurring in 20% of presentations in the UK prospective study. Opsoclonus may also be completely absent (Herman and Siegel 2009; Krug et al. 2010). Common misdiagnosis includes Guillain-Barre syndrome, cerebellar ataxia, and epileptic seizures (Haden et al. 2009; Tate et al. 2005).

A set of diagnostic criteria has been proposed by international collaborators (Matthay et al. 2005). These criteria specify that three out of four features must be present for a diagnosis of OMS: (1) opsoclonus, (2) ataxia and/or myoclonus, (3) behavioral changes or sleep disturbances, and (4) a diagnosis of neuroblastoma (Matthay et al. 2005). Given the atypical presentations, only three out of four features must be present to make the diagnosis.

3   Etiology

Opsoclonus-myoclonus syndrome has been associated with viral infections, toxins, metabolic disorders, and neoplastic diseases but is in most instances idiopathic. In adults, about 14% of patients with opsoclonus are found to have an underlying cancer (Klaas et al. 2012). It affects 2–3% of children with neuroblastoma (Pranzatelli et al. 2017a) (Table 1).

Table 1 Etiology of opsoclonus myoclonus

Structural brain lesions
Midbrain or thalamic hemorrhages
Hydrocephalus
Parainfectious
Viral
Epstein-Barr virus
Coxsackievirus B2
Enterovirus
Human herpes virus (HHV-6)
Influenza
West Nile virus
Varicella
Cytomegalovirus
Human immunodeficiency virus[a]
Hepatitis C
Bacterial
Mycoplasma pneumoniae
Salmonella
Streptococcus
Note: Postimmunization associations have been reported following varicella, measles, and diphtheria-pertussis-tetanus vaccine administration
Toxins
Phenytoin, diazepam, lithium, thallium, amitriptyline, organophosphates, cocaine, toluene
Metabolic: Celiac disease, hyperosmolar coma
Autoimmune
Guillain-Barre syndrome (anti-GQ1b antibodies)
Sarcoidosis
Myasthenia gravis
Multiple sclerosis
Head trauma
Post-allogeneic hematopoietic stem cell transplantation
Complication of pregnancy
Transient phenomenon of normal infants
Idiopathic

[a]Note: The presence of a "viral infection" should not preclude the search for a tumor

There are several well-described paraneoplastic syndromes that have first clinically presented with opsoclonus-myoclonus syndrome (OMS). In childhood, neuroblastoma, which usually presents before age 3, is the most commonly associated malignancy (Pranzatelli et al. 2017a). In adults, the most frequently involved tumor is lung cancer, particularly small cell lung cancer, followed by breast and gynecological cancers (Klaas et al. 2012).

4   Neuroblastoma

The prevalence of neuroblastoma in OMS was 8% in the 1970s, 16% in the 1980s, 38% in the 1990s, and 43% in the 2000s, with tumors being mainly low grade (Pranzatelli et al. 2017a; Brunklaus et al. 2011). It typically presents between 6 months and 3 years of age. CT/MR imaging of the chest and abdomen is the most accurate test to detect occult neuroblastoma and should be the gold standard in the workup (Brisse et al. 2011).

Two thirds of patients will develop typical symptoms of OMS including opsoclonus, myoclonus, ataxia, and behavioral changes. However, up to one-third of cases have an atypical presentation (Pranzatelli et al. 2017a). Dysarthria, expressive, and receptive vocabulary impairment as well as attention deficit issues are common. Delayed diagnosis has shown a higher correlation with short-term memory deficits and motor disabilities. There are often chronic neurological sequelae in about 50–70% of patients, most notably in the cognitive and behavioral domains (De Grandis et al. 2009). In one study, borderline low IQ or mental retardation was found in 62% of children with opsoclonus-myoclonus syndrome related to neuroblastoma (Klein et al. 2007).

Patients with a neuroblastoma-associated OMS may have a better survival compared to patients with neuroblastoma alone (Cooper et al. 2001). Autoantibodies in OMS, against both intracellular and surface epitopes to cerebellar neurons, are predominantly IgG3. This effect is not caused by a difference in total serum IgG3. An autoimmune pathogenesis of pediatric OMS is supported not only by the detection of autoantibodies but also by the response to immunosuppressive treatment and inflammatory changes in the CSF.

Other paraneoplastic antibodies associated with OMS are anti-Hu, anti-Ro, anti-Ri, anti-Ma1 and anti-Ma 2, anti-Ta-Ma2, and anti-CRMP-5/anti- CV-2; however, there continues to be additional antibodies identified in case reports. Anti-Yo antibody, or Purkinje cell cytoplasmic antibody type 1 (PCA1), is the most common variant of paraneoplastic cerebellar degeneration (PCD) (Venkatraman and Opal 2016). The majority of cases reported have been women with pelvic or breast tumors. Anti-Ri has been associated with lung, breast, cervix, and bladder cancer. Anti-Ri antibody reacts with 55-kd and 80-kd proteins (Grant and Graus 2009). ANNA-1 is a marker of cytotoxic T-cell-mediated neuronal injury that has been associated with small cell lung carcinoma and neuroblastoma (Grant and Graus 2009). Antineurofilament antibodies (NF210K antibody), anti-Purkinje cell antibodies, and immunoglobulin G autoantibodies binding to the surface of isolated rat cerebellar granular neurons have been isolated (Hero and Schleiermacher 2013). GABA$_B$R antibody and amphiphysin antibody are seen in limbic encephalitis. One reported case showed opsoclonus that developed months before limbic encephalitis. Clinical improvement has shown some correlation with B-cell reduction in the cerebrospinal fluid (CSF) (Armangue et al. 2014).

N-methyl-D-aspartate receptor (NMDAR) antibodies are associated with systemic teratomas.

Epidemiologic studies have shown that teenagers and young adults, particularly females, are more likely to have a teratoma if presenting with acute-subacute onset OMS with CSF pleocytosis (Armangue et al. 2014). Voltage-gated potassium channel (VGKC) complexes are plasma membrane proteins that primarily function as regulators for axonal conduction, neurotransmitter release, and control of neuronal excitability. VGKC complexes become dysfunctional when in contact with pathogenic antibodies. Another similar target for VGKC complex antibodies have been described in anti-contactin-associated protein-like 2 (CASPR2) or leucine-rich glioma-inactivated 1 (LGI1).

CSF studies in paraneoplastic syndromes are often positive for oligoclonal bands, which is a marker for inflammation and neuronal injury. There is often lymphocytic pleocytosis (white blood cell count >5/µL) and elevated protein (>45 mg/dL) (Pranzatelli et al. 2004).

There continues to be additional paraneoplastic syndromes discovered for several neurological disorders. Treatment is not antibody-specific; however, in paraneoplastic syndromes, patients typically improve after immunotherapy such as intravenous immunoglobulin (IVIG) and/or plasma exchange (Pranzatelli and Tate 2017) (Table 2).

Table 2 Common paraneoplastic antibodies associated with opsoclonus

Antibody	Neurologic signs/symptoms	Cancer type
Anti-Yo (Purkinje cell)	Cerebellar	Breast, pelvic tumors
ANNA-2 (anti-Ri)	Cerebellar	Lung, breast, cervix, bladder cancers
ANNA-1 (anti-Hu)	Encephalitis, cerebellar, sensory neuropathy	SCLC, neuroblastoma
N-methyl-D-aspartate receptor (NMDAR)	Limbic encephalitis	Systemic teratomas
Anti-ta-Ma2	Encephalitis, lateral medullary syndrome	Testicular germ cell tumors
Anti-Ma1	Encephalitis, cerebellar	Lung cancer

5 Pathophysiology

The exact pathophysiology remains unknown; however, several hypothesis have been proposed. The most supported theories include the following (1) disinhibition of the fastigial nucleus in the cerebellum, (2) damage to afferent projections to the fastigial nucleus, and/or (3) a delay in the saccadic burst neuron circuitry in the brain stem (Brunklaus et al. 2011).

One of the most accepted hypotheses involves disinhibition of the fastigial nucleus of the cerebellum. Disruption of Purkinje cells in the dorsal vermis and/or their inhibitory projections to the fastigial nucleus results in the disinhibition of the fastigial nucleus in the cerebellum. The fastigial nucleus outflow pathway passes contralaterally through the uncinate fasciculus, near the opposite fastigial nucleus, to the brain stem. If there is a lesion in one fastigial nucleus it could possibly affect the contralateral fastigial nucleus or its efferents, causing bilateral saccadic overshoot dysmetria. Research supporting this hypothesis includes the following:

1. A functional magnetic resonance imaging (MRI) study demonstrating bilateral activation (i.e., disinhibition) of the fastigial nucleus in two patients with opsoclonus, not observed in healthy controls while performing high frequency saccades (Brisse et al. 2011).
2. Histopathological examination of a patient with opsoclonus revealing damage to afferent projections to the fastigial nucleus (De Grandis et al. 2009).
3. Single-photon emission computed tomography identified the area of dysfunction to the cerebellar vermis, where Purkinje cells normally exert inhibitory control over the fastigial nucleus in two patients with opsoclonus (Klein et al. 2007; Cooper et al. 2001).
4. Nova-2 is a neuronal-specific RNA binding protein that is an autoimmune target in patients with paraneoplastic opsoclonus. Defective Nova-2, which normally contributes to inhibitory synaptic transmission or synaptic plasticity, may be responsible for reduced inhibitory control of movements seen in opsoclonus-myoclonus syndrome (Venkatraman and Opal 2016).

Another hypothesis of the pathophysiology of opsoclonus involves the instability and delay in brain stem circuitry. Opsoclonus occurs when saccadic oscillations of both the horizontal and vertical saccadic pulse generators become unstable (Grant and Graus 2009; Hassan et al. 2008). Saccadic oscillations can arise both from intrinsic lesions of the pulse generator and from abnormal inputs to the pulse generator including those from the midbrain, the diencephalon, and probably the cerebellum.

There is likely two distinct mechanisms for saccadic oscillation, one which causes large oscillations, such as opsoclonus or flutter, and another which causes small oscillations such as voluntary nystagmus or microsaccadic flutter (Grant and Graus 2009; Hassan et al. 2008).

Intrinsic biological delays along neural pathways that are involved in the propagation of impulses lead to oscillations. As the delay is increased, the amplitude of the oscillation increases and the frequency decreases. If the delay is large enough, oscillations can be sustained even when the level of pause cell tone is normal. Also, when the delay is set between the normal value and the value required for sustained oscillation, damped/smaller oscillations can be produced after a saccade terminates. This may explain why flutter often occurs immediately after a voluntary saccade (Grant and Graus 2009).

Compared to microsaccadic flutter, the oscillations of flutter and opsoclonus are lower in frequency and larger in amplitude than those produced by removing the pause cell input alone. To simulate flutter and opsoclonus, one must not only decrease the level of pause cell activity but also increase the delay in the feedback signal by the saccadic pulse generator to terminate the saccade (Grant and Graus 2009).

Microelectrode studies in pontine reticular formation of monkeys have identified three types of premotor neurons related to saccadic eye movements: burst, tonic, and pause cells.

Omnipause neurons include Purkinje cells, granular cells, and the dentate nuclei of the cerebellum. Omnipause neurons inhibit horizontal saccadic burst neurons in the pontine paramedian reticular formation (PPRF). They also inhibit vertical saccadic burst neurons in the rostral interstitial nucleus of the medial longitudinal fasciculus (MLF). Omnipause neurons are found between the abducens nuclei in the nucleus raphe interpositus (Armangue et al. 2014). Inputs into the omnipause neurons arise in the superior colliculus, frontal eye fields, and mesencephalic reticular formation. The sources of excitatory and inhibitory input to the pause neurons remain unclear. These pause cells tonically discharge except during saccades. When they pause, they inhibit burst cells and prevent saccadic oscillations. Omnipause neuronal dysfunction affects saccades in all directions of gaze.

The neural integrator is comprised of neurons that provide a tonic signal proportional to eye position and provide innervation needed to maintain eccentric position of gaze after a saccade is completed. The neural integrator for horizontal movements is found just below the abducens nucleus in the nucleus prepositus hypoglossi and medial vestibular nuclei. Most of the neural integrator neurons discharge in proportion to saccadic velocities. The neural integrator is thought to be distributed over a network of burst-tonic neurons (Pranzatelli et al. 2004).

6 Pathology

Pathological reports of OMS show a variety of dysfunction. Purkinje cell loss and gliosis in the cerebellum, lesions in the periaqueductal gray, and olivary nucleus as well as perivascular inflammatory infiltrates. Granular cell layer loss in the cerebellum is also common. Lymphocytic infiltrates can be seen in the neocortex, pons, and cerebellum. Despite these abnormalities, approximately half of OMS patients are found with no significant central nervous system pathology (Stefanowicz et al. 2008).

7 Clinical Investigations

Those developing acute or subacute opsoclonus myoclonus in which an immediate underlying cause, i.e., viral infection or drug intoxication, cannot be identified, should promptly be worked up, including in particular serum examination for paraneoplastic autoantibodies. Both antibody-positive and antibody-negative patients need cancer workup. Even in patients whom the initial search for cancer is negative, they should be considered tumor suspects and investigated at frequent, appropriate intervals.

8 Laboratory Investigations

A diagnostic approach to OMS is provided in Fig. 1.

- CSF oligoclonal banding and flow cytometry. Oligoclonal bands were found in 35% of pediatric OMS patients in a prospective study of 135 patients, with the highest frequency in severe cases. Neuroblastoma detection, duration of disease, or relapse history did not differ. In untreated patients, there was a 75% reduction of bands after immunotherapy, indicating that it could serve as a marker for response to therapy. Flow cytometry is however a more sensitive marker of B-cell infiltration (Pranzatelli et al. 2011).
- Antigliadin antibodies of immunoglobulin A subtype, anti-endomysial antibodies, and anti-CV2 antibodies (celiac disease). Confirmatory diagnosis of celiac disease includes a duodenal biopsy which shows villous atrophy. Eight to ten percent patients with celiac disease have neurological manifestations (Deconinck et al. 2006).
- Check levels of urinary catecholamine metabolite. Homovanillic acid (HVA) and vanillylmandelic acid (VMA) are useful tumor markers for neuroblastoma. These assays are reportedly elevated in 95% of cases of neuroblastoma and are useful for confirmatory purposes (Wong 2007). There are, however, cases of falsely negative results, and therefore these urinary assays alone are inadequate for assessing for the presence or absence of neuroblastoma.
- Infectious workup.
- Serum and cerebrospinal fluid paraneoplastic screening.

9 Radiological Investigations

- Magnetic resonance imaging (MRI) with thin cuts through the brain as well as MRI or CT imaging of the neck, thorax, abdomen, and pelvis is recommended for evaluation for possible neuroblastoma.
- Mammography and colonoscopy in adults.
- Whole body 18F-fluoro-2-deoxyglucose positron emission tomography (FDG-PET) scan should be considered if above is negative.

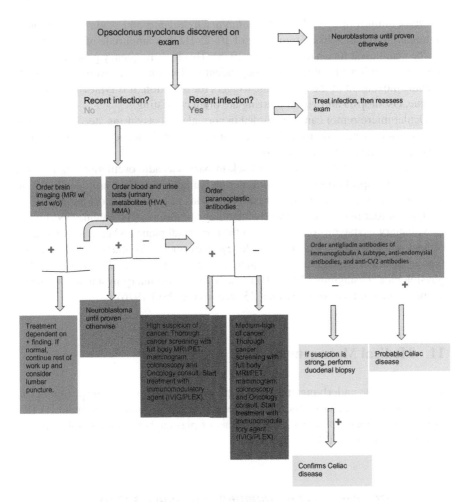

Fig. 1 Stepwise schematic workup for opsoclonus myoclonus

10   Clinical "Mimics" of Opsoclonus

Ocular flutter (OF) is considered a subtype of opsoclonus and is typically described as infrequent, rapid, conjugate, and horizontal oscillations occurring in bursts. Eye measurements reveal bursts of fast phases without inter saccadic intervals (Anagnostou et al. 2013). Frequencies of OF are typically higher than opsoclonus. OF has been described in patients with essential tremor (ET) (Anagnostou et al. 2013). The pathophysiology of OF remains controversial; however, it is hypothesized that abnormal cerebellar Purkinje cell activity leads to reduced inhibition on the fastigial nucleus. This therefore leads to enhanced inhibition of brain stem neurons. This is clinically demonstrated by saccadic burst neurons that become free to oscillate. Studies showing Purkinje cell loss in neuropathological investigation in essential tremor patients support the underlying cerebellar pathology (Anagnostou et al. 2013).

Other mimics of opsoclonus are *voluntary nystagmus* and *ocular microtremor*. Voluntary nystagmus has a frequency of 10–25 Hz, has amplitude of up to 60, and can be maintained for up to 35 s (Zahn 1978). This high-frequency pendular nystagmus can be initiated voluntarily independent of fixation. A convergence or wide opening movement of the eyelids are triggers for initiation. It is typically limited by fatigue. Oscillopsia and visual blurring are often coexisting.

Ocular microtremor can be recorded in normal subjects during steady fixation. It is a continuous near sinusoidal oscillation that has a frequency of 100 Hz and a mean amplitude of 30 (Zahn 1978).

Microsaccadic flutter consists of back-to-back saccadic oscillations, not visible by unaided inspection but seen with an ophthalmoscope. They are usually conjugate, horizontal, and symmetric in both directions of gaze. For most patients it is a benign disorder not associated with a neurological disease (Ashe et al. 1991).

Voluntary nystagmus only occurs in the horizontal plane and consists of a series of back-to-back and to-and-fro saccadic eye movements. Voluntary nystagmus is a type of pendular nystagmus. Its frequency is approximately 5–28 Hz, slower than physiological nystagmus (30–90 Hz) (Zahn 1978). Voluntary nystagmus typically cannot be maintained for more than 25 seconds (Aschoff 1976).

11 Medical Management

Treatment of the underlying etiology of opsoclonus myoclonus is the primary goal. Often, immunomodulatory therapies such as intravenous immunoglobulin (IVIG), rituximab, cyclophosphamide, azathioprine, or plasmapheresis are combined with corticosteroids or ACTH.

11.1 Corticosteroids and Immunomodulatory Agents

Corticosteroids are usually given orally as prednisolone or prednisone at a starting dose of 2 mg/kg/day. If symptoms improve, prednisone is slowly tapered starting at 2–3 months over a 9–12-month period. Recent studies have shown favorable outcomes after intravenous or oral pulses of 20 mg/m2/day of dexamethasone for 3 days given monthly (Rostasy et al. 2006). If a relapse should occur, then a higher dose of prednisone would be initiated.

ACTH: ACTH is normally secreted by the pituitary gland, and stimulates the adrenal gland to produce cortisol. It is given intramuscularly starting at 75 IU per dose twice daily × 40–52 weeks with a slow taper (Tate et al. 2012). The combination of IVIG and corticosteroids has been tried in recent years. ACTH alone and corticosteroids alone have only shown short-term benefits. Side effects of long-term therapy ACTH can include cushingoid features, fluid retention, psychological effects, cardiovascular effects, gastric ulcers, skin changes, osteoporosis, infection, and diabetes mellitus. Most are related to the treatment duration and dose. Growth suppression is common but reversible.

A treatment study with 74 children with OMS compared combination therapies of corticotropin alone, corticotropin in combination with IVIG, corticotropin in combination with IVIG and Rituxan, corticotropin in combination with IVIG and Cytoxan, and Rituxan plus chemotherapy. Fifty-five percent had adverse events (corticosteroid excess), more so with multiagents, and 10% had serious adverse events. The results showed better long-term clinical improvement in the groups with the three-agent and four-agent multimodal treatments compared to conventional treatment with corticotropin alone. This study demonstrated greater efficacy of corticotropin-based multimodal therapy compared with conventional therapy, greater response to corticotropin than corticosteroid-based therapy, and overall tolerability. The treatment responses did not differ significantly between patients whom had tumors discovered (Tate et al. 2012).

Intravenous immunoglobulins (IVIG): IVIG binds to autoantibodies and target cells and decreases levels of circulating immune complexes. IVIG is a useful treatment when given as 1–2 g/kg/day for 1 or more days then a monthly maintenance dose of 1 g/kg. There are typically fewer side effects than chemotherapy or corticosteroids. Potential side effects include headache, nausea, fever, or flu-like symptoms but are usually mild and self-limited. Renal impairment and aseptic meningitis can occur. Some strategies to prevent these side effects include reducing the rate of the infusion and/or concomitant administration of Benadryl and acetaminophen intravenously 30 min prior to IVIG infusion. Preliminary blood work includes an IgA level and kidney function tests (Pless and Ronthal 1996).

There may not be any response until 10–14 days from the first infusion. If there is a good response, IVIG should be given monthly. If there is lack of efficacy, a higher dose should be used the following month. If there is no significant improvement after the second infusion, IVIG should be discontinued.

When a child does not respond to IVIG or has side effects, a different brand should be tried.

Some IVIG subtypes: Cytomegalovirus IVIG, Gamimune N S/D 10%, Flebogamma, Gamunex, Privigen, Gammagard, Panglobulin, Polygam solvent/detergent treated, normal immunoglobulin (human) 12% (120 g/L) intravenous injection (Sandoglobulin), Venoglobulin-S (5% or 10%), and $Rh_o(D)$ immune globulin intravenous (Hartung 2008).

Azathioprine (Imuran): Dose is typically 50 mg per day for 6 months. Prior blood work is necessary before starting Imuran which includes peripheral leukocyte count, platelet count, and liver function tests. The delay to onset of therapeutic effect is 6–12 months. Potential adverse effects are fever, rash, flu-like symptoms, and nausea.

Approximately 10% of patients develop an idiosyncratic flu-like reaction precluding its use. Another important adverse effect is bone marrow suppression.

Rituximab (Rituxan): Rituximab, a chimeric anti-CD20 monoclonal antibody that depletes circulating B cells, has been shown to be well tolerated and beneficial. It is given by intravenous infusion once a week for total of four doses. Its effect can last 6–9 months. Rituximab can be associated with more serious hypersensitivity reactions such as low blood pressure (hypotension), breathing difficulties (bronchospasm), and sensation of the tongue or throat swelling (angioedema). Epinephrine, antihistamine, and corticosteroids should be on hand for treatment of hypersensitivity. Complete blood counts (CBC) and platelet counts should be monitored at intervals. Twelve immunotherapy-naïve children with opsoclonus-myoclonus syndrome and CSF B-cell expansion received rituximab, ACTH, and IVIG. Motor severity lessened 73% by 6 months and 81% at 1 year (Pranzatelli et al. 2010). Additional maintenance rituximab may be necessary to prevent relapses (Toyoshima et al. 2016; Leen et al. 2008).

Cyclophosphamide (Cytoxan): Cytoxan can be given in different forms, tablet, injectable, or oral solution. Typical pediatric dose is 1–5 mg per kg of body weight. It is taken over a period of 60 to 90 days. Frequent monitoring for leukopenia as well as hematuria is suggested.

Potential adverse effects include loss of appetite, nausea, vomiting, constipation, temporary hair thinning or brittleness, and increased risk of infections. Combination therapy of cyclophosphamide and dexamethasone pulses showed improvement in two girls with prolonged clinical course with many relapses, even after 18 months of treatment (Wilken et al. 2008).

6-mercaptopurine (6-MP): 6-MP, a known immunosuppressant, has been shown to normalize cerebrospinal fluid (CSF) lymphocyte frequencies in opsoclonus-myoclonus syndrome and function as a steroid sparer. 6-MP is difficult to titrate and may elevate liver transaminases, which are typically mild and reversible. Possible side effects include anemia, risk of bruising or bleeding, infection, nausea, sour taste, and increased uric acid levels (Pranzatelli et al. 2017b).

Therapeutic Apheresis: Five or six exchanges are usually required. Improvement may be rapid and last for up to 2 months. One of the main limitations in small children is that plasmapheresis is technically infeasible. Compared to IVIG, the main disadvantages include reduced blood volume, which may induce hypotension, increased immunosuppression, and the placement of a large bore central venous catheter, leading to more frequent and serious side effects (Yiu et al. 2001) (Table 3).

12 Prognosis

In adults younger than 40 years of age, the most likely cause of OMS is idiopathic or parainfectious, both of which have a better prognosis than paraneoplastic-related OMS or OMS in children (De Grandis et al. 2009). A Mayo case series of adult-onset OMS showed a favorable outcome with an idiopathic (presumed parainfectious) disorder of short duration, with full recovery after 4 to 6 weeks of treatment. Of the 19 patients, OMS remitted in 13 patients and improved in 3. Nonetheless, in the Mayo Clinic patients, cancer was an important cause in (14%), and a few patients also had a relapsing course or a poor outcome (Klaas et al. 2012).

Table 3 Treatment options for opsoclonus myoclonus

Agent	Dosing	Potential adverse effects	Other
Corticosteroids	Starting oral dose of prednisone 2 mg/kg/day with slow tape starting at 2–3 months over a 9–12-month period	Cushingoid symptoms, impaired glucose levels, insomnia, hair growth, high blood pressure, leg swelling, muscle weakness, easy bruising	
ACTH	Intramuscularly starting at 75 IU per dose twice daily x 40–52 weeks with a slow taper	Growth suppression, Cushing's syndrome, hair growth, acne	
IVIG	Intravenous injection of 1–2 g/kg/day x 1 or more days and then a monthly maintenance dose of 1 g/kg	Headache, nausea, fever, or flu-like symptoms	Preliminary blood work includes an IgA level and kidney function tests
Azathioprine (Imuran)	50 mg per day x 6 months	Fever, rash, flu-like symptoms, nausea, idiosyncratic flu-like reaction Bone marrow suppression	
Rituximab (Rituxan)	4–5 IV infusions of 375 mg/m^2	Low blood pressure, breathing difficulties, and sensation of the tongue or throat swelling (angioedema)	CBC and platelet counts should be monitored
Cyclophosphamide (Cytoxan)	1–5 mg per kg of body weight x 60 to 90 days	Loss of appetite, nausea, vomiting, constipation, temporary hair thinning or brittleness, and increased risk of infections Hemorrhagic cystitis	
6-mercaptopurine (6-MP)	Unclear off-label use. Oral 100–200 mg daily x 2 months	Anemia, risk of bruising or bleeding, infection, nausea, sour taste, increase uric acid levels, elevated liver transaminases	Monitor CBC, LFTs
Therapeutic apheresis	5–6 exchanges every other day	Reduces blood volume and therefore may induce hypotension, immunosuppression	Requires the placement of a large bore central venous catheter

Opsoclonus can remit spontaneously or have a relapsing-remitting course. Even after resolution, formal testing typically still reveals abnormalities of eye movement pursuits and to a lesser extent saccadic eye movements.

For the pediatric population, younger age and severity upon presentation seem to have a worse .prognosis for developing neurologic sequelae. Relapses should be treated. Usually children who responded initially to immunotherapy will do so again, even to a single agent.

Although childhood neuroblastoma carries a significant mortality rate, neuro-blastomas associated with OMS tend to be low grade and have a more favorable outcome (De Grandis et al. 2009). Overall, the survival prognosis of children with OMS secondary to neuroblastoma is very favorable. One series of patients with neuroblastoma reported that 90% of patients with OMS presented with nonmeta-static disease, whereas nonmetastatic disease was present in only 35% of patients without OMS. In the same series, OMS patients were estimated to have a 3-year survival rate of 100%, whereas those without OMS had a survival rate of 77% (Rudnick et al. 2001). Despite this excellent survival prognosis, children with OMS have a more guarded neurological prognosis. Mitchell et al. reported that only a minority of patients followed a monophasic course and that those patients generally had a more favorable neurologic prognosis than patients who followed a more chronic, relapsing course (Mitchell et al. 2005). Russo and colleagues reported that 69% of children with OMS suffered from long-term neurological problems. Interestingly, ten of the patients received chemotherapy for their underlying neuro-blastoma, and six of those ten patients had no neurological sequelae. In contrast, 19 children did not receive chemotherapy, and only 3 of them were free of neurological sequelae. This may suggest that more intense immunosuppression with chemother-apy may help improve neurological outcomes in OMS patients (Russo et al. 1997). The reported expansion of autoreactive B cells has been thought to correlate with severity and duration of OMS; however, it remains controversial if this is an ade-quate biomarker of disease. B-cell activating factor (BAFF), a key molecule neces-sary for B-cell survival, seems to be involved in the intrathecal B-cell expansion (Armangue et al. 2014). Some studies have followed serum and CSF B lymphocytes after treatment with rituximab, and it appears that lower values correlate with better prognosis (Armangue et al. 2014).

Acknowledgments Authors thank Dr. Aasef Shaikh for helpful suggestions with the manuscript.

References

Anagnostou, E., Kararizou, E., & Evdokimidis, I. (2013). Ocular flutter in essential tremor: Clinical course and response to primidone. *Journal of Neurology, 260*, 2672–2674.

Armangue, T., Titulaer, M. J., Sabater, L., Pardo-Moreno, J., Gresa-Arribas, N., Barbero-Bordallo, N., Kelley, G. R., Kyung-Ha, N., Takeda, A., Nagao, T., Takahashi, Y., Lizcano, A., Carr, A. S., Graus, F., & Dalmau, J. (2014). A novel treatment-responsive encephalitis with frequent ops-oclonus and teratoma. *Annals of Neurology, 75*, 435–441.

Aschoff, E. (1976). Voluntary nystagmus in five generations. *Journal of Neurosurgery, 44*.

Ashe, J., Hain, T. C., Zee, D. S., & Schatz, N. J. (1991). Microsaccadic flutter. *Brain, 114*(Pt 1B), 461–472.

Brisse, H. J., McCarville, M. B., Granata, C., Krug, K. B., Wootton-Gorges, S. L., Kanegawa, K., Giammarile, F., Schmidt, M., Shulkin, B. L., Matthay, K. K., Lewington, V. J., Sarnacki, S., Hero, B., Kaneko, M., London, W. B., Pearson, A. D., Cohn, S. L., Monclair, T., & International Neuroblastoma Risk Group Project. (2011). Guidelines for imaging and staging of neuro-blastic tumors: Consensus report from the International Neuroblastoma Risk Group Project. *Radiology, 261*, 243–257.

Brunklaus, A., Pohl, K., Zuberi, S. M., & de Sousa, C. (2011). Outcome and prognostic features in opsoclonus-myoclonus syndrome from infancy to adult life. *Pediatrics, 128*, e388–e394.

Cooper, R., Khakoo, Y., Matthay, K. K., Lukens, J. N., Seeger, R. C., Stram, D. O., Gerbing, R. B., Nakagawa, A., & Shimada, H. (2001). Opsoclonus-myoclonus-ataxia syndrome in neuroblastoma: Histopathologic features-a report from the Children's Cancer Group. *Medical and Pediatric Oncology, 36*, 623–629.

De Grandis, E., Parodi, S., Conte, M., Angelini, P., Battaglia, F., Gandolfo, C., Pessagno, A., Pistoia, V., Mitchell, W. G., Pike, M., Haupt, R., & Veneselli, E. (2009). Long-term follow-up of neuroblastoma-associated opsoclonus-myoclonus-ataxia syndrome. *Neuropediatrics, 40*, 103–111.

Deconinck, N., Scaillon, M., Segers, V., Groswasser, J. J., & Dan, B. (2006). Opsoclonus-myoclonus associated with celiac disease. *Pediatric Neurology, 34*, 312–314.

Eberhardt, O., & Topka, H. (2017). Myoclonic disorders. *Brain Sciences, 7*.

Grant, R., & Graus, F. (2009). Paraneoplastic movement disorders. *Movement Disorders, 24*, 1715–1724.

Haden, S. V., McShane, M. A., & Holt, C. M. (2009). Opsoclonus myoclonus: A non-epileptic movement disorder that may present as status epilepticus. *Archives of Disease in Childhood, 94*, 897–899.

Hartung, H. P. (2008). Advances in the understanding of the mechanism of action of IVIg. *Journal of Neurology, 255*(Suppl 3), 3–6.

Hassan, K. A., Kalemkerian, G. P., & Trobe, J. D. (2008). Long-term survival in paraneoplastic opsoclonus-myoclonus syndrome associated with small cell lung cancer. *Journal of Neuro-Ophthalmology, 28*, 27–30.

Herman, T. E., & Siegel, M. J. (2009). Ataxia without opsoclonus: Right lumbar sympathetic trunk neuroblastoma. *Clinical Pediatrics (Phila), 48*, 336–340.

Hero, B., & Schleiermacher, G. (2013). Update on pediatric opsoclonus myoclonus syndrome. *Neuropediatrics, 44*(06)m 324–329.

Smith, J. L., & Walsh, F. B. (1960). Opsoclonus—ataxic conjugate movements of the eyes. *Archives of Ophthalmology, 64*, 244–250.

Kinsbourne, M. (1962). Myoclonic encephalopathy of infants. *Journal of Neurology, Neurosurgery, and Psychiatry, 25*, 271–276.

Klaas, J. P., Ahlskog, J. E., Pittock, S. J., Matsumoto, J. Y., Aksamit, A. J., Bartleson, J. D., Kumar, R., McEvoy, K. F., & McKeon, A. (2012). Adult-onset opsoclonus-myoclonus syndrome. *Archives of Neurology, 69*, 1598–1607.

Klein, A., Schmitt, B., & Boltshauser, E. (2007). Long-term outcome of ten children with opsoclonus-myoclonus syndrome. *European Journal of Pediatrics, 166*, 359–363.

Krug, P., Schleiermacher, G., Michon, J., Valteau-Couanet, D., Brisse, H., Peuchmaur, M., Sarnacki, S., Martelli, H., Desguerre, I., & Tardieu, M. (2010). Opsoclonus-myoclonus in children associated or not with neuroblastoma. *European Journal of Paediatric Neurology, 14*, 400–409.

Leen, W. G., Weemaes, C. M., Verbeek, M. M., Willemsen, M. A., & Rotteveel, J. J. (2008). Rituximab and intravenous immunoglobulins for relapsing postinfectious opsoclonus-myoclonus syndrome. *Pediatric Neurology, 39*, 213–217.

Leigh, R. J., & Zee, D. S. (2015). *The neurology of eye movements.* Oxford University Press.

Matthay, K. K., Blaes, F., Hero, B., Plantaz, D., De Alarcon, P., Mitchell, W. G., Pike, M., & Pistoia, V. (2005). Opsoclonus myoclonus syndrome in neuroblastoma a report from a workshop on the dancing eyes syndrome at the advances in neuroblastoma meeting in Genoa, Italy, 2004. *Cancer Letters, 228*, 275–282.

Mitchell, W. G., Brumm, V. L., Azen, C. G., Patterson, K. E., Aller, S. K., & Rodriguez, J. (2005). Longitudinal neurodevelopmental evaluation of children with opsoclonus-ataxia. *Pediatrics, 116*, 901–907.

Pless, M., & Ronthal, M. (1996). Treatment of opsoclonus-myoclonus with high-dose intravenous immunoglobulin. *Neurology, 46*, 583–584.

Pranzatelli, M. R., Travelstead, A. L., Tate, E. D., Allison, T. J., Moticka, E. J., Franz, D. N., Nigro, M. A., Parke, J. T., Stumpf, D. A., & Verhulst, S. J. (2004). B- and T-cell markers in opsoclonus-myoclonus syndrome: Immunophenotyping of CSF lymphocytes. *Neurology, 62*, 1526–1532.

Pranzatelli, M. R., Tate, E. D., Swan, J. A., Travelstead, A. L., Colliver, J. A., Verhulst, S. J., Crosley, C. J., Graf, W. D., Joseph, S. A., & Kelfer, H. M. (2010). B cell depletion therapy for new-onset opsoclonus-myoclonus. *Movement Disorders, 25*, 238–242.

Pranzatelli, M. R., Slev, P. R., Tate, E. D., Travelstead, A. L., Colliver, J. A., & Joseph, S. A. (2011). Cerebrospinal fluid oligoclonal bands in childhood opsoclonus-myoclonus. *Pediatric Neurology, 45*, 27–33.

Pranzatelli, M. R., & Tate, E. D. (2017). Opsoclonus myoclonus syndrome. In K. Swaiman, S. Ashwal, D. M. Ferriero, N. F. Schor, R. S. Finkel, & A. L. Gropman, et al. (Eds.), *Swaiman's pediatric neurology: principles and practice (Chap. 120)* (pp. 938–944). London, UK: Elsevier.

Pranzatelli, M. R., Tate, E. D., & McGee, N. R. (2017a). Demographic, clinical, and immunologic features of 389 children with opsoclonus-myoclonus syndrome: A cross-sectional study. *Frontiers in Neurology, 8*, 468 https://doi.org/10.3389/fneur.2017.00468.

Pranzatelli, M. R., Tate, E. D., & Allison, T. J. (2017b). 6-Mercaptopurine modifies cerebrospinal fluid T cell abnormalities in paediatric opsoclonus-myoclonus as steroid sparer. *Clinical and Experimental Immunology, 190*, 217–225.

Rostasy, K., Wilken, B., Baumann, M., Muller-Deile, K., Bieber, I., Gartner, J., Moller, P., Angelini, P., & Hero, B. (2006). High dose pulsatile dexamethasone therapy in children with opsoclonus-myoclonus syndrome. *Neuropediatrics, 37*, 291–295.

Rudnick, E., Khakoo, Y., Antunes, N. L., Seeger, R. C., Brodeur, G. M., Shimada, H., Gerbing, R. B., Stram, D. O., & Matthay, K. K. (2001). Opsoclonus-myoclonus-ataxia syndrome in neuroblastoma: Clinical outcome and antineuronal antibodies-a report from the Children's Cancer Group Study. *Medical and Pediatric Oncology, 36*, 612–622.

Russo, C., Cohn, S. L., Petruzzi, M. J., & de Alarcon, P. A. (1997). Long-term neurologic outcome in children with opsoclonus-myoclonus associated with neuroblastoma: A report from the Pediatric Oncology Group. *Medical and Pediatric Oncology, 28*, 284–288.

Shawkat, F. S., Harris, C. M., Wilson, J., & Taylor, D. (1993). Eye movements in children with opsoclonus-polymyoclonus. *Neuropediatrics, 24*, 218–223.

Stefanowicz, J., Izycka-Swieszewska, E., Drozynska, E., Pienczk, J., Polczynska, K., Czauderna, P., Sierota, D., Bien, E., Stachowicz-Stencel, T., Kosiak, W., & Balcerska, A. (2008). Neuroblastoma and opsoclonus-myoclonus-ataxia syndrome--clinical and pathological characteristics. *Folia Neuropathologica, 46*, 176–185.

Tate, E. D., Allison, T. J., Pranzatelli, M. R., & Verhulst, S. J. (2005). Neuroepidemiologic trends in 105 US cases of pediatric opsoclonus-myoclonus syndrome. *Journal of Pediatric Oncology Nursing, 22*, 8–19.

Tate, E. D., Pranzatelli, M. R., Verhulst, S. J., Markwell, S. J., Franz, D. N., Graf, W. D., Joseph, S. A., Khakoo, Y. N., Lo, W. D., Mitchell, W. G., & Sivaswamy, L. (2012). Active comparator-controlled, rater-blinded study of corticotropin-based immunotherapies for opsoclonus-myoclonus syndrome. *Journal of Child Neurology, 27*, 875–884.

Toyoshima, D., Morisada, N., Takami, Y., Kidokoro, H., Nishiyama, M., Nakagawa, T., Ninchoji, T., Nozu, K., Takeshima, Y., & Takada, S. (2016). Rituximab treatment for relapsed opsoclonus–myoclonus syndrome. *Brain and Development, 38*, 346–349.

Venkatraman, A., & Opal, P. (2016). Paraneoplastic cerebellar degeneration with anti-Yo antibodies – A review. *Annals of Clinical Translational Neurology, 3*, 655–663.

Wilken, B., Baumann, M., Bien, C., Hero, B., Rostasy, K., & Hanefeld, F. (2008). Chronic relapsing opsoclonus-myoclonus syndrome: Combination of cyclophosphamide and dexamethasone pulses. *European Journal of Paediatric Neurology, 12*, 51–55.

Wong, A. (2007). An update on opsoclonus. *Current Opinion in Neurology, 20*, 25–31.

Yiu, V. W., Kovithavongs, T., McGonigle, L. F., & Ferreira, P. (2001). Plasmapheresis as an effective treatment for opsoclonus-myoclonus syndrome. *Pediatric Neurology, 24*, 72–74.

Zahn, J. R. (1978). Incidence and characteristics of voluntary nystagmus. *Journal of Neurology, Neurosurgery, and Psychiatry, 41*, 617–623.

Eye Movement Disorders in Patients with Epilepsy

Macym Rizvi and Fareeha Ashraf

Abstract A broad spectrum of eye movement disorders can be observed in patients with epilepsy, many of which are consequences of antiepileptic medications but can also occur secondary to epileptic discharges. In this chapter, we will review clinical phenomenology and mechanistic underpinning of the eye movement deficits in epilepsy patients due to both etiologies.

Keywords Eye movements · Epilepsy · Nystagmus · Antiepileptic drugs

1 AED-Induced Eye Movement Disorders

Antiepileptic medications reduce neuronal depolarization in order to prevent the abnormal rhythmic and synchronous discharges seen in seizures. Such mode of action to depress neuronal activity can induce a wide range of eye movement disorders. In this section we will discuss AEDs that are known to cause eye movement disorders, and for each AED discussed, we will review the mechanistic underpinning of the induced ocular motor deficit.

Multiple neuroanatomical circuits will be discussed in further detail below. As an overview, the main components involved in AED-induced eye movement disorders include the horizontal neural integrator, the vertical neural integrator, and the neuroanatomical components of smooth pursuit and saccades. The role of the neural integrator is to integrate information regarding eye velocity and eye position and to

M. Rizvi · F. Ashraf (✉)
Department of Neurology, University Hospitals of Cleveland, Cleveland, OH, USA

Department of Neurology, Louis Stokes Cleveland VA Medical Center, Cleveland, OH, USA
e-mail: Fareeha.Ashraf@va.gov

© Springer Nature Switzerland AG 2019   487
A. Shaikh, F. Ghasia (eds.), *Advances in Translational Neuroscience of Eye Movement Disorders*, Contemporary Clinical Neuroscience,
https://doi.org/10.1007/978-3-030-31407-1_24

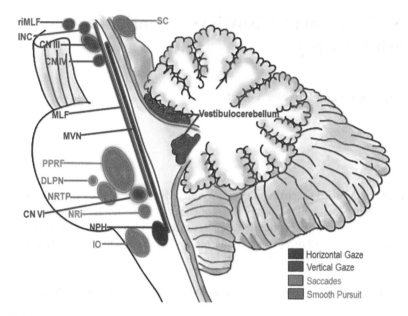

Fig. 1 Neuroanatomical networks. Horizontal gaze network (red) includes oculomotor nucleus (CN III), medial longitudinal fasciculus (MLF), medial vestibular nucleus (MVN), abducens nucleus (CN VI), and nucleus prepositus hypoglossi (NPH) and fined-tuned by the vestibulocerebellum. Vertical gaze network (blue) includes rostral interstitial medial longitudinal fasciculus (riMLF), interstitial nucleus of Cajal (INC), and trochlear nucleus (CN IV) and fine-tuned by the vestibulocerebellum. Saccade network (green) includes paramedian pontine reticular formation (PPFR), and nucleus raphe interpositus (NRi); of note burst neurons for vertical and torsional saccades lie within the rostral interstitial medial longitudinal fasciculus (riMLF, labeled in blue). Smooth-pursuit network (orange) includes dorsolateral pontine nucleus (DLPN), nucleus reticularis tegmenti pontis (NRTP), inferior olive (IO), and vestibulocerebellum. Vestibulocerebellum is comprised of the flocculus and nodulus, known as the flocculonodular lobe

provide signals to extraocular motoneurons to maintain or attain a new position of the eyes. The horizontal neural integrator is comprised of the medial vestibular nucleus and nucleus prepositus hypoglossi located in the caudal pons and the rostral medulla, while the vertical neural integrator is comprised of the interstitial nucleus of Cajal and the rostral interstitial medial longitudinal fasciculus located in the rostral midbrain. Both the horizontal and vertical neural integrators are fine-tuned by projections from the vestibulocerebellum (Sanchez and Rowe 2016). In addition to the neural integrators, AEDs can impair smooth-pursuit and saccade generation networks. The network for smooth pursuit includes the vestibulocerebellum, dorsolateral pontine nucleus, nucleus reticularis tegmenti pontis, and inferior olive, whereas the network for saccade generation includes the superior colliculus, omnipause neurons, and excitatory burst neurons.

AEDs can impair the normal functioning of these networks introduced here and can manifest as different eye movement disorders observed in patients with epilepsy to be discussed in detail below (Fig. 1).

2 Sodium Valproate

Sodium valproate, a commonly used anticonvulsant, blocks voltage-gated sodium channels and increases the levels of gamma-aminobutryic acid (GABA). Sodium valproate possibly inhibits GABA degradative enzymes, such as GABA transaminase and succinate-semialdehyde dehydrogenase, and inhibits GABA reuptake thus raising cerebral and cerebellar levels of this inhibitory neurotransmitter. The eye movement abnormalities associated with valproate include gaze-evoked nystagmus and vertical gaze palsies.

2.1 Gaze-Evoked Nystagmus

Gaze-evoked nystagmus is an oscillation of the eyes that occur when attempting to maintain extreme eye positions. Gaze is unable to be held in an extreme position, and the eyes drift back toward the null point (which is often straight-ahead gaze), and a corrective saccade is generated to move the eyes back to the eccentric position. This process repeats which is seen as a rhythmic oscillation of the eyes. The classic description of the gaze-evoked nystagmus is that it is caused by a deficiency of the neural integrator which is comprised of a cellular network converting the eye velocity commands into an eye position signal. The nucleus prepositus hypoglossi is a major contributor for the horizontal neural integrator with a minority of inputs being from the medial vestibular nucleus. Both nuclei are located within the region of the caudal pons and rostral medulla. The vertical and torsional neural integrator is in the interstitial nucleus of Cajal, within the midbrain. If there is minimal abnormality in integrator function, gaze-evoked nystagmus will manifest in extreme angles of gaze. If there is major dysfunction in integrator function, gaze-evoked nystagmus will appear as soon as the eyes deviate from the primary gaze (Suzuki et al. 2003).

Many antiepileptic drugs, including sodium valproate, will cause gaze-evoked nystagmus, and it is likely due to its effects on the neural integrator. These medications can cause decreased firing of the neural integrator network, affecting its ability to effectively transform velocity to position. This phenomenon, termed a leaky neural integration, causes the eyes to drift while attempting to fixate in an eccentric position. Then a corrective saccade is made to attempt to maintain the eccentric position. This cycle will then continue to repeat, and a gaze-evoked nystagmus will be observed (Glassauer et al. 2003; Rett 2007).

The brainstem neural integrators receive fine tuning by the cerebellum, in particular the floccular and parafloccular Purkinje cells within the vestibulocerebellum. These Purkinje cells are known for their inhibitory nature and influence the brainstem neural integrators via the cerebellar peduncles (Sanchez and Rowe 2016). Sodium valproate also leads to dysfunction at the level of cerebellar Purkinje neurons with prolonged use resulting in cerebellar atrophy (Zaccara et al. 2004). Lack of optimized Purkinje neuron function results in deficient feedback to the integrator, hence its dysfunction, and subsequent gaze-evoked nystagmus (Zee et al. 1976).

2.2 Vertical Gaze Palsy

Sodium valproate's effect on GABA may be the cause for selective vertical gaze palsy seen when there is sodium valproate toxicity. Vertical gaze control centers are the rostral interstitial medial longitudinal fasciculus and the interstitial nucleus of Cajal (INC) which are housed in the midbrain. The INC in particular has abundant GABAergic neurons. Studies in the macaque monkey discovered GABAergic projections from the interstitial nucleus of Cajal to the contralateral superior oblique and inferior rectus motoneurons. During upward eye movements, these commissural GABAergic projections inhibit the superior oblique and inferior rectus motoneurons and premotor down-burst-tonic neurons (Horn et al. 2003). The proposed mechanism for selective vertical gaze palsy with sodium valproate toxicity is based on increased levels of GABA. At high concentrations of sodium valproate, GABA transaminase is inhibited and the degradation of GABA is prevented. Subsequently, there will be elevated levels of GABA and deactivation of the vertical burst neurons producing a selective vertical gaze palsy in valproate toxicity (Gosala Raja et al. 2013).

3 Phenytoin/Fosphenytoin

The mechanism of action of phenytoin is thought to involve blockade of voltage-gated sodium channels by preferentially binding to the fast inactive state. This blocks high-frequency firing of action potentials. There are various eye movement abnormalities that have been reported with phenytoin and fosphenytoin. These include gaze-evoked nystagmus, impaired smooth pursuit, impaired VOR suppression, downbeat nystagmus, periodic alternating nystagmus, and pendular nystagmus. It should be noted that the phenytoin serum concentration does not necessarily correlate with the appearance of phenytoin-induced eye movement disorders (Riker et al. 1978).

3.1 Gaze-Evoked Nystagmus

Gaze-evoked nystagmus secondary to phenytoin is thought to be due to a leaky neural integrator with a similar mechanism discussed above with sodium valproate. The neural integrator's role is to mathematically integrate eye and head velocity signals to provide an accurate eye position signal to maintain gaze. Much of the eye and head velocity signals are provided from vestibular and cerebellar connections. It is also believed that phenytoin likely impairs the central vestibular and cerebellar connections required by the neural integrator. This pharmacologic impairment of the neural integrator can manifest clinically as a gaze-evoked nystagmus in addition to ataxia (Riker et al. 1978).

3.2 Smooth Pursuit

Smooth-pursuit movements are slower tracking movements of the eyes to keep a moving target on the fovea. Cerebral cortical areas concerned with visual motion are relayed to brainstem gaze-holding network which integrates eye and head velocity signals, vestibular, and pursuit eye movements into eye position commands. In addition to brainstem networks, the cerebellum will provide information regarding initiation and programming of the pursuit movements. The three components of the cerebellum involved in smooth pursuit are the vestibulocerebellum (paraflocculus), dorsal vermis and its projections to the fastigial nucleus, and the ansiform lobule (VII) of the cerebellar hemisphere. Pharmacological inactivation of the paraflocculus and flocculus in the macaque monkey substantially impairs smooth pursuit and VOR cancellation (Belton and McCrea 2000). The mechanism in which phenytoin impairs smooth pursuit is likely due to its disrupting effects on the flocculus and paraflocculus and is discussed further below.

3.3 Vestibulo-Ocular Reflex Suppression

The vestibulo-ocular reflex (VOR) maintains gaze on stationary object while the head is moving. Consequent eye movements are generated with equal speed (as head movement) but in the opposite direction, with a goal to maintain visual fixation. Smooth pursuit is used to track moving objects while the head is stationary. An instance where the VOR and smooth-pursuit systems come into conflict is when the head is moving with a moving target. Here the VOR drives the eyes in the opposite direction while the pursuit system attempts to overcome this by driving the eyes with the target. This mechanism to overcome the VOR is termed VOR suppression.

Impairments in both smooth pursuit and VOR suppression can be seen with phenytoin. It may be related to phenytoin's effects on cerebellar Purkinje cells and impairing the functions of the flocculus and paraflocculus. The abnormalities in smooth pursuit and VOR suppression can be seen in patients who do not have any clinical or biochemical evidence of anticonvulsant toxicity. This has been seen in patients treated with phenytoin or phenobarbital, and marked improvements in smooth pursuit and VOR suppression is seen after discontinuation of those antiepileptic medications (Bittencourt et al. 1980).

3.4 Downbeat Nystagmus

Downbeat nystagmus results from a defective vertical gaze holding causing a pathologic updrift of the eyes and a corrective downward saccade. Downbeat nystagmus can be seen in dysfunction of Purkinje cells within the floccular lobe of the vestibulocerebellum. The anterior semicircular canal provides vestibular inputs which are normally inhibited by the floccular lobe. Floccular dysfunction can result in disinhibition of the anterior semicircular canals and give rise to an upward drift of the eyes with corrective downward saccades. This is seen clinically as downbeat nystagmus. The presence of downbeat nystagmus seen with phenytoin is likely related to its effects on the vestibulocerebellum (Berger and Kovacs 1982).

3.5 Periodic Alternating Nystagmus

Periodic alternating nystagmus (PAN) is a horizontal nystagmus that predictably oscillates in direction, frequency, and amplitude. For example, a leftward beating nystagmus will develop increasingly larger amplitudes and higher frequencies until a point in which it will then progressively diminishes. There will then be a brief null period in the horizontal nystagmus where downbeat nystagmus or square wave jerks can be seen. Then a reversal in the direction of nystagmus begins where a right beating nystagmus of progressively larger amplitudes and higher frequencies is seen until a point where it will begin to wane. This cycle continues to repeat in a crescendo-decrescendo pattern and reverses direction approximately every 2 min.

PAN can be acquired from disruption of vestibulocerebellum's nodulus and uvula. These cerebellar structures play a role in the time constant of rotational velocity storage. Pharmacological evidence suggests that the nodulus and uvula maintain inhibitory control on the vestibular rotational responses. Dysfunction can result in an oscillatory shifting of the null point (Cohen et al. 1987). Phenytoin can disrupt the vestibulocerebellum's normal inhibitory control on rotational vestibular inputs thus precipitating a PAN (Campbell 1980; Schwankhaus et al. 1989).

3.6 Pendular Nystagmus

Pendular nystagmus when acquired is a type of ocular oscillation which often causes continuous oscillopsia. There are typically horizontal, vertical, and torsional components. Depending on whether the horizontal and vertical oscillatory components are in phase or out of phase, the resultant trajectories can either be oblique, elliptical, or circular. Also when comparing the oscillations of each eye, the nystagmus may be conjugate or disconjugate with different sizes of oscillations.

The etiology for acquired pendular nystagmus lies within the neural substrates for gaze holding, particularly the neural integrator. The neural integrator involves a network within the brainstem and cerebellum. This network will mathematically integrate vestibular, optokinetic, saccadic, and pursuit eye velocity signals. This network then takes these signals and provides information for eye position to maintain gaze. If there is instability in the neural integrator, pendular nystagmus can arise.

Phenytoin causes slowing of neural conduction. This slowing can delay feedback between the cerebellum and the oculomotor neural integrator. Since the neural integrator network's role is to mathematically integrate eye and head velocity signals to provide an accurate eye position signal to maintain gaze, a slowing of neural conduction can lead to its instability. This instability is owed to the neural integrators inability to quickly and accurately provide the correct eye position based on eye and head velocity signals. The delayed feedback from slowed neural conduction will disrupt gaze holding due to errors in eye position from the integrator. This erroneous eye positioning for gaze holding can then manifest as pendular nystagmus. The instability of the neural integrator from pharmacologically slowed neural conduction is the likely etiology of phenytoin-induced pendular nystagmus (Shaikh 2013).

4 Carbamazepine

Carbamazepine binds preferentially to the voltage-gated sodium channels in their inactive conformation, preventing repetitive and sustained neuronal depolarization. Its anticonvulsant efficacy is due to the inhibition of sodium channel activity.

Impairments of eye movements associated with carbamazepine include gaze-evoked nystagmus, slowing of saccade velocity, impaired smooth pursuit, downbeat nystagmus, jerk seesaw nystagmus, oculogyric crisis, and ophthalmoplegia.

4.1 Gaze-Evoked Nystagmus

As commonly seen with other antiepileptic use, the gaze-evoked nystagmus is also seen in patients who take carbamazepine. The mechanism of gaze-evoked nystagmus secondary to carbamazepine also involves the vestibulocerebellum (flocculonodular lobe) or its connections (Umeda and Sakata 1977; Remler et al. 1990). Further details regarding neuroanatomy and mechanism are discussed above under sodium valproate-induced gaze-evoked nystagmus.

4.2 Slow Saccades

Saccades are ballistic movements of the eyes that abruptly change the point of fixation. Generation of saccades include the eye fields in cerebral cortex, superior colliculus, and the brainstem saccadic pulse generator network. Within the brainstem network that generates premotor saccade commands are two types of neurons: burst neurons and omnipause neurons. For horizontal saccades, the saccadic burst neurons are within the paramedian pontine reticular formation within the caudal pons. For vertical and torsional saccades, burst neurons are within the rostral interstitial nucleus of the medial longitudinal fasciculus. There are two types of burst neurons: excitatory burst neurons and inhibitory burst neurons (Büttner Ennever and Büttner 1988). The excitatory burst neurons when activated will cause tonic firing of the associated extraocular muscle to perform a saccade. The inhibitory burst neurons' roles include silencing activity in antagonist extraocular muscles during saccades. Another role may be to help end the saccade when the eye is on target (Tedeschi et al. 1989). The duration of most saccades is less than 100 milliseconds and have a peak velocity of approximately 500 degrees/second. Carbamazepine may decrease peak saccade velocity. This may be due to carbamazepine causing dysfunction neuronal depolarization of the brainstem saccade pulse generator system in the paramedian pontine reticular formation (Tedeschi et al. 1989).

4.3   Smooth Pursuit

Smooth-pursuit movements are slower tracking movements of the eyes to keep a moving target on the fovea. Neuroanatomical structures involved for smooth pursuit include cerebral cortical areas, the neural integrator, and vestibulocerebellum. Further details are discussed above under phenytoin and smooth pursuit.

The mechanism in which carbamazepine impairs smooth pursuit can be presumed to be similar to phenytoin as they have similar inhibitory actions on sodium channels. Phenytoin impairs smooth pursuit via its effects on the vestibulocerebellum. The vestibulocerebellum provides information regarding initiation and programming of the pursuit movements. If dysfunction of the vestibulocerebellum occurs, impairments in smooth pursuit may occur. Thus it may be carbamazepine's disrupting effects on the vestibulocerebellum that give rise to smooth-pursuit impairments (Reilly et al. 2008).

4.4   Downbeat Nystagmus

Downbeat nystagmus results from a defective vertical gaze holding causing a pathologic updrift of the eyes and a corrective downward saccade. Downbeat nystagmus can be present in dysfunctions of the vestibulocerebellum. Further information regarding downbeat nystagmus is discussed above under phenytoin and downbeat nystagmus. In summary, floccular dysfunction can result in disinhibition of anterior semicircular canals, giving rise to an upward drift of the eyes with a corrective downward saccade.

The induction of downbeat nystagmus by carbamazepine may reflect disruption of the cerebellum and its control of vestibular, pursuit, or otolith-ocular reflexes (Chrousos et al. 1987).

4.5   Seesaw Nystagmus

Seesaw nystagmus consists of a half cycle with elevation and intorsion of one eye and synchronous depression and extorsion of the other; during the next half cycle, the vertical and torsional movements reverse. Jerk nystagmus will have an alteration in slow and quick phases. Factors that contribute to jerk seesaw nystagmus include imbalance and miscalibration of vestibular responses that normally optimize gaze during head rotations in roll. The vestibular nuclei will provide inputs to oculomotor and trochlear nuclei via the medial longitudinal fasciculus (MLF) to maintain gaze. Discoordination and miscalibration between these structures may manifest as jerk seesaw nystagmus. Oxcarbazepine may disrupt the coordination between vestibular inputs and normal gaze to cause a jerk seesaw nystagmus (Adamec et al. 2013).

4.6 Oculogyric Crisis

Oculogyric crisis is characterized by prolonged involuntary upward gaze deviation.

In an oculogyric crisis, affected patients have great difficulty in looking downward as the eyes become fixed in an upward gaze. It is often associated as an acute dystonic reaction due to neuroleptic drug treatment. But it can rarely be seen in carbamazepine use. This may be related to an imbalance of the neural integrator and it's vertical gaze-holding mechanism (Berchou and Rodin 1979).

4.7 Ophthalmoplegia

One study described ophthalmoplegia induced by carbamazepine (Noda and Umezaki 1982). It is possible that global weakening of burst generation or profound and sustained inhibition by the omnipause neurons have caused the deficit. The omnipause neurons are located within the nucleus raphe interpositus which is located within the dorsal caudal pons. They exert a tonic inhibition upon horizontal and vertical saccadic burst neurons during fixation or smooth-pursuit eye movements. Before a saccade is generated, the high level of tonic inhibition by the omnipause neurons is removed. The release of inhibition by omnipause neurons disinhibits the saccadic burst neurons which are then able to perform a saccade (Horn et al. 1994). It may be that carbamazepine impairs the normal release of inhibition by omnipause neurons which would lead to a constant high tonic inhibition upon eye movement networks that may manifest as an ophthalmoplegia.

5 Lamotrigine

The mechanism of action of lamotrigine is not fully elucidated as it may have multiple actions that likely contribute to its broad clinical efficacy. In regard to its antiepileptic effect, lamotrigine is a member of the sodium channel blocking class. Lamotrigine blocks voltage-activated sodium channels which decreases presynaptic release of excitatory neurotransmitters such as glutamate.

5.1   Downbeat Nystagmus

Lamotrigine toxicity can cause downbeat nystagmus. The presence of downbeat
nystagmus supports the involvement of the vestibulocerebellum, particularly the
floccular lobe as it contains gaze-velocity Purkinje cells with a downward on-
direction for smooth pursuit and participates in gaze holding. The flocculus normally inhibits the central vestibular pathways from the anterior but not from the
posterior semicircular canals. Disinhibition of the anterior semicircular canals due
to floccular dysfunction would give rise to upward drift of the eyes and result in
downbeat nystagmus. Thus, lamotrigine may give rise to a pharmacologically
induced transient dysfunction of the vestibulocerebellum resulting in downbeat nystagmus (Oh et al. 2006; Alkawi et al. 2005).

6   Benzodiazepines and Barbiturates

Lorazepam, diazepam, clobazam, and others are among the class of benzodiazepine
antiepileptic drugs. They enhance the effect of the inhibitory neurotransmitter
GABA at the $GABA_A$ receptor to decrease the excitability of neurons.
Benzodiazepines binding acts as a positive allosteric modulator and hyperpolarizes
the neuronal membrane potential by increasing the chloride ion conductance.
Barbiturates, such as phenobarbital and primidone, have a similar mechanism of
action as benzodiazepines as they too potentiate the effects of GABA. Like benzodiazepines, barbiturates bind as a positive allosteric modulator to the $GABA_A$ receptor. At higher doses, they will act as $GABA_A$ agonists. Benzodiazepines increase the
frequency of chloride channel opening, whereas barbiturates will increase the duration of opening. In addition, barbiturates block the excitatory glutaminergic AMPA
and kainate receptors. Taken together, barbiturates potentiate inhibitory $GABA_A$
receptors and inhibit excitatory AMPA receptors. Similar eye movement abnormalities arise in patients taking benzodiazepines or barbiturates. This may be due to their
GABAergic properties. They include increase duration and decreased velocity of
saccades and impaired smooth pursuit.

6.1   Saccade Duration and Velocity

During a saccade, a high-frequency burst of activity occurs in ocular motoneurons
and its agonist ocular muscle. This burst of activity is the *saccadic pulse of innervation* and generates the force necessary to overcome the orbital viscous drag for the
eyes to move. After the pulse of innervation to perform the saccade, the eyes are
held in the new position by the orbital muscles countering the orbital elastic restoring forces. To hold the eyes in this new position after a saccade, a higher level of
tonic innervation of the ocular motoneurons and its agonist orbital muscles is

required and is termed the *saccadic step of innervation*. Meanwhile, reciprocal inhibitory changes occur in the antagonist orbital muscles and motoneurons to perform the saccade.

The neuroanatomy for saccade initiation includes the caudal pons (burst neurons within the paramedian pontine reticular formation) for horizontal saccades and rostral midbrain (burst neurons within the rostral interstitial nucleus of the medial longitudinal fasciculus) for vertical and torsional saccades. Omnipause neurons (OPNs) within the nucleus raphe interpositus in the caudal PPRF exert continuous inhibition upon all burst neurons for saccade generation in any direction. For saccade initiation, the high tonic inhibition of OPNs is released. This disinhibition of the burst neurons allows activation of ocular muscles resulting in a saccade. In addition, the superior colliculus has projections to regulate the OPNs and thus has strong control over saccade generation.

Saccades necessitate a fine balance between inhibitory and excitatory activity as well as close interplay between structures throughout the brainstem, cerebellum, and cortical eye fields. Antiepileptics can impair the neural network that generates saccades. This can result in reduced velocity, increased latency, and increased duration of saccades in the setting of barbiturate or benzodiazepine use. The velocity of saccades is strongly related to the discharge rate of burst neurons in the PPRF. A factor that influences burst neuron discharge rate in the PPRF includes a person's level of sedation as slower saccades occur in drowsiness. Benzodiazepines and barbiturates have sedating properties which alter the tonic firing of saccadic OPNs (Keller 1977).

In addition to sedation decreasing the velocity of saccades, there can also be increased saccade latencies. Diazepam was found to increase latencies of saccades in saccadic reaction time testing. This finding may be owed to the vigilance-lowering properties of diazepam and its effects on attention (Fafrowicz et al. 1995; Rothennberg and Selkoe 1981; Masson et al. 2000). Another study found that serum concentrations of benzodiazepines, such as diazepam, also resulted in decreased peak horizontal saccadic velocity. This demonstrated a relationship between serum benzodiazepine concentration and its slowing and sedating effect on brainstem reticular formation function by measuring peak saccade velocity (Bittencourt et al. 1981).

In conjunction with their sedative effects, the $GABA_A$ agonist activity of benzodiazepines may also influence the neural networks of saccades through the OPNs. OPNs have GABAergic, glycinergic, and glutaminergic afferents. The inhibitory $GABA_A$ agonist activity of benzodiazepines may impair normal OPNs activity upon burst neurons and play a role in increased saccade latency and reduced peak saccade velocity (Jürgens et al. 1981).

6.2   Smooth Pursuit

Smooth-pursuit eye movements allow clear vision of objects by reducing retinal slip from the fovea as it moves within the visual environment. Descending pathways serving smooth visual tracking include cortex, pons, and cerebellum which then project to ocular motoneurons. Within the cerebellum, the dorsal vermis and para-flocculus receive inputs from the pons (dorsolateral pontine nucleus and nucleus reticularis tegmenti pontis) and from the inferior olive. Then the dorsal vermis and paraflocculus project to the fastigial nucleus and vestibular nucleus which then provide outputs to the ocular motoneurons to perform smooth-pursuit eye movements.

The dorsal vermis and caudal fastigial nucleus contribute most to the onset of smooth pursuit, while the vestibulocerebellum contributes most to steady-state tracking. Purkinje cells in the paraflocculus and flocculus modulate their discharge according to gaze velocity and position during smooth pursuit. The caudal fastigial nucleus receives inputs from Purkinje cells of the dorsal vermis and axon collaterals from pontine nuclei and discharge to initiate onset of smooth pursuit.

Impairment of smooth pursuit by antiepileptic medications may be due to its effects on the cerebellum. There are multiple brain regions where benzodiazepines will bind, one of which being the vestibulocerebellum. Benzodiazepines bind densely to vermal and floccular cerebellar areas. Binding sites within these cerebellar areas with $GABA_A$ receptors include the granular layer (houses granule cells and Golgi cells) and the molecular layer (houses Purkinje cell dendrites, parallel fibers, stellate cells, and basket cells).

These inhibitory GABAergic effects on the cerebellum may be the mechanism in which benzodiazepines impair smooth pursuit. Particularly, diazepam can reduce the amplitude (gain) of smooth pursuit eye tracking as well as decrease smooth-pursuit velocity (Rothenberg and Selkoe 1981; Bittencourt et al. 1983).

7 Propofol and Midazolam

Propofol and midazolam are intravenous anesthetics commonly employed in the management of patients in status epilepticus. Both propofol and midazolam act upon $GABA_A$ receptors. Propofol increases GABA-mediated inhibitory tone in the central nervous system by decreasing the rate of dissociation of GABA from the $GABA_A$ receptor. This results in hyperpolarization of cell membranes by increasing the duration of GABA-activated opening of chloride channels. At higher concentrations, propofol behaves as a receptor agonist by directly activating $GABA_A$ receptors in the absence of GABA. Midazolam also enhances the effect of GABA on the $GABA_A$ receptor by increasing the frequency of chloride channel opening resulting in neuronal hyperpolarization. Midazolam does not activate $GABA_A$ receptors directly, differing from propofol at higher concentrations.

7.1 Saccade Latency and Velocity

The described eye movement abnormalities associated with both propofol and midazolam include increased latency and decreased peak velocity of saccades. Propofol has also been found to reduce ocular microtremor which is a small high-frequency random tremor of the eyes linked to neural activity in the brainstem and reticular formation. The sedative effect upon saccadic metrics can be attributed to their suppression of the brainstem reticular formation, which is supported by alterations in the ocular microtremor observed in anesthetized patients. There are also effects upon the OPNs that feature a high baseline firing rate and briefly stop during saccades and blinks. They exert a powerful inhibitory action upon saccadic burst neurons and have glycine as a neurotransmitter. Although OPNs use glycine, they receive GABAergic, glycinergic, and glutaminergic afferents. Anesthetics affect alertness via suppression of the brainstem reticular formation and will also alter the tonic firing of the saccadic OPNs by altering the tone of their GABAergic afferents (Busettini and Frolich 2014; Bojanic et al. 2001).

8 Ictal Eye Movement Disorders

Eye and head movement abnormalities are common clinical manifestations of epileptic seizures. They include horizontal gaze deviations seen in versive seizures and epileptic nystagmus.

9 Versive Seizures

Versive movements are defined by lateral head and eye deviations to a position that is involuntary, sustained, and unnatural. The eye movements precede the head movements in which the eyes first deviate laterally followed by head turning. Also, the gaze deviation of the eyes can either appear to move horizontally in a stepwise fashion, or the eyes can appear to move horizontally in a smooth and continuous manner.

The direction of gaze deviation and head turning in a versive seizure can aid localization of the seizure focus. The direction of version typically occurs contralateral to the seizure focus but can also be ipsilateral. Version can also be seen in frontal, temporal, or occipital lobe foci. One distinguishing factor to aid localization of the seizure focus is if awareness is preserved during version. If awareness is maintained during versive movements, the focus originates in the contralateral frontal lobe. Loss of awareness during versive movements are more likely to occur with temporal lobe foci. In addition to loss of awareness in temporal lobe foci, the direction of version can either be ipsilateral or contralateral to the temporal lobe focus. This is in contrast to frontal lobe seizure foci which are typically always contralateral to the direction of version.

Frontal eye field activation causes the contralateral horizontal gaze deviation. The frontal eye fields are located in the frontal lobe and are located near the intersection of the caudal end of the middle frontal gyrus and the precentral gyrus. During the initiation of eye movements, such as voluntary saccades and pursuit movements, the frontal eye field is activated and provides input to the contralateral paramedian pontine reticular formation (PPRF) via the frontopontine fibers. From there, the medial longitudinal fasciculus (MLF) system is activated which coordinates horizontal eye movements by stimulating the oculomotor nucleus and its innervation to the medial rectus and the abducens nucleus and its innervation to the lateral rectus. In summary, a seizure can stimulate the ipsilateral frontal eye field which will cause contralateral horizontal gaze deviation via descending frontopontine fibers to activate the PPRF-MLF network.

A seizure focus in the right frontal lobe stimulates the right frontal eye field which send excitatory inputs to the left PPRF via the frontopontine fibers and ultimately activates the medial rectus of the right eye and lateral rectus of the left eye. Clinically this is seen as a left horizontal gaze deviation typical of versive seizures. In this example, awareness is preserved. In contrast, versive seizures seen in temporal lobe epilepsy first has loss of awareness and later can stimulate the frontal eye field as the seizure spreads from the temporal lobe to the frontal lobe. In addition, version can be contralateral or ipsilateral to the temporal lobe seizure foci. The ipsilateral version may be due to seizure spread to the contralateral medial temporal lobe and further spread to cortical and subcortical structures of the frontal and parietal lobes. This spread may then stimulate the frontal eye field contralateral to the temporal lobe seizure focus and generate an ipsilateral versive seizure.

A similar mechanism of version is also seen in parietal and occipital lobe epilepsies. In these cases, the direction of version is typically contralateral, similar to frontal lobe seizures. It is also likely due to parietal eye field activation from seizure spread from parietal and occipital lobe foci (McLachlan 1987).

It was mentioned earlier in this section that the eye movements seen in versive seizures can either be stepwise or smooth and continuous. This difference is dependent on the area within the frontal eye field that is activated. Stimulation studies performed on the frontal eye field resulted in different patterns of eye version. When the posterior portion of the frontal eye field is stimulated near the precentral gyrus, a contralateral smooth-pursuit eye movement is seen. When the frontal eye field is stimulated more anteriorly, contralateral stepwise saccades are seen (Schall 2009). In addition to the stepwise or smooth-pursuit eye movements seen with stimulation of different portions of the frontal eye field, there can also be a difference in the size of saccades as well. Stimulation of the ventrolateral portion of the frontal eye field generates shorter saccades, whereas stimulation of the mediodorsal frontal eye field generates larger saccades (Milea et al. 2002) (Fig. 2).

10 Epileptic Nystagmus

Epileptic nystagmus is due to epileptic activity which causes rapid, repetitive eye movements. If observed, it is a reliable sign lateralizing the epileptogenic zone to the contralateral hemisphere to the fast phase of nystagmus. The duration of epilep-

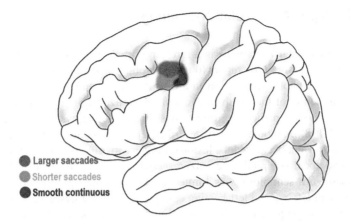

Fig. 2 Frontal eye field. Located at the caudal end of the middle frontal gyrus and anterior to the precentral gyrus. Posterior frontal eye field (blue) activation results in a contralateral smooth-pursuit eye movement. Activation of the anterior frontal eye field results in contralateral stepwise saccades. If the anterior ventrolateral region of the frontal eye field (orange) is activated, shorter saccades are seen. If the anterior mediodorsal region of the frontal eye field (red) is activated, larger saccades are seen

tic nystagmus is short and persist less than 1 min (Ma et al. 2015). There are two proposed mechanisms for epileptic nystagmus. One involves epileptic activation of a cortical saccade region (frontal eye field, supplementary eye field, posterior parietal cortex) causing contraversive quick phases combined with a defect in the gaze-holding system such as a leaky neural integrator. This allows the eyes to drift back toward the midline. In this situation, an epileptic discharge would cause a contraversive quick-phase eye movement followed by a slow phase caused by the leaky neural integrator (which is based upon an assumption of impaired neural integrator function secondary to AED use and is discussed further in the next paragraph) that would bring the eyes toward the midline of the orbit. Another mechanism involves epileptic activation of a cortical pursuit region (primary visual areas, temporo-parietal-occipital junction) generating an ipsiversive slow phase followed by a reflexive saccade as the eye reaches the far eccentric position in the orbit (Kaplan and Tusa 1993).

The proposed neural mechanism for epileptic nystagmus is based on the assumption of a leaky neural integrator. Many antiepileptic medications such as phenytoin, phenobarbital, carbamazepine, and benzodiazepines can cause a leaky neural integrator. Also in patients found to have epileptic nystagmus, many had seizure activity within the temporo-parieto-occipital cortex and had high-frequency epileptic discharges greater than 10 Hz. Epileptic nystagmus may be rarely seen as these combinations of factors are uncommonly met (Kaplan and Tusa 1993; Kellinghaus et al. 2008).

11   Epileptic Monocular Nystagmus

Epileptic monocular nystagmus is a rarely reported phenomenon with an unclear mechanism of action. One report found epileptic monocular nystagmus provoked by intermittent photic stimulation in a cognitively impaired 15-year-old girl with poor visual acuity of the involved eye and generalized epilepsy. The authors concluded was due to a pathologic form of flash-induced after nystagmus (Jacome and FitzGerald 1982). Another report described epileptic monocular nystagmus in a patient with a right occipital lobe lesion secondary to Sturge-Weber syndrome. This patient had focal occipital lobe seizures contralateral to the involved eye. That patient's seizure semiology was limited to altered visual perceptions of the left hemifield and left eye and left-beating epileptic monocular nystagmus. The authors provided a few proposed mechanisms. One being that the seizure discharge activated a cortical saccade region and caused simultaneous supranuclear inhibition of ipsilateral eye movement. Another proposed mechanism relies on the theory of von Helmholtz in that premotor eye movement commands are monocular and that the focal seizure discharge creates a monocular eye movement command at the cortical or brainstem level thus manifesting as epileptic monocular nystagmus (Grant et al. 2002). Another study described ictal monocular nystagmus of the left eye secondary to a right frontal focal cortical dysplasia type 2b. In addition, there was also interictal monocular nystagmus of the right eye. The proposed hypothesis included not only focal cortical activation but also an irregular brainstem imbalance that overrides the regular binocular cortical and subcortical eye movement control mechanisms. The finding of interictal right eye monocular nystagmus may be due to a dysfunction of the right brainstem with possibly a "Todd's paralysis" of the analogous left brainstem nuclei due to high seizure frequency. In summary, this proposed hypothesis states that the simultaneous control of eye movements was transiently overridden by epileptic activity of the right frontal cortical focus exciting the right frontal eye field, stimulating cranial nerve nuclei III and VI in the ictal state, resulting in monocular nystagmus of the left eye. Then during the interictal state there was a monocular nystagmus of the right eye due to a presumed Todd's paralysis-like phenomenon of the left brainstem (Schulz et al. 2013).

12   Ictal Eye Blinking

Ictal eye blinking is thought to occur from ictal cortical activity with descending activation of brainstem trigeminal fibers resulting in eye blinks (Saporito et al. 2017). Ictal unilateral blinking is a lateralizing sign in epilepsy to the ipsilateral hemisphere, typically being frontal but has been described secondary to ictal activity in other brain regions as well (Kalss et al. 2013a). The pathway and mechanism for unilateral eye blinking are not well understood. Ipsilateral precentral, postcentral, temporal, and cerebellar regions as well as trigeminal fibers are thought to play a role in ipsilateral eye blinking. One proposed hypothesis is that ictal activity may pathologically inhibit the contralateral eye from blinking with descending inputs to trigeminal fibers causes blinking of the ipsilateral, non-inhibited eye (Kalss et al. 2013b). Another proposed theory regarding ipsilateral eye blinking is based on stimulation studies of the dura. This hypothesis is based upon ictal activity stimulating trigeminal fibers coursing through and innervating subdural structures and pial vessels of the ipsilateral hemisphere which then unilaterally stimulates blinking of the ipsilateral eye (Falsaperla et al. 2014) (Table 1).

Table 1 Summary of mechanism and neuroanatomical site of action of each anti-epileptic drug and the resulting eye movement abnormality

AED	Mechanism of action	Eye movement abnormality	Mechanism of abnormality
Valproate	Na channel blocker	Gaze-evoked nystagmus	Neural integrator
	Increases GABA	Vertical gaze palsy	Deactivation of vertical burst neurons
Phenytoin	Na channel blocker	Gaze-evoked nystagmus	Neural integrator
		Impaired smooth pursuit	Vestibulocerebellum – flocculus and paraflocculus
		VOR suppression	Vestibulocerebellum – flocculus and paraflocculus
		Downbeat nystagmus	Vestibulocerebellum – flocculus
		Periodic alternating, nystagmus (PAN)	Vestibulocerebellum – nodulus and uvula
		Pendular nystagmus	Neural integrator
Carbamazepine	Na channel blocker	Gaze-evoked nystagmus	Neural integrator
		Slowed saccades	PPRF (burst neurons)
		Impaired smooth pursuit	Vestibulocerebellum
		Downbeat nystagmus	Vestibulocerebellum – Flocculus
		Seesaw nystagmus	Vestibular inputs and gaze holding
		Oculogyric crisis	Neural integrator and vertical gaze holding
		Ophthalmoplegia	Omnipause neurons
Lamotrigine	Na channel blocker	Downbeat nystagmus	Vestibulocerebellum – flocculus
Benzodiazepines	Enhances GABA	Slowed saccades	PPRF (burst neurons) and OPNs
Barbiturates	Enhances GABA, AMPA Antagonist	Impaired smooth pursuit	Vestibulocerebellum – vermis and flocculus
Propofol midazolam	Enhances GABA	Slowed saccades	PPRF and OPNs

References

Adamec, I., Nankovic, S., Zadro, I., et al. (2013). Oxcarbazepine-induced jerky see-saw nystagmus. *Neurological Sciences, 34*, 1839–1840.

Alkawi, A., Kattah, J. C., & Wyman, K. (2005). Downbeat nystagmus as a result of lamotrigine toxicity. *Epilepsy Research, 63*, 85–88.

Belton, T., & McCrea, R. A. (2000). Role of the cerebellar flocculus region in cancellation of the VOR during passive whole body rotation. *Journal of Neurophysiology, 84*, 1599–1613.

Berchou, R. C., & Rodin, E. A. (1979). Carbamazepine-induced oculogyric crisis. *Archives of Neurology, 36*, 522–523.

Berger, J. R., & Kovacs, A. G. (1982). Downbeat nystagmus with phenytoin. *Journal of Clinical Neuro-Ophthalmology, 2*, 209–211.

Bittencourt, P. R., Gresty, M. A., & Richens, A. (1980). Quantitative assessment of smooth-pursuit eye movements in healthy and epileptic subjects. *Journal of Neurology, Neurosurgery, and Psychiatry, 43*, 1119–1124.

Bittencourt, P. R., Wade, P., Smith, A. T., et al. (1981). The relationship between peak velocity of saccadic eye movements and serum benzodiazepine concentration. *British Journal of Clinical Pharmacology, 12*, 523–533.

Bittencourt, P. R., Wade, P., Smith, A. T., et al. (1983). Benzodiazepines impair smooth pursuit eye movements. *British Journal of Clinical Pharmacology, 15*, 259–262.

Bojanic, S., Simpson, T., & Bolger, C. (2001). Ocular microtremor: A tool for measuring depth of anaesthesia? *British Journal of Anaesthesia, 86*, 519–522.

Busettini, C., & Frolich, M. A. (2014). Effects of mild to moderate sedation on saccadic eye movements. *Behavioural Brain Research, 272*, 286–302.

Büttner Ennever, J. A., & Büttner, U. (1988). The reticular formation. In J. A. Büttner-Enever (Ed.), *Neuroanatomy of the oculomotor system* (pp. 119–176). New York: Elsevier.

Campbell, W. W. (1980). Periodic alternating nystagmus in phenytoin intoxication. *Archives of Neurology, 37*, 178–180.

Chrousos, G. A., Cowdry, R., Schuelein, M., et al. (1987). Two cases of downbeat nystagmus and oscillopsia associated with carbamazepine. *American Journal of Ophthalmology, 103*, 221–224.

Cohen, B., Helwig, D., & Raphan, T. (1987). Baclofen and velocity storage: A model of the effects of the drug on the vestibulo-ocular reflex in the rhesus monkey. *The Journal of Physiology, 393*, 703–725.

Fafrowicz, M., Unrug, A., Marek, T., et al. (1995). Effects of diazepam and buspirone on reaction time of saccadic eye movements. *Neuropsychobiology, 32*, 156–160.

Falsaperla, R., Perciavalle, V., Pavone, P., et al. (2014). Unilateral eye blinking arising from the ictal ipsilateral occipital area. *Clinical EEG and Neuroscience, 47*(3), 243–246.

Glassauer, S., Hoshi, M., Kempermann, U., et al. (2003). Three-dimensional eye position and slow phase velocity in humans with downbeat nystagmus. *Journal of Neurophysiology, 89*, 338–354.

Gosala Raja, K. S., Srinivas, M., Raghunandan, N., et al. (2013). Reversible vertical gaze palsy in sodium valproate toxicity. *Journal of Neuro-Ophthalmology, 33*, 202–203.

Grant, A. C., Jain, V., & Bose, S. (2002). Epileptic monocular nystagmus. *Neurology, 59*, 1438–1441.

Horn, A. K., Büttner-Enever, J. A., Wahle, P., & Reichenberger, I. (1994). Neurotransmitter profile of saccadic omnipause neurons in nucleus raphe interpositus. *The Journal of Neuroscience, 14*, 2032–2046.

Horn, A. K. E., Helmchen, C., & Wahle, P. (2003). GABAergic neurons in the rostral mesencephalon of the macaque monkey that control vertical eye movements. *Annals of the New York Academy of Sciences, 1004*, 19–28.

Jacome, D. E., & FitzGerald, R. (1982). Monocular ictal nystagmus. *Archives of Neurology, 39*, 653–656.

Jürgens, R., Becker, W., & Kornhuber, H. H. (1981). Natural and drug-induced variations of velocity and duration of human saccadic eye movements: Evidence for a control of the neural pulse generator by local feedback. *Biological Cybernetics, 39*, 87–96.

Kalss, G., Leitinger, M., Dobesberger, J., Granbichler, C. A., Kuchukhidze, G., & Trinka, E. (2013a). Ictal unilateral eye blinking and contralateral blink inhibition - a video-EEG study and review of the literature. *Epilepsy & Behavior Case Report, 1*, 161–165.

Kalss, G., Leitinger, M., Dobesberger, J., Granbichler, C. A., Kuchukhidze, G., & Trinka, E. (2013b). Ictal unilateral eye blinking and contralateral blink inhibition - a video-EEG study and review of the literature. *Epilepsy & Behavior Case Report, 1*, 161–165.

Kaplan, P. W., & Tusa, R. J. (1993). Neurophysiologic and clinical correlations of epileptic nystagmus. *Neurology, 43*, 2508–2514.

Keller, E. L. (1977). Control of saccadic eye movements by midline brain stem neurons. In R. Baker & A. Berthoz (Eds.), *Control of gaze by brain stem neurons* (pp. 327–336). Amsterdam: Elsevier.

Kellinghaus, C., Skidmore, C., & Loddenkemper, T. (2008). Lateralizing value of epileptic nystagmus. *Epilepsy & Behavior, 13*, 700–702.

Ma, Y., Wang, J., Li, D., & Lang, S. (2015). Two types of isolated epileptic nystagmus: Case report. *International Journal of Clinical and Experimental Medicine, 8*(8), 13500–13507.

Masson, G. S., Mestre, D. R., Martineau, F., et al. (2000). Lorazepam-induced modifications of saccadic and smooth-pursuit eye movements in humans: Attentional and motor factors. *Behavioural Brain Research, 108*, 169–180.

McLachlan, R. S. (1987). The significance of head and eye turning in seizures. *Neurology, 37*(10), 1617–1619.

Milea, D., Lobel, E., Lehéricy, S., et al. (2002). Intraoperative frontal eye field stimulation elicits ocular deviation and saccade suppression. *Neuroreport, 13*(10), 1359–1364.

Noda, S., & Umezaki, H. (1982). Carbamazepine-induced ophthalmoplegia. *Neurology, 32*, 1320.

Oh, S. Y., Kim, J. S., Lee, Y. H., et al. (2006). Downbeat, positional, and perverted head-shaking nystagmus associated with lamotrigine toxicity. *Journal of Clinical Neurology, 2*, 283–285.

Reilly, J. L., Lencer, R., Bishop, J. R., Keedy, S., & Sweeney, J. A. (2008). Pharmacological treatment effects on eye movement control. *Brain and Cognition, 68*(3), 415–435. https://doi.org/10.1016/j.bandc.2008.08.026.

Remler, B. F., Leigh, R. J., Osorio, I., et al. (1990). The characteristics and mechanisms of visual disturbance associated with anticonvulsant therapy. *Neurology, 40*, 791–796.

Rett, D. (2007). Gaze-evoked nystagmus: A case report and literature review. *Optometry, 78*, 460–464.

Riker, W. K., Downes, H., Olsen, G. D., et al. (1978). Conjugate lateral gaze nystagmus and free phenytoin concentrations in plasma: Lack of correlation. *Epilepsia, 19*, 93–98.

Rothenberg, S. J., & Selkoe, D. (1981). Specific oculomotor deficit after diazepam. II. Smooth pursuit eye movements. *Psychopharmacology (Berlin), 74*, 237–240.

Rothennberg, S. J., & Selkoe, D. (1981). Specific oculomotor deficit after diazepam. I. Saccadic eye movements. *Psychopharmacology (Berlin), 74*, 232–236.

Sanchez, K., & Rowe, F. J. (2016). Role of neural integrators in oculomotor systems: A systematic narrative literature review. *Acta Ophthalmologica, 96*(2), e111–e118.

Saporito, M. A. N., Vitaliti, G., Pavone, P., et al. (2017). Ictal blinking, an under-recognized phenomenon: Our experience and literature review. *Neuropsychiatric Disease and Treatment, 13*, 1435–1439.

Schall, J. D. (2009). Frontal eye fields. In *Encyclopedia of neuroscience* (pp. 367–374). Elsevier BV. Academic Press. ISBN: 978-0-08-045046-9. https://doi.org/10.1016/B978-008045046-9.01111-6

Schulz, R., Tomka-Hoffmeister, M., Woermann, F. G., et al. (2013). Epileptic monocular nystagmus and ictal diplopia as cortical and subcortical dysfunction. *Epilepsy & Behavior Case Reports, 1*, 89–91.

Schwankhaus, J. D., Kattah, J. C., Lux, W. E., et al. (1989). Primidone/phenobarbital-induced periodic alternating nystagmus. *Annals of Ophthalmology, 21*, 230–232.

Shaikh, A. G. (2013). Fosphenytoin induced transient pendular nystagmus. *Journal of the Neurological Sciences, 330*, 121–122.

Suzuki, Y., Kase, M., Hashimoto, M., et al. (2003). Leaky neural integration observed in square-wave jerks. *Japanese Journal of Ophthalmology, 47*, 535–536.

Tedeschi, G., Casucci, G., Allocca, S., et al. (1989). Neuroocular side effects of carbamazepine and phenobarbital in epileptic patients as measured by saccadic eye movements analysis. *Epilepsia, 30*, 62–66.

Umeda, Y., & Sakata, E. (1977). Equilibrium disorder in carbamazepine toxicity. *The Annals of Otology, Rhinology, and Laryngology, 86*, 318–322.

Zaccara, G., Cincotta, M., Borgheresi, A., & Balestrieri, F. (2004). Adverse motor effects induced by antiepileptic drugs. *Epileptic Disorders, 6*(3), 153–168.

Zee, D. S., Yee, R. D., Cogan, D. G., Robinson, D. A., & Engel, W. K. (1976). Ocular motor abnormalities in hereditary cerebellar ataxia. *Brain, 99*(2), 207–234.

Index

© Springer Nature Switzerland AG 2019
A. Shaikh, F. Ghasia (eds.), *Advances in Translational Neuroscience of Eye
Movement Disorders*, Contemporary Clinical Neuroscience,
https://doi.org/10.1007/978-3-030-31407-1